P9-AGI-362

ENCYCLOPEDIA OF

Mathematics and Society

ENCYCLOPEDIA OF

Mathematics and Society

Sarah J. Greenwald
Jill E. Thomley
Appalachian State University

VOLUME 1

Salem Press

The paper used in these volumes conforms to the American National Standard for Permanence of Paper for Printed Library Materials, X39.48-1992 (R1997).

LIBRARY OF CONGRESS CATALOGING-IN-PUBLICATION DATA

Encyclopedia of mathematics and society / Sarah J. Greenwald , Jill E. Thomley, general Editors.

p. cm.

Includes bibliographical references and index.

ISBN 978-1-58765-844-0 (set : alk. paper) -- ISBN 978-1-58765-845-7 (v. 1 : alk. paper) -- ISBN 978-1-58765-846-4 (v. 2 : alk. paper) -- ISBN 978-1-58765-847-1 (v. 3 : alk. paper)

1. Mathematics--Social aspects. I. Greenwald, Sarah J. II. Thomley, Jill E.

QA10.7.E53 2012

303.48'3--dc23

2011021856

First Printing

PRINTED IN THE UNITED STATES OF AMERICA

Produced by Golson Media

President and Editor J. Geoffrey Golson

Senior Layout Editor Mary Jo Scibetta

Author Manager Joseph K. Golson

Copy Editors Carl Atwood, Kenneth Heller, Holli Fort

Proofreader Lee A. Young

Indexer J S Editorial

Contents

Publisher's Note

The *Encyclopedia of Mathematics and Society* (three volumes) explains how mathematics is at the root of modern civilization, from measuring temperature on a frigid day to driving a car to using a digital camera; enthusiasts might say applied mathematics rules the world. The set includes 478 articles, all of which were written specifically for the work.

Scope of Coverage

The *Encyclopedia of Mathematics and Society* is designed to provide students at the high school and undergraduate levels with a convenient source of information on the fundamental science and the mathematics behind our daily lives, explaining to students how and why mathematics works, and allowing readers to better understand how disciplines such as algebra, geometry, calculus, and others affect what we do every day. This academic, multiauthor reference work serves as a general and nontechnical resource for students and teachers to understand the importance of mathematics; to appreciate the influence of mathematics on societies around the world; to learn the history of applied mathematics; and to initiate educational discussion brought forth by the specific social and topical articles presented in the work.

The articles in the set fall into one or more of the following broad categories: architecture and engineering (35 articles); arts, music, and entertainment (41): business, economics, and marketing (32); communication and computers (22); friendship, romance, and religion (18); games, sport, and recreation (42); government, politics, and history (43); history and development of curricular concepts (63); mathematics around the world (21); mathematics culture and identity (27); medicine and health (34); school and society (19); space, time, and distance (25); travel and transportation (18); and weather, nature, and environment (35).

Rationale for Choice of Topics

Mathematics is a fundamental part of society, yet many people may not be aware of the interconnections between what they have learned in school and their everyday lives. In its Curriculum Guide (MAA, 2004), the Mathematical Association of America's Committee on the Undergraduate Program in Mathematics (CUPM) recommends that mathematics programs lead people "to learn mathematics in a way that helps them to better understand its place in society: its meaning, its history, and its uses." In keeping with this philosophy, the editors chose topics for inclusion based on one or more of the following criteria:

- The topic is timely and likely to remain so.
- The topic can be tied to mathematical concepts that people likely have been exposed to.
- The topic is related to concepts and connections that professional mathematical organizations have suggested are important.
- The topic is one that the general public has expressed interest in.
- The topic is one we have successfully used or that we know has been successfully used in other contexts.

Article Length and Format

Articles in the encyclopedia range in length from 500 to 3500 words. Each is first presented with the category to which it belongs (for example, architecture and engineering), an article summary, and fields of study for the article. The fields of study include the following:

- Algebra
- Calculus
- Communication
- Connections
- Data Analysis and Probability
- Geometry
- Measurement
- Number and Operations
- Problem Solving
- Reasoning and Proof
- Representations

Each article is then followed by "See Also" cross-references to other relevant articles and "Further Reading" sources that include bibliographic citations. Many articles are richly illustrated with photos and captions, and charts, graphs, and tables. Finally, each article is signed by the contributor to the encyclopedia.

Frontmatter and Backmatter

Volume 1 of the *Encyclopedia of Mathematics and Society* begins with "About the Editors" and then presents their introduction to the encyclopedia. The "List of Articles," repeated in all three volumes, features all the articles in alphabetical order with page numbers as they are listed in the encyclopedia. A "Topic Finder" shows all the articles organized by category to enable readers to find related article by topic. The "List of Contributors" presents all the writers for the encyclopedia along with their academic or institutional affiliations.

The backmatter of the encyclopedia at the end of Volume 3 has the "Chronology of Mathematics," a timeline of major milestones in the discoveries and development of mathematics. Next is the "Resource Guide" for further research that includes books that are major works in the history of mathematics as well as current editions of new works, journals in the mathematics field, and Internet sites that pertain to mathematics. A "Glossary" provides mathematical definitions for terms encountered in the articles. Lastly, a comprehensive subject index references all concepts, terms, events, persons, places, and other topics of discussion.

Online Access

Salem provides access to its award-winning content both in traditional printed form and online. Any school or library that purchases this three-volume set is entitled to complimentary access to Salem's online version of the content through our Salem Science Database. For more information about our online database, please contact our online customer service representatives at (800) 221-1592.

The advantages are clear:

- Complimentary with print purchase
- Fully supported
- Unlimited users at your library
- Full access from home or dorm rooms
- Immediate access via online registration
- A simple, intuitive interface
- User profile areas for students and patrons
- Sophisticated search functionality
- Complete content, including appendixes
- Integrated searches with any other Salem Press product you already have on the Salem Science platform
- E-books are also available

About the Editors

Sarah J. Greenwald is a professor of mathematics and a women's studies core faculty member at Appalachian State University in Boone, North Carolina. She obtained her Ph.D. in mathematics from the University of Pennsylvania in 1998 and since then has published more than 35 articles. Her areas of expertise include Riemannian geometry, popular culture as it pertains to mathematics, and women and minorities in mathematics. Dr. Greenwald has discussed the impacts of scientific popular culture representations on NPR's *Science Friday.* She has spoken all over the country, and her interactive mathematics lecture appears on 20th Century Fox's *Futurama* movie *Bender's Big Score.*

Dr. Greenwald has won numerous teaching awards including a Mathematical Association of America Alder Award for Distinguished Teaching and an Appalachian State University Wayne D. Duncan Award for Excellence in Teaching in General Education. Dr. Greenwald has also been active in professional service as a member of the Joint Policy Board for Mathematics Advisory Panel for Mathematics Awareness Month and as the associate editor for the Association for Women in Mathematics, just to name a few. Her husband, Joel Landsberg, is the bassist for the Kruger Brothers.

Jill E. Thomley is an associate professor of statistics in the Department of Mathematical Sciences at Appalachian State University. Her education and scholarly interests are diverse, generally focusing on mathematics and science applications. She earned a Ph.D. in Decision Sciences from Rensselaer Polytechnic Institute and was awarded the Del and Ruth Karger Dissertation Prize and a Rensselaer Founders Award of Excellence. Additional degrees include an M.S. in industrial/organizational psychology from Rensselaer and an A.B. in psychology from Harvard University.

Along with teaching, Dr. Thomley consults on statistical design and analysis of scientific research and evaluates the results of federal education grants. Areas of focus include computational science, a discipline arising from the intersection of science, mathematics, and computer science, and the adoption and diffusion of educational innovations in mathematics and science. She presented at the first Science in Society Conference in 2009 and was published in *The International Journal of Science in Society*. Additional interests include history of statistics and statistics in popular culture.

Introduction

Mathematics is pervasive in modern society, and on some level we all use mathematics in our daily lives. At the same time, many people are not fully aware of the diverse interactions and connections between mathematics and society. Mathematics takes a readily apparent starring role in highly technological fields like engineering, computer science, and the natural sciences. Outside these fields, however, there are countless ideas, inventions, and advances that cannot be fully realized without the involvement of mathematics.

Organizations like the National Council of Teachers of Mathematics and the Mathematical Association of America recommend that mathematics be explored in the context of contemporary society. To examine these connections, we approach them from different angles. We can look at mathematics through the lens of larger societal structures like nations, cultures, and educational systems, or we can turn this method around to explore the societal structures within mathematics, such as the culture of mathematicians and notions of proof, certainty, and success.

Connections are also found in the countless applications of mathematics to society. Overall, definitions and applications of mathematics are inherently dependent on context: the socio-historical events during which they developed; the people who created or discovered concepts, who built upon the work of others, or who passed their knowledge on to the next generation; the fundamental connections to daily tasks of living; the ethics, controversies, and philosophies surrounding mathematics; the public's perceptions of mathematics and mathematicians; the way current society uses mathematics to solve problems and educate its citizens; and the way mathematics draws from society in order to grow and evolve.

Mathematics shapes the world in which we live. In the twenty-first century it is almost impossible to find an academic field of study that does not use mathematics, either directly or via tools and technology in which mathematics plays a vital role. The world in turn shapes the discipline of mathematics by inspiring mathematicians to formulate new questions, solve new problems, develop new theories, and use new technologies. Each successive generation of mathematicians brings fresh perspectives, expectations, and ways of thinking and working into the culture of mathematics. These mathematicians are influenced by the home, school, and play environments in which they were raised.

However, despite the mathematics all around us, people's exposure may be limited. Representations in the media or in popular culture may portray mathematics and mathematicians in highly stereotypical ways that do not reflect the true depth, breadth, diversity, and culture of the mathematics community.

The goal of *The Encyclopedia of Mathematics and Society* is to weave multilayered connections between society, history, people, applications, and mathematics. These connections address both mathematical concepts that our readers likely have been exposed to at school, work, or through other sources, as well as advanced topics that are built upon these fundamental ideas. The articles in the *Encyclopedia*, which were contributed by a broad spectrum of authors in many fields, also include connections to multiple disciplines within and outside of mathematics.

In general, the articles do not teach or present detailed mathematical theory, derivations, and equations. There is already a vast array of textbooks and other works better able to accomplish that important task. Instead, we intended them to serve as a foundation and jumping-off point for additional explorations. As mathematics professor and educator Art Johnson has noted in other settings, we hope that this type of contextualization helps people to "see mathematics as a discipline that transcends culture, time, and gender, and as a discipline for everyone, everywhere."

In keeping with this focus on linkages and interdisciplinarity, we have organized the articles not by mathematics topic but according to various connecting themes. For example, there are few stand-alone articles about individual people within the encyclopedia. Instead, we encouraged our authors to include significant mathematical contributors within the associated context of one or more topics or applications. The people we did choose to include as stand-alone articles serve to highlight the diversity of individuals who have produced great achievements with mathematics.

Further, our intent was to discuss, via these individuals and other articles in the Mathematics Culture and Identity theme, the community of mathematicians today: who mathematicians are, as professionals and people; the type of work mathematicians do; the different ways in which mathematicians describe mathematics and where their ideas come from; and mathematicians' personal processes when working with mathematics. We also wanted to address in these articles how the mathematics community perceives itself and how it is in turn perceived by society.

Articles within the History and Development of Curricular Topics theme highlight many of the earliest known uses, both ancient and modern advances, and people who have contributed to the development and spread of the concept or field. In contrast, the articles within the School and Society theme examine the importance of broad fields inside and outside of school, primarily in the United States. These articles showcase, for example, what jobs use particular skills and why the field is a fundamental part of current school curricula and society. The Mathematics Around the World theme extends the discussion of cross-cultural attitudes and perspectives on mathematics, with geographic regions grouped according to current United Nations standards. Other themes that center on mathematics application are Games, Sport, and Recreation; Government, Politics, and History; and Space, Time, and Distance.

Why did we choose to focus on connections? In modern society, widespread Internet access has placed data about a broad spectrum of people, objects, and events essentially at our fingertips, yet mathematics content may be buried among other discussions rather than brought to the forefront.

Both Internet and other types of library searches can result in a potentially overwhelming number of results, many of which contain almost nothing of mathematical relevance, though important connections exist. Too often, regardless of the amount of data or sources returned, connections between mathematics, people, objects, and events are missing, or they are presented in isolation from their broader historical context. Such connections are critical components of knowledge acquisition, creation, and dissemination. They are what allow people to extrapolate from what they already know to new situations, to create new knowledge or new applications, to overcome existing negative stereotypes about mathematics, and to fully understand the timeline of human events from multiple perspectives.

Even several hundred articles cannot provide an exhaustive examination of mathematics and society. At best, we can perhaps provide a snapshot of the history, people, applications, and mathematical connections as they exist at the time of publication, with some discussion of the rich history and speculations about future directions. Hopefully, this encyclopedia is a representative sampling of articles that, with the accompanying further readings, will allow a reader to follow the path to related topics of interest.

In making the very difficult decision regarding what topics to include, given that time and space were not unlimited, we used an array of selection criteria, such as: the topic was timely and likely to remain so for a reasonable period of time; the topic was tied to mathematical concepts to which people likely have been exposed; the topic was related to concepts and connections that professional mathematical organizations suggested are important; the topic is one that the general public has expressed interest in; or the topic was one that we ourselves have successfully used or that we know has been successfully used in other educational or professional contexts.

When embarking on this work, we already knew in a general sense how pervasive mathematics is in society, and we were eager to share these ideas with others. However, even though we are mathematicians with diverse interdisciplinary backgrounds, research and teaching interests, we were surprised to discover so many interesting and amazing connections. We learned more than we ever imagined we would. It was regrettably impossible to include everything we thought was interesting or important, and we have accumulated a long list of items that we want to explore in the future, on our own or with our students and colleagues.

The creation of this encyclopedia has been an intellectual pleasure and a profound learning experience, and we hope that our readers find the same kind of enjoyment and wonder that we experienced.

Sarah J. Greenwald
Jill E. Thomley
General Editors

List of Articles

Topic Finder

The following list is provided for readers to find articles related by topic.

Business, Economics, and Marketing

Communication and Computers

Friendship, Romance, and Religion

Games, Sport, and Recreation

Government, Politics, and History

Medicine and Health

School and Society

Space, Time, and Distance

List of Contributors

Stephen Abbott
Middlebury College
John G. Alford
Sam Houston State University
Micah Altman
Harvard University
Mohamed Amezziane
DePaul University
Or Syd Amit
Boston College
Jim Austin
Independent Scholar
Sukantadev Bag
University College Cork
Zenia C. Bahorski
Eastern Michigan University
Hyungryul Baik
Cornell University
Thomas E. Baker
University of Scranton
Ron Barnes
University of Houston, Downtown
Eric Barth
Kalamazoo College
John Beam
University of Wisconsin, Oshkosh
Judith E. Beauford
University of the Incarnate Word
Linda Becerra
University of Houston, Downtown
Robert A. Beeler
Eastern Tennessee State University
Kimberly Edginton Bigelow
University of Dayton
Bonnie Ellen Blustein
West Los Angeles College
Norma Boakes
Richard Stockton College of New Jersey
Matt Boelkins
Grand Valley State University
Mark Bollman
Albion College
Vladimir E. Bondarenko
Georgia State University
Casey Borch
University of Alabama at Birmingham
Sarah Boslaugh
Washington University School of Medicine
Marek Brabec
Academy of Sciences of the Czech Republic
Murray R. Bremner
University of Saskatchewan

David Brink
University College Dublin, Ireland
Patrick L. Brockett
University of Texas at Austin
John N. A. Brown
Independent Scholar
Chris D. Cantwell
Imperial College London
Peter J. Carrington
University of Waterloo
Jen-Mei Chang
California State University, Long Beach
Darrah Chavey
Beloit College
John T. Chen
Bowling Green State University
Diana Cheng
Middle Tennessee State University
Ka-Luen Cheung
The Hong Kong Institute of Education
Jason L. Churchill
Cleo Research Associates
Loren Cobb
University of Colorado, Denver
Shirley Coleman
Newcastle University
Dogan Comez
North Dakota State University
Justin Corfield
Geelong Grammar School
Beth Cory
Sam Houston State University
Vesta Coufal
Gonzaga University
Kumer Pial Das
Lamar University
Richard De Veaux
Williams College
Marilena Di Bucchianico
Rutgers University
Daniel Disegni
Columbia University
Maria Droujkova
Natural Math
Leigh H. Edwards
Florida State University
Steven R. Edwards
Southern Polytechnic State University
Caleb Emmons
Pacific University
Jonathan L. Entin
Case Western Reserve University
Gisela Ernst-Slavit
Washington State University, Vancouver
Amy Everton
Independent Scholar
Jonathan David Farley
University of Oxford
Lee Anne Flagg
University of Alabama at Birmingham
Francesco Flammini
University of Naples Frederico II
Robert D. Foreman
University of Oklahoma
Daniel J. Galiffa
Penn State Erie, The Behrend College
Angela Gallegos
Occidental College
Catherine C. Galley
Independent Scholar
Joseph A. Gallian
University of Minnesota, Duluth
Joaquim Alves Gaspar
Universidade de Lisboa
Sommer Gentry
United States Naval Academy
Mark Ginn
Appalachian State University
Darren Glass
Gettysburg College
Deborah L. Gochenaur
Shippensburg University
Christopher Goff
University of the Pacific
Lidia Gonzalez
City University of New York
Jeff Goodman
Appalachian State University
Rick Gorvett
University of Illinois at Urbana-Champaign
Judith V. Grabiner
Pitzer College
Michael K. Green
State University of New York, Oneonta
Sarah J. Greenwald
Appalachian State University

William Griffiths
Southern Polytechnic State University
Alexander A. Gurshtein
Mesa State College
Juan B. Gutierrez
University of Miami
Simone Gyorfi
O. Goga High School, Jibou, Romania
Gareth Hagger-Johnson
The University of Leeds
Thomas W. Hair
Florida Gulf Coast University
Anthony Harkin
Rochester Institute of Technology
Ziaul Hasan
University of Illinois, Chicago
Deborah J. Hilton
Independent Scholar
Holly Hirst
Appalachian State University
Calli A. Holaway
University of Alabama
Liang Hong
Bradley University
Brian Hopkins
Saint Peter's College
Linda Hutchison
University of Wyoming
Yih-Kuen Jan
University of Oklahoma
Jerry Johnson
Western Washington University
Pete Johnson
Eastern Connecticut State University
Phillip Johnson
Appalachian State University
D. Keith Jones
University of Southampton
Ugur Kaplan
Kadir Has University, Istanbul
David I. Kennedy
Shippensburg University of Pennsylvania
Cathy Kessel
Independent Scholar
Michael "Cap" Khoury
University of Michigan, Ann Arbor
Christine Klein
Independent Scholar
Michael Klucznik
St. Bonaventure University
Rick Kreminski
Colorado State University, Pueblo
Matt Kretchmar
Denison University
Bill Kte'pi
Independent Scholar
Maria Elizete Kunkel
University of Ulm, Germany
Konnie G. Kustron
Eastern Michigan University
Alistair Kwan
Yale University
James Landau
Independent Scholar
Carmen M. Latterell
University of Minnesota, Duluth
Michele LeBlanc
California Lutheran University
Stephen Lee
Mathematics in Education and Industry
Eddie Leung
Hong Kong Institute of Education
Fuyuan Liao
University of Oklahoma
Silvia Liverani
University of Bristol
Michael G. Lovorn
University of Alabama
Chad T. Lower
Pennsylvania College of Technology
Margaret MacDougall
University of Edinburgh Medical School
Yiu-Kwong Man
The Hong Kong Institute of Education
Philip McCartney
Northern Kentucky University
Elizabeth A. McMillan-McCartney
Northern Kentucky University
Liliana Monteiro
Affiliation TK
Mariana Montiel
Georgia State University
Deborah Moore-Russo
State University of New York, University at Buffalo
Ashwin Mudigonda
Universal Robotics Inc.

Andrew Nevai
University of Central Florida
Samuel Obara
Texas State University
Eoin O'Connell
Deakin University
Serkan Ozel
Bogazici University
Zeynep Ebrar Yetkiner Ozel
Fatih University
Julian Palmore
University of Illinois at Urbana-Champaign
Robert W. Peck
Louisiana State University School of Music
Josipa G. Petrunic
University College London
Thomas J. Pfaff
Ithaca College
Biljana Popovic
University of Nis
Zoran Petrovic
University of Belgrade
Michael Qaissaunee
Brookdale Community College
Matina J. Rassias
University College of London
Gregory Rhoads
Appalachian State University
Mark Roddy
Seattle University
Maria Elizabeth S. Rodrigues
University of Ulm, Germany
David C. Royster
University of Kentucky
Douglas Rugh
Independent Scholar
Karim Salim
Independent Scholar
Alun Salt
University of Leicester
Kady Schneiter
Utah State University
Richard Schugart
Western Kentucky University
Carl R. Seaquist
Texas Tech University
Dorry Segev
Johns Hopkins University School of Medicine
Abhijit Sen
Suri Vidyasagar College
Padmanabhan Seshaiyer
George Mason University
Shahriar Shahriari
Pomona College
Barbara A. Shipman
University of Texas at Arlington
Kevin L. Shirley
Appalachian State University
Lawrence H. Shirley
Towson University
Daniel Showalter
Ohio University
Jorge Nuno Silva
University of California, Berkeley
Florence Mihaela Singer
University of Ploiesti, Romania
Kelli M. Slaten
University of North Carolina, Wilmington
David Slavit
Washington State University
Mark R. Snavely
Carthage College
Henrik Sorensen
Aarhus University
Ravi Sreenivasan
University of Mysore
Christopher J. Stapel
University of Kentucky
Catherine Stenson
Juniata College
Kristi L. Stringer
University of Alabama at Birmingham
Stephen Szydlik
University of Wisconsin, Oshkosh
Tristan Tager
Indiana University
Courtney K. Taylor
North Greenville University
Jill E. Thomley
Appalachian State University
Todd Timmons
University of Arkansas, Fort Smith
Elena Toneva
Eastern Washington University
Marcella Bush Trevino
Independent Scholar

Juliana Utley
Oklahoma State University
K. G. Valente
Colgate University
Daniela Velichova
Slovak Technical University in Bratislava
Carlos J. Vilalta
Center for Economic Research and Teaching
Eliseo Vilalta-Perdomo
Tecnologico de Monterrey
Jiri Wackerman
Institute for Frontier Areas of Psychology and Mental Health
Karen Doyle Walton
DeSales University
Christopher J. Weinmann
Independent Scholar
Matthew West
University of Alabama at Birmingham
Bethany White
University of Western Ontario
Sharon Whitton
Hofstra University
Connie Wilmarth
Northwest Christian University
Elizabeth L. Wilmer
Oberlin College
Daniel P. Wisniewski
DeSales University
Todd Wittman
University of California, Los Angeles
Qiang Zhao
Texas State University
Linda Reichwein Zientek
Sam Houston State University

A

Accident Reconstruction

Category: Travel and Transportation.
Fields of Study: Algebra; Data Analysis and Probability; Measurement.
Summary: Accidents can be mathematically reconstructed to model accident risk and to improve safety equipment designs.

Accident reconstruction is important for understanding how accidents happen and for preventing accidents in the future. Principles and techniques from physics, mathematics, engineering, and other sciences are used to quantify critical variables and calculate others. For example, the initial speed of a suddenly braking vehicle can be determined by mathematically analyzing tire skid and yaw marks. The length of skid marks is a function of vehicle velocity and the amount of friction between the wheels and the road surface. In the case of yaw or circular motion, the radius of the yaw mark is also a factor in the calculation, as well as the elevation of the road. Speed can also be calculated from the trajectories, angles, and other characteristics of objects struck by a speeding vehicle, or between two or more colliding vehicles. Investigators may use distances and angles to determine the original positions of passengers who have been ejected from a vehicle. For more complex modeling, mathematicians, engineers, and other accident reconstructors rely on principles and equations from physics, such as those governing energy and momentum, as well as vehicle specifications, mechanical failure analyses, geometric characteristics of highways, and quantification of visibility, perception, and reaction. Data from both real accidents and staged collisions, along with statistically designed safety analyses and other methods such as stochastic modeling, are often used to construct accident simulations and visualizations for use in a wide variety of contexts, including legal proceedings. Actuaries use accident data to model accident risk, which in turn influences insurance rates and public policy, such as seat belt and helmet laws.

Modeling Accident Reconstructions

Accidents related to travel and transportation can have a variety of negative consequences including personal injury and death. The analysis of accidents can lead to improved designs of vehicles and reduced fatalities as well as warning travelers about potential risks of travel. In reconstructing accidents, evidence from photographs, videos, eyewitnesses, or police reports is collected. Decision trees are used to ask questions at each stage of reconstruction and help decide the closest accident scenario dictated by the available evidence. In such reconstructions, probability must be assigned for the likely cause of the accident and for the particular accident type among the possible accident scenarios based on the available evidence. Stochastic modeling is used to help solve such problems in accident reconstructions.

Uses of Accident Reconstructions

Another important aspect of accident reconstructions is to estimate the probability of occurrence of various types of injuries one may suffer in accidents. Such probability estimates are used to help calculate travel insurance. By nature, accidents happen randomly and—since the types of injuries suffered in accidents also vary randomly—it is important to model accident types and predict the kinds of injuries one may suffer in different accident types. Such models can help prepare communities with the optimal number of emergency services and also help doctors prepare for any unique types of injuries they are likely to deal with.

A typical problem is determining the types of special medical facilities that should be established to deal with travel-related accidents in a city. Such problems require stochastic modeling based on past data, which will help in simulating different types of accidents. Simulations help in planning emergency services to deal with accidents. Accident reconstructions may also help in forecasting the number of accidents of different types likely to happen in the near future, which may lead to better planning of the health, emergency, and disaster management facilities in the city.

Safety and Design Using Accident Reconstructions

Accident reconstructions also may help in improving vehicle design. Incorporating safety devices in vehicles is also a very important aspect of design. Safety devices, which help in avoiding severe injuries to passengers because of accidents, are designed with the help of accident reconstruction and are always a matter of high priority. Simulations can be used to develop sensors that can give an early warning about impending accidents or reduce the speeds of vehicles—thereby reducing the severity of an accident. In creating such designs, mathematical optimization methods are used to determine the optimal cost and space to be allotted. Another crucial application of accident reconstruction and accident modeling is driver training. Sophisticated simulators can be used to simulate different accident scenarios and train drivers to react appropriately to each situation in real time. These simulators are based on algorithms and use random number generators to simulate accident situations. Well-developed algorithms that closely simulate real accidents are needed to reduce—or even eliminate—major accidents.

Further Reading

Brach, Raymond, and R. Matthew Branch. *Vehicle Accident Analysis and Reconstruction Methods.* Warrendale, PA: SAE International, 2005.

Franck, Harold, and Darren Franck. *Mathematical Methods for Accident Reconstruction: A Forensic Engineering Perspective.* Boca Raton, FL: CRC Press, 2009.

Ravi Sreenivasan

See Also: Animation and CGI; Crime Scene Investigation; Data Mining; Insurance; Probability.

Accounting

Category: Business, Economics, and Marketing.
Fields of Study: Algebra; Number and Operations.
Summary: Accounting applies mathematics to the recording and analysis of a business's financial status.

Accounting is the recording, interpretation, and presentation of financial information about a business entity, typically with the goal of producing financial statements that describe the business's economic resources in standardized terms. Formal accounting began with the work of Franciscan friar Luca Pacioli, who introduced accounting techniques in his 1494 mathematical work *Summa de Arithmetica, Geometria, Proportioni et Proportionalita*. During the Industrial Revolution, Josiah Wedgwood introduced cost accounting, a technique to ensure a profit margin by calculating the costs of materials and labor at every stage of production and setting the price accordingly. The needs of stockholders and other interested parties within the business, and an increasingly complex business environment, have increased the need for financial record-keeping techniques that are thorough and produce useful financial statements. Modern accounting is assisted by a variety of software packages, but the accountant must still be well-versed in mathematics in order to interpret the information. The fundamental accounting equation can be stated as the following:

Assets = Liabilities + Owners' Equity.

For any given company, assets can be thought of as what the company owns. This includes cash (actual cash and bank accounts), money that is owed to the business (called accounts receivables), inventory, buildings, land, equipment, and intangibles like patents and goodwill. Liabilities are what the company owes. This includes money owed to a bank (notes payable), suppliers (accounts payable), or the government (taxes payable). Owners' equity can take several forms depending on who the owners are: a single person (sole proprietor), a few people (partnership), or shareholders (corporation). Each method of ownership has advantages and disadvantages, but regardless of the method, the owners' equity can be thought of as a net asset since it can be found by subtracting liabilities from assets.

Accounting is the process of keeping track of the operations and financial status of a business.

Accounting as Record Keeping

Whenever a financial transaction takes place, it must be recorded in at least three locations. First, it will be recorded in the general ledger (a book of entry summarizing a company's financial transactions). When recorded, the entry should contain the date of the transaction, a brief description of the transaction, and the monetary changes to all accounts affected (which will be at least two).

From there, the transaction gets recorded a second time in a secondary (or subsidiary) ledger for each of the accounts affected. When the amounts are recorded, they are put into the left (debit) column or the right (credit) column of the ledger. (In bookkeeping, "debit" and "credit" mean left and right, respectively; they are not related to debit or credit cards in this situation.) The total of each column of the general ledger record must add to the same sum. In that manner, all money can be accounted for as going into or out of an account.

In order to determine whether to credit or debit an account, a general rule that works for most accounts is to first look at the fundamental accounting equation. Since assets are listed on the left, to increase assets, the transaction is recorded in the left column (debiting the account) and to decrease assets, the transaction is recorded in the right column (crediting the account). Similarly, since liabilities and owners' equity are listed on the right-hand side of the equation, to increase liabilities and owners' equity, the transaction is recorded in the right column (crediting the account) and to decrease liabilities and owners' equity, the transaction is recorded in the left column (debiting the account). For example, suppose a company needed to purchase $100 worth of office supplies. Furthermore, suppose the company pays $40 with cash and puts the remaining $60 on account (store credit). The general ledger may look like the following:

Figure 1. Purchased office supplies.

04-31-2017	Office supplies	$100	
	Cash		$40
	Accounts payable		$60

In Figure 1, notice that both the right and left columns add up to $100; this shows that no money was lost in the process. Office supplies are considered an asset, so since the company increased the amount of office supplies, that account was recorded on the left—in other words, it debited office supplies for $100. Cash is also an asset, but the company decreased the amount of cash it had. As a result, cash was credited (the transaction was recorded on the right for that account). Accounts payable is a liability the company owes to the retailer it purchased the products from. Since the company increased the amount it owed the retailer, that account was recorded on the right as an increase to the company's liabilities—accounts payable was credited.

Once this transaction was recorded in the general ledger, the company would also need to record this transaction in the Office Supplies ledger, the Cash ledger, and the Accounts Payable ledger. Accounts are debited or credited in their specific ledgers in the exact same manner that they are debited or credited in the general ledger. In a similar manner, the retailer who sold the office supplies would need to record this same transaction into his or her general and secondary ledgers. However, the retailer's transaction would use the opposite side to denote the sale as follows:

Figure 2. Sold office supplies.

04-31-2017	Cash	$40	
	Account Receivable	$60	
	Inventory		$100

Again, the right and left columns add up to the same amount. Contrary to the purchasing company, the receiving company lists three assets to record the transaction. Cash and accounts receivable are both being increased, so debited. The asset "inventory" is being decreased and results in a credit to inventory. If this were a large company, rather than record each individual transaction, the retailer would most likely record an entire day's transactions as a single entry at the end of each business day. Once the general ledger has been recorded, the secondary ledgers need adjusting entries as well to denote the transaction(s).

Accounting as Record Sharing

In addition to keeping records of transactions for a business, accounting is responsible for creating reports that summarize the journals to share with others. To learn about the reports and how to create reports intended for people outside the business (such as shareholders, creditors, or government agencies), a person can take a class in financial accounting. To learn about the reports and how to create reports intended for people inside the business (such as managers), a person can take a class in managerial accounting.

The most common reports created for people outside the business are balance sheets, income statements, cash flow statements, and retained earnings statements. Of the four statement types, the balance sheet is written as a snapshot of the company at a point in time. In contrast, the other three statements are created to show what happened over a period of time such as a month, quarter, or year. When creating these reports, the income statement is usually completed first. As its name implies, the income statement is created to determine the company's income during a specific time period. The income statement is also known as a profit and loss statement (P&L) or earnings statement. Information from the income statement is then used to create the retained earnings statement. Finally, the information from the retained earnings statement is used on the balance sheet.

The balance sheet first lists all of the company's assets in order of liquidity (the ability to turn the asset into

Benford's Law

Benford's law, named after physicist Frank Benford, gives the probability with which the numbers 1 through 9 will occur as the first digit in many types of real-life data. For example, in a list of actual bank account deposits in a given day, about 30% of the time the first digit of the deposit amount will be a 1. Fraudulent data that has been created by people often does not match the expected probabilities.

In very large modern data sets, highly focused tests use this principle to find deviations in selected subsets; for example, the occurrence of a suspiciously large frequency of $24 receipts submitted in a company that has a $25 maximum meal allowance.

cash easily) from the most liquid to the least liquid. The assets are then added together to find the total assets of the company. The balance sheet next lists all of the company's liabilities in order of due date from the soonest due to the latest due. Below the liabilities is listed the owners' equity (which includes retained earnings from the retained earnings statement). The liabilities and owners' equity are added together. Referring back to the fundamental accounting equation, both of these amounts (the total assets and the sum of the liabilities and owners' equity) should equal one another.

Reports created for internal users vary widely depending on the reasoning and the need for the report. Internal reports are usually created and specifically designed for making decisions within the company. For example, manufacturers could use internal reports to determine the optimal price of their product.

Manufacturers may also use internal reports to determine if it is more cost effective to create a needed part or to purchase the part from another company. They may need to consider continuing or eliminating a division of their company. Managerial accounting is also responsible for budgeting and forecasting.

Mathematical Models

Many areas in financial accounting rely on mathematical models for explanation and prediction. For example, models have played important roles in applications such as understanding the consequences of public disclosure, formalizing market efficiency or competition, measuring income, and evaluating equilibrium pricing for goods and services. Some important mathematical techniques used in accounting models include linear regression, systems of simultaneous equations, equilibrium notions, and stochastic analysis. In the latter, random rather than constant inputs are used to model scenarios where decisions must be made under realistic conditions of uncertainty. The data used in these models may be cross-sectional (representing a single snapshot in time) or longitudinal (one or more variables are measured repeatedly to detect trends and patterns). Probability theory is also used to detect instances of accounting fraud.

Further Reading

Davis, Morton D. *The Math of Money: Making Mathematical Sense of Your Personal Finances.* New York: Copernicus, 2001.

Hoyle, Joe Ben, Thomas F. Schaefer, and Timothy S. Doupnik. *Fundamentals of Advanced Accounting.* New York: McGraw-Hill, 2010.

Kimmel, Paul D., Jerry J. Weygandt, and Donald E. Keiso. *Financial Accounting: Tools for Business Decision Making.* 5th ed. Hoboken, NJ: Wiley, 2009.

Mullis, Darrell, and Judith Handler Orloff. *The Accounting Game: Basic Accounting Fresh From the Lemonade Stand.* Naperville, IL: Sourcebooks, 2008.

Verrecchia, Robert. "The Use of Mathematical Models in Financial Accounting." *Journal of Accounting Research* 20 (1982).

Weygandt, Jerry J., Paul D. Kimmel, and Donald E. Keiso. *Managerial Accounting: Tools for Business Decision Making.* 4th ed. Hoboken, NJ: Wiley, 2008.

Chad T. Lower

See Also: Budgeting; Payroll; Shipping.

Acrostics, Word Squares, and Crosswords

Category: Games, Sport, and Recreation.
Fields of Study: Geometry; Number and Operations; Problem Solving.
Summary: Mathematics and symmetry come into play in creating and solving word puzzles.

Acrostics, word squares, and crossword puzzles are the most common forms of word puzzles in English. Acrostics and word squares are over 2000 years old and call for the solver to discover words hidden either covertly (acrostics) or overtly (word squares). The crossword puzzle premiered in 1913 and is similar to a word square expanded onto a larger grid, with gaps. Word puzzles have been used as mnemonics, ciphers, literary devices, educational exercises, and as simple games. Their construction, especially in the case of crossword puzzles, is informed by geometry; their solution can be pursued through probability theory. In a sense, the construction and solving of word puzzles provide pleasures very similar to those of doing mathematics.

Historic Examples

The earliest examples of acrostics are in the Old Testament of the Bible. The Lamentations of Jeremiah and 12 Psalms are arranged so that the first letters of each verse spell out the Hebrew alphabet.

In Greece in 400 B.C.E., Dionysius forged a Sophoclean text titled *Parthenopaeus* with the intention of mocking his rival, Heraclides. Having declared the author to be Sophocles, Heraclides was referred to in one of the several acrostics that Dionysius had included, which read, "Heraclides is ignorant of letters."

In more contemporary times, novelist Vladimir Nabokov enjoyed chess problems, and one can find acrostics, number puzzles, cryptic references, and puns in several of his novels and stories. The last paragraph of his 1951 short story "The Vane Sisters," for example, can be read both as the narrator's confusion and acrostically (taking the first letter of each word) as a message from the dead sisters.

Acrostics are often found in poetry because of its greater flexibility in syntax and phrasing. Former U.S. President George Washington is known to have constructed at least one acrostic when he was 15—a love poem for a girl about whom nothing is known other than her name, Frances.

Another good example of an acrostic poem is to be found at the end of Lewis Carroll's 1871 book *Alice Through the Looking Glass*; each letter of the name Alice Pleasance Liddell begins a new line in the poem about childhood innocence.

Word Squares

If the first acrostics appeared in the Old Testament, word squares were not far behind. One of the most well known is a Latin word square from about 2000 years ago:

S A T O R
A R E P O
T E N E T
O P E R A
R O T A S

This word square is called a 5-by-5 symmetric word square because there are five words that can be read either down or across. The words "TENET," "OPERA," and "ROTAS" will be familiar to speakers of languages descended from Latin. SATOR is a Latin word for planter or creator. AREPO is a contentious word; it can be assumed that it was at some time used in Latin. This particular word square is unique in another way—SATOR reversed is ROTAS, AREPO is OPERA reversed, and TENET is palindromic (reads the same forward and backward).

Below is an example of an ordinary symmetrical 4-by-4 word square using English words

B A S E
A W A Y
S A L E
E Y E S

Many 5-by-5 and 6-by-6 squares exist in English. There are even a few 9-by-9 word squares, though many of the constituent words are extremely unfamiliar.

Those with an interest in algebra will notice that symmetry in word squares is equivalent to symmetry in matrices. If one transposes—swaps the rows and columns—a symmetrical word square, the resulting word square is the same as the original. A non-symmetrical word square does not have this property. A 4-by-4 double word square, like the one below, is not symmetrical. It is a double word square because it contains twice the number of words of a 4-by-4 symmetrical square, that is, eight:

D A R T
O B O E
C L A M
K E M P

Crosswords

Word squares can be entertaining in themselves. However, simply by expanding a word square onto a larger grid and using gaps to section long words into shorter ones, one can create a puzzle of an altogether different kind. By doing so, puzzle creator Arthur Wynne turned the largely esoteric practice of crafting word squares into a puzzle for the masses—the crossword.

The first published crossword appeared in December 1913 in the newspaper *New York World*. Wynne wrote definitions for each of the words he had used to complete a diamond-shaped grid, and it was up to the solvers of the newspaper's puzzle page to fill in the blanks.

Wynne's grid was almost fully "checked," which means that most letters were part of two words—a white square is "unchecked" when it is part of only one word. In U.S. crosswords, it remains the norm to have very heavily—if not fully—checked grids. For other crossword types, particularly cryptic crosswords, grids may be only 50% to 60% checked. Having a fully checked grid means that it is possible to complete the crossword by entering only the across (or down) words. As the number of unchecked squares increases, however, the ability to build on one's correct answers decreases. Most crosswords have a 15-by-15 grid and twofold rotational symmetry (they look the same after 180 degrees of rotation), but differences in the number of checked squares can produce as many as 80 words or as few as 30.

PROVERB, a computer program designed to solve crosswords, relies on the heavily checked nature of American-style grids. Computer scientist Michael Littman and others report that PROVERB averaged more than 95% correct answers in less than 15 minutes per puzzle on a sample of 370 puzzles. This result is better than average human solvers but not better than the best. If nothing else, the complexity of the PROVERB program serves to highlight the vast computing power humans naturally possess.

Instinctively, many people may not be aware that the five most frequently used letters in the English language are E, T, A, O, and I. Crosswords setters (and PROVERB), on the other hand, are acutely aware of this and aim to use letters in their longer words that will be easy to intersect with the shorter ones. It is therefore worth bearing in mind that, for example, "Erie" and "Taoist" will appear in crosswords much more often than "jazz" and "Quixote." Incidentally, the five least frequently used letters are K, J, X, Z, and Q.

Estimates suggest that fewer than 100 people construct crossword puzzles for a living in the United States. Mathematician Byron Walden has been called "one of the best" by a *New York Times* crossword editor. For some, he may be most well known for writing the puzzle that was used in the championship round of the American Crossword Puzzle Tournament, later featured in the film *Wordplay*. He has also analyzed and given talks on symmetry and patterns associated with conventional crossword construction, with the aim of helping people become more skilled puzzle solvers.

Mathematician Kiran Kedlaya is also a well-known puzzle solver and creator. He believes that the brain processes required for computer science, mathematics, music, and crossword puzzles are similar, and he pursues all of these activities professionally and recreationally. One puzzle he created was published on the well-known *New York Times* crossword page, and he regularly contributes mathematics puzzles to competitions like the USA Mathematical Olympiad. He has been quoted as saying, "It's important to tell kids who are interested in math as a career that there are many venues to do it, not just in the academic area within math departments."

Further Reading

Balfour, Sandy. *Pretty Girl in Crimson Rose (8).* Sirlingshire, UK: Palimpset Book Production, 2003.

Littman, M., et al. "A Probabilistic Approach to Solving Crossword Puzzles." *Artificial Intelligence* 134 (2002).

MacNutt, Derrick Somerset. *Ximenes on the Art of the Crossword*. London: Methuen & Co., 1966.

Eoin O'Connell

See Also: Literature; Poetry; Puzzles; Religious Writings; Sudoku.

Actors

See *Writers, Producers, and Actors*

Addition and Subtraction

Category: History and Development of Curricular Concepts.

Fields of Study: Communication; Connections; Number and Operations; Representations.

Summary: Addition and subtraction are binary mathematical operations, each the inverse of the other, and are among the oldest mathematical concepts.

Addition can be thought of as a process of accumulation. For example, if a flock of 3 sheep is joined with a flock of 4 sheep, the combined flock will have 7 sheep. Thus, 7 is the sum that results from the addition of the numbers 3 and 4. This can be written as $3 + 4 = 7$ where the sign "+" is read "plus" and the sign "=" is read "equals." Both 3 and 4 are called addends. Addition is commutative; that is, the order of the addends is irrelevant to how the sum is formed. Subtraction finds the remainder after a quantity is diminished by a certain amount. If from a flock containing 5 sheep, 3 sheep are removed, then 2 sheep remain. In this example, 5 is the minuend, 3 is the subtrahend, and 2 is the remainder or difference. This can be written as $5 - 3 = 2$ where "−" is read "minus." Subtraction is not commutative and therefore the ordering of the minuend and subtrahend affects the result: $5 - 3 = 2$, but $3 - 5 = -2$.

The concept of addition can be extended to have meaning for fractions, negative numbers, real numbers, measurements, and other mathematical entities. The algorithms used for computing the sum or difference, some of which have been taught for millennia, ultimately depend on the representation used for the numbers. For example, the approach used for adding Roman numerals is different from that used to add Hindu-Arabic numbers. Computers perform subtraction using the same circuits they use for addition.

History and Development of Addition and Subtraction

Human beings' ability to add and subtract small whole numbers is probably innate. Some of the earliest descriptions of techniques for handling large numbers come from ancient China during the Warring States period (475–221 B.C.E.), when arithmetic operations were performed by manipulating rods on a flat surface that was partitioned by vertical and horizontal lines. The numbers were represented by a positional base-10 system. Some scholars believe that this system—after moving westward through India and the Islamic Empire—became the modern system of representing numbers.

The Greeks in the fifth century B.C.E., in addition to using a complex ciphered system for representing numbers, used a system that is very similar to Roman numerals. It is possible that the Greeks performed arithmetic operations by manipulating small stones on a large, flat surface partitioned by lines. A similar stone tablet was found on the island of Salamis in the 1800s and is believed to date from the fourth century B.C.E. The word "calculate" was derived from the Latin word for "little stone."

The Romans had arithmetic devices similar in appearance to the typical Chinese abacus. It is difficult to use modern paper-and-pencil techniques for adding and subtracting Roman numerals (with I as one, II as two, V as five, X as ten, L as fifty, C as one hundred, D as five hundred, M as one thousand)—but it worked well in its time, since it was devised for use with an abacus.

During the Middle Ages, counting boards were used to perform arithmetic. A counting board consisted of a series of actual or virtual horizontal lines that were labeled from the bottom by I, X, C, M, and so on. The system borrowed the symbols used for core numbers from the Roman system. The spaces between the lines were labeled starting from the bottom by V, L, and D. A number like MMDCCXXXVIIII (2739) would be represented by placing the appropriate number of counters on each line. The line labeled M would have 2 counters (for 2000, or two thousands). The space just below, labeled D, would have 1 counter (500, or one five-hundreds); the line labeled C, 2 counters (200, or two hundreds); the space labeled L, 0 counters; the line labeled X, 3 counters (for 30, or three tens); the line labeled V, 1 counter (5); and the line labeled I, 4 counters (4, or four ones). The total of all these numbers is 2739. Note that accountants used VIIII (denoting five plus four) to represent 9, whereas stonemasons used "IX"(denoting 10 less 1). To compute the sum MMDCCXXXVIIII + MCLXI, a person would simply transcribe the numbers to the counting board and then combine the counters following rules of carrying to ensure that no more than 4 counters were on any line and 1 counter on any space. This representation was then easily transcribed back into Roman numerals.

Many early books on arithmetic claim that this method of performing arithmetic was especially preferred by women, who at times had the responsibility for keeping the books for small family businesses. Hindu-Arabic numerals and paper-and-pencil methods for performing arithmetic began to appear in Europe in the twelfth century and replaced Roman numerals and the counting board by the nineteenth century.

Two Methods for Subtracting by Hand

Two popular methods for handling "borrowing" that are taught today are shown below. The method shown in the figure below on the left is popular in Italy, England, and the United States, while the one on the right is popular in Spain, France, and parts of Latin America. The example is to compute 3047 − 1964. Starting with the method on the left, first begin with the rightmost column and subtract 4 from 7. Write the result, 3, below the 4. Moving one column to the left, try to subtract 6 from 4, which cannot be done without using negative numbers. The method is thus to attempt to "borrow" 1 from 0, which is the digit to the left of the 4. Again, this cannot be done without using negative numbers. Therefore, the method is to borrow 1 from 3, which is the digit to the left of the 0 resulting in crossing out the 3 and replacing it with a 2. Then the zero becomes a 10, and it in turn can be replaced by a 9 so the borrowed 1 can be placed in front of the 4 to make it 14. Now, one can subtract 6 from 14 to get 8, which is written below the 6. Moving left to the next column, one can subtract 9 from 9 to get a 0, which is written below the 9. Finally, 1 is subtracted from 2 to get a 1, which is written below.

```
 2 9 1
 3 0 4 7          3 0 4 7
                    1 1
-1 9 6 4        - 1 9 6 4
--------        ---------
 1 0 8 3          1 0 8 3
```

To solve the problem using subtraction with carry, use the example on the right. The carrying numbers (the small 1s) affect the numbers on a diagonal, as shown in the example. The number 1 adds 10 to the integer in the top row and adds 1 to the integer in the bottom row. Starting from the rightmost column, 4 is subtracted from 7, resulting in 3, which is written below. Then, try to subtract 6 from 4, which cannot be done, so insert a small 1 to the left of the space between the 4 and the 6. This is interpreted to mean that the 4 has become 14. Subtract 6 from 14 and record the answer, 8, below. Move left to the next column containing 0 and 9. The small 1, written above and to the right of the 9, is added to the 9 to get 10. Attempt to subtract the 10 from the 0 above, which cannot be done. Instead, write a small 1 just to the left of the space between the 0 and 9, and interpret this to mean that the 0 has become a 10. Now, 10 minus 10 is 0, which is written below. Move left to the next column. The small 1, written above and to the right of the 1, is added to the 1 giving 2, which is subtracted from 3 resulting in 1, which is written below.

Adding and Subtracting on a Computer

At the most basic level, whole numbers are represented in a computer in base-two by a sequence of the binary states "Hi" and "Lo" interpreted as "1" and "0." The circuits that perform addition are implemented by sequences of logical gates. Typically a "1" in the leftmost bit indicates that the number is negative, with the remaining bits indicating the magnitude of the number. Subtraction can be performed by the same circuits that perform addition. Two popular approaches are designated as "one's complement" and "two's complement." "One's complement" can best be explained by performing subtraction in base-10 using "nine's complement." Assume a computation of 3047 − 1964. To find the "nine's complement" of 1964, subtract each digit from 9 to obtain 8035. This is added to 3047 resulting in 11,082. The leftmost 1 is viewed as a "carry" and brought around and added to the rightmost digit in an operation called "end-around carry" to obtain the final result: 1083.

Generalizing Addition and Subtraction

The sum of two fractions a/b and c/d is defined to be

$$\frac{ad + bc}{bd}.$$

The sum of irrational numbers (numbers that cannot be represented as fractions of whole numbers) can be approximated only by adding their approximating rationals. The exact sum of two irrational lengths, a and b, can be found exactly using geometry by first extending the segment representing a sufficiently on one end so that the length b can be marked off from that end with a compass.

Addition can be generalized to other mathematical objects, such as complex numbers and matrices. One of these objects, typically called the additive identity and denoted by "0," has the property such that if "a" is any object then the sum of 0 and a is a. The additive reciprocal of an object a is denoted by $-a$ and is defined to the object so that the sum $a + (-a)$ is 0. The difference $a - b$ is defined to be $a + (-b)$.

Further Reading

Flegg, G. *Numbers: Their History and Meaning.* New York: Schocken Books, 1983.

Karpinski, L. C. *The History of Arithmetic.* New York: Russell & Russell, 1965.

Pullan, J. M. *The History of the Abacus.* New York: F. A. Praeger, 1969.

Rafiquzzaman, M. *Fundamentals of Digital Logic and Microcomputer Design.* Hoboken, NJ: Wiley, 2005.

Yong, L. L., and A. T. Se. *Fleeting Footsteps.* Singapore: Word Scientific Publishers, 2004.

CARL R. SEAQUIST
CATHERINE C. GALLEY

See Also: Multiplication and Division; Number and Operations; Number and Operations in Society.

Advertising

Category: Business, Economics, and Marketing.
Fields of Study: Algebra; Data Analysis and Probability; Number and Operations.

Summary: Mathematics is used to weigh the costs and gains of advertising and to profile and target consumers.

Advertising delivers product information from suppliers to consumers—suppliers may be manufacturers, hospitals, software developers, educators—and is critical to the success of a business in marketing development. Advertising media may be traditional (such as television, newspapers, and posters) or technological (via Internet and e-mail), as well as commercial (to sell products for profit) or noncommercial (in political campaigns or for religious purposes). The annual advertising cost in the United States amounts to more than $100 billion.

Advertising includes two stages: the planning stage for marketing strategies, whose goal is business development, and the analysis stage of cost analysis involved with the forms and the contents of communication between suppliers and potential customers. Mathematics and statistics play critical roles in both stages of advertising.

Market Shares

In the planning stage, the analysis of market shares for advertising necessitates matrix operations and multivariate probability inequalities to portray the dynamics of market shares over time. The following is an example of matrix operations, which bridge advertising with market shares. Consider the market shares of General Motors (GM) and Ford in the U.S. automobile industry. Assume that the current market shares distribute as follows:

General Motors:	21%
Ford:	17%
Other Manufacturers:	62%

If GM starts an advertising campaign with the goal of increasing the market share to 29% in three years, GM may count on customers to switch from Ford or other manufacturers to GM. However, in reality, some of the GM customers may switch to Ford or to other manufacturers.

Let a_1, a_2, a_3 be the percentages of original GM users who, at the end of the advertising campaign, remain with GM, who switch to Ford, and who switch to other manufacturers, respectively. Let b_1, b_2, b_3 be the percentages of original Ford users who switch to GM, who remain with Ford, and who switch to other manufacturers, respectively. Let c_1, c_2, c_3 be the percentages of the other customers who switch to GM, who switch to Ford, and who remain with their manufacturers, respectively. Then, the market shares x_{GM}, x_{Ford}, and x_{Others} at the end of the three years are determined by the following simple matrix equation:

$$\begin{bmatrix} x_{\text{GM}} \\ x_{\text{Ford}} \\ x_{\text{Others}} \end{bmatrix} = \begin{bmatrix} a_1 & b_1 & c_1 \\ a_2 & b_2 & c_2 \\ a_3 & b_3 & c_3 \end{bmatrix} \begin{bmatrix} 21\% \\ 17\% \\ 62\% \end{bmatrix}.$$

If GM intends to increase x_{GM} to 29%, GM should advertise specifically to different groups of customers. This is mathematically equivalent to manipulating the elements in the 3×3 matrix above within plausible ranges of the elements.

The foregoing scenario is a simplified example to illustrate the role of matrix operations in advertising. In reality, the story is more complex. For example, the 3×3 matrix above will become an $n \times n$ matrix, where n is the number of competing suppliers in the market. Also, the stochastic feature of the supply-demand market, the market shares, and the corresponding elements for the $n \times n$ matrix change constantly under the influence of the advertising campaign.

Thus, it is more appropriate to treat the market shares as a vector consisting of random variables. In this case, one of the convenient approaches to evaluating the market shares is the method of multivariate probability inequalities in conjunction with the construction of Hamilton-type circuits.

Advertising Costs and Effects

The analysis stage examines costs and effects associated with various communication channels and advertising media. For instance, in Internet advertising, typical cost considerations are cost per mile (CPM), cost per click (CPC), and conversion rate. These terms have strong connections with mathematics and statistics.

For Web advertising, CPM usually refers to the cost for every thousand visits to the publisher's Web site. For example, assume that an ad network offers a $5 CPM for a banner, which was put on three Web sites for three months. If the total page views for the three Web sites are 80,000, 110,000, and 140,000 during the three-month period, the total cost of Web advertising for the ad network is

$$\$5\,\frac{80{,}000}{1000} + \$5\,\frac{110{,}000}{1000} + \$5\,\frac{140{,}000}{1000} = \$1{,}650$$

In general, if an ad is posted in n Web sites, the total cost is

$$\sum_{i=1}^{n} \text{CPM} \times (W_i / 1000)$$

where W_i is the number of Web impressions (visits) to the ith publisher's Web site for the same period of time.

Consider that the number of Web impressions on each publisher's Web site depends on many continuously changing factors; then W_i is a random number. Let $E(W_i)$ be the expected value of W_i, which measures the long-term average of the number of Web impressions of the banner on the ith publisher's Web site. The long-term average cost is

$$\sum_{i=1}^{n} \text{CPM} \times (E(W_i)/1000).$$

CPC refers to the amount that the advertiser pays for each click generated from the Web publisher. For example, if the cost per click is $0.04, and three Web publishers generate 1700, 1600, and 900 clicks in three months, the cost of Web advertising is

$$\$0.04(1700) + \$0.04(1600) + \$0.04(900) = \$168$$

In general, if a Web ad is posted in m Web sites, the total cost is

$$\sum_{i=1}^{m} \text{CPC} \times C_i$$

where C_i is the number of clicks generated on the ith publisher's Web site for a given period of time.

Consider the fact that the number of clicks on each publisher's Web site depends on various unexpected factors: C_i is actually a random variable. Let $E(C_i)$ be the expected value of C_i, which measures the long-term average of the number of clicks generated from the ith publisher's Web site over a given period of time. The long-term average ad cost is then

$$\sum_{i=1}^{m} \text{CPC} \times E(C_i).$$

The foregoing two concepts, CPM and CPC, measure the potential impact of the internet ad only in terms of clicks or Web visits. However, these two concepts are unable to provide the advertiser with information regarding whether the Web impression has been transferred into the desired action (such as buying the advertised product). A useful measurement in Web advertising to help account for the advertising effect is the "conversion rate" (or CR, the average number of people taking the action encouraged by the ad per 100 visits to the publisher's Web site). For example, if out of 2000 clicks on an ad posted on a publisher's Web site, 12 people end up buying the product, the conversion rate of the ad for this Web site is then

$$\left(\frac{12}{2000}\right) \times 100 = 0.6\%.$$

Being highly associated with key factors such as the design of the publisher's Web site, the conversion rate is an index that directly measures the final impact of the ad for the Web site.

Since the conversion rate directly reflects the performance of the Web site, it can be used to compare

Figure 1.

	May	June	July	August
Google AdSense	5%	6.1%	4.3%	7.5%
Chitika	7.3%	5.2%	5.7%	6.4%

advertising effects of two or more Web sites. However, it is risky to compare conversion rates directly. The example in Figure 1 helps illustrate this point. Consider two Web sites: Google AdSense and Chitika. If the conversion rates of the two Web sites are as follows in the past four months, it is impossible to claim which site has better performance on Web advertising.

In fact, the raw values shown in Figure 1 include the stochastic influence of many online factors. In this case, to evaluate the monthly advertising effect of different Web sites accurately, statistical data analysis is needed.

Because of random effects, the expected value of the conversion rate of each Web site should be considered when comparing two or more publishers' Web sites in terms of the conversion rates. Given a set of historical data involving all the Web sites of interest, one of the statistical estimation approaches is the method of "simultaneous confidence intervals," which compares the ranges of expected conversion rates with a prespecified confidence level. For example, with a set of data for the conversion rates of three Web sites over a period of time, if a 95% simultaneous confidence interval reads

$$0.5\% < CR_{\text{Google}} - CR_{\text{Chitika}} < 2\%$$

and

$$1.3\% < CR_{\text{Google}} - CR_{\text{Yahoo}} < 3.4\%$$

it means that at 95% confidence level, the advertising performance (in terms of conversion rate) of Google is better than that of Chitika and Yahoo.

To enhance the accuracy of the simultaneous confidence ranges, or to improve the power of testing multiple advertising effects, the two-stage estimation procedure can be considered. When the underlying distribution of the monthly conversion rates is skewed, the two-stage estimation procedure can be used with nonparametric tests to make inferences on the performance of multiple Web sites.

Data Mining and Advertisements

Masses of personal data being collected every day about consumers, via mechanisms like credit card applications, consumer discount cards, and product views and ratings on shopping Web sites are poised to revolutionize the field of advertising. Data mining is the mathematical and statistical method for sifting through large volumes of data to find patterns and create prediction models, in this case of consumer behavior. In 2009, the online video rental company Netflix awarded a $1 million prize to the winners of its three-year contest to develop a better algorithm to predict what movies its users would prefer, based on ratings data provided by the company.

Finally, mathematics is used not only to decide when, where, and how to advertise products and services but also to determine what to emphasize within the advertisements themselves: discounts on pricing or the number of calories per serving, just to name two. However, it is often difficult to verify those numbers. Many will remember Trident Gum's 1960s slogan, "Four out of five dentists surveyed would recommend sugarless gum to their patients who chew gum." Although the statement was popular at the time, its legitimacy was later questioned, since it came from a survey whose details have never been released.

IBM has initiated a Smarter Planet campaign focused on dispersed or cloud computing (Internet-based computing). Its "Smarter Math Builds Equations for a Smarter Planet" commercial cites mathematics as the universal language and gives a number of ways in which mathematics will be used to create a "smarter planet."

Further Reading

Baines, Paul. "A Pie in the Face." *Alternatives Journal* 27, no. 2 (2001).

Graydon, Shari. *Made You Look—How Advertising Works and Why You Should Know.* Toronto: Annick Press, 2003.
Kotabe, Masaki, and Kristiaan Helsen. *Global Marketing Management.* Hoboken, NJ: Wiley, 2004.
Laermer, Richard, and Mark Simmons. *Punk Marketing.* New York: HarperCollins, 2007.
Murray, David, Joel Schwartz, and S. Robert Lichter. *Ain't Necessarily So: How the Media Remake Our Picture of Reality.* New York: Penguin, 2002.
Russell, J. Thomas, and W. Ronald Lane. *Kleppner's Advertising Procedure.* Upper Saddle River, NJ: Prentice Hall, 1999.

John T. Chen

See Also: Expected Values; Market Research; Matrices.

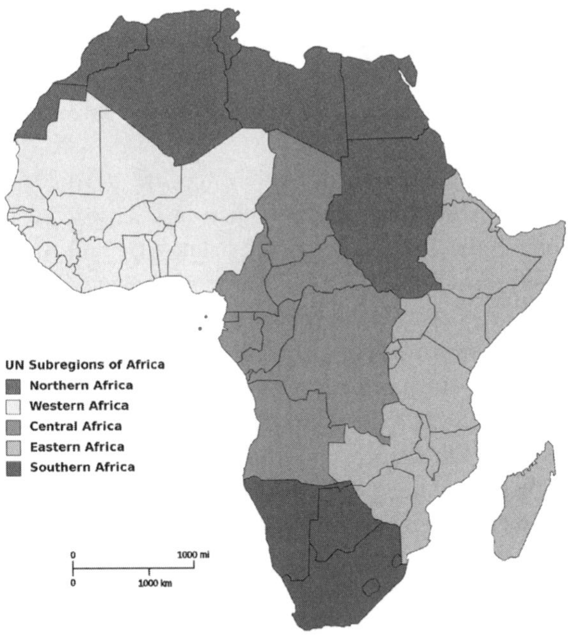

Central Africa is composed of eight countries and is shown in the medium gray shaded area.

Africa, Central

Category: Mathematics Around the World.
Fields of Study: All.
Summary: Central African contributions include counting games and decorative geometric patterns.

Central Africa comprises Angola, the Central African Republic, Chad, Congo, the Democratic Republic of the Congo, Equatorial Guinea, Gabon, and Sao Tome and Principe. Mathematical concepts developed in central Africa include variations of the counting game Mancala and the sophisticated geometric patterns used in traditional art. These patterns, in sand art and pottery, woven into mats and baskets, and displayed in tattoos, include complex symmetries and fractals. Some educators have advocated incorporating these indigenous African manifestations of mathematics into school curriculums.

Mancala

As with much of Africa, variations of the mathematical counting game Mancala were played throughout the region. The mathematics of Mancala games are discussed in more detail in the entry "Africa, East," but some description here is warranted. *The Complete Mancala Games Book* gives rules for 28 different versions of this game played in central Africa. These variations arise throughout much of central Africa but especially in Cameroon and the Congo. While the version of Mancala best known in the United States is a two-row version (also called Wari or Oware), many of the variations played in the Congo have four rows, which adds substantially to the complexity of the game, as well as the complexity of the arithmetic calculations and logical thinking required to play them well. Even with the two-row version, the Congolese variation Mbele uses a complicated game board (a two-row version with many holes in each row, with the rows pinched together near the ends). Again, this adds mathematical complexity to the game.

Geometric Patterns

Many of the most interesting mathematics developed by the peoples of central Africa have been geometric in nature. A significant part of African art traditions include quite complex—and mathematically sophisticated—geometric patterns. These patterns include symmetries in various combinations, between different elements, and between various colors. Claudia Zaslavsky writes: "If one wanted to survey the whole field of geometric design in Africa, one would have to catalogue almost every aspect of life." In central Africa, such geometric patterns are found on pottery, cloths, mats, carvings, baskets, bowls, tattoos, and other objects of daily use.

The Kuba people of the Congo are particularly famous for such art, especially their raffia embroidered cloth. Both *Africa Counts* and *Geometry From Africa* show many examples of Kuba artwork, along with artwork of other African peoples. The woven mats of the Yombe women of the Congo are another example of complex geometric design. Paulus Gerdes has studied these mat designs as an interplay between cultural values and mathematics.

The art of the Chokwe people of the Congo and Angola includes a mathematically challenging art form called "sona," usually drawn in the sand. These drawings are made with a single line continuously weaving through an arrangement of dots, such as the "Lion With Cubs" drawing of the accompanying figure. The heads and tails of the animals are added after the principal line is drawn. These drawings represent stories, morals, or values of the Chokwe, or just an animal or object from their environment. The techniques for determining which dot arrangements will generate such one-line drawings are fundamentally mathematical in nature. Drawings that can be done in a single line, without retracing, are a mathematics topic known as Eulerian Graphs. This artwork of the Chokwe is strongly connected to this mathematical idea, and was being investigated by the Chokwe artists about the same time that the idea was first studied by European mathematicians in the mid-eighteenth century.

The geometric patterns of central Africa extend to include fractal designs. Fractals are a mathematical structure that can be viewed as a repetition of the same shapes at many different sizes or scales. For example, trees have branches, each with smaller branches, and then even smaller branches. Western architecture often has rectangular blocks with rectangular houses, but rarely are such shapes repeated at more than two scales, and rarely is this a conscious shape imitation. African fractals often use circular, oval, or diamond shapes at several scales, with smaller shapes inside or around the larger shapes. There is substantial evidence that at least some of these fractal designs are a conscious choice of the artists and builders, and not accidental. *African Fractals* shows several Cameroonian examples of fractal designs in cities and villages, and even in hair braiding. This book also shows a similar style of pattern, using increasingly smaller but otherwise identical shapes in the art of the Mangbetu people of the Congo.

Education

Several African educators have suggested incorporating these traditional mathematical elements into their schools. The Cameroonian educator A. N. Boma writes: "In African traditional education, the curriculum was organized holistically rather than in discipline areas such as mathematics, history. . . .Education for all cannot afford the luxury of isolating education in terms of disciplines, rather it should take the holistic approach in developing a total person. . . ." The ideas described here integrate mathematics with cultural,

"Lion With Cubs" drawing made with a single line weaving through an arrangement of dots.

artistic, and other elements to achieve this holistic approach. Unfortunately, the schools in Central Africa cannot easily incorporate such ideas. The 2009 *Mathematics in Africa* report describes low percentages of the population attending schools, high student-to-teacher ratios, heavy use of recycled European mathematics textbooks, and few prepared teachers in most of central Africa outside of Cameroon. All of these facts make it difficult to customize mathematics education for African students. Cameroon has a more developed education system, but at the college level it is struggling with filling the mathematics faculty positions that have been approved, and most mathematics teaching there is done in large classes by low-level staff. Nevertheless, with more than half of the central African Ph.D.s in mathematics, Cameroon may become a leader in mathematics education for the region.

Further Reading

Boma, A. N. "Some Lessons From Traditional Practices for Present-Day Education in Africa." In *African Thoughts on the Prospects of Education for All.* Dakar, Senegal: United Nations Educational, Scientific and Cultural Organization (UNESCO)-United Nations Children's Fund (UNICEF), 1990.

Eglash, Ron. *African Fractals: Modern Computing and Indigenous Design*. Piscataway, NJ: Rutgers University Press, 1999.

Gerdes, Paulus. *African Doctorates in Mathematics: A Catalogue*. Maputo, Mozambique: Research Centre for Mathematics, Culture and Education, 2007.

———. *Geometry From Africa*. Washington, DC: Mathematical Association of America, 1999.

Gerdes, Paulus, and Ahmed Djebbar. *Mathematics in African History and Cultures: An Annotated Bibliography*. Cape Town, South Africa: African Mathematical Union, 2004.

International Mathematical Union. "Mathematics in Africa: Challenges and Opportunities." 2009. http://www.mathunion.org/publications/reports-recommendations.

Russ, Laurence. *The Complete Mancala Games Book*. New York: Marlowe Co., 1999.

Zaslavsky, Claudia. *Africa Counts: Number and Pattern in African Cultures*. 3rd ed. Chicago: Chicago Review Press, 1999.

Darrah Chavey

See Also: Africa, Eastern; Africa, Southern; Africa, West; African Mathematics; Board Games; Graphs.

Africa, Eastern

Category: Mathematics Around the World.
Fields of Study: All.
Summary: East African contributions include Mancala, logic games, and games similar to Tic-Tac-Toe.

Eastern Africa is the birthplace of the human species, and includes Burundi, Comoros, Djibouti, Eritrea, Ethiopia, Kenya, Madagascar, Malawi, Mauritius, Mayotte, Mozambique, Reunion, Rwanda, Seychelles, Somalia, Tanzania, Uganda, Zambia, and Zimbabwe. Mancala, an ancient counting game with many variations throughout the continent, originates in East Africa, which is also home to complicated geometric patterns in woven art and a number of logic puzzles and other mathematical games. The quality of mathematics education continues to be a serious concern.

Mancala

Eastern Africa is home to an impressive variety of mathematically based games. The most well known are the many variations of Mancala, often called the "African national game." Although there are hundreds of variations, the general idea is: (1) stones or seeds are placed in pits laid out with two to four rows and several pits per row; (2) players collect the seeds from one pit and "sow" them one at a time into other pits around the board; (3) under some circumstances, the player picks up the seeds from the final pit and continues sowing those seeds; (4) when the move ends, the player will, in some cases, capture seeds from his or her opponent. These games generally involve a substantial amount of counting, adding, and subtracting (for example, to determine where the final seed will land), as well as consideration of multiple possibilities, analysis to calculate where an opponent can move afterward, strategy, and logic. It is no wonder that some leaders (including Tanzanian president Julius Nyerere) were first noticed as good Mancala players. It is uncertain where the game originated, but the oldest dated game boards come

The game of Mancala has many variations, one of which is played on this board in Zanzibar.

from Ethiopia and Eritrea about 1300 years ago. The game is surely older than that, possibly as much as 3300 years old. *The Complete Mancala Games Book* includes 61 different variations of this game played in eastern Africa, including variations specific to every country except Burundi.

Other Puzzles and Games

Logic puzzles come in many forms. One puzzle type common to eastern Africa is the river-crossing puzzle. For example, a man with a wolf, a goat, and a cabbage must use a boat to cross a river except that (1) he can take only one item across at a time; and (2) the goat cannot be left alone with the wolf (who would eat it) or the cabbage (which it would eat). These kinds of puzzles are mathematical because, as Marcia Ascher writes, "A stated goal must be achieved under a given set of logical constraints." Variations of this puzzle, with different logical constraints, appear in Ethiopia, Zambia, and Mozambique.

Several "three-in-a-row" games, related to Tic-Tac-Toe, are played in eastern Africa. In Shisima, from Kenya, players start with an octagonal board, the eight corners, a center point, and lines connecting opposite corners through that center. Players start with three stones each, on the corners closest to them. During a turn, players move one of the stones to one of the nine points (eight corners and the center) connected to it, if it is empty. The goal is to get three stones in a row (a straight line), which must include the center and two corners opposite each other. *Africa Counts* describes two other three-in-a-row games from Zimbabwe, each of which begins like Tic-Tac-Toe where players place stones on points on the board, then continues like Shisima with players moving their stones to get a triple. In Tsoro Yematatu, the board has seven spots, each player has three stones, and one spot is always empty. In African Morris, there are 24 spots, and each player has 12 stones. Here, it could happen that the board becomes filled, but if there is a three-in-a-row during that stage, the player does not win; instead, the player captures an opponent's stone. Hence, the game usually continues into the second phase. These three-in-a-row games are logic puzzles and are examples of games of position, which have been widely studied in mathematics.

Geometric Patterns

The geometric patterns of art from eastern Africa contain a great deal of mathematical and geometric structure and symmetry. Some of the most well known of such crafts are the woven *sipatsi* baskets of Mozambique, and other types of woven baskets and mats from Mozambique, Kenya, Tanzania, Uganda, and Madagascar. This artwork contains varied types of symmetries and dramatic patterns. Paulus Gerdes writes that this art "reveals the force of the imagination and the artistic and geometric creativity of the women and men who weave [these baskets]." Examples exist in the Ba-ila settlement in Zambia and in Ethiopian processional crosses.

Mathematical Education

Mathematical education in eastern Africa shares many of the challenges that exist throughout the continent, especially the lack of prepared teachers at the secondary level. As the South African mathematics educator Jan Persons writes, "At the departure of the Portuguese from Mozambique in the early 1970s, there were only a handful of qualified secondary mathematics teachers. In general, starving the local population of decent and effective education was used as a weapon to halt or, at least, retard development."

This issue has been a major problem in eastern and central Africa, which combined have 48% of Africa's population but have produced less than 8% of Africa's mathematics Ph.D.s. Kenya has a strong college-level mathematics program, having produced nearly half of all Ph.D.s in eastern Africa. Unfortunately, as also happens in central Africa, most of the mathematics students are attracted into professions other than teaching because of the low salaries for teachers. There are several efforts in place to improve mathematics education in these countries, but much work on the educational structures remains to be done throughout this region.

Further Reading

Ascher, Marcia. "A River-Crossing Problem in Cross-Cultural Perspective," *Mathematics Magazine* 63, no. 1 (1990).

Eglash, Ron. *African Fractals: Modern Computing and Indigenous Design*. Piscataway, NJ: Rutgers University Press, 1999.

Gerdes, Paulus. *African Doctorates in Mathematics: A Catalogue*. Maputo, Mozambique: Research Centre for Mathematics, Culture and Education, 2007.

———. *Geometry From Africa*. Washington, DC: Mathematical Association of America, 1999.

Gerdes, Paulus, and Ahmed Djebbar. *Mathematics in African History and Cultures: An Annotated Bibliography*. Cape Town, South Africa: African Mathematical Union, 2004.

International Mathematical Union. "Mathematics in Africa: Challenges and Opportunities." 2009. http://www.mathunion.org/publications/reports-recommendations.

Russ, Laurence. *The Complete Mancala Games Book*. New York: Marlowe Co., 1999.

Zaslavsky, Claudia. *Africa Counts: Number and Pattern in African Cultures*. 3rd ed. Chicago: Chicago Review Press, 1999.

Darrah Chavey

See Also: Africa, Central; African Mathematics; Basketry; Board Games; Mathematical Puzzles; Tic-Tac-Toe.

Africa, North

Category: Mathematics Around the World.
Fields of Study: All.
Summary: North Africa has been a major contributor to mathematics, particularly in ancient Egypt and the Islamic Golden Age.

North Africa, comprised of Algeria, Egypt, Libya, Morocco, Sudan, Tunisia, and Western Sahara, has long been geographically and culturally distinct from the rest of the continent because of the Sahara desert (which includes most of the region) and the proximity to southern Europe and the Middle East. The mathematics of ancient Egypt is among the oldest known mathematics traditions, and the Egyptian city of Alexandria was an important center of learning in the ancient world. Centuries later, Egyptian mathematicians were among the contributors to the Islamic Golden Age, translating classical works, which also helped bring about the Renaissance and Age of Enlightenment.

Mathematics historians and teachers have explored a variety of historical mathematics in the area, such as string figures and precolonial mathematics in Sudan, or the work of Gaston Julia, who was born in Algeria at the end of the nineteenth century and is known for his investigations on dynamical systems. The Julia set is named for him. Modern mathematicians and scholars in North Africa continue to take part in mathematics research and teaching.

Ancient Developments

Papyrus scrolls predating 1500 B.C.E. have been found in Egypt that discuss mathematical topics. One of the more famous is the Ahmes scroll (after the name of the scribe to whom it is attributed), currently held in the British Museum, which describes many problems in algebra and geometry and demonstrates their solutions. It is of particular interest for its use of unit fractions (fractions with a numerator of 1, such as 1/8) and for demonstrating a method of calculating circular areas.

In the Hellenistic period (c. 323–146 B.C.E.), and in the Roman period that followed, the city of Alexandria in Egypt was a center of learning, and the Great Library of Alexandria was the most important library in the ancient world. Euclid (c. 300 B.C.E.), a Greek mathematician who worked in Alexandria, is best known for his treatise *Elements*, which formed the basis for how geometry has been understood and taught for more than 2000 years. Eratosthenes of Cyrene (276–194 B.C.E.) was born in what is now Libya. He estimated the circumference of Earth and is known for the Sieve of Eratosthenes, which is useful in number theory.

One of the best-known Egyptian mathematicians from the Roman period was Ptolemy (c. 90–168 C.E.), a Roman citizen who lived in Egypt. One of his well-known works is the *Almagest*, the most comprehensive surviving ancient treatise on astronomy. Hypatia (c. 350–415), a Greek who lived in Alexandria, was a female mathematician who wrote commentaries and was also known as a teacher of astronomy and philosophy.

Islamic Period

Mathematics flourished during the Islamic Golden Age (c. mid-eighth to mid-thirteenth century). One impetus to this development was the translation of classical Greek works, such as Ptolemy's *Almagest* and Euclid's *Elements*. These translations were often the only surviving copies and their preservation by Islamic scholars allowed them to be reintroduced into Western thought. Besides the appreciation of knowledge for its own sake, the development of mathematical sciences had practical uses in the Islamic world; for instance, knowledge of astronomy was required to understand the phases of the moon and thus correctly observe Islamic holy days, while algebraic notation was developed in part to solve problems relating to the laws of inheritance. Geometric motifs are very common in Islamic art and design, in part because, for religious reasons, Islamic artists did not create representational art, such as portraits. Instead, complex patterns such as tessellation figures (tilings) were developed for artistic use.

Many mathematicians worked in Egypt during the Islamic Golden Age. Ahmed ibn Yusuf (c. 835–912) was born in what is now Iraq but moved to Egypt and died in Cairo. He worked with his father, Yusuf ibn Ibrahim, on mathematics and wrote a book on ratio and proportion, which commented on Euclid's *Elements* and was translated into Latin in the twelfth century. Abu Kamil Shuja ibn Aslam (c. 850–930) was a mathematician who made important contributions to the study of real numbers, irrational numbers, and combinatorics, and some of whose techniques were adopted by the thirteenth-century Italian mathematician Fibonacci. Ibn Yunus (c. 950–1009) was an Egyptian astronomer and mathematician whose most famous work is a handbook of astronomical tables, which is notable for the accuracy of his observations and for his meticulous description of numerous planetary conjunctions and lunar eclipses. Abu Ali al-Hasan ibn al-Hasan ibn al-Haytham (c. 965–1039) was born in Persia but lived primarily in Egypt and died in Cairo. He worked as an engineer, reportedly attempting to develop a method to dam the Nile River, and made important contributions to optics and to the development of the scientific method. Al-Marrakushi ibn Al-Banna (c. 1256–1321) lived in Morocco and may have been born there. He worked on Euclid's *Elements* and texts on algebra and arithmetic operations.

Besides being an Egyptian mathematician, Ptolemy was also an astronomer, a geographer, and an astrologer.

Modern Developments

In the early twenty-first century, mathematical study and research continues in North Africa. Mathematicians belong to professional organizations like the Association Mathématique Algérienne, the Egyptian Mathematical Society, the Tunisian Mathematical Society, and the Société des Sciences Naturelles et Physiques du Maroc. Egypt and Tunisia are members of the International Mathematical Union, which is a worldwide organization designed to promote mathematics. North African countries have participated in the International Mathematical Olympiad, an annual competition held since 1959 for high school students. Algeria first participated in 1977, Morocco in 1983, and Tunisia in 1981.

Further Reading

Gerdes, Paulus. *African Doctorates in Mathematics: A Catalogue*. Maputo, Mozambique: Research Centre for Mathematics, Culture and Education, 2007.

Gerdes, Paulus, and Ahmed Djebbar. *Mathematics in African History and Cultures: An Annotated Bibliography*. 2nd ed. Cape Town, South Africa: African Mathematical Union, 2007.

Joseph, George Gheverghese. *The Crest of the Peacock: Non-European Roots of Mathematics*. Revised ed. Princeton, NJ: Princeton University Press, 2000.

Kani, Ahmad. "Arithmetic in the Pre-Colonial Central Sudan." In *Science and Technology in African History With Case Studies From Nigeria, Sierra Leone, Zimbabwe, and Zambia*. G. Thomas-Emeagwali, ed. Lewiston: Edwin Mellin Press, 1992.

Sarah Boslaugh

See Also: Africa, West; African Mathematics; Arabic/Islamic Mathematics; Egyptian Mathematics.

Africa, Southern

Category: Mathematics Around the World.
Fields of Study: All.
Summary: Southern Africa is the home of ancient mathematical artifacts and modern mathematical innovations.

Southern Africa comprises the five nations of the Southern African Customs Union: Botswana, Lesotho, Namibia, South Africa, and Swaziland. Colonization led to significant European populations, especially in South Africa and Namibia.

The oldest known mathematical artifact is the Lebombo bone, discovered in a rock shelter in the Lebombo Mountains near the South Africa/Swaziland border. There is evidence of the cave having been inhabited continuously beginning some 200,000 years ago, and the bone itself is estimated to be 35,000 years old. The Lebombo bone is a fragment of baboon fibula with 29 notches, most likely used as a tally stick—a notched object used to keep track of quantities. In this case it may have been a menstrual calendar.

Historically, the Dutch and British were particularly influential in this region. For example, the nineteenth-century Boer (also known as Afrikaner) community established the Boer States, including Transvaal and the Orange Free State. It has been documented that the Boer farmers, who were largely descendants of Dutch and some other European settlers, relied heavily on education at home. The migration of large numbers of predominantly British settlers into South Africa in the nineteenth century saw the establishment of more schools and later, universities in the European style. The mathematics heritage of southern Africa reflects both the diversity of the native cultures and the effects of this European colonialism.

South African Mathematicians

One early South African mathematician was Francis Guthrie (1831–1899), who proposed the Four Color Problem. It stemmed from a problem he first explored as a student in which only four colors could be used to denote the counties of England, and no two counties sharing a border could have the same color. Guthrie was born in London but immigrated to South Africa, where he worked as both a mathematician and a botanist. Mathematician Stanley Skewes (1899–1988), who was a faculty member at the University of South Africa and grew up near Johannesburg, postulated his Skewes number, which is an important concept in number theory.

Within South Africa, one well-known mathematician is Chris Brink, who grew up in a town on the edge of the Kalahari Desert and studied at Johannesburg. He earned a degree in mathematics before earning a scholarship to Cambridge University in England, where he completed his doctoral thesis on algebraic logic.

Returning to South Africa, he worked on Boolean modules and was vice-chancellor of the University of Stellenbosch from 2002 until 2007. Outside of the country of South Africa, another early mathematics Ph.D. from the southern Africa region is Abraham Busa Xaba. He was born in Swaziland in 1938 and earned his Ph.D. in 1984. His doctoral dissertation was titled "Maintaining an optimal steady state in the disturbances."

During the latter years of the twentieth century, some South African mathematicians also became known for their work overseas. For example, Lionel Cooper (1915–1979) left the country for political reasons. He grew up in Cape Town and won a Rhodes scholarship to study mathematics at Oxford University. Afterward, he served as a lecturer at Birkbeck College, London, and at Cardiff University, then became head of the Mathematics Department at Chelsea College, London. Abraham Manie Adelstein (1916–1992) was born in South Africa but left to live in England in 1961, where he became a leading medical statistician.

Organizations

As well as these important role models, there have been many attempts to encourage collaboration and development of mathematics in the southern African region. The Southern Africa Mathematical Sciences Association was founded in 1981 and is headquartered in Botswana. Its serves as a forum for the sharing of mathematical ideas for the countries in southern Africa as well as some neighboring countries that may be more broadly defined as being in the southern portion of the African continent.

The African Institute for Mathematical Sciences was founded in 2003 as a partnership of six universities: Cambridge University (England), University of Cape Town (South Africa), Oxford University (England), Université Paris-Sud XI (France), Stellenbosch University (South Africa), and University of the Western Cape (South Africa). Its three primary goals are: promoting mathematics and science in Africa; recruiting and training talented students and teachers of science and mathematics; and building capacity for educational, research, and technological initiatives in Africa. The South African Mathematics Olympiad is held each year for high school students, and teams from southern Africa have participated in the International Mathematical Olympiad since 1992.

Further Reading

Gerdes, Paulus. *African Doctorates in Mathematics: A Catalogue*. Maputo, Mozambique: Research Centre for Mathematics, Culture and Education, 2007.

———. "On Mathematics in the History of Sub-Saharan Africa." *Historia Mathematica* 21, no. 3 (1994).

Gerdes, Paulus, and Ahmed Djebbar. *Mathematics in African History and Cultures: An Annotated Bibliography*. Cape Town, South Africa: African Mathematical Union, 2004.

Simkins, C. E. W., with Andrew Paterson. *Learner Performance in South Africa: Social and Economic Determinants of Success in Language and Mathematics*. Cape Town, South Africa: HSRC Press, 2005.

Vithal, Renuka, Jill Adler, and Christine Keitel. *Researching Mathematics Education in South Africa: Perspectives, Practices and Possibilities*. Cape Town, South Africa: HSRC Press, 2005.

Justin Corfield

See Also: Africa, Central; Africa, Eastern; Africa, West; African Mathematics.

Africa, West

Category: Mathematics Around the World.
Fields of Study: All.

Summary: Mathematics has long been used in west African art, architecture, industry, and music.

The peoples of west Africa have a long history of using mathematics. Everyday uses were similar to mathematics in other traditional societies around the world. Farmers measured their fields and counted their crops, anticipating the production figures. Fishers designed boats to carry them off the coast and prepared nets for catching fish. For both, there were processes to handle their products, either for immediate consumption or—with additional mathematics—for sale in local or distant markets. Markets served as centers of trade and also as centers of mathematical calculations of quantities and sizes, profits and losses. Everyone designed and built houses, often round in shape, which calculus shows to provide the maximum area for a given perimeter. As larger societies and governing units grew beyond the villages, mathematics played a role in governments, from taxation and salaries to the design of palaces and warehouses.

Mathematics in West African Art

West Africa has long been known for its art, textiles, music, and dance. Mathematics is central to the creative and performing arts. Some particular west African examples include carved sculptures, wall paintings, tie-dyed textiles, and woven cloth. Sculptures often show symmetries, not only of human features but also of geometrical designs and proportions of animals, village scenes and daily life, and abstractions of circles, rhombi, stars, and repeating patterns. Often, the palaces of chiefs or emirs became sites of art, especially with designs on the walls or in the architecture of the structure—all incorporating geometrical designs.

Throughout west Africa, textiles have been a central part of culture. From the multicolored patterns in Sierra Leone to the deep blues and indigos of the Hausas, the techniques of dyeing cloth have been popular, especially with tied or sewn folds of the cloth to yield intricate patterns of dyed and nondyed areas of the material. Often, the use of symmetries and Euclidean geometric constructions is necessary to produce the desired circular, radial, rhombic, and zigzag patterns. Woven cloth includes the brightly colored *kente* of Ghana, the metallic shine of the Okenne cloth of western Nigeria, and others. Weaving requires engineering mathematics to design and build looms, and then careful planning so that the strips of material that come

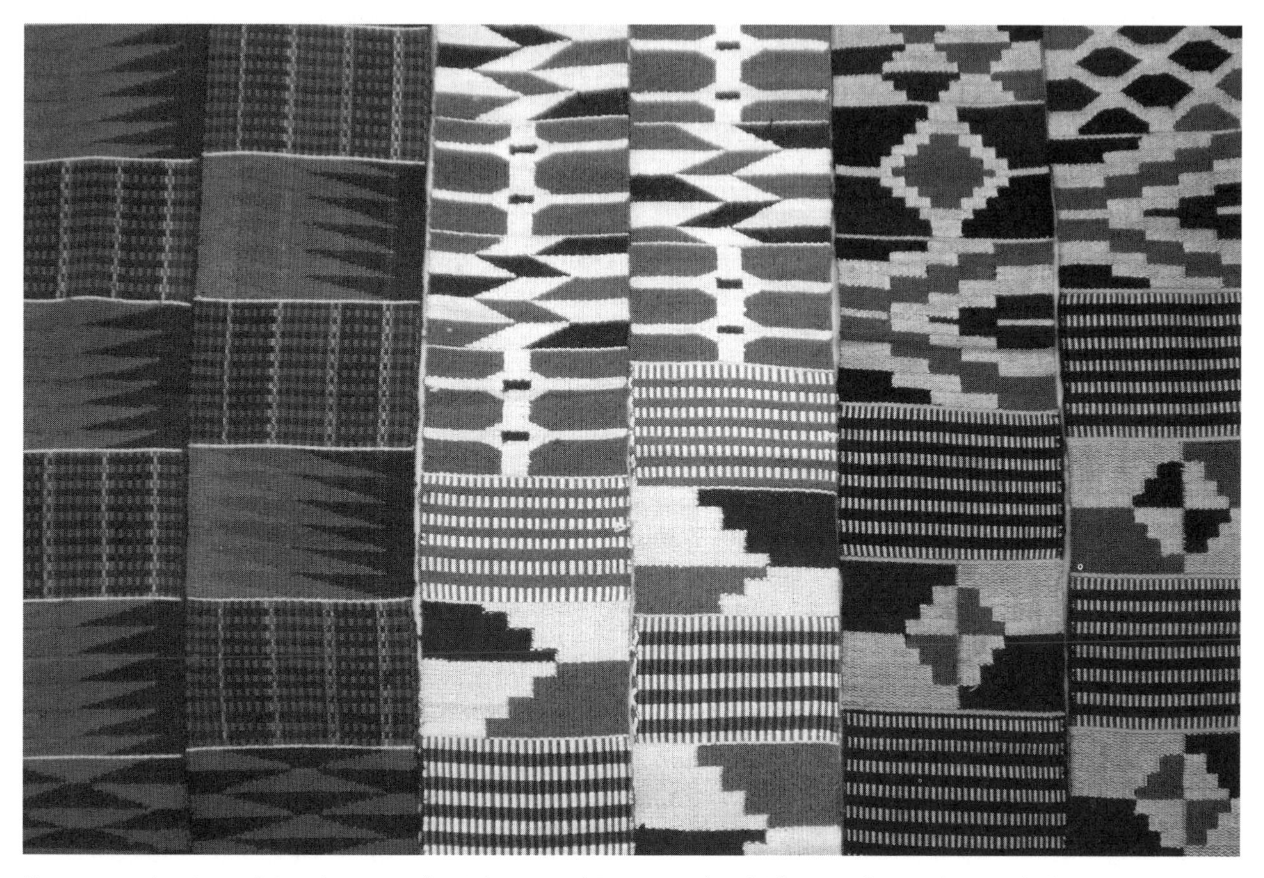

Kente *weaving is traditional among the Ashanti and Ewe people of Ghana, Africa. A* kente *cloth is sewn together using narrow strips of brightly colored cloth with different geometric patterns.*

off the looms will fit together in two-dimensional symmetrical arrangements.

These traditional artistic products have been carried into the present day. Traditional designs are now seen in modern buildings throughout the region. Fashionable textiles sometimes use new materials or printed cloth but continue the geometric traditions. *Kente* has become a popular material not only in Ghana but also in the United States, especially the symmetrical strips used as wraps and ties. Recent studies by ethnomathematician Ron Eglash have demonstrated a variety of uses of fractal patterns in traditional west African arts, ranging from repeating smaller patterns in the geographical arrangement of savannah villages, to necklaces and bracelets, carvings from Mali of increasingly small antelopes, and even corn-row hair braids that repeat smaller shapes as the pattern goes from the forehead and temples to the rear of the head.

Music and dance from west Africa are famous to both ethnomusicologists and jazz aficionados—and to ethnomathematicians. The rhythm patterns, especially from complex drumming structures, often involve unusual time signatures and alternations of loud and soft sounds. The three-dimensional movements of dance, like the carvings and textiles, show complex symmetries and geometrical arrangements of the dancers.

Early in the second millennium, Islam was introduced in west Africa, along with Islamic mathematical studies. This introduction added to the original practical base of west African mathematics, as west Africans adapted Islamic counting methods, reflected not only in the languages of west Africa but also in theoretical mathematics studied at scholarly centers such as Timbuktu (in modern Mali) and Katsina (in modern Nigeria).

Mathematics and West African Development

Since gaining independence, mostly in the 1960s, west African countries have moved rapidly to modernize. In the process, they have shown a dynamic use of mathematics—on a smaller scale than but similar to the technical mathematics of the developed world. Oil production in Nigeria, gold mining in Ghana, and diamond mining in Sierra Leone all use modern mathematical techniques, including those employed by geological surveys, sophisticated industrial equipment design, accounting, marketing, and business management. New businesses are being established to work with cell phones, the Internet, automatic teller machines, television and film production, and other industries that rely on technical mathematics and engineering. Modern freeways connect the larger cities and are designed by civil engineers and urban planners.

Education and West African Mathematics

Education throughout west Africa has grown dramatically since independence—universal primary education remains elusive, but the percentage of children attending school is approaching that goal in several countries. Political independence also brought educational independence, including national curricula offered by the Ministries of Education, the West African Examinations Council's system of standardized examinations, and locally produced textbooks and teaching materials, using familiar names, places, and situations in examples. Local researchers are studying their own cultures, seeking examples of ethnomathematics in traditional life, often with the goal of using these findings to strengthen the content of school mathematics curricula. With only a few universities in existence at the time of independence, west African countries now have numerous universities. These are often managed by the national governments—though some states of Nigeria operate their own universities and research centers, and the number of private universities is growing. These universities offer degree programs in mathematics, the sciences, engineering, and computer science, all with curricula based on the accepted world standards of these fields. Most countries have professional and scholarly organizations of mathematicians and mathematics educators, and periodically there are regional and continent-wide conferences, such as the meetings of the African Mathematical Union (AMU). The AMU's activities include the Commissions on Mathematics Education in Africa, Women in Mathematics in Africa, the African Mathematics Olympiads, and publishing the journal *Afrika Matematica*. Thus, even as west Africa maintains its traditional uses of mathematics in the arts and music, it has also become a part of the modern world mathematics community.

Further Reading

Eglash, Ron. *African Fractals: Modern Computing and Indigenous Design*. New Brunswick, NJ: Rutgers University Press, 1999.

LaGamma, Alisa, and Christine Giuntini. *The Essential Art of African Textiles: Design Without End*. New York: Metropolitan Museum of Art, 2008.
Mendonsa, Eugene L. *West Africa: An Introduction to Its History, Civilization and Contemporary Situation*. Durham, NC: Carolina Academic Press, 2002.
Zaslavsky, Claudia. *Africa Counts: Number and Pattern in African Culture*. 3rd ed. Chicago: Lawrence Hill Books, 1999.

Lawrence H. Shirley

See Also: African Mathematics; Arabic/Islamic Mathematics; Textiles.

African Mathematics

Category: Government, Politics, and History.
Fields of Study: Connections; Geometry; Measurement; Number and Operations; Representations.
Summary: Throughout African history, mathematics has been used in the arts, in engineering and business dealings, and in games.

As in all societies, mathematics has always been a part of the cultures and daily life of people in Africa. One difficulty of studying African mathematics is that for much of the history of Africa, the societies were non-literate, relying on oral traditions to pass their stories to the coming generations.

The wet tropical and subtropical climates of most African civilizations destroyed whatever records may have been kept—or at least hid them from the eyes of future probing historians. Hence, when assertions like the first African mathematical achievement are proclaimed, there should be a caveat that this is the first "that we know of," for similar earlier achievements may well have been lost to history. The African Mathematical Union hosts a Commission on History of Mathematics in Africa—and readily recognizes the difficulty of its charge when even details of the social, political, and military history of precolonial Africa remain difficult to find. Discovering the history of African mathematics is an even greater challenge. Hence, much mathematics history in Africa remains speculative, based on general understandings of how mathematics works in other societies past and present, and fitted into the growing framework of bits and pieces of the history of Africa and African society.

Modern Western mathematics (now used around the world) has indeed come from the developments in the European academy, but it is only the formalized structures of pure theoretical mathematics and their applications in science, industry, and technology that grew from this theoretical work. However, mathematical thinking is much broader than the tightly logical structures of academic mathematics. Everyone who thinks about counting, arranging, or designing—anyone who makes strategic plans for achieving a goal—is thinking in mathematical terms. These examples of mathematics have occurred in Africa as much as anywhere else in the world.

Development of African Mathematics

Before recorded history, Africans herded their animals, planted and harvested crops, and built homes and other structures. All these activities required mathematics. Farming required finding the best time to plant and the appropriate time for harvest. Over time, it is likely that this led to formal or informal calendars, so the farmers would be prepared to do their tasks at the right time. They applied measurements and design as they laid out their fields, including sorting out boundary disputes with neighbors. Anthropologists have even studied the variations in the arrangements of fields in farming communities. When the time for harvesting came, several other mathematical issues arose. Initially, there would be a need for containers and storage bins for the produce, requiring geometrical design.

Later, business mathematics would be used in the markets—even those using barter systems—to determine the comparative values of the products, the gains and losses, and the purchases of other products. Some societies developed currencies—a famous example is the use of strings and bundles of cowry shells by the Yorubas. This probably contributed to the complex numeration system of the Yoruba language, which can handle very large numbers. It has even been suggested that the use of higher numbers came as a result of inflation requiring higher prices. Also, the use of strings and bundles easily flows into the grouping used in place-value of counting systems.

Village life also measured the times of human life, from the diurnal movement of the sun and language of timekeeping, to the much longer periods of milestones of maturity—birth, initiation as an adult, old age, and death. These time markers sometimes went beyond the individual and family, such that entire age cohorts measured time and followed the appropriate customs of their ages together. Kinship relations sometimes were built into mathematical structures, attempting to avoid disputes and maintain a smoothly functioning society.

Over the past one to two millennia, villages grew and coalesced into larger units. As societies grew beyond the size of villages, the mathematics correspondingly grew. The savanna of west Africa saw Songhai, ancient Mali and Ghana, and the Hausa States. The Swahili civilization grew along the coast of the Indian Ocean in east Africa. Also, trading links reached to increasingly distant targets across the Sahara and along distant stretches of ocean coastline. Although few records have survived, it is acknowledged that large governmental and trading organizations required complex record keeping and accounting. A trader would certainly want to keep careful records of items being traded to avoid being cheated by faraway customers. Governments had to handle administrative and logistical details of the equivalent of civil servants and the king's retinue, and, especially, of armies. Longer trade routes required the design of stronger boats for coastal travel and navigational skills for caravan travel across empty desert landscapes. Also, the needs for currencies grew far beyond those of local markets, as traders had to convert the prices of the sellers to those of the buyers and still control costs and profits.

A string of cowry shells on display in a museum. The Yorubas used these shells as currency.

Reaching out from local roots also put Africans in contact with others—and often, the reverse happened as outside groups came into Africa. Either way, this led to a mixing of culture and a growth of experience. Mathematical ideas jumped from culture to culture, contributing a growth of power and sophistication of mathematics. It is reported that when king Mansa Musa of Mali accepted Islam and traveled across the Sahara to make the hajj pilgrimage to Mecca in 1324–1325 C.E., he brought so much of the golden riches of his empire that he upset the economy of Egypt as he passed through! The flow of the Arabs into both west Africa and east Africa brought the intellectual riches of Islamic mathematics. Even in the terminology of counting words, Arabic influence can be seen in the words for the decade numbers (20, 30, 40, and so on) in both the Hausa language of west Africa and Swahili of east Africa. Arab mathematics, which would later also make fundamental contributions to European mathematics, was taught in Qur'anic schools, and scholarly centers were established in various place including Timbuktu and Mombasa. One of the few documented examples of precolonial history of mathematics in west Africa was the work of Muhammad ibn Muhammad, who worked in Katsina—now in northern Nigeria—in the early 1700s. Interestingly, part of his work became controversial—his calculations of "magic squares," which some of the Islamic authorities considered as flirting with the occult. The astronomical calculations required to maintain the calendar of Islamic festivals led to a growth of formalized geometry and trigonometry.

Mathematics in Egypt

In addition to the mathematics of subsistence, daily life, government, and trade, there was also considerable mathematics used in the arts and recreation. Probably the most famous and spectacular mathematics of the arts and architecture on the African continent is the mathematics of early Egypt. Beyond the famous hieroglyphic mathematics of ancient Egyptian numerals and the arithmetic of the problems found in rolls of papyrus, the mathematics of Egyptian architecture reached the level of "wonders of the world." Notably, the famous pyramids are built with precise lengths, angles, and alignments. They fit into near-perfect geometrical shapes—all the more impressive given their massive size and the belief they were actually constructed by uneducated laborers working under the supervision of masters of labor. The mathematical history questions remain: Who did the design work? How were the designs communicated to the individual laborers?

Mathematics in Sub-Saharan Africa

In sub-Saharan Africa such spectacular wonders are not often seen, but the mathematics of the arts remains impressive. Other architectural examples include the massive structure of the Zimbabwe fortress as well as decorative design in chiefs' palaces and public structures throughout the continent. Walls are often decorated with geometrical patterns—some to be washed off for new work when a new king would arrive.

On a smaller scale, many parts of Africa are known for their textile designs. Sierra Leone has intricate tie-and-dye patterns in cloth. Akan weavers in Ghana produce long strips of woven *kente* cloth in bright colors of red, blue, green, and gold, and then align them side by side to create broad sheets used as toga-like robes in traditional dress. Okenne weavers also make cloth, often with metallic threads giving a shiny appearance to the design. All of these patterns require mathematics in their design—especially considerations of symmetry. Tie-and-dye requires careful planning of the ties so that the resulting dye pattern reflects the design pattern. *Kente* and *Okenne* cloth show symmetry both along the initial woven strips and also across the strips in the full cloth of the robe.

The sculptures from many parts of Africa contributed to some of the designs of modern Western art. They show much use of symmetry, scale distortion, and even repetitive fractal-like patterns. Similarly, African music and dance, especially from west Africa, show mathematically complex rhythm structures in drumming and in the use of a variety of plucked and strummed musical instruments. Like African art, the music of Africa has contributed much to Western music, especially via the music the African slaves brought to the Americas, which formed the roots of jazz.

Beyond the arts, recreational mathematics is seen in numerous African games and pastimes. The best example is the many varieties of the mancala games (known under various names in different countries), which involve sharing seeds into pits in a game board, trying to capture the seeds of the opponent. There are many variations of the rules but all require a careful strategy of play and mathematical problem solving. Some game experts have listed mancala among the great games of the world.

Further Reading

Gerdes, Paulus. *Geometry From Africa: Mathematical and Educational Explorations.* Washington, DC: Mathematical Association of America, 1999.

Zaslavsky, Claudia. *Africa Counts: Number and Pattern in African Culture.* 3rd ed. Chicago: Lawrence Hill Books, 1999.

Lawrence H. Shirley

See Also: Africa, Central; Africa, Eastern; Africa, Southern; Africa, West; Arabic/Islamic Mathematics; Egyptian Mathematics.

AIDS

See *HIV/AIDS*

Aircraft Design

Category: Architecture and Engineering.
Fields of Study: Geometry; Number and Operations.
Summary: Mathematics plays a pivotal role in designing, manufacturing, and enhancing aircraft components and launch platforms.

Achieving flight has been a dream of mankind since prehistory, one never abandoned. As early as Leonardo da Vinci, mathematics—the cornerstone of engineering and physics—was recognized as the key to realizing the dream. Da Vinci's 1505 "Codex on the Flight of Birds," for instance, is a brief illustration-heavy discussion attempting to discover the mechanics of birdflight in order to replicate those mechanics in manmade flying machines. Da Vinci considered not simply the wingspan and weight of birds but a fledgling notion of aerodynamics. He was the first to note that in a bird in flight, the center of gravity—the mean location of the gravitational forces acting on the bird—was located separately from its center of pressure where the total sum of the pressure field acts on the bird. This fact would be important in later centuries when aircraft were designed that are longitudinally stable. Today, mathematics is used in the study of all aspects of flight, from launch platform design to the physics of sonic booms.

Complex Analysis and the Joukowski Airfoil

Abstract mathematics can find its place in physical applications people experience quite often. For example, complex analysis and mappings play a vital role in aircraft. In layman's terms, complex analysis essentially amounts to reformulating all the concepts of calculus using complex numbers as opposed to real numbers. This formulation leads to new concepts that cannot be achieved with only real numbers. In fact, the very notion of graphing complex functions, rather than real functions, is quite different—mathematicians often call the graphing of complex functions a "mapping." Taking a simplistic geometric figure, like a circle, and then applying a complex function transforms the figure into a more complicated geometric structure. One figure that results from such a transformation looks like an airplane wing. Furthermore, one can consider the curves surrounding the circle as fluid flow, that is, air currents, and we obtain a rudimentary model of airflow around an airplane wing. This transformation is entitled the Joukowski Airfoil, which is named after the Russian mathematician and scientist Nikolai Joukowski (1847–1921), who is considered a pioneer in the field of aerodynamics. Variations of this transformation have been utilized in applications for the construction of airplane wings.

Nature-Inspired Algorithms

An example of how various fields of mathematics, science, and engineering coalesce is epitomized at the Morpheus Laboratory, where applications of methods and systems found in nature are applied to the study and design of various types of aircraft. For example, biologically inspired research is conducted by studying an assortment of details related to the mechanics of birds in flight.

Birds are an example of near perfection in flight, a fact that humans have long observed. Birds have been evolving for millions for years and have adapted to various environmental changes, thus altering their flight mechanics accordingly. By studying the mathematical properties related to their wing morphing, surface pressure sensing, lift, drag, and acceleration, among other aspects, the researchers at Morpheus Laboratory can use the knowledge they have gleaned and apply it to several different types of aircraft. In order to accomplish this feat, mechanical models of actual birds are constructed and analyzed. Morpheus researchers utilize an assortment of mathematics and physics, including fluid mechanics (the study of air flow in this case) and computer simulations, to analyze the data that result from studying the mechanical birds in flight. The analysis, in turn, results in novel perspectives in flight as well as the design of innovative types of planes.

In addition, many of the problems that arise regarding the machinery and components that comprise an aircraft carrier can also be potentially solved via Darwinian-inspired mathematical models. For example, the structural components of aircraft are constantly being optimized, as numerical performance is attempted to be maximized while cost is minimized.

The managing of cabin pressurization has made it possible for aircraft to fly safely under various weather conditions and landscape formations. This ability is due in large part to devices known as "pressure bulkheads," which close the extremities of the pressurized cabins. Because of the wealth of physical phenomena that influence the stability of these bulkheads, such as varying pressures, it has been a challenge to optimize their design. In the early twenty-first century, it was proposed that the bulkheads should have a dome-like shape, as apposed to a flat one, which was suggested by both mathematical and biological evidence. Interestingly, these two structures demonstrate completely dissimilar mechanical behaviors, which lead researchers to consider different approaches to modeling the dome-like bulkheads.

The dome-like structured bulkheads are analogous to biological membranes and can be mathematically modeled in a similar fashion. In addition to the imple-

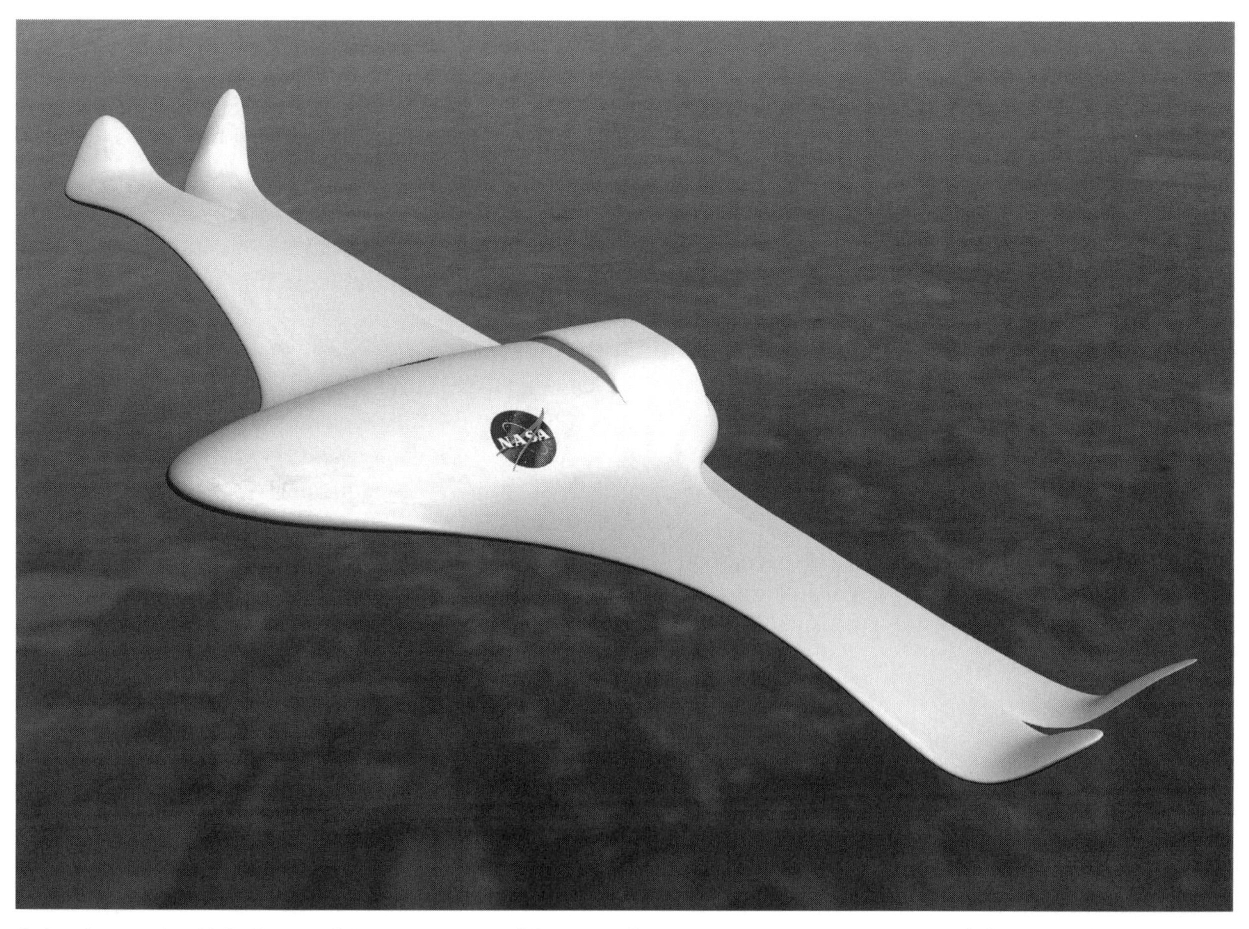

Scientists at the U.S. National Aeronautics and Space Administration (NASA) envisioned this design as a twenty-first-century aerospace vehicle. The "Morphing Airplane" is part of NASA's vision for aircraft of the future.

mentation of these membrane-like designs, the minimization of the cost of their construction and the assurance of their durability is mathematically modeled.

Simulating Sonic Booms

Every time an aircraft travels faster than the speed of sound, a very loud noise is produced called a "sonic boom." The boom itself results when an aircraft travels faster than the speed of the corresponding sound waves. The boom is a continuous event, as opposed to an instantaneous sound, which is a result of the compression of the sound waves. Other fast-moving projectiles like bullets and missiles also produce sonic booms.

Mathematically, this concept means that the velocity of an aircraft (v_a) exceeds the wave velocity of sound (v_s). The Mach Number (M), named after the Austrian physicist and philosopher Ernst Mach (1838–1916), is defined as the ratio of the velocity of an aircraft to the velocity of sound. This ratio is expressed mathematically as

$$M = \frac{v_a}{v_s}.$$

When $v_a < v_s$, $M < 1$, the object is moving at what is often referred to as "subsonic speed." If $v_a = v_s$, $M = 1$, and the object is moving at what is frequently called "sonic speed." Whenever $v_a > v_s$, $M > 1$, and the object is moving at what is titled "supersonic speed." Furthermore, whenever $v_a > v_s$, a shock wave is produced.

The shock waves from jet airplanes that travel at supersonic speeds carry a great amount of concentrated energy resulting in great pressure variations. In fact, two booms are often produced when jets fly at supersonic speeds. Usually, these two booms coalesce into an N-shaped sound wave that propagates in the atmosphere toward the ground. Although shock waves are exceedingly interesting, they can be unpleasant to the human ear and can also cause damage to buildings including the shattering of windows.

However, there is increasing economical interest in designing aircraft carriers that can travel at supersonic speeds with a low sonic boom. To demonstrate, the flight time for a trip from New York to Los Angeles can essentially be cut from 10% to 50% if the plane flies at a supersonic cruise speed instead of subsonic speed. Therefore, physicists are currently developing adaptive methods that model sonic booms in order to ultimately develop aircraft that can travel at supersonic speeds without causing structural damage—aircraft that create a low sonic boom. Aspects such as near-field airflow as well as pressure distribution have been analyzed in these models by utilizing techniques of mathematical analysis.

Aircraft Carriers

Airplanes were a major evolution in modern warfare. World War II aircraft carriers that moved airplanes closer to targets that would otherwise be well beyond their fuel ranges proved to be pivotal to many battles, especially in the Pacific. They continue to be a key component of many countries' navies for rapid deployment of aircraft for surveillance, rescue, and other military uses. Launching from and landing airplanes on aircraft carriers is considered one of the most challenging pilot tasks because of the restricted length of the deck and the constant motion of the deck in three dimensions. A catapult launch system gives planes the added thrust they need to achieve liftoff and requires calculations that take into account mass, angles, force, and speed. Similar issues apply to the tailhook capture system that stops planes when they land.

There are also significant scheduling issues for multiple aircraft on a carrier, fuel use, weapons logistics, and radar systems used to monitor both friendly and enemy planes. Aircraft carriers are like large, self-contained floating cities. Mathematicians work in the nuclear or other power plants that provide electricity for the massive aircraft carriers of the twenty-first century and in many other logistics areas beyond direct flight launch and control. They also help design and improve aircraft carriers. For example, mathematician Nira Chamberlain modeled the lifetime running costs of aircraft carriers versus operating budgets to develop what are known as "cost capability trade-off models," which were used to help make decisions about operations. He also worked on plans for efficiently equipping ships to optimize speedy access to spare components. Some of the mathematical methods he used include network theory, Monte Carlo simulation, and various mathematical optimization techniques.

Further Reading

Alauzet, Frederic, and Adrien Loseille. "Higher-Order Sonic Boom Modeling Based on Adaptive Methods." *Journal of Computational Physics* 229 (2010).

Balogh, Andres. "Computational Analysis of a Boundary Controlled Aircraft Wing Model." *Sixth International Conference on Mathematical Problems in Engineering and Aerospace Sciences.* Cambridge, England: Cambridge International Science Publishing, 2007.

Freiberger, Marianne. "Career Interview With Nira Chamberlain: Mathematical Modelling Consultant." http://plus.maths.org/content/career-interview-mathematical-modelling-consultant.

Morpheus Laboratory. http://www.morpheus.umd.edu.

Niu, Michael Chun-Yung, and Mike Niu. *Airframe Structural Design: Practical Design Information and Data on Aircraft Structures.* Granada Hills, CA: Adaso Adastra Engineering Center, 1999.

Viana, Felippe, et al. "Optimization of Aircraft Structural Components by Using Nature-Inspired Algorithms and Multi-Fidelity Approximations." *Journal of Global Optimization* 45 (2009).

Yong, Fan, et al. "Aeroservoelastic Model-Based Active Control for Large Civil Aircraft." *Science China: Technological Sciences* 53 (2010).

Daniel J. Galiffa

See Also: Airplanes/Flight; Mathematics, Applied; Numbers, Complex; Weightless Flight.

Airplanes/Flight

Category: Travel and Transportation.
Fields of Study: Algebra in Society; Geometry in Society; Number and Operations.

Summary: Aerodynamics is necessary to understanding the flight of objects through three-dimensional space and the forces acting upon them.

Human flight involves moving in a three-dimensional environment within the atmosphere in a stable, controlled way. Aerodynamics is the study of forces and the resulting motion of objects through air. It comes from Greek *aerios*, meaning "air," and *dynamis*, meaning "force." Mathematics is fundamental to understanding flight and in the design of different flying devices and machines, including kites, balloons, helicopters, and airplanes. From Orville and Wilbur Wright's initial experiments with gliders at the beginning of the twentieth century, to the breaking of the sound barrier in the middle of the century, to the development of suborbital craft at the start of the twenty-first century, airplanes have been constructed in many different forms.

However, the ability to fly for all fixed-wing aircraft ultimately depends on a differential movement of air above and below the wings to generate positive lift. Control depends on three parameters, known as "pitch," "yaw," and "roll," that are angles of rotation in three dimensions or axes about the plane's center of mass. Mathematicians and others continue to study flight in order to more fully understand the mathematical and scientific principles that keep heavier-than-air craft in the air and to produce designs that are faster, safer, and more efficient. They also explore related issues in air travel, such as optimal strategies for loading passengers onto planes and the scheduling of aircraft flight crews.

Mathematical History

Stories from many cultures around the world suggest that humans have been interested in flight for thousands of years. There is evidence that the Chinese used kites well before the first century C.E. Leonardo da Vinci

The Father of Aviation

Engineer George Cayley (1773–1857), working in the eighteenth and nineteenth century, is often called the "father of modern aviation" for his research, which helped identify the aerodynamic forces of flight: weight, lift, drag, and thrust. Though Cayley experimented with manned gliders, modern heavier-than-air flight is generally traced to the 1903 launch of the Wright Flyer, a twin propeller biplane with a single motor to provide thrust and mechanisms so that the pilot could control for pitch, roll, and yaw.

Their design helped overcome previous obstacles to sustained stable and controlled flight by adding ailerons to the wings, elevators to the tail surfaces, and rudders to the fuselage to manage airflow. By common convention, roll is motion about the longitudinal axis of the plane. Yaw is movement about the vertical body axis. Pitch is movement about an axis that is perpendicular to the longitudinal plane of symmetry. Pilots require a firm grasp of this three-dimensional geometry to navigate aircraft and follow directional headings.

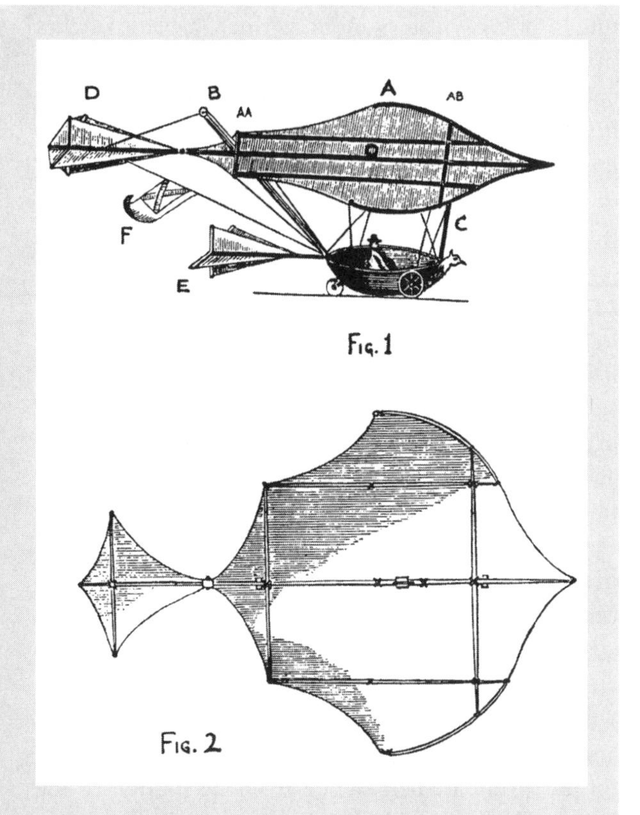

George Cayley's "Governable Parachute" design was printed in Mechanics' Magazine *in 1857.*

recorded his studies of flight in the fifteenth century with more than 100 drawings, including his theoretical ornithopter. Air is a fluid, and so much of the mathematics of flight science derives from fluid force studies, such as those performed by mathematician Daniel Bernoulli in the seventeenth century. Bernoulli's principle is one foundation of flight mechanics.

Mathematical models for flight rely on the Navier-Stokes equations, named for mathematicians Claude-Louis Navier and George Stokes, which are fundamental partial differential equations describing fluid flow. They have many extensions. The Darcy–Weisbach equation, derived by dimensional analysis and named for engineer Henry Darcy and mathematician Julius Weisbach, is important to understanding the dissipation of energy because of friction, such as drag. Working in the early twentieth century, mathematician Otto Blumenthal studied the theory of complex functions, which he also applied to problems such as stress in airplane wings. Mathematician Selig Brodetsky studied equations of airplane motion, including three-dimensional phugoids, which are extensions of common, undesirable oscillatory motions where a plane pitches up and climbs, then pitches down and descends, with changes in airspeed. Peter Lax studied a class of nonlinear equations that can develop singularities, which have applications in aerodynamics that are related to phenomena like the shock waves that result from breaking the sound barrier.

Principles of Flight

Balloons are an example of lighter-than-air craft that use buoyancy to ascend and descend within the atmosphere, and hot air balloons are known to have been explored and used in the eighteenth century. There is also evidence that miniature hot air balloons were used in China for several centuries.

Heavier-than-air craft use the principle of lift to overcome gravity. There have been various mathematical and physical theories posed regarding how lift in airplane wings is accomplished. Aerodynamicists have analyzed how the motion of the air over an airplane wing creates circulation and differential pressure above and below the wing, which creates lift. Lifting forces on the airfoil are perpendicular to the motion of the lifting surface through the air and, in level flight, they counteract gravity. An observable example is the "sing" or hum that occurs in telephone wires in a steady wind, which is a repeating pattern of swirling vortices. This effect is because of the oscillations induced by a phenomenon called "vortex shedding," which causes the wires to oscillate perpendicular to the wind flow.

Studies and models suggest that an airfoil produces circulation in a similar manner. Airfoils can be optimally designed to take advantage of this effect by allowing a smooth flow to develop over the surface of the airfoil, called "laminar flow." The Reynolds number, named for mathematician Osborne Reynolds, quantifies laminar flow. Without laminar flow over an airfoil, turbulence is produced and vortex shedding occurs. Others suggest that aircraft lift is a Newtonian reaction force, named for Isaac Newton, coupled with the Coandă effect, named for engineer Henri Coandă, which is the tendency of a fluid to be attracted to a surface, like an airplane wing. The wing pushes the air down, so the air pushes the wing up.

Lift and Thrust

In general, a pilot taking off from the ground initially accelerates directly into oncoming wind whenever possible, since there is agreement based on observation and mathematics that relative forward motion of the plane's wings with respect to the air is required for flight. Usually, the plane itself is in motion, though a strong wind over a stationary wing can also generate some lift. To maintain a steady, level flight path after takeoff, without any added acceleration, two mathematical relations must be maintained: thrust = drag and lift = weight. Early aircraft engines were powered by gasoline, similar to automobile combustion engines. A fundamental problem of weight, which inhibited lift, was solved by using aluminum as a construction material. Although oxygen is needed to burn gasoline, it is not carried by the aircraft but extracted from the atmosphere so that it does not add to the mass of the aircraft. Jet engines compress and discharge a fast-moving jet of air to generate thrust, using the same principles of fluid dynamics that govern other aspects of aircraft flight, according to Newton's third law of motion. In contrast, a rocket must carry propellants, both fuel and oxidizer, and can thus fly outside of the atmosphere. The added force helps compensate for the extra weight.

Flight Speed

The types of speeds of flight are typically classified as slow subsonic flight, fast subsonic flight, trans-sonic flight, and supersonic flight. The Bell X-1 rocket-propelled airplane is credited as the first piloted aircraft in the world to break the sound barrier, under control of test pilot Charles Yeager. Other planes have been thought to have broken the sound barrier during steep dives, which many do not consider flight. The joint United Kingdom and France plane known as the *Concorde*, which flew from the 1970s until its retirement in 2003, was the only commercial supersonic aircraft. Commercial jets of the early twenty-first century typically achieve speeds in the range of 80% to 85% of the speed of sound, the slower end of trans-sonic flight.

The design speeds tend to avoid compressibility effects in air, which occur above roughly 80% of Mach 1. The Mach number is a ratio of the speed of the aircraft to the speed of sound at the aircraft's altitude. Supersonic flight requires much more energy to sustain, and generally only military aircraft conduct sustained supersonic flight within the atmosphere. The Prandtl-Glauert equation, named for scientists Ludwig Prandtl and Hermann Glauert, is used to help correct computations of fluid flow at high speeds a function of compressibility, while the Prandtl-Glauert singularity is observed as a visible cloud of vapor that results from air pressure changes around a trans-sonic airplane. The pressures can be modeled as an N-wave, named because a plot of pressure versus time resembles the letter N.

A mode of atmospheric flight explored with experimental aircraft at the beginning of the twenty-first century is hypersonic flight, which starts at speeds approximately 5–10 times the speed of sound. Special engines must be developed to make this speed possible. Previously, the Lockheed Aircraft SR-71 held the speed record at greater than Mach 3. It was powered by a special fuel and was air breathing. In 1974, the SR-71 set a speed record flying across the Atlantic from Beale Air Force Base in Louisiana to London in less than two hours. This flight occurred many decades after aviator Beryl Markham's speculations about flying the Atlantic in an hour. Hypersonic aircraft flying at speeds greater than Mach 5 likely will be powered by different forms of air breathing propulsion systems, such as turbine-free engines known as "scramjets," which at very high speeds use ram air compression to ignite a fuel in the engine. In principle, such designs have the capability of going at very high speeds at high altitude and form a transition to spaceflight.

Further Reading

Anderson, David, and Scott Eberhardt. *Understanding Flight*. 2nd ed. New York: McGraw-Hill, 2009.

Tennekes, Henk. *The Simple Science of Flight: From Insects to Jumbo Jets*. Revised and expanded ed. Cambridge, MA: MIT Press, 2009.

Julian Palmore

See Also: Aircraft Design; Skydiving; Weightless Flight; Wheel; Wind and Wind Power.

Algebra and Algebra Education

Category: History and Development of Curricular Concepts.

Fields of Study: Algebra; Communication; Connections; Geometry.

Summary: Algebra and algebra education have undergone many radical changes and remain highly adaptable mathematical disciplines with many real-world applications.

When they hear the term "algebra," many people may think only of solving an equation for an unknown variable x. In reality, algebra is a broad mathematical discipline that includes a range of theories and methods and which has no single agreed-upon definition. Even young children may engage in algebraic reasoning, such as understanding the relationships between quantities or manipulating symbols, without referring to it by name. For much of human history, computations were likely performed using a variety of words and symbols to meet needs such as accounting, taxation, and planting. There is evidence of algebraic problem solving in Egypt and Babylonia. Their techniques appear to have relied a great deal on spoken rhetoric rather than symbol manipulation, though the Babylonians solved quadratic equations using methods similar to those taught in the twenty-first century.

Algebraic thinking is also found in works from ancient China. Greeks, Hindus, Arabs, Persians, and Europeans all made advances and contributions to algebra, and the term is derived from an Arabic word. In the nineteenth century, mathematicians began to expand the notions of algebraic form and structure to encompass more types of mathematical objects such as vectors and matrices as well as operations that could be carried out upon these objects. Also, algebra was not constrained to the ordinary systems of numbers, and noncommutative algebras emerged. The discipline of abstract or modern algebra has grown even further to encompass concepts like groups, rings, and fields.

Concurrently, algebra has become increasingly more important in education at all levels. One of the perceived advantages of algebra and algebraic thinking is that problem solving can be accomplished by symbolic manipulation rules without constant reference to meaning, and these generalized problem-solving skills are viewed as advantageous for students in a wide range of life and occupations skills. This notion has led to a somewhat controversial "algebra for all" approach in many K–12 educational systems in which all students must take an algebra course before graduating, and basic algebraic concepts are introduced as early as the primary grades.

Early History

Even in the classical and ancient period, people had started to use numerals such as 1, 2, and 3 (or I, II, and III, for example) to represent quantities. Numerals, however, bore a direct relation to the quantity being counted. The numeral 1, for instance, only ever referred to a quantity of one. In ancient Egypt, some mathematicians had started to use other symbols, called *ahau*, to represent unknown quantities. These symbols are called "variables" in the twenty-first century because the quantity or number they represented could vary. But the variation in quantity that the Egyptians allowed for was much more restricted than what is allowed for in modern algebra. For example, the symbol x can refer to any number (whole, integer, or other) depending on the mathematical context in which it is used.

Thus, while ancient Egyptians and mathematicians in other ancient civilizations may have used symbols to represent quantities, they did not use symbols in the generalized way in which they are used today. In fact, it was only in the third century C.E. that a Greek mathematician, Diophantus of Alexandria, first used letters of the alphabet to stand in for numbers. It is because of Diophantus's works that mathematicians started to express "an unknown quantity" using symbols such as x and y rather than written words.

Al-Khowarizmi

Diophantus's symbolical technique was not widespread, however. In fact, the term "algebra" actually stems from a period much later than that of Diophantus. It comes from the work of the eighth-century Muslim scholar Muhammad Ibn Musa Al-Khowarizmi (there are various spellings of his name). Al-Khowarizmi worked as a scholar and intellectual during the reign of the Caliph al-Ma'mun (r. 813–833 C.E.).

Al-Khowarizmi was a prominent member of the Bayt al-Hikma, the "House of Wisdom," which the

Pioneers in algebra include Muslim scholar Muhammad Ibn Musa Al-Khowarizmi, Italian mathematician and friar Luca Pacioli, and French mathematicians René Descartes and Évariste Galois.

Caliph had created as an academy and library to promote science. Al-Khowarizmi's book, *Al-Kitab al-mukhtasar fi hisab al-jabr w'al-muqabalah* (an abridged book on the operations of al-jabr and al-muqabalah) is the oldest surviving Arabic book on mathematics. Al-Khowarizmi was also one of the first algebra teachers, as he taught algebra within the Bayt al-Hikma as a subject on its own. Although the ancient Egyptians and Babylonians did produce texts on arithmetic, algebraic, and geometric problems as early as 2000 B.C.E., Al-Khowarizmi was among the first to teach algebra as a science on its own rather than as a subbranch of other branches of mathematics.

The word *al-jabr*, from which the modern-day term "algebra" is derived, first appeared in the title of Al-Khowarizmi's book. Some historians have interpreted it to mean "the restoration of a broken bone" or, in mathematical terms, "the removal of the negative quantity from the equation," while the word *al-muqabalah* has often been interpreted to mean the removal of positive quantities. Gandz has contested these interpretations, however, to argue that a better translation of *al-jabr* is simply "the science of equations."

Europeans

Europeans first became acquainted with Al-Khowarizmi's works through Latin translations by Gerhard of Cremona (1114–1187) and Robert of Chester (c. 1150), both of which first appeared in the twelfth century. Historians have often accredited these Latin translations of Arabic mathematics with the origins of European algebra. One of the first European treatises on algebra to emerge in the Renaissance period was written by the Italian mathematician and friar Luca Pacioli in 1494. Other Italians worked on varied algebraic problems in subsequent years, including Scipione del Ferro (1465–1526), who was able to derive the solution to a cubic equation in the early sixteenth century. The Italian mathematician, Niccolò Tartaglia (1499–1557), derived a general solution to cubic equations a few years later.

In the same century, the French mathematician René Descartes (1596–1650) began to combine algebra (and algebraic rules) with geometry. Descartes was the first to apply algebra to the study of geometric curves. In 1637, he published a work in which he represented curves by means of algebraic equations. Descartes' innovation was to study curves in their algebraic form rather than in their geometric form. The result was a field of mathematics known as "analytic geometry" (also called "geometric analysis") according to some eighteenth and nineteenth practitioners. Analytic geometry allowed mathematicians to use symbols, along with the rules that govern the combination and interaction of symbols, to solve problems related to the motion of bodies in space and the behavior of geometrical objects, such as circles, parabolas, and hyperbolas.

Solving Equations

Algebra could therefore be used to find solutions to linear equations such as $ax + by = 0$, which describe lines in space; quadratic equations, such as $ax^2 + bx + c = y$, which describe parabolas in space; cubic equations, such as $ax^3 + bx^2 + cx + d = 0$, which describe cubic relations in space; and other higher-order equations, such as $a_n x^n + a_{n-1} x^{n-1} + \cdots + a_0$, which describe various curves. The upshot of the Cartesian use of algebra in geometry was that algebraic manipulations could be used to also solve "systems of equations," such as

$$\begin{aligned} ax + by &= c \\ dx + by &= f. \end{aligned}$$

Another outcome of the rise of analytic geometry was the development of the calculus in the seventeenth century. However, although calculus uses the tools of algebra—including symbolic representation and algebraic manipulation—to compute its solutions, it is not the same as algebra. Algebra is generally understood to include only those expressions that possess a finite number of terms and factors. This means that the computation of solutions to algebraic equations terminates after a certain number of steps. In calculus, on the other hand, the concept of a "limit" means that the process of differentiation can be repeated *ad nauseam* and therefore never terminate.

Modern Period

Over the course of the past 1000 years, algebra has thus expanded from a basic use of symbols in simple numerical reasoning to the analysis of structures called algebraic "fields" and "groups" in the nineteenth and twentieth centuries. In fact, the "modern" period in algebra is typically understood as having begun in the early nineteenth century with the work of mathematicians such as the French mathematicians Joseph

Louis Lagrange (1736–1813) and Évariste Galois (1811–1832), as well as the Norwegian mathematician Niels Henrik Abel (1802–1829). Galois, for instance, worked on the concept of an algebraic field. Though Galois died prematurely young at the age of 20 (as the result of a duel in which he was shot), his work later culminated in what is called Galois Theory.

Another important change in the field of algebra occurred in the mid-nineteenth century with the algebraic-geometric work of the Irish natural philosopher Sir William Rowan Hamilton (1805–1865). Hamilton started to work on couples (number-pairs that can be represented as (x, y) on a Cartesian graph) to understand the algebra that could be used to describe their behavior.

In trying to extend the algebra of couples to the algebra of triplets (numbers that could be represented by the point (x, y, z) on a three-dimensional axis system), Hamilton generated an interesting mathematical operator known as the "quaternion." The quaternion can be represented as $w + xi + yj + zk$, where w, x, y, and z represent real numbers, and i, j, and k represent imaginary numbers. To get his quaternion algebra to work, however, Hamilton had to manipulate the standard rules of algebra as they were conceived of at the time. While it is the case in normal arithmetic that $1 \times 2 = 2 \times 1$, such that the order of the numbers does not affect the outcome of the operation of multiplication, Hamilton's quaternions did not follow this rule. Hamilton found that when numbers are represented as directed lines in space (called "vectors"), the order in which the numbers are multiplied with one another does matter. In Hamilton's algebra, therefore, $1 \times 2 \neq 2 \times 1$. Rather, $1 \times 2 = -(2 \times 1)$.

Hamilton is often seen as a pioneer in the study of algebras. Based on his work in quaternion algebra, other mathematicians developed the idea that by changing the rules of the game—by playing around with the standard rules of algebra and arithmetic, such as the commutative principle in multiplication—one could generate new algebraic systems in which the component parts—the variables being manipulated and the objects they represent—do not necessarily follow the same rules as normal algebra.

Another mathematician who developed a similarly new algebraic system was Hermann Grassmann (1809–1877). However, Grassmann's works were largely unknown across Europe until the mid-nineteenth century, by which point Hamilton had already published his major works on quaternions. A British mathematician who attempted to extend Hamilton and Grassmann's new algebraic systems to n-dimensional space was William Kingdon Clifford (1845–1879). Clifford died young, and, as a result, it took many years for his bi-quaternion algebraic operator to become widely known, understood, or used.

Fermat's Theorem

The history of algebra is therefore replete with breakthroughs. In the seventeenth century, a French mathematician, Pierre de Fermat, worked on a problem in number theory that he had picked up while studying the works of Diophantus. Fermat was interested in studying the Pythagorean numbers. Pythagorean numbers are sets of three numbers, such as a, b, and c, which satisfy the equation $a^2 + b^2 = c^2$. Students often learn about Pythagorean numbers through the Pythagorean theorem, which describes the length of sides in right-angle triangles in geometry. Fermat, however, was not interested in triangles so much as he was interested in the consequences of slight manipulations to the Pythagorean theorem.

He attempted to determine the consequence of manipulating the exponents in the Pythagorean numbers from 2 to n. In so doing he wrote, "I have discovered a truly remarkable proof" Fermat explained that when the Pythagorean theorem is made to read $a^n + b^n = c^n$, the new equation has no integer solutions for any value of n greater than 2. In other words, it is impossible to find numbers a, b, and c that satisfy the equation $a^5 + b^5 = c^5$. Fermat never offered a full proof of this claim and mathematicians ever since have struggled to generate it. This bit of algebra is still called a "theorem" to indicate that, although it is believed to be true, one cannot be sure that it actually holds true for all integer values of n.

Mathematicians who have tried to prove Fermat's Theorem over the years have been led to develop other branches of algebra along the way. One example is Eduard Kummer (1810–1893), who created the concept of "ideals" in algebra. The theory of ideals remains an important tool in algebraic systems. An "ideal" A is a (nonempty) subset of a ring R whenever the sum of two elements of A is an element of A as well. In addition, if a is any element of the subset A, and r is any element of the ring R, the products ar and ra are both in the subset A.

An example of this is the integer 3. All of the multiples of 3 form an "ideal" in the ring of the set of integers.

Later Developments

By the late nineteenth century, mathematicians in Europe, Great Britain, and the United States also became interested in studying the structure of certain algebraic equations. Rather than concerning themselves with particular solutions to individual equations, these mathematicians wanted to identify the axioms (or laws) that governed the behavior of differing algebraic equations. These mathematicians focused on the structure of algebraic systems, where a system consists of a set of elements and a set of operations that abide by certain axioms (or rules). The simplest example of an algebraic structure is called a "group." The French mathematician Camille Jordan (1833–1922), the German mathematician Felix Klein (1849–1925), and the Norwegian Sophus Lie (1842–1899) studied groups and did much to establish this area of algebraic research, although older mathematicians such as the eighteenth-century German mathematician Leonhard Euler and the nineteenth-century German mathematician Carl F. Gauss (1777–1855) had already developed some foundational notions that related to abstract groups. Groups are a fundamentally nineteenth-century idea. By the mid-twentieth century, the notion of a group had become widely accepted and had even come to form the core of abstract algebra.

Throughout the nineteenth and twentieth centuries, mathematicians who worked on various aspects of algebraic structure included people such as Benjamin Pierce, Eduard Study, Karl Weierstrass, Richard Dedekind, Theodor Molien, Élie Cartan, Emil Artin, and the twentieth-century female mathematician Emmy Noether (and the entire "school" of mathematicians that she fostered). Some of the "groups" that they helped to define, use, and develop include semigroups, loops, rings, integral domains, fields, lattices, modules, Boolean algebras, and linear algebras, among others.

In the twentieth and twenty-first centuries, abstract algebra has come to include a wide variety of subject topics, including negative and complex numbers, proportions, theory of exponents, finite arithmetic progression, geometric progression, mathematical induction, the binomial theorem, permutations and combinations, the theory of equations, partial fractions, inequalities, and determinants.

Algebra Instruction

Algebra developed because of the need to solve real-life questions and as an extension of mathematical investigations, but in the eighteenth century, mathematicians such as Colin Maclaurin and Euler thought of algebra as a universal arithmetic, and education focused on solving equations for unknown quantities by symbol assignment and manipulation. The focus on symbol manipulation and transformational activities such as collecting like terms, factoring, and simplifying equations continued in school algebra until the mid-1960s, when educators experimented with ways to make algebra more meaningful to students. By the early 1990s, generational activities that included algebra as a way to describe numerical or geometric patterns replaced transformational activities in some countries. Teachers also investigated the effectiveness of a wide variety of teaching strategies such as computer algebra software, historical perspectives, or active learning methodologies, and there were also many algebra survival books marketed such as *Hot X: Algebra Exposed* by actress Danica McKellar, who majored in mathematics.

Teachers continue to experiment with ways to help students understand algebraic equations and models as well as the process of manipulating them. There is also a long history of debate about when to begin teaching algebra. Before 1700, algebra was not routinely part of the U.S. curriculum at any level of schooling, though evidence suggests it was taught in some places, such as Harvard University, in the early part of the 1700s. By 1820, Harvard required algebra for admission, and several other Ivy League schools adopted this standard over the next three decades. Massachusetts also passed a law in 1827 requiring algebra to be taught in many high schools. As early as the first part of the twentieth century, some educators such as Claude Turner suggested that algebra should be taught in eighth grade to help students understand concepts like cube roots. Some educators pointed to developmental theories such as Jean Piaget's theory of cognitive development in order to resist teaching algebra any earlier than eighth grade. In the twenty-first century, many states in the United States have adopted an "algebra for everyone" approach to teaching, and several states require students to pass an algebra test to graduate from high school. This emphasis is due in part to the increased focus on problem-solving skills believed to develop a wide range of life and occupational skills.

Further Reading

Barton, Bill, and Victor Katz. "Stages in the History of Algebra With Implications for Teaching." *Educational Studies in Mathematics* 66, no. 2 (2007).

Cooke, Roger. *Classical Algebra: Its Nature, Origins, and Uses.* Hoboken, NJ: Wiley-Interscience, 2008.

Derbyshire, John. *Unknown Quantity: A Real and Imaginary History of Algebra.* New York: Plume, 2007.

Greenes, Carole, and Rheta Rubenstein. *Algebra and Algebraic Thinking in School Mathematics: 70th Yearbook.* Reston, VA: National Council of Teachers of Mathematics, 2008.

Stacey, Kaye, Helen Chick, and Margaret Kendal. *The Future of the Teaching and Learning of Algebra.* New York: Springer, 2004.

Varadarajan, V. S. *Algebra in Ancient and Modern Times.* Providence, RI: American Mathematical Society, 1998.

Josipa G. Petrunic

See Also: Algebra in Society; Cubes and Cube Roots; Polynomials; Squares and Square Roots; Wiles, Andrew.

Algebra in Society

Category: School and Society.
Fields of Study: Algebra; Connections.
Summary: Algebra provides tools for orderly thinking and problem solving, applicable across a spectrum of pursuits.

Among the many discussions in his 1961 book *The Realm of Algebra*, science fiction author and biochemist Isaac Asimov described the real-life uses of algebra; explored the role it played in the discoveries of scientists and mathematicians such as Galileo Galilei and Sir Isaac Newton; and suggested the idea that "the real importance of algebra, and of mathematics in general, is not that it has enabled man to solve this problem or that, but that it has given man a new outlook on the universe." This notion underlies many of the perspectives on algebra in the twenty-first century.

Knowledge of algebra is seen as important not only for scientific research and the workplace but also for teaching general logical thinking and for making decisions that are important to personal well-being and society as a whole. For example, some functional relationships among people's day-to-day activities that may affect personal decisions include the relationship between how much food a person eats and weight; the amount of exercise and weight loss; and calculations for loans, interest, and other financial matters.

Some would say that the ramifications of these relationships and a lack of understanding of them mathematically are found in the housing crisis of the early twenty-first century and the increase in obesity. Algebra is reported as being a challenging subject for some people.

Many consider algebra to be a major gateway into higher mathematics in both high school and college, and it is thus critical to careers in engineering, science, mathematics, and other disciplines that require advanced mathematics training. Performance of primary and secondary students on algebra tests is one common comparison measure used to evaluate the relative standing of countries with regard to education. Professional organizations like the National Council of Teachers of Mathematics (NCTM) continue to examine the role of algebra in society and make recommendations. Some of the numerous careers that have been cited as requiring algebra include architecture, banking, carpentry, dentistry, civil engineering, nursing, pharmacy, and plumbing.

How Is Algebra Useful?

In 2003, the RAND Corporation's Mathematics Study Panel underscored the key role of algebra in education by choosing it as one of the panel's main areas of focus, explaining their decision in part by saying, "Algebra is foundational in all areas of mathematics because it provides the tools (i.e., the language and structure) for representing and analyzing quantitative relationships, for modeling situations, for solving problems, and for stating and proving generalizations." In algebra, there are general laws or algebraic models that can be used to represent a given scenario.

Algebra is sometimes noted as a type of language that provides answers to all cases at all times and models the relationships between quantities, reducing the need for repeated or inefficient computation. For example, in order to determine the savings in an interest-bearing account after a given period of time, one could compute the savings each month or year by

Clifford Ho, 2010 Asian American Engineer of the Year in Sandia National Laboratory's solar power heliostat field. Studying algebra is critical to pursuing careers in engineering, science, and mathematics.

multiplying by the interest rate. However, this computation is cumbersome after many compounding cycles. Instead, the algebraic formula

$$A = P(1+r)^t$$

can be applied directly, where P is the initial investment, r is the interest rate per period, t is the number of periods, and A is the amount of money in the bank after t periods. People may want to know if it is profitable to leave money in a bank subjected to the stated formula. On the other hand, people may want to determine the present and future value of the money they have invested because of the effect of inflation. In other instances, such as taking a car or home loan, similar algebraic laws exist. These laws help people know how much money, for instance, they may save if they pay off their loan earlier than the due date. In the eleventh century, scholar, poet, and mathematician Omar Khayyam explained the following:

> ... Algebra is a scientific art. The objects with which it deals are absolute numbers and measurable quantities which, though themselves unknown, are related to "things" which are known, whereby the determination of the unknown quantities is possible. ... What one searches for in the algebraic art are the relations which lead from the known to the unknown. ... The perfection of this art consists in knowledge of the scientific method by which one determines numerical and geometric unknowns.

Early History

Algebra definitions and applications have evolved over time, though many aspects of algebraic thinking and methods that are taught in twenty-first-century schools can be traced back to antiquity. The Babylonians and Egyptians used algebraic techniques to solve problems directly related to the everyday needs of society, such as dividing land and keeping financial records. One such example from Babylonian mathematics is an alternative method for solving cubic equations of the form $x^3 + x^2 = b$, via tabulated numerical values of squares and cubes. The Babylonians were able to solve this polynomial by using the table that gave the values of $x^3 + x^2$ or $x^2(x+1)$. They constructed the table to solve: $x^2(x+1) = 1;30$ in sexigesimal notation. The "periods" below are used to represent multiplication.

x	$x^3 + x^2$
1	1.2 = 2
2	4.3 = 12
3	9.4 = 36
4	16.5 = 80
5	25.6 = 150
6	36.7 = 252
7	49.8 = 392
8	64.9 = 576
9	81.10 = 810
10	100.11 = 1100
.	
30	900.31 = 27900

The algorithm used by the Babylonians to find the roots of cubic equations is different from the modern approach, although it can be explained using modern language.

For example, in modern notation, in solving the equation $x^3 + 2x^2 - 3136 = 0$ set $x = 2y$. Then the equation can be rewritten as the following:

$$(2y)^3 + 2(2y)^2 - 3136 = 0$$
$$8y^3 + 8y^2 - 3136 = 0$$
$$y^3 + y^2 = 392.$$

From the table, $y = 7$. Since $x = 2y$, then $x = 14$.

Topics that are viewed as algebra in contemporary mathematics were often numerical or geometric in nature. The Pythagorean theorem, named for Pythagoras of Samos, can be expressed in terms of the algebraic equation that relates the sum of the sides surrounding a right angle in a triangle squared to the square of the hypotenuse. However, historically, there is evidence that the Babylonians explored numerical versions of the theorem, while the Greeks examined the areas of the geometric squares that sat on the edges of the triangle.

The Pythagorean theorem can be found in twenty-first-century algebra classrooms, and it is useful in setting right angles in constructions and in measuring distance in flat objects. Symbolic notation for algebra was developed in India and became popular in Europe in the seventeenth and eighteenth centuries. Historical methods reflect the unique construction of understanding, indicative of the localized culture at that time. Algebraic methods have also been found in some ancient Chinese works.

Greeks, Hindus, Arabs, Persians, and Europeans all contributed to the development of algebra. The term itself comes from the Arabic word *al-jabr*, which has several translations including "the science of equations." The word appears in the title of the early algebra text written by Muhammad Ibn Musa Al-Khowarizmi in the ninth century.

Applied Algebra

For a long time, one major emphasis in algebra was solving polynomial equations, but in the eighteenth century, algebra went through a transformation that broadened the field to include study of other mathematical structures. Around that time, textbooks defined algebra in many different ways. According to mathematician Colin Maclaurin, "Algebra is a general method of computation by certain signs and symbols which have been contrived for this purpose, and found convenient.

It is called an universal arithmetic, and proceeds by operations and rules similar to those in common arithmetic, founded upon the same principles." Leonhard Euler defined algebra as: "The science which teaches how to determine unknown quantities by means of those that are known." As the concept of variables was further developed, many physical properties, including time, mass, density, pressure, temperature, charge, and energy, were expressed algebraically.

For instance, Albert Einstein's equation relates energy to mass times the speed of light squared. In the twenty-first century, defining algebra commonly requires a broader approach. First, one could say that early or elementary algebra is essentially the study of equations and methods for solving them; and second, that modern or abstract algebra is the study of various mathematical structures. High school algebra textbooks typically contain a breadth of topics, such as polynomials and systems of linear equations. These are important in modeling many relationships in society. For example, parabolas represent the paths of ball or bullet trajectories, and systems of linear equations and matrices give rise to digital images. At the college level, students continue their study of algebraic equations in virtually every mathematics and statistics class. Students in a broad range of majors, including the sciences and mathematics, may further their understanding of systems of linear equations and their applications in a linear algebra class.

Mathematics majors in modern or abstract algebra study topics like groups, rings, and fields, and graduate students further explore these and other algebraic structures. These concepts have been useful in chemistry, computer science, cryptography, crystallography, electric circuits, genetics, and physics. Algebra is a core area from the middle grades and high school to undergraduate and graduate mathematics. Research fields include the connections of algebra with other subdisciplines, like algebraic geometry, algebraic topology, or algebraic number theory, and the abstract structures and notions in pure algebra have been applied in many contexts. Some algebraists work for the National Security Agency and others work as professors.

In general, mathematicians and scientists often algebraically derive laws for a given scenario or relationship from patterns. For example, consider a triangle number pattern. It is fairly simple to find the next number recursively but finding larger values such as the 1000th triangular number without a general rule can be more challenging. (See Figure 1 and Table 1.)

Figure 1.

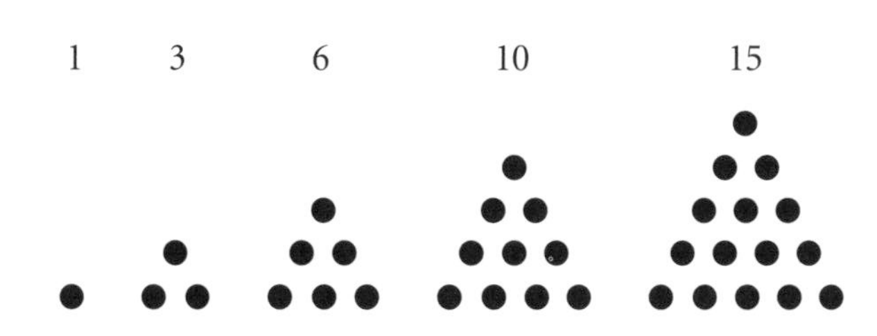

Table 1.

1st term	2nd term	3rd term	4th term	5th term	nth term
$1 = 1$	$3 = 1 + 2$	$6 = 1 + 2 + 3$	$10 = 1 + 2 + 3 + 4$	$15 = 1 + 2 + 3 + 4 + 5$	$= 1 + 2 + 3 + \cdots + n$

Algebra can be used to generalize the preceding case and derive that

$$1 + 2 + 3 + \cdots + n = \frac{n(n+1)}{2}$$

so the general law will be

$$a_n = \frac{n(n+1)}{2}.$$

Hermann Weyl noted "The constructs of the mathematical mind are at the same time free and necessary. The individual mathematician feels free to define his notions and set up his axioms as he pleases. But the question is will he get his fellow mathematician interested in the constructs of his imagination. We cannot help the feeling that certain mathematical structures which have evolved through the combined efforts of the mathematical community bear the stamp of a necessity not affected by the accidents of their historical birth. Everybody who looks at the spectacle of modern algebra will be struck by this complementarity of freedom and necessity."

Many algebraic equations are used in everyday life to meet societal needs. For example, the area of a rectangle is given by the length times the width. There are algebraic equations like finding the area of a square or circle, and also finding volume, which are used in applications like home decorating, cooking, landscaping, and construction. Building houses and fences, determining amounts of material needed for a project, and completing everyday chores use algebra to make work accurate and efficient. Economists use algebraic laws to project business profits or losses and to advise investors and other decision makers. In other instances such as taking a car or home loan similar algebraic laws help people know how much money, for instance, they may save if they pay off their loan earlier than the due date.

Many formulas are easy to use and can easily be entered in a hand calculator or computer to generate the required result. Such formulas have been adapted to Web-based applets and software like spreadsheets to track financial records, making them widely accessible and often easy to use.

Mathematician Roger Cooke explained, "Algebra provided more than just a compact notation for writing down relations among variables. Its rules made it possible to manipulate those laws on paper and derive some of them from others. For example, a consequence of Kepler's third law is that the ratio . . . of the square of a planet's period to the cube of its distance from the sun is the same for all planets . . . Kepler's third law and Newton's law of gravitation are equivalent statements, given certain basic facts of mechanics." Kepler's laws were named for Johannes Kepler and Newton's for Sir Isaac Newton. The ability to express algebraic relationships using variables and rates of change in calculus increased the applicability of equations in a wide variety of contexts.

The U.S. Bureau of Labor Statistics highlights the importance of coursework in algebra for numerous careers, including brickmasons, blockmasons, and stonemasons; carpenters; computer control programmers and operators; construction and building inspectors; engineers and engineering technicians; line installers and repairers; machine setters, operators and tenders in metal and plastic; machinists; opticians; physical therapist assistants; power plant operators, distributors and dispatchers; sheet metal workers; surveyors, cartographers, photogrammetrists, and surveying and mapping technicians; radiation therapists; tool and die makers; and veterinarians.

Algebra's Role as a Gateway

Some would argue that in the United States, mathematics achievement has not met the same standards of excellence as in other developed countries, and that, as a result, students may not be prepared to enter college. Some historians trace the growing need for mathematics education to the turn of the twentieth century or the Industrial Revolution, when there were debates about the appropriate level of mathematics for high school education. Historically, popular opinion was often against algebra as a subject of widespread study in secondary schools, since many did not see clear connections between algebra and real-world needs. Mathematics educator W. D. Reeve cited one newspaper editorial as an example of such an attitude in a 1936 National Council of Teachers of Mathematics report, *The Place of Mathematics in Modern Education*, saying the following:

> Quite frankly, I see no use for algebra except for the few who will follow engineering and technical lines. . . . I cannot see that algebra contributes one iota to a young person's health or one grain of inspiration to his spirit. . . . I can see no use for it in the home as an aid to a parent, a citizen, a producer, or a consumer."

The same report noted deficits in algebra skills even among graduate students and relatively high failure rates for algebra students in some high schools, such as in New York City, which were used by some as additional arguments against algebra's broad inclusion in the high school curriculum. With regard to who should and should not study algebra, Reeve countered: ". . . no one, I think, has the wisdom to decide who will profit most by its study or predict who the future Newtons and Einsteins are to be."

Mathematician and philosopher Alfred Whitehead stated the following:

> Quadratic equations are part of algebra, and algebra is the intellectual instrument which has been created for rendering clear the quantitative aspects of the world. There is no getting out of it. Through and through the world is infected with quantity. To talk sense, is to talk in quantities. It is no use saying that the nation is large. . . . How large? It is no use saying that radium is scarce. . . . How scarce? You cannot evade quantity. You may fly to poetry and to music, and quantity and number will face you in your rhythms and your octaves. . . . This question of the degeneration of algebra into gibberish, both in word and in fact, affords a pathetic instance of the uselessness of reforming educational schedules without a clear conception of the attributes which you wish to evoke in the living minds of the children. . . . First, you must make up your mind as to those quantitative aspects of the world which are simple enough to be introduced into general education; then a schedule of algebra should be framed which will about find its exemplification in these applications.

Other newspapers like the *Columbus Dispatch* supported broad high school mathematics education during that time period, asserting that schools should provide the "mathematical key" to the "gateways of a larger life."

Algebra eventually became commonplace in high schools and some middle schools, with basic algebraic concepts often introduced even in the primary grades, and yet questions about how to teach algebra continued. Algebra is usually a prerequisite for all higher mathematics courses in both high school and college, and, in some cases, it is required for high school graduation. Students will not advance in many majors or career paths unless they pass algebra, and the result is that some students change majors or abandon education altogether. Students requiring remediation courses at the college level are fairly common.

The result is that in the twenty-first century, algebra is still viewed by many as a major gatekeeper to educational and career advancement, and learning algebra has been promoted as a civil rights issue for every U.S. citizen, though many of the same arguments from past decades continue to be debated. In the latter twentieth century, algebra education became a renewed topic of discussion from local school districts all the way to the White House. The RAND Corporation's panel further explained its decision to focus on algebra by saying, "Without proficiency in algebra, students cannot access a full range of educational and career options, and this curtailment of opportunities often falls most directly on groups that are already disadvantaged."

At the same time, naysayers continue to publish counterpoints regarding algebra's lack of utility. One 2006 *Washington Post* article about a student named Gabriella, who purportedly dropped out of high school after failing her algebra course many times, asserted that writing teaches logical reasoning more effectively than algebra and stated that many students will "never need to know algebra" in the real world, since most mathematics can now be done by computer or calculator. It concluded that having an algebra requirement for high school graduation is potentially more detrimental than helpful because it may spur students to drop out who otherwise might have graduated.

This article spurred many further discussions, and it appeared to reflect the author's own difficult experiences with algebra, a phenomenon that has been reported by many educational researchers and that drives further curricular revisions. Authors of algebra textbooks and self-help books have explored different ways to help students connect to algebra. For instance, actress Danica McKellar has written algebra readiness and algebra books that include stories and characters in order to express equations and solutions in contextual situations. Some educators incorporate mnemonics, songs, or other memory techniques such as First, Outside, Inside, Last (FOIL) in order to teach the multiplication of two binomials. Other authors highlight real-life applications, historical connections, or solutions using technology.

Many national reports have indicated that education in the United States is in a critical period, and some would say particularly in mathematics and science. Educators and politicians have proposed changes to the mathematics education curriculum to prepare U.S. students. The number of students entering college and requiring courses that enable them to be effective in the workplace is rising. Further, engineering and other technical fields that were once seen as elite or remote are increasingly a part of daily life, including computing, electronics, business, and architecture. Technology is changing every day, which has changed society, including mathematics. As a result, there is an increased need for people who can adapt to the changes and continue being effective in society. In this context, there has been a movement to reform algebra education so that it can be more readily accessed by everyone. The "algebra for all" movement has been a central point within the reform initiatives. National standards such as those published by the NCTM have stressed the need to make algebra more accessible to students, and they often outline both the content to be covered and instruction expectations. Some research has shown that students who take algebra by eighth or ninth grade are more likely to pursue higher mathematics, though this cannot be interpreted as a cause-and-effect relationship.

Further Reading

Cohen, Richard. "What Is the Value of Algebra?" *Washington Post* (February 16, 2006). http://www.washingtonpost.com/wp-dyn/content/blog/2006/02/15/BL2006021501989.html.

Cooke, Roger. *Classical Algebra: Its Nature, Origins, and Uses*. Hoboken, NJ: Wiley-Interscience, 2008.

Edwards, Edgar. *Algebra for Everyone*. Reston, VA: National Council of Teachers of Mathematics, 1990.

Eves, Howard. *An Introduction to the History of Mathematics*. Philadelphia: Saunders College, 1990.

The Futures Channel. "Algebra in the Real World." http://www.thefutureschannel.com/algebra_real_world.php.
Gallian, Joseph. *Contemporary Abstract Algebra.* 7th ed. Belmont, CA: Brooks Cole, 2009.
McKellar, Danica. *Hot X: Algebra Exposed: Word Problems, Polynomials, Quadratic Equations and More.* New York: Hudson Street Press, 2010.
Rappaport, Josh. *Algebra Survival Guide: A Conversational Guide for the Thoroughly Befuddled.* Santa Fe, NM: Singing Turtle Press, 1999.
United States Bureau of Labor Statistics. "Occupational Outlook Handbook." http://www.bls.gov/oco.
Zaccaro, Edward. *Real World Algebra.* Bellevue, IA: Hickory Grove Press, 2001.

Samuel Obara

See Also: Algebra and Algebra Education; Equations, Polar; Exponentials and Logarithms; Function Rate of Change; Polynomials; Squares and Square Roots; Vectors; Wiles, Andrew.

Analytic Geometry

See *Coordinate Geometry*

Anesthesia

Category: Medicine and Health.
Fields of Study: Algebra; Measurement.
Summary: Anesthesia dosages must be precisely determined and the patient must be monitored for signs of too high or low a dosage.

The word "anesthesia" was coined from the Greek word *anaisthesis* meaning "insensibility." As early as 4200 B.C.E, opium poppies were used as an herbal remedy in Samaria and, later, in Cyprus, India, and China. The three main types of anesthesia are local (loss of sensation in a small area of the body by the blockage of nerve signals), regional (loss of sensation in a larger area of the body), and general (loss of consciousness), are used to relieve the feeling of pain during medical and dental procedures. Anesthesia uses mathematics in variety of ways including the calculation of appropriate drug dosages, the monitoring of patients under general anesthetics during surgery and recovery, and the design and use of anesthetic equipment including vaporizers, ventilators, and pressure gauges.

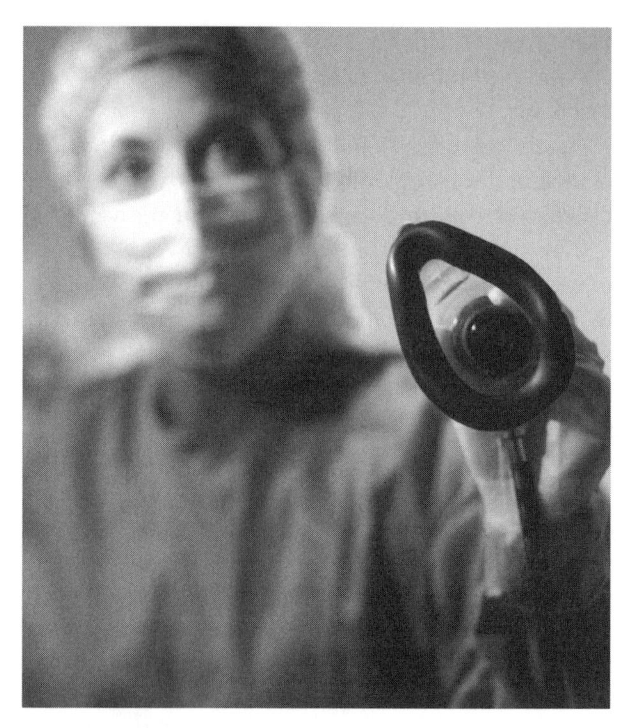

Anesthesiologists uses mathematics for tasks from calculating drug dosages to monitoring patients.

Non-pharmacological anesthetic techniques historically have included local anesthetics such as ice and rum. Nitrous oxide ("laughing gas"), ether, and chloroform were used as general anesthetics in the 1800s during childbirth and surgery.

Applications of Mathematics

Anesthetic drug dosages per minute are based on milligrams of drug per kilogram weight of the patient. The rate of elimination of drugs from the body per unit of time is proportional to the amount of drug in the body. The time taken for the drug concentration in the plasma to be reduced by 50% is called the elimination half life.

The measurements monitored while a patient is under general anesthetics can include data such as temperature; heart rate via ECG (electrocardiogram);

oxygen saturation via pulse oximetry; ratio of oxygen, carbon dioxide, and nitrous oxide from the patient's inspired and expired gases; urine output; arterial blood pressure; central venous pressure; pulmonary artery occlusion pressure; cerebral activity via EEG (electroencephalogram); and neuromuscular function.

A major concern is keeping the patient at the appropriate level of anesthesia. The cardiovascular system is threatened if the anesthetic is too deep, but if the anesthetic is too light, the patient may experience pain or regain consciousness.

Researchers have attempted to measure the depth of anesthesia by monitoring on a graph and on a time scale EEG signals generated by electrical discharges of neurons near the brain surface. One method is to administer a gaseous anesthetic drug and hypothesize that the concentration of the drug in the expired air is proportional to the blood-plasma concentration. An alternative research technique is to observe respiratory sinus arrhythmia (RSA), which is the variation in heart rate during a breathing cycle. The heart rate increases during inhalation and decreases during exhalation. On the graph of an ECG, each heartbeat is referred to as an R peak. The difference between two consecutive R peaks is an RR interval, which is shortened during inspiration and lengthened during expiration.

Anesthesia Providers' Educational Backgrounds

The academic and clinical preparation for an anesthesiologist in the United States consists of four years of college, four years of medical school, one year of internship, and three years of anesthesiology residency. A description of Steven Cruickshank's 1998 book *Mathematics and Statistics in Anaesthesia* states that anesthesia residents are required to study and understand pharmokinetics (the study of what the body does to a drug) and statistics as "a core part of their training." In addition to physician anesthesiologists, anesthesiologist assistants or Certified Registered Nurse Anesthetists (CRNAs) can apply anesthesia or sedation while working with healthcare professionals. CRNAs complete four years of college, at least one year of acute-care nursing, and a 24- to 36-month master's degree program before passing the required certification examination. Anesthesiologist assistants (AAs) with master's degrees may practice under the supervision of an anesthesiologist in several states.

Further Reading

Chen, Z., et al. "Linear and Nonlinear Quantification of Respiratory Sinus Arrhythmia During Propofol General Anesthesia." http://www.ncbi.nlm.nih.gov/pmc/articles/PMC2804255.

Cruickshank, Steven. *Mathematics and Statistics in Anaesthesia*. New York: Oxford University Press, 1998.

Neligan, Pat. "Pharmacokinetics." http://www.scribd.com/doc/39115053/Pharmacokinetic.

Widman, G., T. Schreiber, B. Rehberg, A. Hoeft, and C. E. Elger. "Quantification of Depth of Anesthesia by Nonlinear Time Series Analysis of Brain Electrical Activity." http://arxiv.org/PS_cache/nlin/pdf/0007/0007027v1.pdf.

Karen Doyle Walton

See Also: Drug Dosing; LD50/Median Lethal Dose; Medical Imaging; Surgery.

Animals

Category: Weather, Nature, and Environment.
Fields of Study: Algebra; Communication; Geometry; Measurement.
Summary: Principles of engineering, physics, and mathematics are demonstrated by the physiology, movement, and behavior of animals.

Animals, including human beings, are living organisms that belong to the domain Eukaryota (having complex cellular structures enclosed with membranes) and the kingdom Animalia. Within this taxonomy, the kingdom is defined by several characteristics, including internal digestion of food (called "heterotrophism") and the ability to move using its own energy in at least some stages of life (called "motility"). Some say that what distinguishes humans from other animals is mathematical ability. However, researchers have studied a diverse range of mathematical concepts as they relate to animal behavior and have found evidence of abilities such as symbolic calculation, efficiency in locomotion, and synchrony. There are questions about whether these findings are biased perceptions of mathematical significance. Many mathematical patterns and symmetry can also be found in the structure of animals, ranging from their cellular tissue to their coat patterns. Some of the motivation behind the development of many statistical measures and methods, such as standard deviation and regression, was to characterize natural variability and associations in animal species.

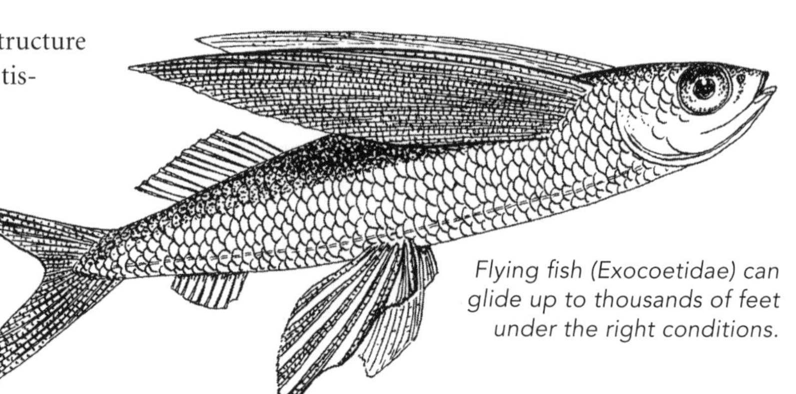

Flying fish (Exocoetidae) can glide up to thousands of feet under the right conditions.

Biological Systematics, Set Theory, and Logic

Biological systematics is the field that describes and names living organisms, provides their classifications and keys for identification, and situates classes of organisms within evolutionary history and modern adaptations. In particular, classification of organisms (called "taxonomy") is an empirical science, where description of classes is the final step in the discovery and description of organisms. Existing biological classifications may include the ranks of domain, kingdom, phylum, class, order, family, genus, and species.

The definition of the kingdom Animalia is intensional definition—it specifies necessary and sufficient conditions for belonging to the set of animals. The particular subclass of definitions used in systematics to define animals is called definition by genus and differentia. Such definitions rely on a structure of sets, subsets, and supersets as well as their differentiating conditions. For example, defining negative numbers as the set of rational numbers that are less than zero, mathematicians use the superset of rational numbers (defined elsewhere) and the differentiating condition of being less than zero. Animalia is one of several kingdoms (subsets) of the domain Eukaryota, differentiated from other kingdoms by particular conditions.

Careful decisions are made in the organization of kingdoms and in defining differentiated conditions. For example, if only the conditions of internal digestion of food and motility were used, the Venus flytrap would be considered an animal rather than a plant. However, plants are also differentiated by the sufficient condition of having plastids, such as chloroplast, in their cells. Internal digestion of food and motility are necessary but not sufficient conditions for declaring an organism an animal. There are historical and modern systems defining anywhere from two to eight kingdoms of living organisms, depending on the necessary and sufficient conditions used for definitions.

Animal Tissue Structures

All animal cells have extracellular matrix, the boundary that can serve many functions, including exchanging substances between cells, segregating tissues, and anchoring cells. Animal cells typically form tissues; groups of cells carry out particular functions within animal bodies. There are four types of animal tissues, defined by their functions: muscle, nervous, epithelial, and connective. Cells within tissues and tissues within organs may be tessellated (filling space or surface infinitely, without gaps).

Tissue engineering is an interdisciplinary field combining biology, material science, chemistry, and engineering to re-create, change, or replace tissues. It pays special attention to the mechanical and structural properties of tissues, often modeled mathematically before being implemented in the lab.

Technological Metaphors and Models

Beginning in the Renaissance, it was common for people to conceptualize living organisms in terms of human-made machines. This phenomenon worked both ways, since human constructions were informed by new understandings and observations of nature. During the Renaissance, animal tissues and organs were seen as combinations of relatively simple mechanisms such as levers. Attempts were made to imitate some functions of animals in construction, such as making bird-like wings. This analytic approach informed the development of scientific methods in biology—in contrast with a holistic view of living things as having a completely different nature from human-made mechanisms. In the seventeenth century, this philosophical approach of modeling animals on machines was supported by such influential scientists as Galileo Galilei, René Descartes, and Isaac Newton.

Engineering and mathematics developed along with explanations in biology. Developments in steam technology introduced the ideas of energy and work, which, in turn, led to the analysis of gas and liquid pressures as explanations of the interaction of tissues and organs in animal bodies. The metaphors of heart or cellular structures as pumps—or kidneys and the liver as filters—persist to this day. When electricity and magnetism were first discovered, there were numerous attempts to apply them directly to explanations of animal bodies, but many of these early models were discarded later. In the twentieth century, animal processes are often conceptualized as computer entities, such as nervous system as a computational network. Likewise, animal brains are observed for the purpose of building the artificial intelligence. Mathematical models in biology developed from simple measurements of weight, length, and proportion to those incorporating calculus, differential equations, statistics, computational science, and other areas of modern mathematics.

Animal Motility, Field Perception, and Gradients

Animals can move under their own power. Animals movement in response to external stimuli or gradients of stimuli is called taxis. In calculus, the gradient is a vector field; its vectors point in the direction of the greatest rate of increase in a variable and have the magnitude equal to that rate. Depending on the nature of the variable in the gradient, animals or animal cells can exhibit different types of taxis, such as thermotaxis along temperature gradients or phototaxis along light gradients. Mathematical models of taxis are based on calculus, differential equations, and statistics.

Chemotaxis is the movement along the gradient in a chemical substance. Animal cells may have multiple chemical receptors around their boundaries, allowing the cell to determine the direction of chemical gradient vectors. Animal cells can move toward chemoattractors, such as immune cells arriving where they need to be, or away from chemorepellents. The development of animal embryos involves the movement of cells and is regulated by gradients in signal chemicals. Sperm movement occurs because of chemotaxis and thermotaxis.

Magnetoperception (the ability to detect magnetic fields) is observed in migrating birds, sharks, rays, honeybees, and other animals. It is an important factor in regulating animal movement and navigation—for example, during bird migrations. Experiments and applications in magnetoperceptions usually involve attaching magnetic substances to animals and observing effects. For example, cows and deer grazing under power lines orient themselves differently. The mechanisms of magnetoperception continue to be actively investigated.

Animal Locomotion

The way animals move, in addition to being a matter of biological interest, is a source of engineering ideas. Until the twentieth century, the main source of data on animal movement was observation and, sometimes, experiments with animals or their body parts. Photography and videography added details to the observation. Animals may be equipped with miniature devices that track their positions in space, as well as the electric activity within muscles, the contraction of muscles, or the forces exerted by muscles. These devices allow the development of detailed models of animal bodies during movement.

Every type of locomotion has been modeled in physics, with a variety of relevant equations. There are three major types of terrestrial locomotion (movement on

Snakes move with slithering terrestrial locomotion by undulating in one of several patterns.

solid surfaces): legged movement, slithering, and rolling. Legged animals may have from two to 750 legs, with the geometry of leg and joint position defining posture, and the pattern and pace of leg use defining gaits. Snakes move by undulating in several patterns, such as sidewinding, or by lifting parts of their belly slightly off the ground, moving them forward relative to their ribs, and then pulling the body to them (rectilinear motion). These movements on land are described by kinematic equations, in water by hydrodynamic equations. Rolling animals, such as pangolins, can briefly achieve great speed, usually by forming a wheel or a ball out of their whole body and using gravity to escape predators.

Swimming is accomplished by body movement propulsion in fish, jet propulsion in mollusks, undulation in several types of animals, and limb movement in some birds and mammals. Jet propulsion requires relatively high energy but can provide animals with an occasional burst of speed. Models of swimming include such measures as buoyancy and are modeled with fluid dynamics and mechanics.

Gliding, soaring, and flying are energy-efficient ways of locomotion, and attract much interest in biomechanics and aerodynamics. Scientists study concepts like lift and drag as well as ratios of wing measurements such as loading (weight to area). Animals use different types of motion through the air, which are defined by a combination of timing and geometry. For example, falling with increased drag forces that prolongs the fall can be either parachuting (when the angle to earth is more than 45 degrees) or gliding (when the angle is less than 45 degrees). Gliding animals such as fish and squirrels have aerodynamic adaptations including streamlining. The variable glide ratio is the ratio between the horizontal and the vertical speed components (lift to drag). A flying squirrel has a glide ratio of about two, and a human in a glider windsuit modeled after gliding animals has a glide ratio of about two and a half. Soaring birds glide during parts of their flight.

The properties of winged flight in birds and bats depend on proportions of the animal's body. Wingspan is the distance between wingtips, and the mean wing chord is the average of the distances between the front and the back edge of the wing, found using calculus. Aspect ratio of a wing is the ratio of wingspan to mean chord. Fast birds such as falcons have pointy short wings with high aspect ratio (narrow wings). Long wings with high aspect ratios such as the wings of albatrosses, on the other hand, can produce slow soaring and gliding flight. Wide, rounded wings with medium aspect ratios can be used for a variety of flight types, for example, in storks or sparrows.

Biophysicists first attempted to explain insect flight using bird flight mechanics. They found that the resulting forces were several times less than what would be needed to lift and to propel an insect. Current theories of insect flight are still controversial. The theories use computational differential equations to model effects such as vortexes created in front of wings. When wings flap with high enough frequency, such a vortex can provide significant additional suction force.

Relatively rare types of animal locomotion depend on surface tension and capillary forces for walking on the water surface, or moving faster over released liquid (Marangoni effect). These forces are studied in fluid dynamics and thermodynamics.

Researchers debate why the wheel, which provides several mechanical advantages in terrestrial locomotion, has never evolved in any animal. The relevant mathematical model is a graph measuring fitness of organisms to the environment, called fitness landscape. Fitness peaks are stable states, with genetic modifications meaning worse fitness. While wheel locomotion may be a fitness peak, it is surrounded by fitness valleys too deep to be crossed by evolutionary means.

Migrating Animals

Many animals migrate—periodically travelling among habitats—sometimes over long distances. Models of migration take into account the time of each leg of the journey as well as the full period of migration. These times can be synchronized with seasonal milestones, developmental stages in the life of each animal, and other natural events. Because migrations can take place across international boundaries, they can help promote international efforts in research and conservation. The Convention on the Conservation of Migratory Species of Wild Animals, for example, covers several endangered species of birds and fish, as well as migratory bats and turtles.

About a fifth of all bird species in the world migrate. Typically, birds migrate closer to the equator in winters, and farther from the equator in summers. Mathematical models of bird migration include the overall patterns for particular populations such as migration corridors as well as random events such as irrup-

The long wings of albatrosses produce slow soaring and gliding flight.

tions (large numbers of birds) migrating farther after population explosions. Biophysics involved in bird migration includes theories of energy efficiency, and various mechanical effects, such as wear on feathers that necessitates periodic molting synchronized with the migration period.

During migration, birds navigate by using the landscape clues they learn while young, orienting by the sun, or using magnetoperception. In some bird species, navigating is mostly a learned behavior; in others, it is mostly coded genetically. Sometimes the coding goes wrong, reversing the migration direction 180 degrees, thus causing birds to reverse-migrate in the opposite direction from the majority of their flock. Bird species that learn their migration routes from their elders, such as cranes, can be taught to use safer routes by following light aircraft of animal preservation specialists.

Shorter migration routes also exist. For example, many fish species rise to the water surface to feed at night—a type of diel vertical migration. Many fish species high in the food chain migrate to follow their prey, with varying times and lengths of migration journeys.

Because many insects are relatively short-lived, their migrations may involve multiple generations being born along the route. In these cases, none of the individual insects travels the full migration route. Some migrating insects, such as locusts, swarm for the purpose of migration. A swarm can be modeled using a system of differential equations where pairs of individuals move closer if they are too far, move away if they are too close, and orient themselves toward the same direction. However, studies of insects, including locusts, show complex mechanisms that include chemoregulation, physiological change in response to overcrowding (measured in contacts per unit of time), emission, and responsiveness to sounds and other variables involved in swarming.

Herds of animals, schools of fish, and flocks of birds can be modeled as groups of particles, with interactions among individuals determined by differential equations with some fixed and some random parameters to account for individual behavior variations. Such mathematical models (called "interacting particle models") can describe flock behavior or predict school migration routes. To observe animal migration, researchers use tracking devices, satellite observation, and echolocation for marine species.

The Physics of Winged Flight

The physics of winged flight in birds, bats, and extinct dinosaurs focuses on the balance of four forces: lift, drag, thrust, and gravity. Air moving around wings produces lift because of speed and pressure differences in the airflow on either side of the wing—a complex process still being studied in aerodynamics and modeled with systems of differential equations.

Drag comes from air resistance to the flying body, and by air turbulence created by wing movements. Newton's second and third laws of physics explain thrust (the force created when a wing flaps). The vector of thrust points in the direction opposite to where the wing is moving. In other words, the flapping wing propels the animal forward by by pushing against the air. The force of gravity is proportional to the mass of the animal.

Food Webs

Food webs and food chains map food relationships in ecosystems. The key measurement of the position within the food web is called "trophic level." Autotrophs (producers) are at trophism level one. Autotrophs are organisms that do not consume other organisms or carbon produced by them, and therefore are not animals. Two mechanisms of autotrophism are photosynthesis in plants, and chemosynthesis in archae and bacteria. The first organisms to evolve on Earth used chemosynthesis. A third mechanism, radiotrophy, is being researched in fungi in high-radiation areas. All food chains within all food webs on Earth start with level one autotrophs. Predator species that no other species predate upon are called apex predators.

More specifically, classes of organisms are named according to the flow chart with three branchings. The first branching determines the source of energy, either light (photo-) or chemical (chemo-). The second branching determines the source of extra electrons in reduction-oxidation reactions, either organic (-organo-) or inorganic (-litho-). The third branching defines the source of carbon, either organic (-heterotroph) or carbon dioxide (-autotrophs). For example, fungi are chemoorganotrophic. All eight combinations resulting from these three branchings exist in nature. Heterotrophic organisms that break down other dead organisms into simpler organic or inorganic compounds are called decomposers. Consumer organisms use other living organisms as their source of energy. Simplistically, the second trophic level comprises primary consumers that eat plants (herbivores) or chemosynthesizing creatures. The third trophic level, secondary consumers or predators, consists of animals that eat primary consumers. Animals that eat those at the third trophic level are said to have the fourth trophic level, and so on. However, most existing animal species obtain energy from several sources. For example, foxes eat rabbits and berries; chickens eat grains and insects.

To address the complexity of food chains, the trophic level of an animal is determined by the formula of adding all products of levels of its food by the fraction of that food in the animal's diet, and adding 1. For example, if a chicken's diet consists of 30% worms (level 2) and 70% grain (level 1), its trophic level is equal to

$$0.3(2) + 0.7(1) + 1 = 2.3.$$

Statistical analysis is used to determine the mean trophic level of a species in a particular ecosystem.

Changes in any part of the food web affect all other parts. For example, the effect of introducing predators that reduce the numbers of the prey and cause abundance in the next trophic level down is an example of a "trophic cascade event." The ability of an ecosystem to withstand disturbances is measured by an index called ascendency, and is derived by formulas from the information theory field of mathematics. Variables in ascendency formulas include both the amounts of energy and matter circulated within an ecosystem and the information shared among members of the system. Low ascendancy values make ecosystems internally unstable; high ascendancy values make ecosystems oversensitive to external disturbances. Ascendency values corresponding to stable systems are called "the window of vitality."

Ascendency is an example of using multiple indices and metrics to model, evaluate, and predict changes in food webs. For example, consider energy or biomass transfer from one feeding level to the next feeding level. The efficiency of this transfer is a measure of an ecosystem called ecological efficiency. For example, in a food chain that consists of four levels, with mean ecological efficiency of 1/10, the apex predator has the ecological efficiency of converting sunlight into its biomass of

$$\left(\frac{1}{10}\right)^4 = 0.0001.$$

Ecological efficiency restricts the number of possible trophic levels.

Fantastic Animals, Hybrids, and Genetic Chimeras

A variety of cultures describe fantastic animals or humanoids with animal traits. These animals—especially those invented before the nineteenth century—are used in mathematics education to help students understand concepts related to combinatorics because they are made by combining parts of existing animals. For example, ancient Greeks invented a chimera that had the body of a lion, the heads of a goat and a lion, and a snake for the tail. In genetics, chimeras are animals that have genetic material from more than one zygote—from four or more parents. Chimeras of different animals of the same species happen naturally when several eggs in one female are fertilized by sperm from different males and then fused. They may also happen artificially, in which case different animals species can be used. For example, a goat-sheep chimera called "geep" was first produced in the 1970s.

Hybrid animals are different from chimeras in that they have two parents, but the parents are of different species. Hybridization has been recognized and used for millennia. For example, humans have produced large populations of mules since ancient times. The mathematics of hybrids involves tracking the amount of genetic material from each species through generations, and calculating the probabilities of achieving particular traits in offspring. For example, a single-cross hybrid has 50% genetic material from either line of parents. Crossing such hybrids with the line of one of the parents (called backcrossing) produces hybrids with roughly 75% genetic material from that parent's species—averaged across a species, as individuals will have either pure or half-and-half genetic material.

Symmetry and Fractals

Most animal bodies exhibit either rotational (radial) or reflection symmetry. Animals with bilateral reflection symmetry (having a plane separating bodies into roughly reflected halves) form the taxon Bilateria. Observation of symmetry is a major tool of evolutionary theory. For example, it is hypothesized that all Bilateria animals evolved from a common ancestor species, Urbilaterian, that lived around six hundred million years ago. This makes Bilateria a clade (a group of animals that come from a common ancestor). Bilaterians have the front end with the mouth and the back end with the anus, defined by the plane of symmetry.

Rotationally symmetric animals such as sea anemones and sea stars usually have the mouth on the axis of the symmetry. When animals have a certain number of body regions positioned around the axis symmetrically, they are called by the number of regions. For example, five-armed stars exhibit pentamerism, and many coral polyps exhibit hexamerism, or six-part rotational symmetry.

Combinations of reflections, rotations, and translations can produce repeated geometric patterns called tessellations or wallpaper groups in plane and crystallographic groups in space. There are 17 types of wallpaper groups and 230 types of crystallographic groups described by the area of mathematics called group theory. Wallpaper and crystallographic groups can be found in colonies of animals such as corals or in arrangements of animal body parts such as fish scales.

Fractals are shapes that can be split into parts that are copies of the whole. Fractals frequently occur in the living nature. For example, feathers are fractal-like structures of the tree type, with three or four levels. Nervous systems and lungs of mammals are also tree-type fractals. Beyond the literal meaning as a geometric shape, the idea of a fractal as a self-repeating structure is applied to many areas related to animals to describe patterns within systems behavior, evolution, migration, and development.

Further Reading

Adam, John. *A Mathematical Nature Walk*. Princeton, NJ: Princeton University Press, 2009.

Ahlborn, Boye. *Zoological Physics: Quantitative Models of Body Design, Actions, and Physical Limitations of Animals*. New York: Springer, 2004.

Ball, Philip. *The Self-Made Tapestry: Pattern Formation in Nature*. New York: Oxford University Press, 2001.

Cheng, Ken, and Nora Newcombe. "Geometry, Features, and Orientation in Vertebrate Animals: A Pictorial Review." Animal Spatial Cognition: Comparative, Neural & Computational Approaches, 2006. http://www.pigeon.psy.tufts.edu/asc/Cheng/Default.htm.

Devlin, Keith. *The Math Instinct: Why You're a Mathematical Genius (Along With Lobsters, Birds, Cats, and Dogs)*. New York: Basic Books, 2005.

Murray, James. "How the Leopard Gets Its Spots." *Scientific American* 258, no. 3 (1988).

Noonan, Diana. *Animal Investigations: Collecting Data*. Huntington Beach, CA: Teacher Created Materials, 2008.

Maria Droujkova

See Also: Airplanes/Flight; Bees; Genetics; Joints; Nervous System; Symmetry; Synchrony and Spontaneous Order.

Animation and CGI

Category: Arts, Music, and Entertainment.
Fields of Study: Geometry; Measurement; Representations.
Summary: Animators have become adept at creating realistic products with the help of mathematics.

Animation is the process of creating the illusion of fluid movement from a series of static images. When these images are viewed in sufficiently fast succession, the human eye sees them as continuous motion rather than a sequence of discontinuous still images. From the earliest mechanical devices, through hand-drawn, stop-motion, and computer-assisted film techniques of the twentieth century, up to the latest computer-generated imagery (CGI), the quest has been to create interesting representations of movement and action.

Early Animation Devices

Historically, there have been several mechanical devices that were developed to simulate movement using still pictures. The Phenakistoscope, invented in 1832 by the Belgian physicist Joseph Plateau, consisted of a spindle with two mounted discs, one with slots around its edge and the other with pictures of successive action. With the discs spinning in unison and the picture side facing a mirror, the view through the slots appeared to show a moving drawing. In 1834, a British mathematician named William Horner produced the Zoetrope, a cylinder cut with vertical slits. Pictures of successive action were positioned on the inside opposite the slits. With the cylinder rotating, the image seen through the slits appeared to be in motion. As the Zoetrope used more pictures, and could be rotated more quickly, this gave a better illusion of movement. Even in the early twenty-first century, the Zoetrope is used to illustrate the basic idea of animation.

Animation Principles

By the start of the twentieth century, these mechanical devices were superseded by animated films. The principal technique was to hand-draw each frame. In the 1930s, animators at Walt Disney Studios developed what became known as the "12 principles of animation," many of which remain pertinent in an era of CGI. To illustrate, consider someone throwing a ball so that it bounces along the ground. A thrown ball is known to follow a parabolic path, a form of arc. The "arc" principle of animation is that almost all actions follow some form of arc. Arcs, as the Disney animators were well aware, give animation a more natural appearance. Another principle, "slow in and slow out," relates to the ball taking time to accelerate and decelerate. The animation looks most realistic if there are more frames near the beginning and end of a movement, and fewer in the middle. The flight of the ball and its bounce involves the principle of "squash and stretch." As the ball falls, animating a slight stretch gives the impression of the ball having speed. Dilation is the mathematical transformation for stretching and shrinking. Animating a squash to the ball as it bounces gives the impression of weight. For the ball to seem real, the animator uses the principle of "solid drawing" by taking into account the form of the ball in three-dimensional space as well as using the geometry of light and shadow.

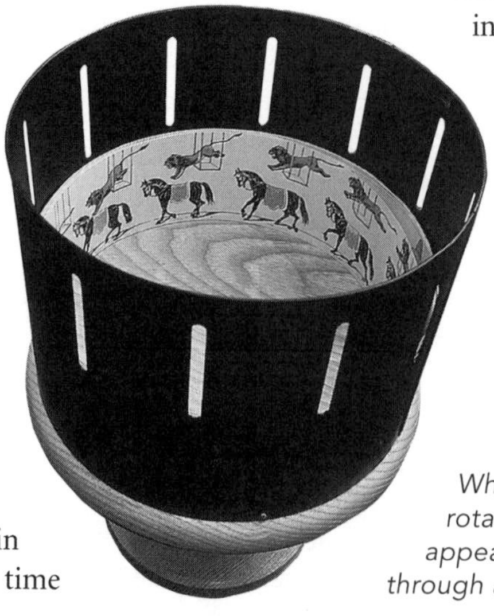

CGI and Mathematics

CGI is even more mathematically based than hand animation because the images must be mathematically represented in order to be manipulated in the computer environment. Oscar-winning computer scientist Tony DeRose, who has worked for Pixar Animation Studios, said, ". . . different kinds of mathematics are used for different aspects of a film, from the simulation of how light bounces around in an environment (integral calculus) to obtaining smooth surfaces efficiently (subdivision surfaces) and making characters move in a realistic fashion (harmonic coordinates)." Trigonometry and vector algebra are widely used in CGI algorithms for creating and manipulating images. Matrices are a standard algebraic way of representing various transformations. Dilation makes objects larger or smaller in addition to stretching; translations move objects; and rotations turn objects.

One classic CGI method for creating three-dimensional animated objects involves using polygonal meshes, which are collections, or grids, of polygons. This method makes use of the geometry of smooth surfaces. Like animated motion, this method relies on the human eye's tendency to smooth discon-

When the cylinder of a Zoetrope rotates, the images on the inside appear to be moving when viewed through a succession of slits.

tinuous regions. Locally, smooth surfaces look flat, so they can be approximated with small, flat polygons such as triangles or quadrilaterals. Basic three-dimensional shapes such as cubes, cylinders, spheres, and cones may be joined to form composite three-dimensional objects. Interpolation is also used. More complex and smoother-looking three-dimensional objects can be modeled using sophisticated mathematics like spline patches and non-uniform rational basis splines, where a spline is a mathematical function defined piecewise by polynomials. Such techniques have become standard practice in CGI. The mathematical representation of three-dimensional shapes, including layout and materials, is used to compute a two-dimensional image from a given viewpoint, a process called "rendering." This process entails addressing issues such as visibility from selected viewer angles (including which parts of objects in the scene are hidden) and appearances, and how objects look different as the lighting varies. Finally, the motion of each object in the scene has to be specified.

Lucasfilm LTD and animator Kecskemeti B. Zoltan of Ste-One provided mathematician Timothy Chartier with digital models of Yoda from the *Star Wars* movies to explore in linear algebra classrooms. One of the models had 53,756 vertices, 4040 triangles, and 49,730 quadrilaterals, illustrating that realistic images and their transformations have many more data points and matrix multiplications than is typical as classroom examples. Chartier noted, "More recently, computer animation produced the character's movement, which required mathematical concepts from such areas as linear algebra, calculus, differential equations, and numerical analysis. Drawing on these popular culture ties in appropriate coursework can pique students' curiosity and compel further learning."

Despite the many available mathematical techniques and advances in the computational and visualization power of computer systems, convincing simulation of some physical features, like hair, continues to be challenging. Pixar noted that it took up to 12 hours to render a single frame of the character Sulley in the 2001 movie *Monsters, Inc.* because of his nearly 3 million individually animated hair strands. Each hair was mathematically modeled as a series of springs connected via hinges.

CGI has come a long way since the 1976 movie *Futureworld*, which many acknowledge as the first use of three-dimensional computer imagery. Even though the first CGI film to win an Oscar was Pixar's short movie *Tin Toy* in 1988, the 1995 movie *Toy Story* was the first full-length, fully CGI feature film. Many challenges and problems remain to be solved in the quest for photorealism in CGI. Examples include more accurate modeling and representation of physical actions, such as swallowing, as well as textural and other properties of materials like skin, including wrinkles. Animators also seek to better differentiate faces for people of varying ages, such as children or the elderly.

Further Reading

Chartier, Timothy. "Using the Force: Star Wars in the Classroom." *PRIMUS* 17, no. 1 (2007).

Mathematical Association of America. "Pixar's Tony DeRose Illuminates the Mathematics of Animation." http://www.maa.org/news/101509derose.html.

McAdams, A., S. Osher, and J. Teran. "Crashing Waves, Awesome Explosions, Turbulent Smoke, and Beyond: Applied Mathematics and Scientific Computing in the Visual Effects Industry." *Notices of the American Mathematical Society* 57, no. 5 (2010). http://www.ams.org/jackson/fea-mcadams.pdf.

D. Keith Jones
Deborah Moore-Russo

See Also: Digital Images; Movies, Making of; Optical Illusions; Painting; Transformations.

Apgar Scores

Category: Medicine and Health.
Fields of Study: Algebra; Measurement; Representations.
Summary: The Apgar score is a simple prognostic device for neonatal care.

Throughout the developed world and in many other countries, every newborn baby is assessed according to various factors, each of which is assigned a score that is aggregated to quantify the baby's condition and prognosis. The system was introduced in the 1950s by Dr Virginia Apgar, whose last name has come to serve as a mnemonic for the assessed categories: activity, pulse, grimace (reflex), appearance (skin color), and respiration. The Apgar score indicates the health of the newborn and the likelihood that medical treatment or special intervention will be necessary much more quickly and more accurately than any system that had previously been in use.

In medicine, there are many scoring systems designed to predict and identify clinical situations in which the potential value of intensive care is low, while the burden of therapy is high, providing a numerical prediction of mortality. In gastroenterology, the Child-Pugh score is a scoring system to assess the prognosis of chronic liver disease; for vascular patients, the Eagle score allows estimation of a patient's risk of dying during heart surgery; the probability of pulmonary embolism is estimated by the Geneva score; the Gleason Grading system is used to help evaluate the prognosis of men with prostate cancer; and for pediatric end-stage liver disease, a scoring system exists for prioritizing allocation of liver transplants for children under 12 years of age.

Development and Effectiveness

Dr. Virginia Apgar was the first woman at Columbia University College of Physicians and Surgeons to be named a full professor. She developed a practical method for measuring the status of probable survival of newborn infants. Initially, she listed all objective signs that pertained in any way to the condition of the infant birth. Then she observed that five of these signs could be easily determined one minute after the complete birth of the baby. Depending on if the sign was absent, weak, or present, a rating of 0, 1, or 2 was given for each signal. The signs are heart rate (slow, normal, fast, or irregular), respiratory effort (from normal to distressed), reflex response (from over- to under-reactive), muscle tone, and color (pale, normal, or blue). In this system, infants in poor condition scored 0–2, infants in fair condition scored 3–7, and a score of 10 indicated a baby in the best possible condition.

In 1953, she observed the mortality rates of 2096 newborn infants with low, moderate, and high Apgar scores within 60 seconds after complete birth. This evaluation was rapidly adopted in delivery rooms throughout the United States and elsewhere. In 1959, a study with 15,348 infants established the predictive value of the Apgar score. The death rate among infants scoring 2, 1, or 0 was about 15%, while the rate for infants scoring 10 was about 0.13%. This prediction is especially useful in judging the urgency for resuscitative measures, such as respiratory assistance. It can be used to guide care, including intensive care. The score is generally determined by doctors and nurses at one minute and at five minutes after delivery. The five-minute score is generally accepted as the best predictor for newborn infant survival. A low score on the one-minute test may show that the neonate requires medical attention but is not necessarily an indication that long-term problems will occur, particularly if there is an improvement for the five-minute test.

Prediction

Probability is used to express knowledge or belief that an event will occur or it has occurred. A prediction is a statement that tells what might happen in the future based upon the given information. Prediction methods are important in various fields, including medicine, physics, and finance. Mathematics can be used to develop predictions, which are based on a careful analysis of patterns and collected data. Apgar recognized the patterns related to a baby's health signs and used them as a basis to make subsequent predictions. This example provides a clear idea of a credible prediction that was based on some form of empirical evidence. Thanks to this predictor approach, thousands of babies with special needs get the care they need immediately. Although it is not possible to make a 100% accurate prediction, predictions based on solid data and statistical analysis can increase the likelihood of accuracy.

Before 1952, the way to judge the condition of a newborn baby quickly and accurately shortly after birth was based on "breathing time" and "crying time." Apgar's accurate observations between 1949 and 1952 allowed the development of the automatic method of one-minute observation covering several signs easily. Thus, using some mathematical tools it is possible to transform qualitative values, such as physiological signs of babies, into quantity values—Apgar scores. By making predictions using Apgar scores it is also possible to perform the reverse: using the quantitative values (scores) to predict future qualitative values (health of babies).

Further Reading

Apel, M. A. *Virginia Apgar: Innovative Female Physician and Inventor of the Apgar Score (Women Hall of Famers in Mathematics and Science)*. New York: Rosen Publishing Group, 2004.

Infarom Publishing. "Probability Theory Guide and Applications." http://probability.infarom.ro.
National Institutes of Health: National Library of Medicine. "Changing the Face of Medicine, Dr. Virginia Apgar." http://www.nlm.nih.gov/changingthefaceofmedicine/physicians/biography_12.html.

Maria Elizete Kunkel
Maria Elizabeth S Rodrigues

See Also: Data Analysis and Probability in Society; Diagnostic Testing; Medical Imaging.

Arabic/Islamic Mathematics

Category: Government, Politics, and History.
Fields of Study: Algebra; Connections; Geometry; Measurement; Number and Operations; Representations.
Summary: Arabic and Islamic mathematicians popularized the decimal system and Arabic numerals and also developed algebra.

Mathematicians living in Islamic lands and writing in Arabic have played a central role in the development of mathematics, particularly during the 700-year period from around the year 750 C.E to around 1450 C.E. These scholars lived in an area that not only includes the present-day Middle East but stretches into the western parts of India, the major cities of central Asia, all of northern Africa, and most of the Iberian Peninsula. Most of the influential mathematicians of this seventh-century era were Muslim, and most wrote in Arabic. However, the lands ruled by Muslim rulers included many ethnicities, cultures, languages, and religions. Muslims, Christians, Jews, Zoroastrians, Manichaeans, Sabians, Buddhists, Hindus, Persians, Turks, Sogdians, Mongols, Arabs, Berbers, Egyptians, and many others contributed to a remarkable multiethnic, multicultural civilization. Mathematics was not an exception. The full story of mathematics in this era has yet to be told. Hundreds of manuscripts await examination, translation, and a critical edition. Undoubtedly, in the years to come, our understanding of the extent, the import, and the influence of the mathematics of this period will change dramatically.

While their knowledge of what came before them was incomplete and uneven, the mathematicians of the Islamic era were aware of—and in some ways heirs to—ideas, methods, and points of view that originated in India, Persia, and—especially—Greek Alexandria. A remarkable translation movement coupled with a scholarly tradition of writing commentaries on previous works meant that mathematicians of this era were comfortable with the contents and the methodology of the works of, among others, Euclid, Archimedes, Apollonius, Ptolemy, and Diophantus as well as the basics of Indian decimal arithmetic and trigonometry. They also had access to Persian astronomical tables. They accomplished a great deal with this heritage. What the mathematicians of the Islamic era bequeathed to those who came later was very different in content, style, and approach than what had come before them. (A note on names: names of mathematicians and places can be transliterated to English based on their Arabic, Persian, or Turkish versions. For the most part, we have chosen what is currently most common in English. The one exception is that we have often omitted the Arabic definite article "al" that precedes titles and nicknames.)

The Decimal System and the Concept of Number

For Euclid—the preeminent mathematician of Greek Alexandria—"number" meant a rational number. In his work, irrational numbers were called magnitudes and were treated quite differently from numbers. In fact, Euclid's very influential book *Elements* contains few numbers and hardly any calculations. Starting with Khwarizmi of Khwarizm (c. 780–850 C.E.), the principles of the positional decimal system that had originally come from India were organized and widely disseminated. Hence, with the use of 10 symbols it was possible to carry out all arithmetic operations. Over the following centuries, the methods for these arithmetical operations were improved and included working with decimal fractions and with large numbers. In fact, in the process, the Euclidean concept of number was gradually enlarged to include irrational numbers and their representation as decimal fractions. The mathematician Kashani (c. 1380–1429), also known as al-Kashi, worked comfortably with irrational numbers and, for example, was able to produce an approximation that was correct to 16 decimal places. The Arabic texts on the decimal number system were translated to Latin and were the basis for what are now called the Hindu-Arabic numerals.

Algebra

While it is possible to recognize algebraic problems in ancient mathematics, algebra as a discipline distinct from geometry and concerned with solving of equations was developed during the Islamic period. The first book devoted to the subject was Khwarizmi's *Al-kitab al-muhtasar fi hisab al-jabr wa-l-muqabala* (*Compendium on Calculation by Completion and Reduction*).

In this title, "al-jabr"—the origin of the word "algebra"—means "restoration" or "completion" and refers to moving a negative quantity to the other side of an equation where it becomes positive. *Al-muqabala* means "comparison" or "reduction" and refers to the possibility of subtracting like terms from two sides of an equation. While all algebra problems were stated and solved using words and sentences—symbolic algebra did not arise until much later in the fifteenth century in Italy—an algebra of polynomials was developed by Abu Kamil (c. 850–930), Karaji (c. 953–1029), and Samu'il Maghribi (c. 1130–1180+, also known as al-Samaw'al). Powers, even negative powers, of unknowns were considered and many algebraic equations were classified and solved. Khwarizmi gave a full account of second-degree equations, and Khayyam (1048–1131) gave a geometric solution to equations of degree three using conic sections. Here, we give a problem—translated to modern notation—solved by Abu Kamil. Some 300 years later, this exact same problem appeared in Chapter 15 of the 1202 text *Liber Abaci* by Leonardo Fibonacci. Abu Kamil gave a solution to the following system of three equations and three unknowns:

$$x + y + z = 10$$
$$x^2 + y^2 = z^2$$
$$xz = y^2.$$

Abu Kamil first started with the choice of $x = 1$ and solved the latter two equations for y and z. Since, for the latter two equations, any scalar multiple of the solutions continues to be a solution, he then scaled the solutions so that the first equation was also satisfied. He simplified the answer to get:

$$x = 5 - \sqrt{\sqrt{3125} - 50}.$$

Geometry

Geometrical methods and problems were ubiquitous in the Islamic era. While algebraic problems were solved using the newly developed algebraic algorithms (the word "algorithm" itself is derived from *algorismi*, the Latin version of the name of the mathematician al-Khwarizmi), the justification for the algebraic methods was usually given using geometrical arguments and often relying on a distinctively Euclidean style. Guided by problems in astronomy and geography (for example, finding, from any place on Earth, the direction of Mecca for the purpose of the Islamic daily prayers), spherical geometry was developed.

But new work in plane geometry was also carried out. Khayyam and Nasir al-din Tusi (1201–1274), for example, studied the fifth postulate of Euclid and came close to ideas that much later on led to the development of non-Euclidean geometries in Europe. However, as is the case with much of the mathematics of this era, applications play an important role in the choice of questions and problems.

For example, Abu'l Wafa Buzjani (940–997) reports on meetings that included mathematicians and artisans. A problem of interest to tile makers is how to create a single square tile from three tiles. A traditional mathematician, Abu'l Wafa explains, translates this problem into a ruler and compass construction and gives a method for constructing a square of side $\sqrt{3}$.

While logically correct, this construction is of little use to the tile maker, who is confronted with three actual tiles and wants to cut and rearrange them to create a new tile. Abu'l Wafa also gives the customary practical method that is actually used by tile makers to solve this problem, and proves that their method, while practical, is not precise, and the final object is not exactly a square. While stressing the importance of being both practical and precise, and the virtues of Euclidean proofs, he presents his own practical and correct methods for solving this and related problems.

Trigonometry

The origins of trigonometry begin with the Greek study of chords as well as the Indian development of what is now called the "sine function." Claudius Ptolemy's table of chords and Indian tables of sine values were powerful tools in astronomy. However, a systematic study and use of all the trigonometric functions motivated by applications to astronomy, spherical geometry, and geography begins in the Islamic era. Abu'l Wafa had a proof of the addition theorem for sines and used all six trigonometric functions; Abu Rayhan Biruni (973–1048) used trigonometry to measure the circumference of Earth; and Nasir al-din Tusi gave a systematic treatment in his *Treatise on the Quadrilateral* that helped establish trigonometry as a distinct discipline.

Combinatorics

One of the earlier known descriptions and uses of the table of binomial coefficients (also known as the Pascal triangle) is that of Karaji. While his work on the subject is not extant, his clear description of the triangle survives in the writings of Samu'il Maghribi. Binomial coefficients were used extensively, among other applications, for extracting roots. Kashani, for example, used binomial coefficients to give an algorithm for extracting fifth roots. He demonstrated it by finding the fifth root of 44,240,899,506,197. Other combinatorial questions were treated as well. Ibn al-Haytham (c. 965–1039, also known as Alhazen) gave a construction of magic squares of odd order, and Ibn Mun'im (died c. 1228) devotes a whole chapter of his book *Fiqh al-Hisab* to combinatorial counting problems.

Numerical Mathematics

The prominence of applied problems, the development of Hindu-Arabic numerals and calculation schemes, and the development of algebra and trigonometry led to a blossoming of numerical mathematics. One prime example is Kashani's *Miftah al-Hisab* or *Calculators' Key*. In addition to his approximation of 2π and his extraction of fifth roots, he also gave an iterative method for finding the root of a third-degree polynomial in order to approximate the sine of one degree to as close as an approximation as one wishes.

Further Reading

Berggren, J. L. *Episodes in the Mathematics of Medieval Islam*. New York: Springer Verlag, 1986.

Katz, Victor J. *A History of Mathematics. An Introduction*. 2nd ed. London: Addison Wesley, 1998.

Katz, Victor, ed. *The Mathematics of Egypt, Mesopotamia, China, India, and Islam. A Sourcebook*. Princeton, NJ: Princeton University Press, 2007.

Van Brummelen, Glen. *The Mathematics of the Heavens and the Earth: The Early History of Trigonometry.* Princeton, NJ: Princeton University Press, 2009.

Shahriar Shahriari

See Also: Babylonian Mathematics; Greek Mathematics; Measurement, Systems of; Number and Operations; Numbers, Rational and Irrational; Numbers, Real; Ruler and Compass Constructions; Squares and Square Roots; Zero.

Archery

Category: Games, Sport, and Recreation.
Fields of Study: Algebra; Geometry; Measurement.
Summary: Mathematics is essential in modeling and predicting a bow's performance.

Archery is the practice of propelling an arrow with a bow for the purpose of hunting, warfare, or sport. A bow is a pair of elastic limbs connected at the tips by a string. A bow acts as a spring and stores in the limbs the energy applied by the archer. As the archer releases the string, the arrow is propelled with a force proportional to the tension on the string. The path of the arrow is a parabola whose shape is determined in part by the angle of release from the bow, measured with reference to the ground.

The origins of archery are lost in the beginning of civilization and probably will never be determined with precision. The earliest bows known today were found in the Holmegaard area of Denmark and were made of elm and yew. The Holmegaard bows date form the Mesolithic period (10,000–3000 B.C.E.); however, there is archaeological evidence of projectile wounds—possibly caused by bows—from the Upper Paleolithic (40,000–10,000 B.C.E.) in all continents. It is speculated that archery was first used for hunting and, later, for warfare as social structures became increasingly complex. By the twelfth century B.C.E., archery was a decisive branch of military power. For example, the wall of the Theban temple of Ramses III depicts the Aegean fugitive fleet—driven from Crete by the Greek immigration—engaging in and losing a battle against the Egyptian fleet, whose primary weapon is shown to be archery. Archery remained the weapon of choice in the West for distance combat until the introduction of gunpowder toward the fourteenth century C.E. Archers in the medieval era would fire in a high arc, achieving accuracy by volume rather than deliberate aim. Today, archery is practiced as a precision sport and for hunting. Men's archery was one of the events of the second modern Olympics in 1900. The first Olympic archery event for women was held in 1904.

Mathematical Modeling of Bows

Since the 1930s, engineers and scientists have studied the design of bows. In 1947, C. N. Hickman made the first accurate mathematical model for flat bows, consisting of an idealized representation of two linear elastic hinges and rigid limbs with point mass (an idealized representation of a body used to simplify calculations) at the tip. More recent modeling efforts by B. W. Kooi and C. A. Bergman consider the limbs as beams that store elastic energy by bending.

The Bernoulli-Euler equation, named for Daniel Bernoulli and Leonhard Euler, describes the change in the curvature of a beam as a function of the "bending moment" (tendency to rotate about an axis) and is used to estimate the force in the string. When the archer draws the bow, the force exerted at the middle of the string causes an increase in the bending of the limbs, thus increasing the momentum and storing more energy for the shot. The elasticity modulus of the bow's material—the proportionality constant that relates limb deformation versus energy stored—determines the force with which the limbs recover their original shape after being deformed.

Mathematical modeling is a viable alternative for the evaluation of the performance of old bow models. As time passes, environmental conditions and natural processes cause considerable degradation within the cell structure of the wood used in ancient bows, which prevents a realistic assessment of the original density of the material and precludes direct testing.

An archer's body should be perpendicular to the target and his or her feet should be shoulder-width apart.

Further Reading

Grayson, Charles E. *Traditional Archery From Six Continents*. Columbia: University of Missouri Press, 2007.

Kooi, B. W., and C. A. Bergman. "An Approach to the Study of Ancient Archery Using Mathematical Modelling." *Antiquity* 71, no. 271 (1971).

Miller, Frederic P., Agnes F. Vandome, and John McBrewster, eds. *History of Archery*. Beau Bassin, Mauritius: Alphascript Publishing, 2010.

Slater, Steven. "The Physics of a Wooden Bow: In Traditional Archery, Not All Bows Are Equal." *Suite 101* (July 6, 2009). http://www.suite101.com/content/the-physics-of-a-wooden-bow-a130234.

Soar, Hugh D. H. *The Crooked Stick: A History of the Longbow*. Yardley, PA: Westholme Publishing, 2009.

Juan B. Gutierrez

See Also: Artillery; Firearms; Learning Models and Trajectories; Middle Ages.

Archimedes

Category: Government, Politics, and History.
Fields of Study: Calculus; Connections; Geometry; Measurement; Representations.
Summary: Archimedes contributed to many areas of mathematics, demonstrated the law of buoyancy, and designed a number of marvelous devices.

Archimedes of Syracuse (287–212 B.C.E.) was born and lived most of his life in Syracuse on the Greek island of Sicily in the Mediterranean. Archimedes possessed an incredibly versatile intellect—today he is remembered as one of the most important mathematicians, astronomers, and engineers in history.

He is credited with numerous inventions, such as the mechanical pump known as the Archimedes screw, the compound pulley, and various engines of war (including advanced catapults, ship-destroying mechanical arms, and the famous Archimedes "death ray"). As important as Archimedes' legacy is to engineering and astronomy, perhaps his most important work was in mathematics. His contributions in geometry, conic sections, and number theory along with his work in computational mechanics, his discovery of the law of buoyancy, and his contributions to the field of mathematics that would become known as calculus almost two millennia later, secure Archimedes' place in mathematical history.

Early Life

Most of what is known about Archimedes was written long after his death by Roman historians such as Plutarch. This lack of contemporary sources—coupled with the fact that surviving works of Archimedes himself are copies made many centuries later—make some of the popular stories about the Greek mathematician questionable.

Archimedes' father was an astronomer named Phidias. According to some authors, young Archimedes was sent to Alexandria, Egypt, to study. The library at Alexandria was the center of learning for the Greek world, containing the mathematical and astronomical manuscripts produced by scholars such as Euclid of Alexandria, the "father of geometry." It is very likely that Archimedes studied mathematics with the students of Euclid. While in Alexandria, Archimedes may have produced his first important invention, a pump now known as the Archimedes screw (some historians claim the device was invented by Archimedes at a later date at the request of King Hieron II to be used as a bilge pump for removing water from ships). Whatever Archimedes' motivation for developing it, the Archimedes screw is a simple mechanical device that is used to move water. In fact, the device continues to be used today in various applications.

Although there is no indisputable evidence that Archimedes studied in Alexandria, there are several indicators that he was friendly with the mathematicians there. For instance, the famous Archimedes' Cattle Problem is found in a letter Archimedes sent to the Alexandrian astronomer and mathematician Eratosthenes. Archimedes challenged the mathematicians of Alexandria to solve a complicated mathematical riddle concerning the number of cattle in the herd of the sun god. Written in verse, the problem involves extremely large numbers and was not solved by the mathematicians of Alexandria, or by anyone else for that matter (including Archimedes), until the late nineteenth century.

Archimedes' Inventions

In addition to the Archimedes screw, Archimedes is credited with the invention of the compound pulley. A compound pulley is a system of movable pulleys that provides a substantial mechanical advantage for doing work. Evidently very confident in his knowledge and abilities, Archimedes once asserted, "Given a place to stand on, I can move the Earth." King Hiero II decided to test Archimedes' boast by assigning him the job of

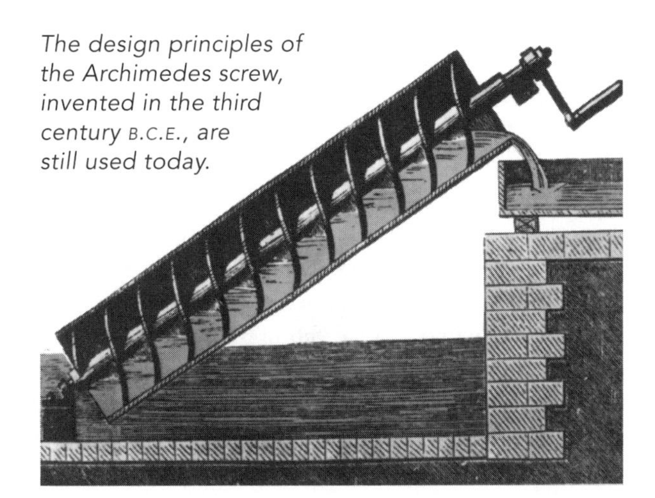

The design principles of the Archimedes screw, invented in the third century B.C.E., are still used today.

The Archimedes Palimpsest

One of the primary duties of historians is to use old letters, manuscripts, and other documents to try to understand a people or culture. In most cases, the existence of these documents is already known to others, and the historian is simply trying to shed new light on the past through the interpretation of the existing manuscripts. Occasionally, a new document is discovered and serves to excite the historical community.

In the tenth century, a manuscript containing several of the works of Archimedes was produced in Greek, the language Archimedes had used to compose the original works thirteen centuries earlier. A few centuries later, this manuscript and several other unrelated manuscripts were reused to produce a Byzantine prayer book.

Since the manuscripts were written on parchments—or animal skins—the words were literally scraped off the page (the word "palimpsest" means "scraped again"), making the parchment ready for its new authors. In this way, the monks saved the expense of new parchment. This prayer book survived the ensuing centuries until the early twentieth century, when a leading Archimedes scholar, John Ludwig Heiberg, determined that underneath the prayers and barely visible in a few places were works of the Greek mathematician Archimedes.

The location of the Archimedes Palimpsest throughout the rest of the twentieth century is a bit of a mystery, but in 1999 a wealthy collector, wishing to remain anonymous, bought the manuscript at auction and entrusted it to the Walters Art Museum in Baltimore for conservation and study. At the museum, experts in various fields have worked with scholars to uncover the hidden text. What they uncovered proved to be one of the most exciting discoveries in the history of mathematics. One of the Archimedean treatises on the palimpsest represents the only known copy of that work in Greek. Two others are manuscripts previously thought lost by scholars.

Although Greek and Roman writers attributed both of these works to Archimedes, no one in the modern age had ever laid eyes on either one—until the secrets of the palimpsest were unlocked.

moving a heavily laden ship from its dock, a job that would have required a great effort from a substantial number of men. Archimedes constructed a system of pulleys by which, with very little effort on his part, he was able to move the heavy ship.

Perhaps the best-known inventions of Archimedes were the engines of war used by the defenders of Syracuse against the invading Roman army. Although the Syracusans were badly outnumbered, Archimedes' ingenious devices kept the attackers at bay, even striking fear into the hearts of the Roman soldiers. In addition to making improvements on existing weapons such as the catapult, Archimedes developed new and frightening methods for defending his home. One of these inventions reportedly involved the use of great mirrors to focus the sun's rays on the Roman ships blockading Syracuse, setting fire to the helpless ships and the soldiers contained in them. This particular story of Archimedes' exploits has been questioned for centuries. In fact, it was a topic of the Discovery Channel television show, *MythBusters*, which concluded that the likely success of such a device was very small.

Archimedes also built, according to Roman historians, gigantic mechanical arms that swung out over the enemy ships. Some of these arms dropped massive stones and other weapons, sinking the ships. Another mechanical device, today known as the Archimedes claw, was used to pluck ships out of the water and bash them against the rocks or simply shake them and drop them from great heights so that they would sink to the bottom of the sea.

Archimedes' inventions were so effective, and so terrifying, that reports claim that the Roman ships and their invading armies would flee in terror at the slightest sound or movement emanating from the Syracuse defenses that hinted at another attack.

Archimedes' Mathematics

Although renowned for his engineering achievements and machines of war, Archimedes was at heart a pure mathematician. His insights and discoveries in many fields of mathematics cause historians today to consider him one of the greatest and most original mathematicians who ever lived. Archimedes was reportedly obsessed with mathematics, and stories abound regarding this obsession. While lounging in the public baths—as was the custom at the time—Archimedes would often draw geometric figures in the chimney embers. This single-mindedness eventually led to his demise. Two stories emerged from biographies of Archimedes by the Roman historian Plutarch, both of which occurred when the Romans finally overran Syracuse. In the first and most popular story (probably because it illustrates the romantic idea of the dedicated but absent-minded scientist), Archimedes is contemplating some geometric figures when a Roman soldier comes upon him and orders Archimedes to come with him. Archimedes' response to the soldier is that the soldier should leave him alone until he has finished the proof to the problem he is contemplating—a response not appreciated by the soldier, as he slew Archimedes with a sword. The other story involves a soldier coming upon Archimedes as he carried various mathematical instruments to General Marcellus, the Roman general in charge of the invasion. The soldier, thinking the instruments were valuable—perhaps even vessels filled with gold—slew Archimedes for the treasure. In either account, General Marcellus is very unhappy, as he had ordered Archimedes brought to him alive.

Today, Archimedes the mathematician is remembered for much, but his discovery of the methods that one day would be called integral calculus is at the forefront. Archimedes used a technique called the method of exhaustion to approximate the area of a circle and thus the value of pi. In this method, Archimedes inscribed a polygon in a circle and calculated the area of the polygon. Inscribing a polygon involves drawing a polygon—a multisided, closed figure such as a pentagon or an octagon—so that each vertex just touches the inside of a circle. He then circumscribed a polygon around the outside of the circle and calculated its area. He knew the actual area of the circle must be somewhere between the areas of the two polygons. By calculating the areas of polygons with more and more sides—eventually using a 96-sided figure—he was able to approximate the area of the circle closely and conclude that the value of π lay somewhere between the following two fractions:

$$3\frac{1}{7} \text{ and } 3\frac{10}{71}.$$

Archimedes used the method of exhaustion to find many other interesting mathematical theorems. Most of these theorems are geometric in nature and address methods for finding areas of plane figures and volumes of three-dimensional solids. For example, Archimedes proved that the surface area of a sphere is four times the area of a great circle.

He also compared a sphere and a cylinder circumscribed around the sphere and found that the sphere's volume and surface area were two-thirds those of the cylinder. Many scholars believe that Archimedes considered his most important work to be in this area. A little more than century after the death of Archimedes, the Roman senator and orator Cicero discovered a grave he believed to belong to Archimedes, marked with sphere inscribed in a cylinder along with the related theorem.

Interestingly, although Archimedes used geometric methods like the method of exhaustion to prove his theorems, he used other methods in the discovery

Archimedes found that a sphere's volume and surface area were two-thirds those of a circumscribed cylinder.

of some of those same theorems. One such method involved the use of infinitesimals. An infinitesimal is an indefinitely small number that proved to be critical to the development of calculus many centuries later. Ironically, Archimedes did not accept the use of infinitesimals—or other mechanical methods he used to uncover interesting mathematical truths—in a rigorous proof. Archimedes accepted the Greek tradition that only pure geometric demonstrations contained the rigor demanded in mathematical proof. According to Archimedes, as translated by British mathematician Thomas Heath, "Certain things first became clear to me by a mechanical method, although they had to be demonstrated by geometry afterwards.... But it is of course easier, when we have previously acquired by the method, some knowledge of the questions, to supply the proof than it is to find it without any previous knowledge."

Archimedes made many other mathematical discoveries. He found methods for calculating the center of gravity of plane figures, methods for summing infinite series, techniques for finding tangents to curves (a forerunner of differential calculus), and a method for finding the weight of a solid body immersed in liquid. In addition to these geometric discoveries, Archimedes found many interesting results in arithmetic, or the theory of numbers. He developed methods for estimating the value of square roots, and in a work called *The Sand Reckoner*, Archimedes invented a new number system capable of representing impossibly large numbers—numbers large enough, according to Archimedes, to count the number of grains of sand in the universe.

Today, we know of Archimedes' great mind through copies of his own works made centuries after his death as well as books from other authors who attribute results to Archimedes. Several works of Archimedes no longer exist, and we know only of their onetime existence from references in other books. Until the twentieth century, an Archimedes work called *The Method* belonged in this category of lost works. Other authors related that *The Method* contained explanations of the techniques used by Archimedes to discover many of his mathematical theorems. Historians and mathematicians alike lamented the fact that this potential insight into the mind of the great man would never be known. A copy of this work has recently come to light and has opened a treasure trove of new information regarding the creative processes used by Archimedes.

Archimedes' Legacy

The esteem with which history holds a figure like Archimedes may be tied to the stories—substantiated or otherwise—that become a part of the folklore surrounding that figure. The most famous story concerning Archimedes is another example of this historical perception. It seems that King Hiero II was concerned that a greedy goldsmith had used a certain amount of silver in a crown that was intended to be pure gold.

The king asked Archimedes to determine the purity of his crown without destroying it. While bathing one day—as the story goes—Archimedes realized that the volume of water displaced by his (or any other) body could be used to calculate the density of the body—a method that could measure the density of the crown and thus its content. Archimedes immediately sprang from his bath and ran naked through the streets of Syracuse yelling, "Eureka!" ("I have found it!")

Although this Archimedean anecdote does not appear in print until several centuries after his death, "Eureka!" remains the first thing that comes to the minds of many modern readers when they encounter the name of Archimedes. If indeed Archimedes did actually solve the problem of the impure crown for his king, it seems much more likely that he used a method that is now called the Archimedes principle. This method, which actually appears in Archimedes' writings, involved weighing an object while it is submerged in water to determine its buoyant force.

In his own lifetime, Archimedes was renowned as an inventor and military engineer rather than as a mathematician. History, however, remembers Archimedes as one of the most brilliant and original mathematical thinkers who ever lived. In today's modern world, "pure" and "applied" mathematics are often scrupulously separated by their practitioners. In his time, however, Archimedes was both a brilliant pure mathematician—whose work involving integral calculus predated Isaac Newton and Gottfried Leibniz by almost 2000 years—as well as a gifted applied mathematician—who used geometric techniques to find, among other things, the center of gravity of solid objects. Certainly, Archimedes is a part of the small pantheon of scientific geniuses like Newton and Albert Einstein whose brilliance changed the way in which we see our world.

Further Reading

Clagett, Marshall. *Archimedes in the Middle Ages.* Madison: University of Wisconsin Press, 1964.

Dijksterhuis, E. J. *Archimedes.* Princeton, NJ: Princeton University Press, 1987.

Gardner, Martin. *Archimedes, Mathematician and Inventor.* New York: Macmillan, 1965.

Hershfeld, Alan. *Eureka Man: The Life and Legacy of Archimedes.* New York: Walker, 2009.

Netz, Reviel, and William Noel. *The Archimedes Codex: How a Medieval Prayer Book Is Revealing the True Genius of Antiquity's Greatest Scientist.* Philadelphia, PA: Da Capo Press, 2007.

Noel, William. "Archimedes Palimpsest." http://www.archimedespalimpsest.org.

Stein, Sherman K. *Archimedes: What Did He Do Besides Cry Eureka?* Washington, DC: Mathematical Association of American, 1999.

Tuplin, Christopher, and T. E. Rihll. *Science and Mathematics in Ancient Greek Culture.* New York: Oxford University Press, 2002.

Todd Timmons

See Also: Calculus and Calculus Education; Greek Mathematics; Infinity; Levers; Limits and Continuity; Mathematics, Applied; Pi; Polygons.

Arenas, Sports

Category: Games, Sport, and Recreation.
Fields of Study: Algebra; Geometry.
Summary: Modern arena designers consult mathematicians to determine the effects of design on play and crowd behavior.

A sports arena is essentially an enclosed area consisting of a large open space where a sport is played, surrounded by seating for spectators. It may also include various facilities for athletes, spectators, and the press. Many sports use specific terms for arenas, like "park" for baseball and "stadium" for football. Some sports arenas are open-air while others are roofed. The word "stadium" comes from *stadion*, an ancient Greek unit of length. Mathematics plays a significant role in the design and maintenance of modern sports arenas, including not only the geometrically shaped playing surfaces but also the optimization of seating, sightlines, acoustics, lighting, spectator traffic flow, and placement of restrooms and concessions. Features such as retractable roofs and convertible forms to accommodate multiple sports require careful design as well. Mathematicians also analyze and model features of sports arenas to determine their potential effect on the game play.

The rules of each sport dictate dimensions for the field of play. Some such as hockey, football, basketball, and soccer specify exact dimensions for the playing surface and delineate areas for specific activities, like the rectangular key in basketball or the half-circle goal crease for amateur hockey. Baseball, on the other hand, standardizes the dimensions of some features such as the distance between bases and the distance between the pitcher and home plate, but the outfield varies depending on the positions of the outfield walls. Further, aspects of game play can be affected by design choices. Fenway Park's outfield wall known as the "Green Monster" is notorious for stopping home runs, yielding more doubles and triples. When the new Yankee Stadium produced a higher rate of home runs, there was speculation about a "wind tunnel" effect. Statistical analyses suggested that curvature and height of the right field wall were more important than wind speeds or patterns. Statistician George "Bill" James developed the concept of park factors, which attempts to measure how park characteristics influence game outcomes.

Robert F. Kennedy Memorial Stadium in Washington, D.C., which opened in 1961, was the first multiuse stadium. It was widely decried for being a "concrete donut." Some critics suggested its wavy shape and curvature optimized it for baseball seating, though the widely replicated design has deficiencies for baseball and football. Some critical seats were too low for football and too high for baseball, resulting in poor sightline angles.

The baseball configuration was also more symmetrical than most baseball-only fields. Modern designers use mathematical techniques and tools (such as Mathcad software), simulations, and three-dimensional modeling for their designs, resulting in unique facilities like The Float in Singapore, which is literally floating on Marina Bay. Similar methods are involved in the design of arena roofs or domes, some of which are retractable. Calculating the amount of material

The Green Monster is the nickname for the left field wall at Fenway Park, home to the Boston Red Sox. The 37-foot-high wall is famous for preventing home runs that would clear the walls of most other ballparks.

needed to construct a curved dome, as well as calculating the weights, forces, and stresses, typically involves the use of calculus. These calculations, in turn, partially determine the type of support required.

Geometry and graph theory also contribute to dome design. R. Buckminster Fuller suggested that domes are strongest when the edges lie along great circles. Triangles are often used to give great strength with minimal weight, while other support structures resemble the latitude and longitude configuration on a globe. Fibonacci sequences and plane tilings also are used in the design of some domes. Veltins Arena in Germany uses features like hinged columns with ball-bearing edges that move in three dimensions. Both Veltins Arena and University of Phoenix Stadium in the United States feature sliding roofs and retractable natural-grass playing surfaces weighing millions of pounds. These were mathematically modeled extensively before construction. Transformative structures of this type have become known as "kinetic architecture."

Mathematicians continue to investigate questions related to sports arenas, some of which have wider applications. Researchers have considered the impact of sports arenas on land values using hedonic regression models. Mathematical analyses of crowd sequence videos (frequently taken from sports venues) benefit research in areas including surveillance, designs of densely populated public spaces, and crowd safety. In some cases, people are conceptualized as a "thinking fluid" to which fluid dynamic and stochastic models may be applied. Unusual events like emergency evacuations are fairly rare, and there are legal barriers to obtaining extensive live footage. As such, computer scientists and mathematicians have developed detailed simulations for both "normal" behavior and unusual crowd events. Some have suggested that topology optimization would be beneficial for investigating arena evacuation plans.

Further Reading

Puhalla, Jim, Jeffrey Krans, and Michael Goatley. *Sports Fields: Design, Construction, and Maintenance.* Hoboken, NJ: Wiley, 2010.

Winston, Wayne. *Mathletics: How Gamblers, Managers, and Sports Enthusiasts Use Mathematics in Baseball, Basketball, and Football.* Princeton, NJ: Princeton University Press, 2009.

Bill Kte'pi

See Also: Baseball; City Planning; Domes.

Artillery

Category: Government, Politics, and History.
Fields of Study: Algebra; Calculus; Measurement; Number and Operations.
Summary Mathematics is essential to the design and firing of artillery pieces.

Mathematics has had numerous military applications, including the development of artillery pieces after the invention of gunpowder in China in the fourteenth century. Mathematical formulas and calculations are critical to the design and use of artillery. The science of ballistics, which relies on mathematical formulas to study the flight paths of projectiles, also plays a major role in artillery development.

Engineer Benjamin Robbins invented the ballistic pendulum and is referred to by some as the founder of modern ballistics. Technological and scientific developments resulted in the modern use of artillery firing tables and computer-based firing calculation programs. Mathematics also plays a significant role in the ability to centralize fire control command centers and the use of indirect fire in which targets are not visible through a weapon's sightlines.

Many mathematicians have worked in places such as the Ballistics Research Laboratory at Aberdeen Proving Ground, such as Gilbert Bliss, who worked on firing tables for artillery.

Early artillery weapons relied on mechanical energy to fire projectiles and were not of uniform design—making them large, cumbersome, and inaccurate. Technological innovations in the eighteenth and nineteenth centuries led to the development of standardized artillery with increased accuracy and mobility.

In the late eighteenth century, British Royal Artillery Lieutenant Henry Shrapnel created a shell (container) that held multiple musket balls and a time fuse that allowed the shell to travel longer distances before exploding, increasing the cannon's range. High-explosive fragmentation shells and improved conventional munitions replaced shrapnel shells beginning in the early twentieth century.

Military scientists, weapon and projectile engineers, and soldiers have utilized the science of ballistics (the study of the flight of projectiles as they exit the weapon, travel through the air, and penetrate the target) since its early development in the fourteenth century to improve the accuracy and range of artillery.

Cannons, which first appeared in the early fourteenth century, spurred the development of ballistics. Early artillery crews used mathematics to determine the optimal angles at which to elevate their weapons for improved accuracy and range. Engineers also used mathematics to determine the angles at which

Artillery Categories

The three main categories of modern artillery are guns (excluding personal weapons such as handguns and rifles), howitzers, and mortars. Guns rely on stored kinetic energy to force a solid projectile through both the weapon's barrel and the air to its intended target, while howitzers and mortars use stored chemical energy and explosive (non-solid) projectiles. Howitzers and mortars generally fire shorter ranges along parabolic arcs while guns fire longer ranges along sightlines. Tanks, anti-aircraft batteries, rockets, and missiles can also be considered part of modern artillery. Artillery can be either self-propelled or towed.

A U.S. M777 Light Towed Howitzer being fired in 2009 in Afghanistan. Howitzers have relatively short barrels and are used to shoot projectiles at high trajectories with a steep angle of descent.

to build fortifications to best defend against artillery bombardments.

Calculations of elevation, distance to target, weather conditions, projectile weight, and flight trajectory are necessary to achieve accuracy. Scientists and mathematicians, beginning with Italian mathematician Niccolo Franco Tartaglia, sought to improve the accuracy and reliability of early artillery pieces through ballistics. Tartaglia's studies on a variety of cannons led to his determination that a 45-degree angle was ideal for firing—with the caveat that external factors such as air drag would affect the results. Tartaglia is also credited with the development of the first ballistics firing tables based on standardized weapons and projectiles.

Other notable mathematical advances in early ballistics included the theories of Galileo Galilei on the effects of the forces of gravity and air drag on the projectile's velocity and flight path, as well as the parabolic nature of ballistic trajectories. In the early eighteenth century, English scientist Benjamin Robbins invented the Ballistic Pendulum, which allowed the measurement of a projectile's velocity and the effects of air drag on that velocity. He also determined that air drag plays a much greater role in affecting a projectile's velocity than gravity does. Sir Isaac Newton is credited with the development of formulas used to calculate aerodynamic drag, which he determined was proportional to air density, the projectile's cross-sectional area, and the approximate square of the projectile's velocity. However, Newton's solution was incomplete, and mathematician Johann Bernoulli produced a more general solution. Mathematician Leonhard Euler integrated the various stages of a projectile's flight to reduce the difficulty of the equations utilized in ballistics.

Artillery projectile designers use ballistics studies that calculate projectile properties, such as mass and diameter, based on the design specifications of the weapon in order to ensure the projectile will fit inside the barrel and generate enough energy to propel the projectile without damaging the weapon. Mathematical formulas are used to determine projectile design based on various input data including the force of aerodynamic drag, the ratio of the projectile's velocity to that of the sound in the medium it will traverse, the properties of the medium, the projectile's caliber (diameter), and the velocity at which it travels.

The mathematics of ballistics can be further broken down into internal, external, and terminal ballistics. Internal ballistics studies the flight properties of a projectile as it travels through the barrel of the weapon. A firing mechanism lights the gunpowder, which creates energy through the pressure generated by expanding gases. The energy is equal to the force times the barrel length. This energy forces the projectile through and out of the barrel. External ballistics studies the flight properties of a projectile as it travels through the air from the weapon to its intended target. Various formulas can be used to determine the kinetic energy of the projectile as it leaves the muzzle. Other calculations are then used to determine ballistic coefficient (a measure of a body's ability to overcome air resistance in flight.). The distance and direction of artillery projectiles is affected by aerodynamic drag caused by a combination of air pressure (the disturbance of air around a projectile creating an area of low pressure behind it) and skin friction (the contact between the air and the projectile's surface). Retardation is the measurement of the degree to which drag will slow a projectile's flight speed and can be calculated by the following formula:

$$R = \frac{D}{M}$$

where R is retardation, D is drag, and M is the projectile's mass. The ballistic coefficient is often used in place of drag because of the greater difficulty in calculating drag, which reduces along the flight path in relation to the decrease in velocity.

External ballistic formulas must also account for the fact that projectiles do not travel along straight flight paths. Physical and meteorological forces must be taken into account when determining or predicting a projectile's flight path. These forces include yaw (caused when the nose of the bullet rotates away from a straight trajectory) and precession (caused when the bullet rotates around the center of mass). Terminal ballistics studies the impact of the projectile as it hits the target. Mathematical calculations can be used to study how a projectile's design and flight features, such as velocity, shape, and mass, will affect its damage and wound capabilities.

Artillery firing requires the use of mathematical equations to determine range, elevation, and deflection, as well as the arc of fire and the probability of hitting the intended target. Artillery equation data also include the projectile's initial velocity, which is further divided into vertical and horizontal velocity components. Calculating the distance a projectile travels is performed by multiplying the time the projectile is in the air by the velocity's horizontal component. The needed angle to achieve a certain distance can then be determined by solving the equation for distance as a function of the angle.

Modern artillery crews rely on indirect fire, a technique developed in the early twentieth century in which a target is fired upon despite not being visible along sightlines. Indirect fire required more complex mathematical formulas and calculations, increasing the importance of specialized trained military personnel. These personnel calculated the range and bearing to the target. New techniques of determining the locations of enemy artillery batteries and subsequent firing data included flash spotting, sound ranging, air photography, and registration point. Indirect fire led to the development of graphical or tabular firing tables and the maintenance of a command center. Technological developments also allowed for greater adjustments to firing data based on such variable conditions as wind speed and weather. Initially, firing data derived from these tables was placed on the weapon's sights.

Use of Computers

Battlefield computers began to appear by the 1960s and were in use by the British and U.S. military by the following decade. Computerized firing tables utilize input data to determine the angle and position of artillery, which weapons will fire, and how many rounds will be fired (although some military forces still rely on older instruments and human calculations as backups).

Firing data such as quadrant elevation, azimuth (an angular measurement in a spherical coordinate system), fuse setting, and projectile properties are inputted into the software program spreadsheets based on established data and standard conditions, which determine ideal firing information. The firing information is then corrected for deviations from standard conditions, such as meteorological conditions. Further technological developments include computer-based surveillance and target acquisition systems, global positioning systems, and laser rangefinders.

Further Reading

Carlucci, Donald, and Sidney Jacobson. *Ballistics: Theory and Design of Guns and Ammunition*. Boca Raton, FL: CRC Press, 2007.

Hackborn, William. "The Science of Ballistics: Mathematics Serving the Dark Side." http://www.augustana.ualberta.ca/~hackw/mp480/exhibit/ballisticsMP480.pdf.

McMurran, Shawnee, et al. "The Impact of Ballistics on Mathematics." Proceedings of the 16th ARL/USMA Technical Symposium, 2010. http://www.math.usma.edu/people/rickey/talks/08-10-25-Ballistics-ARL/08-10-23-BallisticsARL-pulished.pdf.

Marcella Bush Trevino

See Also: Archery; Infantry (Aerial and Ground Movements); Missiles.

Asia, Central and Northern

Category: Mathematics Around the World.
Fields of Study: All.
Summary: The contributions of central Asia have included algebra and its great houses of wisdom.

Throughout history, countries in Asia have had shifting sociopolitical boundaries. The names of some countries have changed, influenced by the Arab and Islamic empires as well as European colonialism in the eighteenth and nineteenth centuries. Though not widely used, Northern Asia sometimes refers to the part of the Asia occupied by the transcontinental country of Russia, which is commonly included in eastern Europe. Central Asia includes the former Soviet satellites of Kazakhstan, Kyrgyzstan, Tajikistan, Turkmenistan, and Uzbekistan. Mongolia, typically considered part of central Asia by historians, is in the modern world classified as part of Eastern Asia by the United Nations. "Northern Asia" is a term that is not commonly used, thus the transcontinental country of Russia is usually thought of as part of Eastern Europe. Knowledge of the contributions of mathematicians around the world is constantly changing as historians discover and translate written materials in many languages. Further, the breakup of the Soviet Union and shifting alliances have given researchers access to documents from decades in which many Eastern Bloc nations kept themselves in isolation, as well as even older works contained in the libraries and educational institutions of these nations. For example, medieval Islamic texts in Uzbekistan have helped shed light on the rich mathematics culture of central Asia. However, the mathematics contributions and achievements of some people from central Asia may be included in the histories of other areas, countries, or cultures.

In the seventh century, the great Library of Alexandria in Egypt was captured by a Muslim army, and there are some historians who believe some contents of the library were taken into Muslim lands. Many cities in central Asia became famous in the medieval period for their own libraries, which contained original works and translations of texts from Greek and Sanskrit, some of which became the only surviving copies of these earlier works. Houses of wisdom provided places for scholars to gather, as well as scientific centers such as the fifteenth-century Samarkand Observatory in what is now Uzbekistan, which was founded by astronomer Muhammad Taragay Ulughbek. This observatory reputedly served as a model for later observatories in India. Astronomer and mathematician Ala al-Din Ali ibn Muhammed, also known as Ali Kushji, later preserved and disseminated some of the knowledge gathered by the observatory when it was destroyed. This catalogue of stars, containing the most accurate mathematical measurements of location known prior to the invention of the telescope, is still studied.

Significant Central Asian Mathematicians

In the same way that mathematicians in central Asia studied and developed many concepts that were first introduced by other cultures, other concepts and techniques in twenty-first-century mathematics were first brought to Europe by mathematicians who worked in or came from central Asia. The word "algorithm" derives from a Latin transliteration of the name of eighth- and ninth-century mathematician Abu Abdallah Muhammad ibn Musa al-Khwarizmi (sometimes written as Al-Khoresmi). The Khwarizm (or Koresm) region included portions of what are now Turkmenistan and Uzbekistan. The word "algebra" comes from the term *al-jabr*, which was found in al-Khwarizmi's treatise on that subject. Another of his mathematical writings, the *Book of Addition and Subtraction by the Indian Method*, helped promote the Hindu base-10 decimal system within the Arabic world. This system spread to Europe and revolutionized mathematics around the world in subsequent centuries.

Historical evidence suggests that tenth-century astronomer and mathematician Abu Mahmud Hamid ibn al-Khidr Al-Khujandi was born in the city of Khudzhand, in what is now Tajikistan. His mural sextant produced some of the most accurate astronomical observations of the day, and he may have contributed to trigonometry. The tenth- and eleventh-century mathematicians Abu Rayhan al-Biruni and Abu Nasr Mansur are also cited as being natives of Khwarizm. Al-Biruni studied a diversity of topics in mathematics and science, including cartography and map projections, trigonometry, combinatorial analysis, ratio theory, algebraic problem solving, geometry, Archimedes of Syracuse's theorems, conic sections, and spherical triangles. Along with his own prolific body of writings, he was also a translator of Sanskrit texts. Abu Nasr Mansur taught and collaborated with al-Biruni—the two frequently cited one another's contributions to their own work.

Many consider Mansur's primary mathematical contributions to be his commentary on Menelaus of Alexandria's *Sphaerica*, his development of trigonometry, and his tables for numerical solutions to problems in spherical astronomy. In the same time period, Abu Ali al-Husain ibn Abdallah ibn Sina, also known by the Latin name Avicenna, wrote on many topics, including medicine and mathematics. Some of his investigations included ruler and compass constructions, areas of circles, and geometric algebra. He also considered music to be a subdiscipline of mathematics, and some believe that his studies led to musical tuning by the method of just intonation, where the note frequencies are related by ratios of small whole numbers, rather than Pythagorean tuning, named for Pythagoras of Samos.

Beginning in about the twelfth century, central Asia underwent a great deal of social and political disruption, and there is often little surviving evidence regarding mathematics and science during those eras. During the Soviet period, mathematicians from Kazakhstan, Kyrgyzstan, Tajikistan, Turkmenistan, and Uzbekistan may have been drawn to some of the central academic centers in Russia and other parts of the Soviet Union. Since the fall of the Soviet Union, these countries are reestablishing themselves as independent nations, and achievement in mathematics continues. For example, students from central Asia have participated in and won numerous medals in the International Mathematical Olympiad, an annual competition for high school students in which individual medals are awarded based on each student's success in solving a set of mathematics problems. Countries send six-member teams.

Kazakhstan, Kyrgyzstan, and Turkmenistan first participated in 1993, Uzbekistan in 1997, and Tajikistan in 2005. In 2010, Kazakhstan hosted the 51st Olympiad in its capital of Astana. Students from 98 countries around the world participated. Professor Askar Dzhumadildayev noted, "Mathematics is one of the most important indexes of the education level in the country. Gathering the best young mathematicians in Astana is a great honor for us." A news report regarding the Olympiad acknowledged the rich history of central Asia: ". . . we should not forget that our country is an heiress of the mathematical school founded by great scientists of the Middle Ages. . . . who greatly contributed to development of mathematics long before the modern countries of the West appeared."

Further Reading

Bobojan, Gafurov. "Al-Biruni: A Universal Genius Who Lived in Central Asia 1000 Years Ago." *UNESCO Courier* (June 1974). http://unesdoc.unesco.org/images/0007/000748/074875eo.pdf.

"The 51st International Mathematical Olympiad." http://www.imo2010org.kz.

Matvievskaya, G. P. "History of Medieval Islamic Mathematics: Research in Uzbekistan." *Historia Mathematica* 20 (1993).

SARAH BOSLAUGH

See Also: Arabic/Islamic Mathematics; Asia, Western; Europe, Eastern.

Asia, Eastern

Category: Mathematics Around the World.
Fields of Study: All.
Summary: Across eastern Asia, mathematics education is given a high priority, with the goal of continuing the region's tradition of excellence.

Eastern Asia is one of the most populated regions of the world, lagging behind only southern Asia, and includes the Chinese cultural sphere once called the "Far eastern civilizations": China, Hong Kong, Macau, Taiwan, Japan, Mongolia, North Korea, and South Korea. The region is by no means homogeneous but has certainly been influenced to varying degrees by China in its writing systems, its cuisine, its architecture, and its religion. These influences are principally historical, cultural exchange being less centralized now, and influences like the Western world and the Soviet Union (in the case of Mongolia and North Korea) having been significant in the recent past. The technology sector is important in much of this region and mathematics education is a priority. Mathematics education in most of east Asia follows the Confucian model.

Number System

The number system in all Chinese-based east Asian languages centers on the same decimal system as the West but with stricter adherence to simple place-value patterns. For example, employing literal translations, the word for the number 12 is *ten-two*, 20 is *two-ten*, 37 is *three-ten, seven*, and 533 is *five-hundred, three-ten, three*. This system, along with the use of an abacus, facilitates the understanding of place value among east Asian elementary students. east Asian countries also follow the Chinese myriad-grouping system, which groups large numbers by ten thousands, rather than thousands. In other words, these languages have single words for the numbers "ten thousand" and "one-hundred million," but not for "million" or "billion."

Educational Philosophy

Historically, public east Asian mathematics classrooms could be generalized as teachers delivering lectures to large classes of students who are expected to master calculations and grasp theory through repetition and memorization. Inherent in this Confucian approach is the assumption among students, parents, and teachers that mathematical success results more from diligent studying than natural talent. Student-centric and practical applications of mathematics are not a primary focus in east Asia, as they sometimes are in the West. This educational philosophy is true not only of the textbooks, which in east Asia are succinct and cover the minimal core set forth by each of the national governments but also of the classrooms, which must closely follow the textbooks. However, since the international test results illuminated relative weaknesses in problem solving, creativity, and practical applications, the east Asian governments have been working to adapt curricula in various ways.

China

Chinese children's task of memorizing thousands of Chinese characters naturally seems to transfer to the subject of mathematics where memorization of formulas and processes is assumed to lead to understanding and discovery.

While mainland China did not participate in some international comparisons, the Chinese team has performed exceptionally well in the annual International Mathematical Olympiad (IMO), a competition among high school students, where it placed first almost every year between 1990 and 2010. But these achievements in mathematics are not limited to Chinese students; two Chinese mathematicians have received the distinguished Wolf Prize in Mathematics: Shiing-Shen Chern in 1983–1984 and Shing-Tung Yau in 2010.

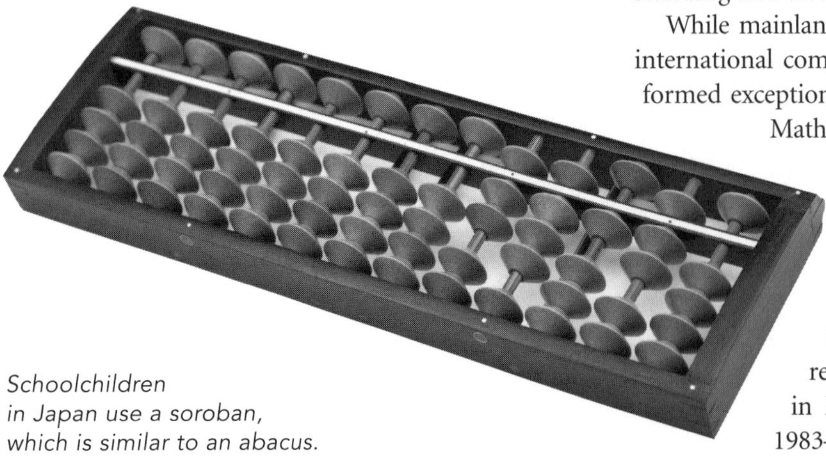

Schoolchildren in Japan use a soroban, which is similar to an abacus.

Hong Kong

The mathematics education system in Hong Kong employs elements both from mainland China and Great Britain. Despite the fact that international test scores ranked students from Hong Kong as years ahead of many Western countries, there is widespread concern about students viewing mathematics as irrelevant beyond testing. This concern has been leading to a curriculum that emulates the Western approach to teaching more mathematics related to problem solving and practical abilities.

Japan

While Japan distinguished itself in mathematics from the other east Asian countries during the Edo period (1603–1868), modern Japanese mathematics carries few remnants of this period. One such remnant is the *soroban*, a Japanese modification of an abacus. Japanese schoolchildren continue to use this beaded calculation device as a means of mastering the decimal system. Like in all east Asian countries, private schools (called *juku*) are attended widely by Japanese students. Japan has produced some of Asia's best mathematicians of the past century, including three winners of the Wolf Prize (Kunihiko Kodaira in 1984–1985, Kiyoshi Ito in 1987, and Mikio Sato in 2002–2003), and three winners of mathematics' most revered award, the Fields Medal (Kunihiko Kodaira in 1954, Heisuke Hironaka in 1970, and Shigefumi Mori in 1990).

Mongolia

Geographically, Mongolia lies between China and Russia. Until the early twentieth century, it was largely under the control of China and was later strongly influenced by Russia and the Soviet Union, adopting a Soviet-style government until 1990. Mongolian teams began participating in the International Mathematical Olympiad in 1964. Ming Antu was a Mongolian mathematician and astronomer, though he has been referred to as Chinese in the past. He worked on infinite series in the eighteenth century, among other accomplishments.

North Korea

While North Korea has the same Confucian background as the other east Asian countries, the former Soviet Union played a significant role in sculpting the modern approach to mathematics education. As do most countries around the world, the North Korean education system upholds mathematics as a central focus for both primary and secondary students, although North Korean story problems tend to be phrased in a nationalistic context. Students who excel in mathematics during their secondary school education may be admitted into the esteemed Kim Il-Sung University. In terms of global rankings, North Korea has sporadically entered a team into the International Mathematics Olympiad, some of which placed in the top 10.

South Korea

From childhood, South Koreans grow up using two separate number systems in their daily lives. The first one, a purely Korean system, is used mainly for counting objects, animals, and people and is no longer used for numbers larger than 99. It is worth noting that the numerals in this Korean system do not follow the same simple place-holding constructions as the number systems rooted in the Chinese language. The Sino-Korean number system, on the other hand, does follow these rules, and is most commonly used with money and large numbers. In school, many South Korean students receive just as much, if not more, of their mathematics instruction from private tutors or *hagwons* (academies) as from the public school environment. This system stems from the inextricable link between a student's mathematics performance on entrance exams and his or her eventual place in society. Some people cite this pressure as an explanation for why South Korean and Japanese students, despite performing exceptionally well on international tests, also rank the highest in their professed dislike for mathematics.

Taiwan

Private mathematics academies in Taiwan are referred to as *buxiban* (cram schools), suggesting their primary, but not exclusive, role of preparing Taiwanese students for entrance examinations. With electronics as a major industry, there has been a recent overhaul of the Taiwanese education system to focus on practical applications of mathematics instead of only theoretical mathematics.

Further Reading

Kennedy, Peter. "Learning Cultures and Learning Styles: Myth-Understandings About Adult (Hong Kong) Chinese Learners." *International Journal of Lifelong Education* 21 (2002).

Lankov, Andrei. *North of the DMZ: Essays on Daily Life in North Korea*. Jefferson, NC: MacFarland & Company, 2007.
Lee, Jihyun. "Universals and Specifics of Math Self-Concept, Math Self-Efficacy, and Math Anxiety Across 41 PISA 2003 Participating Countries." *Learning and Individual Differences* 19 (2009).
Leung, Frederick K. S., Klaus-D. Graf, and Francis J. Lopez-Real, eds. *Mathematics Education in Different Cultural Traditions—A Comparative Study of East Asia and the West*. New York: Springer, 2006.
Rong, Xin. "The General Solution of Ming Antu's Problem." *Acta Mathematica Sinica, English Series* 20, no. 1 (2004).
Usiskin, Zalman, and Edwin Willmore, eds. *Mathematics Curriculum in Pacific Rim Countries—China, Japan, Korea, and Singapore*. Chicago: Informations Age Publishing, 2008.
Yau, Shing-Tung. *S.S. Chern: A Great Geometer of the Twentieth Century*. Boston: International Press of Boston, 1998.

Daniel Showalter

See Also: Asia, Southeastern; Chinese Mathematics; Europe, Eastern.

Asia, Southeastern

Category: Mathematics Around the World.
Fields of Study: All.
Summary: Mathematics in the region has long been intertwined with religion and astrology and in recent generations has been impacted by colonialism.

The United Nations classification of southeastern Asia includes Brunei Darussalam, Cambodia, Indonesia, Lao People's Democratic Republic, Malaysia, Myanmar, Philippines, Singapore, Thailand, Timor-Leste, and Vietnam. Throughout history, the countries of Asia have had shifting political and social boundaries, and the names of many countries have changed over time, especially from the European colonial eras of the eighteenth and nineteenth centuries—when Western historians often began to study and document these countries—into the twenty-first century. For example, Burma became known as Myanmar; Siam became Thailand; Malay or Malaya became Malaysia; the Dutch East Indies or Netherlands East Indies and Java became Indonesia; and French Indochina included Laos, Cambodia, and Vietnam. Singapore was also part of Malaysia for a brief time in the 1960s, and the two regions share many historical developments. China and India, which have long histories of mathematics work and achievement, also had an influence in this region of the world. Therefore, mathematics contributions of some people from southeastern Asia, may be included within the histories of other regions or countries.

Early History

The great architectural feats found in places such as Borobudur, built in the ninth century on the island of Java, now part of Indonesia, and Angkor Wat, constructed three centuries later in Cambodia, suggest to scholars and historians that the architects and the builders must have had considerable mathematics knowledge. Some mathematics was probably brought to the region from India and China, as also happened in Europe and other areas, but there were almost certainly local mathematicians as well. The geometry involved in the design of both Borobudur and Angkor Wat has amazed generations of scholars who have discovered many complex ratios and formulas in the designs. Historians have also discussed the interconnection between religion, astronomy, mathematics, and astrology in southeastern Asia. Often there was little distinction made between mundane and divine matters, and some sequences of numbers (for example, 4, 8, 16, and 32) had religions connotations.

These numbers were used in both government and spiritual matters, such as the number of chiefs and territories in some Malay courts. Numerical systems emerged for the Burmese, Siamese, Cambodian, Laotian, Vietnamese, and Javanese languages. When Europeans began to explore and colonize southeastern Asia, they brought with them their own formal methods of school structure and mathematics teaching, which were documented by historians. Colonial influence saw the Vietnamese language develop a Romanized script, along with Western systems of counting, but the other scripts kept their systems of numerals. The introduction to southeastern Asia of a European-style school education, which replaced previous systems of instruction at pagodas or mosques, was a contributing factor in mathematics education. Much of this education came from the commercial needs of colonial powers to educate boys for work as bookkeepers and businessmen, so Western accounting systems were introduced to these populations—though many merchants continued to use Chinese systems, including the abacus, up through the twenty-first century.

Singapore and Malaysia

Singapore and Malaysia have active mathematics programs. The Raffles Institution in Singapore has a mathematics club whose members compete in events like the Singapore Mathematical Olympiad. The school was established in 1823 and named for (Thomas) Stamford Raffles, who is known as the founder of the British colony in Singapore. The Singapore Mathematical Society was founded in 1952. In the twenty-first century, it organizes participation in events like the national and international mathematics olympiads and the Singapore Mathematics Project Festival, among other educational and professional activities. Singapore first participated in the International Mathematical Olympiad (IMO) in 1988. While many twentieth-century textbooks on mathematics were imported into Singapore, the "Singapore Math Method," first developed in the 1980s and used in the national curriculum in Singapore, is now used in several places in the United States and elsewhere.

One of Singapore's well-known mathematicians is Tony Tan, who completed his doctorate, with a dissertation on "Mathematical models for commuter traffic in cities," at the University of Adelaide, South Australia. He taught mathematics before going into banking, and then into politics, ultimately becoming his country's deputy prime minister. Raffles College in Singapore taught mathematics from the time it started operations in 1928. Relations between Singapore and Malaysia in the twentieth century led to its transformation into the University of Malaya, then the University of Singapore, and the National University of Singapore. Sir Alexander Oppenheim, the vice-chancellor of the University of Malaya 1957–1965, was a prominent mathematician who had taught at Raffles College.

The Malaysian Mathematical Sciences Society, founded in 1970, was formerly known as the Malaysian Mathematical Society. It hosts events like the National Mathematical Olympiad in Malaysia; Malaysia first participated in the IMO in 1995. The Penang Free School, established in Malaysia in 1816, has taught mathematics from its inception. The Institute of Mathematical Sciences at the University of Malaya, founded in 1959 as the Department of Mathematics, continues to provide education for many Malaysian and overseas students and is an important mathematical institute in that country.

Thailand

Historically, Thailand was the only country in southeastern Asia never to be colonized by a foreign nation. Rulers such as the nineteenth-century King Mongkut, the inspiration for the 1946 movie *Anna and the King of Siam* and often called "the father of science and technology," embraced Western innovations. Assumption College, Bangkok, founded in 1885, had an extensive program of mathematics. The Mathematical Association of Thailand publishes the *Thai Journal of Mathematics* and hosts conferences and contests. Thailand has been participating in the IMO since 1989. The Center for Promotion of Mathematical Research of Thailand was established in 1978. Mathematician Yupaporn Kemprasit is an acknowledged world expert on algebraic semigroup theory, ring theory, and algebraic hyperstructure theory.

Cambodia, Laos, and Vietnam

In French Indochina, mathematics was encouraged for commerce. The Quoc Hoc or National Academy, established in 1896, included mathematics in its curriculum, with French as the language of instruction. Until the 1950s, most secondary schools in this region used French and French-language mathematics books—this was done in Cambodia until the early 1970s. Growth in the education system in the late twentieth century produced new native mathematics teachers, including Cambodian Communists Saloth Sar (Pol Pot), Khieu Samphan, and Gaing Kek Ieu (called "Comrade Deuch").

The Vietnamese Mathematical Society was founded in 1965, roughly the same time as one major buildup of American troops during the Vietnam conflict. Many educational institutions were closed for many years because of the war, but the society continued to support regional mathematical research. Vietnam first participated in the IMO in 1974 and hosted the competition in 2007. Mathematics researchers and students from Lao People's Democratic Republic (Laos) also participate in conferences and competitions. For example, in 2010, two high school students won a mathematics prize in a competition that included students from Brunei, Cambodia, Indonesia, Malaysia, the Philippines, Singapore, Vietnam, Thailand, and Laos.

Indonesia

The Dutch in the Netherlands East Indies operated a system of European schools, so-called "native schools," and vocational schools, teaching primarily in Dutch with Dutch-language textbooks. Many of the applied mathematics courses were directed toward engineering. After independence, with the expansion of the education system in Indonesia, there are mathematics departments in all schools and most universities in the country. Indonesia first participated in the IMO in 1988.

Brunei, Myanmar, and the Philippines

Elsewhere in the region there is also mathematical activity. The study of mathematics in the Philippines has been influenced by its close connections with the United States. The Mathematics Society of the Philippines was established in 1973, and the Philippines began participating in the IMO in 1988. Brunei participated in the IMO in 2000. The country of Myanmar has been isolated for much of the period since its independence in 1948. At the start of the twenty-first century, it initiated a 30-year plan for educational reform to address the challenges of the information age. Traditionally, state schools focused on writing, reading, and speaking in Myanmar and English, as well as mathematics, science, and Myanmar geography and history. Newer programs offer increased access to computer skills, as well as courses on information technology, medicine, and engineering, which require more advanced mathematics skills.

Further Reading

Hong, Kho Tek, Yeo Shu Mei, and James Lim. *The Singapore Model Method for Learning Mathematics.* Singapore: EPB Panpac Education, 2009.

Southeast Asian Mathematical Society. http://www.seams-math.org.

Southeast Asian Ministers of Education Organization-Regional Centre for Education in Science and Mathematics. http://www.recsam.edu.my/html/history.html.

Justin Corfield

See Also: Asia, Eastern; Asia, Southern; Chinese Mathematics; Europe, Northern; Europe, Western; Vedic Mathematics.

Asia, Southern

Category: Mathematics Around the World.
Fields of Study: All.
Summary: Southern Asia's history of mathematics reaches back thousands of years and mathematics continues to be a priority.

Southern Asia has a rich tradition in mathematics. Persian, Hindu, and Vedic scholars, among others in this area, contributed to the body of mathematics knowledge. Some of the achievements that have been historically credited to Arabic or Islamic mathematicians may have been influenced by pre-Islamic Persia. From ancient times, the rise and fall of various empires, wars, migration, and colonial influences have resulted in shifting cultural and geographical boundaries. As a result, many countries and regions in southern Asia have changed over time. The United Nations statistical classification for southern Asia contains Afghanistan, Bangladesh, Bhutan, India, Iran (Islamic Republic of), the Maldives, Nepal, Pakistan, and Sri Lanka. In the twenty-first century, these Asian nations continue to make advances in mathematics and mathematics education.

History

Construction of many ancient temples or monuments in southern Asia clearly involved mathematical knowledge, and mathematicians from this time period made various contributions to mathematics. One example is Indian scholar Baudhayana, who lived around 800 B.C.E. and is credited by some with developing the Pythagorean theorem, although others feel he was reflecting Babylonian work. The Vedic priest Katyayana, who lived approximately six centuries later, appears to have been interested in mathematics for religious purposes. Panini (520–460 B.C.E.), born in Shalatula, now part of Pakistan, wrote a scientific theory of Sanskrit. Some historians have theorized that development of algebraic structures and number systems in this region may be tied to the linguistic structure of

Sanskrit. Panini's work also influenced computer languages. Aryabhata (476–550) wrote a mathematical text known as the *Aryabhatiya*. It is composed of 123 metrical stanzas, whose organization has been studied by mathematicians because it differs from later mathematical works in many traditions. Some historians believe that it was influenced by Mesopotamia, while others suggest that it might be an anthology of works by earlier mathematicians. Another text, the *Bakhshali* manuscript, discovered in 1881 near Peshawar in present-day Pakistan, is believed to date from the seventh century, although some experts have dated it to up to eight centuries earlier or five centuries later.

By medieval times, Indian mathematicians had developed the notion of zero as a number, the use of negative numbers, and the definitions of sine and cosine. Some early Indian poetry also shows evidence of the binary number system and the use of decimal numbers. Mathematician Abd Al-Hamid ibn Wasi ibn Turk Al-Jili (c. ninth century) is believed to have been born in Iran, Afghanistan, or Syria. He wrote an algebra book. Persian mathematician, poet, and astronomer Omar Khayyam (1048–1141) wrote books on arithmetic and algebra by the age of 25 and contributed to many mathematical areas. Mathematician Nasir al-Din al-Tusi (1201–1274) was born in the city of Tus, now in Iran. He wrote Arabic translations of several Greek mathematical texts and is also credited with developing planar and spherical trigonometry from what many considered an astronomical tool into a separate mathematical discipline. Ghiyath al-Din Jamshid Mas'ud al-Kashi (1380–1429) was born and worked in Kashan, now in Iran. His *Treatise on the Circumference* included a calculation of π, which exceeded any known precision at the time. He also authored a teaching text called *The Key to Arithmetic*.

Education

Mathematics education has long been a focus in southern Asia. Mathematics was a part of *garakula* residential schools in ancient Nepal and India. From the fourteenth century, what became known as the "Kerala school of astronomy and mathematics" emerged in southern India. There was a flourishing of new discoveries, including the use of calculus long before it was developed by Isaac Newton and Gottfried Leibniz. These developments continued under mathematicians such as Citrabhanu (c. 1530) and Jyesthadeva (c. 1500–1575). English scholar Charles Whish (1794–1833) publicized many of the Kerala achievements to the rest of the world. Even then, the work of Whish—primarily a collector of Sanskrit manuscripts—was largely unknown beyond the scholarly community until the Indian mathematicians K. M. Marar and C. T. Rajagopal were able to demonstrate the advances made in Kerala just prior to the establishment of the European colonial empires in India.

British colonialism brought some European teaching styles into areas of southern Asia, and many universities were founded in the nineteenth century. Also in the nineteenth century, some Nepalese students traveled to India to study, where they were exposed to texts like Bhaskaracharya II's (1114–1185) *Siddhanta Siromani*. French mathematics traditions were introduced to southern Asia by Father Racine (1897–1976), a Jesuit missionary who had previously earned a doctorate in mathematics. With Indian colleagues such as Ramaswamy Vaidyanathaswamy (1894–1960), he promoted "modern" or contemporary mathematics teaching versus solely classical mathematics in the twentieth century. Indo-French collaborations continue to flourish into the twenty-first century and have been cited as contributing to development of areas like algebraic geometry and theoretical partial differential equations in southern Asia. There were other well-known collaborations, such as that between Indian mathematician Srinivasa Ramanujan and British mathematician Godfrey (G. H.) Hardy. In the 1980s, the Maldives introduced a new school curriculum that increasingly emphasized the importance of a variety of subjects, including mathematics.

Researchers in southern Asia have investigated a wide variety of different curricular issues such as gender differences in mathematics in Pakistan. King of Bhutan Jigme Khesar Namgyel Wangchuck noted in 2009:

> In all the countries where progress has been strong in the areas we strive to develop, the strength of the education system has been in Math and Science. In fact in India, the favourite subject for most students is Mathematics. In Bhutan, Mathematics is one of our main weaknesses.

Students from Bangladesh, India, Iran, Pakistan, and Sri Lanka have competed in the International Mathematics Olympiad: Iran since 1985, India since 1989, Sri Lanka since 1995, and Bangladesh and Pakistan since 2005. Mumbai, India, hosted the Olympiad in 1996.

Further Reading

Dauben, Joseph W., and Rohit Parikh. "Beginnings of Modern Mathematics in India." *Current Science* 99, no. 3 (August 10, 2010). http://www.ias.ac.in/currsci/10aug2010/suppl/15.pdf.

Jha, K., P. R. Adhikary, and S.R. Pant. "A History of Mathematical Sciences in Nepal." *Kathmandu University Journal of Science, Engineering and Technology* II, no. 1 (2006). http://www.ku.edu.np/kuset/second_issue/e2/KANAIYA%20JHa-pdf.pdf.

Joseph, George. *The Crest of the Peacock: Non-European Roots of Mathematics*. 3rd ed. Princeton, NJ: Princeton University Press, 2011.

Katz, Victor. *The Mathematics of Egypt, Mesopotamia, China, India, and Islam*. Princeton, NJ: Princeton University Press, 2007.

Waldschmidt, Michel. "Indo–French Cooperation in Mathematics." *Mathematics Newsletter of the Ramanujan Mathematical Society* 19, Special Issue 1 (2010) http://www.math.jussieu.fr/~miw/articles/pdf/IndoFrenchCooperationMaths.pdf.

Justin Corfield

See Also: Arabic/Islamic Mathematics; Asia, Western; Babylonian Mathematics; Vedic Mathematics.

Asia, Western

Category: Mathematics Around the World.
Fields of Study: All.
Summary: The people of western Asia have long studied and influenced mathematics.

Ancient western Asia, including Anatolia, Syria, Mesopotamia, and the Iranian plateau, along with Egypt, is regarded by many as the cradle of civilization. Activities that shaped numerous civilizations are traced historically to this region, including the invention of the wheel, practice of agriculture, first writing systems, and first administrative structures. Many intellectual and scientific disciplines flourished. The development of mathematics followed and was affected by the rise and decline of the civilizations of western Asia. Throughout history, the territory has been settled or invaded by many ethnic groups, including the Babylonian, Persian, Hellenistic, Roman, and Islamic cultures. Some countries were also part of the Soviet Union. It is not always possible to determine the exact origin of historical figures, and, as such, people may be included in the histories of many regions or identified by cultural heritage and the location where they did their work. Further, many of their accomplishments are named for later mathematicians. The twenty-first-century United Nations grouping for western Asia is listed as Armenia, Azerbaijan, Bahrain, Cyprus, Georgia, Iraq, Israel, Jordan, Kuwait, Lebanon, Occupied Palestinian Territory, Oman, Qatar, Saudi Arabia, Syrian Arab Republic, Turkey, the United Arab Emirates, and Yemen.

Babylon

Historical knowledge of Babylonian mathematics is largely limited to translations of the surviving clay tablets that have been unearthed by archaeologists, but even this evidence suggests a rich depth and breadth of mathematics scholarship, largely focused on practical problems. Subsequent cultures that came to the region also left parts of their mathematical legacies. With the emergence of Islam at the end of the sixth century, many of the nomadic tribes living in the Arabian Peninsula joined together to form a significant power.

By the early eighth century, a sociopolitical entity often called the Islamic Empire, which was ruled mostly by a series of government entities known as *caliphates*, spanned from Spain and north Africa to southeastern Anatolia, Persia, and the western portion of central Asia. On the east, the region shared a long border with India, and hence many Muslim intellectuals were also cognizant of Indian culture and mathematical accomplishments. Many local rulers encouraged scholarship, building on the legacy left by the Hellenic and Roman periods.

The House of Wisdom in Baghdad, in what is now Iraq, became the main hub of research and intellectual activity, rivaling Alexandria at its zenith. Works of Hellenistic mathematicians were translated into Arabic—the only surviving copies of certain works. Mathematicians also extended and introduced new ideas and fields. Social factors were another motivating influence in mathematics scholarship in Muslim lands, such as the calculation of the local daily prayer times, the direction of the prayer (toward Mecca), and the determination of the local first day and the end of the holy month of Ramadan. Since the commonly used lunar calendar was 11 days shorter than the solar year, this problem added complexity for numerous peoples and religions in the area. Observing the heavens and predicting the astronomical events was a major field of research for mathematicians and astronomers.

Ottoman Empire and Turkey

Wars brought turmoil to the area, and scholarly activities suffered. Following the conquest of Istanbul in 1453, Ottoman Sultan Mehmed-II built madrasas (buildings used for teaching Islamic theology and religious law, often including a mosque) and encouraged scholars to congregate. However, later events negatively impacted mathematics in western Asia; for example, the destruction of centers of learning such as the Istanbul observatory and the spread of religious scholasticism (a philosophy of teaching that follows a relatively a narrow set of traditional methods heavily influenced by religious teachings), which also occurred in medieval Europe. Some scholars indicate the passage of mathematical leadership over to Europe after about the fifteenth century.

Mathematics was used in Muslim lands to calculate the direction of prayer toward Mecca.

Ottoman Empire efforts of the early nineteenth century reenergized mathematics efforts. Vidinli Hüseyin Tevfik Pasa (1832–1901) contributed to linear algebra and Mehmet Nadir (1856–1927) worked on the theory of Diophantine equations, named for Diophantus of Alexandria. The Ottoman Empire faded after World War I, but the Turkish Republic continued its efforts. A well-known Turkish mathematician is Cahit Arf (1910–1997), known for the Arf invariant in algebraic topology, Arf semigroups, and Arf rings, among others. The *Turkish Journal of Mathematics* is one of the many scientific journals published by the Scientific and Technological Research Council of Turkey. The Turkish Mathematical Society was founded in 1948, and the country is a member of the International Mathematical Union (IMU), a worldwide association that promotes mathematics research and activity. In 1978, Turkey began participating in the International Mathematical Olympiad (IMO), a competition for high school students. Turkey hosted the IMO in 1993.

Israel

Mathematical activity in Israel dates back to antiquity, and it is one of the countries in western Asia with a thriving mathematics community. This fact is due in part to researchers like algebraist Shimshon Avraham Amitsur, who was one of the 1963 founders of the *Israel Journal of Mathematics*. Some other notable Israeli-born mathematicians include Oded Schramm, Saharon Shelah, and 2010 Fields Medal winner Elon Lindenstrauss. The Einstein Institute of Mathematics, named for Albert Einstein, was founded in the 1920s. Israel is a member of the IMU, and Israeli high school students first participated in the IMO in 1979. The Israel Mathematical Union is an organization that offers opportunities for students, teachers, and researchers. In the twenty-first century, there were some calls to boycott Israeli scholars over disputed territories. In response, numerous mathematical organizations worldwide, including the IMU, passed resolutions that stressed the importance of open international scientific exchange.

Other Countries

A revitalization of mathematical activity took place in many other western Asian countries in the twentieth century often connected with professional organizations or national institutes of science. For example, the development of contemporary mathematics in Armenia is tied to the 1944 beginnings of the Institute of Mathematics of the National Academy of Sciences of the Republic of Armenia. The country began participating in the IMO in 1993, the same year as Azerbaijan.

The first issue of the *Azerbaijan Journal of Mathematics* was published in January 2011. The Kuwait Foundation for the Advancement of Sciences supports the Kuwait Mathematics Program at the University of Cambridge, which underscores the relationships between western Asia and universities in other areas of the world. Kuwait began participating in the IMO in 1982.

In 2010, the editor of the *Arab Journal of Mathematics and Mathematical Sciences* was from Jordan. The Cyprus Mathematical Society was founded in 1983 and hosts activities like the Cyprus Mathematical Olympiad. Cyprus began participating in the IMO in 1984, Bahrain in 1990, the United Arab Emirates in 2008, and Syria in 2009. Saudi Arabia first participated in the IMO in 2004. It is also a member of the IMU, and mathematicians gather through the Saudi Association for Mathematical Sciences. Oman is an associate member of the IMU. Countries such as Qatar have developed mathematics standards for grades 1–9. Some countries in western Asia continue to be affected by the area's ongoing sociopolitical volatility. Georgia declared its independence from the Soviet Union in 1991 and is redeveloping many aspects of its national identity. It began participating in the IMO in 1993 and is a member of the IMU through the Georgian National Mathematical Committee. Iraq is also rebuilding itself after the turmoil of the late twentieth century and early twenty-first-century wars.

Some countries in the region participated in the Trends in International Mathematics and Science Study (TIMSS). In 2003, the study included fourth graders from the Republic of Yemen; eighth graders from Bahrain, Israel, Jordan, Lebanon, the Palestinian National Authority, the Syrian Arab Republic, and Saudi Arabia; and both fourth and eighth graders from Armenia and Cyprus. In 2007, even more countries from this region participated, including Armenia, Bahrain, Cyprus, Georgia, Israel, Jordan, Kuwait, Lebanon, Oman, the Palestinian National Authority, Qatar, Saudi Arabia, the Syrian Arab Republic, Turkey, and Yemen. In 2011, Armenia, Azerbaijan, Bahrain, Georgia, Israel, Jordan, Kuwait, Lebanon, Oman, the Palestinian National Authority, Qatar, Saudi Arabia, the Syrian Arab Republic, Turkey, the United Arab Emirates, and Yemen are included with benchmarking participants from this region listed as including Abu Dhabi, UAE, and Dubai, UAE.

Further Reading

Carr, Karen. "West Asian Mathematics." History for Kids. http://www.historyforkids.org/learn/westasia/science/math.htm.

Inonu, Erdal. "Mehmet Nadir: An Amateur Mathematician in Ottoman Turkey." *Historia Mathematica* 33, no. 2 (2006).

Irzik, Gurol, and Güven Güzeldere, eds. *Turkish Studies in the History and Philosophy of Science (Boston Studies in the Philosophy of Science)*. New York: Springer, 2005.

Mathematics in Israel: Historical and Current Affairs. http://imu.org.il/#mathinisrael.

Supreme Education Council. "Qatar Mathematics Standards: Grade 9." http://www.education.gov.qa/CS/en_math/9.pdf.

Trends in International Mathematics and Science Study. http://timss.bc.edu/timss2003.html.

Dogan Comez
Sarah J. Greenwald
Jill E. Thomley

See Also: Africa, North; Arabic/Islamic Mathematics; Babylonian Mathematics; Mathematics and Religion.

Astronomy

Category: Space, Time, and Distance.
Fields of Study: Geometry; Measurement; Number and Operations; Representations.
Summary: Mathematics is used in astronomy to measure and model celestial bodies.

Astronomy is the science that deals with celestial objects. It is divided into two disciplines: positional astronomy (or "astrometry"), which deals with the positions and movements of celestial objects; and astrophysics, which deals with their chemical and physical properties.

Positional astronomy began as a practical science. The first astronomers, before the invention of writing, dealt with such questions as the proper time of the year to plant crops and the proper dates for religious festivals. As their understanding improved over the ages, astronomers tackled other practical problems such as how to predict eclipses, how to tell time within a day, and how to navigate at sea.

Ancient people could take simple observations of the sun and moon and observe the patterns they made. From there it was a short leap to predicting future patterns. They would first record (or, before writing, memorize) the observations, and then perform a mathematical analysis—even if the analysis were nothing more than counting (for example, discovering there were about 365 days between winter solstices).

A much more sophisticated accomplishment was working out the complicated cycles on which lunar and solar eclipses occurred. An eclipse can be terrifying for a people who are not expecting it. If astronomers (many of whom doubled as priests) could predict eclipses, they could warn people in advance and reduce the collective fear.

A number of ancient peoples, including Mayans, Chinese, and Babylonians, developed elaborate calendar systems and tracked the movements of the planets. The Chinese constructed star charts, kept records starting possibly as early as 4000 B.C.E., and developed astronomical instruments. The Babylonians mapped constellations and introduced 60-minute hours and 60-second minutes. Both the Chinese and the Babylonians were able to predict eclipses. By 2500 B.C.E, Egyptians had measured star positions well enough to orient the pyramids to face celestial north. Polynesians traveled throughout the Pacific Ocean using stars as navigational aids.

The Greeks

The ancient Greeks effectively applied mathematics to astronomy. Eratosthenes (c. 200 B.C.E.) used geometry to calculate the size of Earth. Hipparchus (c. 161–126 B.C.E.) discovered the precession of the equinoxes and created the most accurate Greek tables of lunar motion.

Lunar Calendars

Some cultures found it easier to use the moon instead of the sun to tell time. A quick glance shows the phase of the moon, while observing the sun takes careful measurement. There is not an even number of moon cycles in a solar year—making the exclusive use of moon calendars difficult—yet the Babylonians (and others) discovered that there was a 19-year cycle.

In 19 solar years (assuming 365.25 days per year), there are 6939.75 days. The moon takes 29.5306 days to cycle from one new moon to the next. In 235 lunar months, there are 6939.69 days. A lunar calendar based on this cycle has 12 months (six with 29 days and six with 30 days—figuring a lunar month of 29.5 days), adding up to 354 days. Seven times in the 19-year cycle, there is a leap year, with an extra month of 30 days added in—making the leap year 384 days.

$$(12 \times 354) + (7 \times 384) = 6936 \text{ days.}$$

Since the lunar month is slightly more than 29.5 days, a total of four days have to be added during the 19-year cycle, giving 6940 days. This lunar calendar, originated by the Babylonians and refined over the centuries, is still in use today—it sets the dates of Easter and of all Jewish holidays.

Like some other Greek astronomers, he held that Earth revolved around the sun. Claudius Ptolemy (c. 120–150 C.E.) combined observations from Hipparchus and others with his own observations to propose a model of how the solar system worked—assuming Earth was at the center. By using epicycles (circles revolving on circles), he produced what was by far the best model of the heavens until Nicolaus Copernicus.

The Greeks did not only conduct astronomical calculations by hand but used a computer as well. Though not much is currently known about it, a mechanical analog computer was built somewhere in the Greek world about 100 B.C.E., called the "Antikythera mechanism" after the place it was found. This remarkably sophisticated computer was able to show both solar and lunar calendars, track the complicated path of the moon using Hipparchus's results, and predict eclipses for years into the future.

The Renaissance

During the Middle Ages, Arabs, Persians, and Jews, as well as European Christians (after c. 1000 C.E.), continued the work of the Greeks, including making new tables of planetary positions to update Ptolemy's, and keeping track of the precession of the equinoxes. In 1543, Copernicus's book on the solar system was published. Through a mathematical analysis of Ptolemy's work and later observations, Copernicus showed that a system in which the sun was the center of the solar system led to simpler and more accurate analysis than Ptolemy's.

Johannes Kepler used Tycho Brahe's careful naked eye observations of the planets to show that Mars went around the sun in an ellipse, not a circle as the Greeks had assumed. Kepler stated his three laws, which relate the speed of a planet to the shape of its orbit, but he could not explain why these laws worked. Isaac Newton was the first to explain Kepler's laws. He was able to show that any object affected by gravity would move in one of the conic sections: Kepler's ellipse, a line, a circle, a parabola, or a hyperbola. The one exception was the planet Uranus, which did not follow its Newtonian orbit.

It was not until the 1800s that Urbain Leverrier, in France, and John Couch Adams, in England, (unknown to each other) made the assumption that the discrepancies were because of the gravitational pull of an unknown planet. The planet Neptune was discovered in 1846 using Leverrier's prediction. Neptune was found by the consideration of the three components, P_x, P_y, and P_z, of Neptune's position and the three components, V_x, V_y, and V_z, of Neptune's velocity.

Until 1821, Uranus was moving faster in its orbit than expected—more than 4 planetary diameters ahead of its predicted position. After 1821, Uranus moved slower than expected. Obviously, Uranus moved past Neptune around 1821. If one adjusted the coordinate system so that $P_x = 0$ was Uranus's position in its orbit in 1821 and examined how far Uranus was pulled above or below its expected orbit, then one can tell whether Neptune was above or below Uranus in 1821, which gives P_y, and also whether Neptune was moving up or down, which gives V_y. If we have P_z, which represents Neptune's distance from the sun in 1821, then Kepler's laws can be used to find the two remaining parameters: V_x and V_z. Leverrier and Adams used a shortcut to find P_z. Both used Titius-Bode's law, an empirical formula, to predict the next planet beyond Uranus to be 38.8 times Earth's distance from the sun. These predictions were good enough to find Neptune, although Neptune is only 30.1 times Earth's distance from the sun.

Leverrier later examined the orbit of Mercury and found a discrepancy of 43 seconds of arc (which sounds small but is twice the discrepancies used to find Neptune). He computed the orbit of a hypothetical planet, called "Vulcan," which would explain this 43-second variation. Vulcan has never been found, and Einstein's general theory of relativity also explains this discrepancy.

Parallax

The ancient Greeks made attempts using parallax (the difference in the angle to a distant body measured from two different locations, also called triangulation) to find the size of the solar system. Being restricted to naked-eye observations, their results were inaccurate. Using telescopes, a much more accurate measurement was made in 1761 in which observers scattered across Earth found the parallax of Venus when it passed in front of the sun. The observations gave a value of 95.25 million miles from Earth to the sun (the modern estimate is just under 93 million miles). A much more difficult problem was to find the distances of stars by their parallax when viewed from opposite sides of Earth's orbit, first accomplished by Friedrich Bessel in the 1830s. Is space Euclidean or non-Euclidean? If measurably non-Euclidean, this would show up in stellar parallax measurements. No such effect has yet been observed, so one can say—except for relativistic considerations—that space is Euclidean for hundreds of light-years from Earth.

Astrophysics

Astrophysical questions date to the ancient Chinese, who discovered sunspots, and Hipparchus (c. 190–120 B.C.E), who worked on the magnitude (or brightness) of stars. His magnitudes, much refined, are still in use today. However, astrophysics as a discipline can be said to have started with Joseph von Fraunhofer, who in 1815 devised a spectroscope and catalogued the various lines (known as the Fraunhofer lines) that can be seen in the solar spectrum. In the 1850s, Gustav Kirchhoff and Robert Bunsen determined that these lines belonged to different chemical elements. Thus, by examining the spectrum of a star, its chemical composition can be determined. In addition, it was discovered that magnetic fields caused broadening and splitting of Fraunhofer lines, allowing the magnetic fields of stars to also be investigated.

Over the course of the twentieth century, astrophysicists went from studying the spectrum of visible light to studying every frequency of electromagnetic waves—from gamma rays to radio waves. There is now no known radiation from a star that is not being used to help find answers to the questions of what stars are, and how they operate.

Further Reading

Freeth, Tony. "Decoding an Ancient Computer." *Scientific American* 301, no. 6 (December 2009).

Gould Jr., Benjamin Althrop. *Report to the Smithsonian Institution on the History of the Discovery of Neptune.* Washington DC: Smithsonian, 1850.

Hester, Jeff, Bradford Smith, George Blumenthal, and Laura Kay. *21st Century Astronomy*. New York: W. W. Norton, 2010.

James Landau

See Also: Babylonian Mathematics; Chinese Mathematics; Conic Sections; Egyptian Mathematics; Geometry of the Universe; Greek Mathematics; Incan and Mayan Mathematics; Renaissance.

Atomic Bomb (Manhattan Project)

Category: Government, Politics, and History.
Fields of Study: Algebra; Data Analysis and Probability; Number and Operations; Representations.

Summary: The atomic bomb was made possible by Einstein's discovery of energy-mass equivalence.

Influenced by a letter from the famous German-American theoretical physicist Albert Einstein and other prominent scientists, U.S. President Franklin D. Roosevelt authorized the establishment of the Manhattan Project (the code name given to an elaborate effort to design, construct, and detonate an atomic bomb) in mid-1942. The project was directed by physicist J. Robert Oppenheimer, and his group of scientists, mathematicians, and engineers conducted secretive, pioneering research that led to the development of the first nuclear weapons.

Among the scientists who worked on the Manhattan Project were Italian physicist Enrico Fermi, American theoretical physicist Richard Feynman, Hungarian-American mathematician John von Neumann, Hungarian-American theoretical physicist Edward Teller, and Polish-American mathematician Stanislaw Ulam (Einstein also worked as a consultant throughout the project). Notably, several of these scientists, including Einstein and Ulam, were of Jewish decent and eventually resided in America because of Nazi persecution.

Through much trial and tribulation, the first nuclear bomb detonation test titled "Trinity" was successfully conducted on July 16, 1945, in Alamogordo, New Mexico. The Manhattan Project ultimately produced two types of atomic bombs; the plutonium implosion device (the plutonium or implosion bomb), and the uranium bomb (the uranium "gun" bomb). The plutonium bomb was the more difficult of the two to construct and required testing, whereas the uranium bomb was comparatively more simplistic and remained untested until the war.

Following Trinity, the U.S. government attempted to end World War II by detonating its uranium bomb nicknamed "Little Boy" over Hiroshima, Japan, on August 6, 1945. The blast destroyed approximately one-third of the city and caused about 140,000 causalities. Japan's reluctance to surrender prompted the United States to drop its plutonium bomb nicknamed "Fat Man" over Nagasaki, Japan, three days later. This blast killed about 70,000 people, destroyed about one-third of the city, and subsequently ended the war.

The revolutionary science of the Manhattan Project—namely the process of creating atomic explosions—was seemingly insurmountable, and paved the way for significant advancements in physics, chemistry, and mathematics. However, the historical impact of the atomic bombs dropped on Japan, as well as the philosophical and ethical ramifications, is an issue still debated today. In this regard Oppenheimer said, "It is a profound and necessary truth that the deep things in science are not found because they are useful; they are found because it was possible to find them."

The First Nuclear Reactions

Nuclear fission is the splitting of the nucleus of a heavy atom into smaller pieces, which releases a gigantic amount of energy. When this type of reaction is self-sustaining (it stimulates further reactions), it is called a "chain reaction." A critical mass is the minimum mass of fissionable material needed to ensure that a nuclear reaction sustains a chain reaction. Achieving a critical mass and, ultimately, a chain reaction was the essential challenge in developing both Little Boy and Fat Man.

The Little Boy design utilized the "gun method," which was detonated by firing a mass of uranium-235 down a cylinder into another mass of uranium-235 to produce a chain reaction. Fat Man was an implosion-type device that used plutonium-239. In this design, plutonium was placed in the center of a hollowed-out sphere of high explosives, and a number of detonators located on the high explosive's surface were simultaneously fired pressurizing the core and increasing its density—creating an implosion that resulted in a chain reaction. The Trinity test bomb was similar and was nicknamed "The Gadget." Little Boy produced a blast of approximately 12,500 tons of trinitrotoluene (TNT). Fat Man had the explosive power of about 22,000 tons of TNT and The Gadget had a blast yield of around 15–20 tons of TNT.

A tremendous amount of engineering, chemistry, physics, and mathematics was involved in the development and deployment of the atomic bombs. Among these fields was a branch of theoretical physics called "quantum mechanics" (the set of scientific principles that describe the behavior of matter and energy predominating at both the atomic and subatomic levels), which at the time was in its infancy. Quantum mechanics was developed under the assumption that energy is not infinitely divisible but rather composed of quanta (small increments).

Unlike classical or Newtonian mechanics, which describes the motion of objects we encounter every day at the macrocosmic level, quantum mechanics deals with uncertainty in many of its results and is statistical and probabilistic in nature. Although initially this branch of physics was not readily accepted, it nonetheless proved an essential tool in the development of the atomic bomb as it provided many of the insights necessary for its construction. In regard to the science and mathematics utilized in the development of the bomb Stanislaw Ulam said, "It is still an unending source of surprise for me to see how a few scribbles on a blackboard or on a sheet of paper could change the course of human affairs."

The Energy-Mass Equivalence

One of the most imperative concepts in the development of the atomic bomb was the mathematical formulation of the energy-mass equivalence, which was derived by Einstein. He established that mass and energy are, in fact, both different manifestations of the same thing. This idea was a counterintuitive and revolutionary result that spawned from his 1905 special

Explosion resulting from the "Trinity" test of the plutonium bomb.

theory of relativity. Einstein's formulation implied that very minute amounts of mass can be converted into excessively large amounts of energy. For example, this very encyclopedia is, in actuality, a form of energy in storage, which could equivalently be called rest energy or mass. If this encyclopedia could be completely converted into energy, it would yield a gigantic amount of energy indeed. This energy-mass equivalence concept is depicted symbolically through one of the world's most famous equations:

$$E = mc^2.$$

This equation is interpreted as the rest energy E of an object being equivalent to the mass m of the object multiplied by the square of the speed of light c in a vacuum. Alternatively, $E = mc^2$ can be construed as the equation that allows one to determine the amount of mass needed to produce a certain amount of energy—assuming all of the mass can be converted completely into pure energy.

To better understand how this famous simple equation was crucial in the development of the atomic bomb, one needs to understand its nature. First, $E = mc^2$ is a "direct proportion" (E is directly proportional to m), and is symbolically expressed as $E \propto m$. In general, a direct proportion has the form of

$$x \propto y \text{ or equivalently } x = ay$$

where a is the proportionality constant. As a simple example:

$$4 \propto 2 \text{ or equivalently } 4 = 2a.$$

In this case, the proportionality constant is $a = 2$, whereas in the energy-mass equivalence, the proportionality constant is c^2. According to the International Bureau of Weights and Measures, the value of c is 299,792,458 meters per second (m/s), or about 186,282.4 miles per second (mi/s). For computational simplicity, c is often rounded to 300,000,000 m/s (186,000 mi/s), except when performing experiments that require exact values for light speed. Now, taking $c = 300{,}000{,}000 \text{ m/s} = 3 \times 10^8 \text{ m/s}$ one can compute that 1 kilogram of plutonium could theoretically turn into

$$E = 90{,}000{,}000{,}000{,}000{,}000 \text{ kg}\cdot\text{m}^2/\text{s}^2 = 9 \times 10^{16} \text{ kg}\cdot\text{m}^2/\text{s}^2$$

Therefore, one can intuitively understand why a small amount of uranium or plutonium can generate explosions as massive as the ones produced by Little Boy and Fat Man.

It is interesting to note that for the Trinity test, the mushroom cloud expanded to nearly 300 meters (about 984 feet) in .053 seconds.

Further Reading

Bird, Kai, et al. *American Prometheus: The Triumph and Tragedy of J. Robert Oppenheimer.* New York: Random House, 2005.

Coster-Mullen, John. *Atom Bombs: The Top Secret Inside Story of Little Boy and Fat Man.* Self-Published, 2003.

Groves, Leslie M. *Now It Can Be Told: The Story of the Manhattan Project.* New York: Harper, 1962.

Kelly, Cynthia C., et al. *The Manhattan Project: The Birth of the Atomic Bomb in the Words of Its Creators, Eyewitnesses, and Historians.* New York: Black Dog & Leventhal Publishers, 2007.

Rhodes, Richard. *The Making of the Atomic Bomb.* New York: Touchstone, 1986.

Serber, Robert. *The Los Alamos Primer: The First Lectures on How to Build an Atomic Bomb.* Berkeley: University of California Press, 1992.

Sullivan, E. T. *The Ultimate Weapon: The Race to Develop the Atomic Bomb.* New York: Holiday House, 2007.

Daniel J. Galiffa

See Also: Cold War; Einstein, Albert; Radiation; World War II.

Auto Racing

Category: Games, Sport, and Recreation.
Fields of Study: Data Analysis and Probability; Geometry; Measurement.
Summary: Mathematics is essential in the design of race cars and racetracks, and the formulation of race strategy.

Auto racing has taken place for as long as cars have existed. While the early days of racing were related to

Drivers preparing during a practice run for the 2004 Daytona 500 race. An understanding of geometry is critical in determining how to set up cars to handle banking and high speeds.

fairly simple vehicles, it is now a very technical sport that has multiple branches with fans worldwide. Auto racing includes not only cars that are similar to those driven by the average citizen but also cars that are very sophisticated. The different branches of auto racing differ in the specifics of the car but all share a strong relationship to mathematical principles. The design of the car, its tires, the track, and the drivetrain require very careful measurement. The optimal path for a given track and weather condition requires a deep understanding of angles and geometry. Analysis of data to create probability information enables drivers and their teams to make wise decisions for a given set of conditions during a race.

Overview

Auto racing began as soon as the automobile was invented in the late 1800s. Auto racing is a broad term that includes single-seat cars or open wheel cars, which the Indianapolis 500 has made famous. Formula 1 racing is another type of open wheel racing but involves racing around courses that are not oval shaped. The National Association for Stock Car Auto Racing (NASCAR) utilizes cars that are modified from cars that can be bought by the general public. Many successful professional race car drivers began their racing careers with Kart racing, which involves vehicles that look like sophisticated go-karts.

Race Track Design

The racing surface and the track design are significant factors that affect both car design and driving strategy. Race surfaces can include asphalt, concrete, dirt, sand, and (sometimes) ice. Some tracks consist of a very short distance (1/4 mile) and are straight. These tracks are typically used for drag racing, which involves cars trying to go as fast as possible over a short distance. Many track designs have drivers travel in an oval, or near-oval, shape with some banking to help make high-speed turns easier. An understanding of geometry is imperative when determining how best to set up the car to handle the banking and the speeds. Track designs also include road courses in which racers turn both left and right and require a completely different car design to handle banking in both directions. The radius of a turn influences how fast the car can go without losing grip and crashing into the outer wall. The speed affects the size of the down force on the car (caused by spoilers), and, as such, different tracks require car designs.

Car Design

Race car designs evolve in response to technology changes and safety concerns, often as a result of mathematical or statistical analysis. Each branch of auto racing has very strict rules on car design, which are tested before—and sometimes after—each race. The testing includes very careful measurements of various components of the car from the size of various components of the engine, to the car's width, height, and weight. The tests focus on items that affect the car's power (the engine), response to the environment (temperature, air resistance, and gravity), and its influence on forces that are made on the car (width, height, and weight). Because they are such an important part of car performance, tires are supplied to the teams. A large amount of testing by tire companies goes into determining which type of tires will be provided for a particular track. The air and track temperatures often change drastically during a race and can affect how the tires interact with the track surface—providing more or less grip. Likewise, the gas that is put into the car is also provided to drivers. These standardizations provide a more even playing field for the teams so that the driver who wins is, presumably, the one with the greatest skill. Teams can alter the cars slightly during races to modify how the car receives forces from the track and from the air. These modifications include taking out or adding small wedges that alter the angle that the car sits on the track. The impact these small changes make on force is understood using trigonometry.

Race Strategy

Once teams have prepared their car and driver for the race, the issue of strategy plays an important role. Teams use probabilities to determine if and when to stop in the pits to change tires or to add gas. Gas mileage is estimated by using regression involving the number of laps, the speed of the car during the laps run, and the temperature. This estimation is not absolutely exact, and it is not uncommon for drivers to run out of gas near the end of some races because of an error in the team's regression model. Some teams alter the usual pit stop, which involves replacing all of the tires and adding gas, by replacing just some of the tires or just adding gas.

Technology and Safety

Technology is playing a bigger role in auto racing in both car development and car testing. Car teams now use technology to measure a large number of factors that influence their car's performance. For some branches of auto racing, these measurements are made during races. For other branches, the rules prohibit this during races but allow the measurements to occur during practice and research design. Because testing can be so expensive, some tests are done with a few drivers and then shared with all the teams. The use of computer simulation based on mathematical modeling is becoming more prevalent in all branches. It is not unusual for teams to use wind tunnels to test car design, and fluid dynamic modeling has been used to improve the aerodynamic properties of race cars. Off-season drivers use sophisticated driving simulators to hone their skills.

Technology has also been used to make racing safer. Race uniforms, helmets, and car interiors have become much less dangerous because of technological improvements. Additionally, track walls now include what is called a Steel and Foam Energy Reduction (SAFER) barrier, which dissipates the collision energy from a crash so that the impact force felt by the car and driver is smaller and less dangerous.

Further Reading

Beckman, Brian. "The Physics of Racing." http://phors.locost7.info/contents.htm.

Bentley, Ross. *Speed Secrets: Professional Race Driving Techniques.* Osceola, WI: MBI Publishing, 1998.

Genta, Giancarlo. *Motor Vehicle Dynamics: Modeling and Simulation.* Singapore: World Scientific Publishing, 1997.

Gifford, Clive. *Racing: The Ultimate Motorsports Encyclopedia.* Boston: Kingfisher Publications, 2006.

Leslie-Pelecky, Diandra L. *The Physics of NASCAR: The Science Behind the Speed.* New York: Plume, 2009.

Michele LeBlanc

See Also: Diagnostic Testing; Extreme Sports; Highways.

Axiomatic Systems

Category: History and Development of Curricular Concepts.
Fields of Study: Communication; Connections; Geometry; Reasoning and Proof.
Summary: An axiom is a statement that is assumed to be true, and axiomatic systems have a rich and interesting mathematical history.

Axiomatic systems provide a deductive framework for mathematicians to combine related definitions and theorems that make mathematical knowledge systematic and structural. Mathematical theories including number systems, set theory, probability, algebra, and many others are built by using axiomatic systems.

Axiomatic Method and Axiomatic System in Mathematics

To build a deductive mathematical system, one needs to observe two intrinsic limitations in this process.

Limitation 1: Not every mathematical term can be defined. The reason can be seen by the following considerations: To define a term *A*, one needs a term *B*, and possibly some other terms. To define the term *B*, one needs another term, *C*, and so on. One may eventually come back to the term *A*; in which case the definition would be circular as there are a finite number of words. This means that *A* is used to define *A*, which is undesirable.

If the definitions are not to become circular, some terms are needed to start with. The solution is that there will be some terms that will not be defined. These will be called "undefined terms," and will be used to define all the other terms to be considered. One may think that it is strange that this solution can work. How can undefined terms give meaning? This puzzle is partially answered upon consideration of the next limitation.

Limitation 2: Not every mathematical statement can be deduced or proven. The reason is similar to the one in Limitation 1; some statements are needed to start a chain of deduction: if *R*, then *S*; if *S*, then *T*; if *T*, then *U*; and so on. To deal with this limitation, certain statements must be accepted without proof. These statements are called "axioms," and they are the statements that we used to deduce other statements. Actually, the axioms are often statements about the undefined terms. In other words, the axioms often tell us certain properties or restrictions of the undefined terms. Thus, the axioms help provide meaning to the undefined terms. Starting with undefined terms, axioms, and definitions, and by using deductive reasoning to establish important mathematics facts in the form of theorems, the mathematics system so obtained is said to be built by using the "axiomatic method." Such a system that consists of undefined terms, axioms *A*, definitions *D*, statements of the form If *P* then *Q* and proof of such statements is called an "axiomatic system." In an axiomatic system, one does not talk about the validity of *A* or *P*, one talks only about the validity of the proof based on *A* and *D*.

Historical Developments

Historically, Euclidean geometry was the best-known model of an axiomatic system. Around 300 B.C.E., Euclid wrote his 13-volume *Elements*, which contained an axiomatic treatment of geometry. It starts with 23 definitions; Euclid stated 10 axioms. The first five axioms are geometric assumptions, which he called postulates. The last five are more general, which he called common notions. There, Euclid did not use undefined terms.

The most important and fundamental property of an axiomatic system is "consistency" (it is impossible to deduce from these axioms a theorem that contradicts any axiom or previously proved theorem). The Euclidean geometry provides such a consistent axiomatic system. An individual axiom is "independent" if it cannot be logically deduced from the other axioms in the system. The entire set of axioms is said to be independent if each of its axioms is independent. Mathematicians prior to the nineteenth century doubted very much about the independence of the fifth postulate (the parallel postulate). They tried to deduce such a postulate by using the first four postulates. Despite considerable effort devoted to the task, no significant result could be obtained.

Euclid's *Elements* indeed became the most influential book on geometry, as well as the model of logical reasoning and axiomatic system, until the nineteenth century when two fundamental developments took place. First, it was realized that Euclid's logical system was not rigorous enough. A rigorous axiomatic treatment of Euclidean geometry was given by David Hilbert (1862–1943) in his 1899 book *Grundlagen der Geometrie* (*The Foundation of Geometry*). Here, Hilbert used the undefined terms of point, line, lie on, between, and congruent for the geometry system. Second, research results of C. F. Gauss (1777–1855), J. Bolyai (1802–1860), and N. I. Lobachevsky (1793–1856) asserted that the parallel postulate was actually an independent axiom. Non-Euclidean geometry could be developed by replacing the fifth postulate with another independent axiom. The lesson from Euclidean and non-Euclidean geometry is that both are valid axiomatic systems. When studying Euclidean or non-Euclidean geometry, no claims are made on the truth of the axioms about the physical world. One merely claims that if the axioms are valid, then the theorems deduced therein are also valid. Whether the logical system describes the real world is another question.

Current Issues

There are still many issues regarding the axiomatic systems. The set of axioms in an axiomatic system is "complete" if the axioms are sufficient in number to prove or disprove any statement that arises concerning our collection of undefined terms. To determine whether an axiomatic system is complete is by no means an easy question to answer. A great surprise was discovered by Kurt Gödel (1906–1978) in 1931. He proved that in a formal mathematics system that included the integers, there exist statements that are impossible to prove or disprove. This result is called Gödel's incompleteness theorem. Also, to determine whether a given proposition is an axiom has been a very important issue in computer science and is important when one tries to use a computer to do proofs. If the computer cannot recognize the axioms, the computer will also not be able to recognize whether a proof is valid.

Further Reading

Greenberg, Martin Jay. *Euclidean and Non-Euclidean Geometries: Development and History*. 3rd ed. New York: W. H. Freeman and Co, 1993.

Heath, Thomas L. *The Thirteen Books of Euclid's Elements*, Vol. 1–3. 2nd ed. New York: Dover Publications, 1956.

Meyer, Burnett. *An Introduction to Axiomatic Systems*. Boston: Prindle, Weber & Schmidt, 1974.

Venema, G. A. *The Foundations of Geometry*. New Jersey: Pearson Prentice Hall, 2006.

Wallace, E. C., and S. F. West. *Roads to Geometry*. 3rd ed. Upper Saddle River, NJ: Prentice Hall, 2003.

Ka-Luen Cheung

See Also: Geometry and Geometry Education; Greek Mathematics; Mathematical Certainty; Parallel Postulate.

B

Babylonian Mathematics

Category: Government, Politics, and History.
Fields of Study: Algebra; Connections; Geometry; Measurement; Representations.
Summary: Babylon had an advanced utilitarian mathematics from which we inherited sexagesimal timekeeping.

Our knowledge of Babylonian mathematics (2100–200 B.C.E.) is based on extensive mathematical calculations found on clay tablets in the area of Mesopotamia (now Iraq), surrounding the ancient city of Babylon between the Tigris and Euphrates rivers. Because only a fraction of the tablets have survived—and only a small fraction of those have been translated—our knowledge of the depth and breadth of Babylonian mathematics is limited. Mathematics historian Otto Neugebauer likens the situation to tearing a few random pages out of a few textbooks and then trying to reconstruct a representation of modern mathematics. Nonetheless, Babylonian mathematics did involve complicated mathematics, and was used primarily to solve practical problems. These mathematical problems ranged from arithmetic calculations, to algebraic rules, to geometrical formulas, to numerical ideas.

Babylonian Number System

The Babylonian number system was sexagesimal, using both a place value notation based on powers of 60 and a base-10 grouping system for numbers between 1 and 59 within each place value.

Traces of their sexagesimal notation remain today in the recording of time (hours, minutes, seconds) and the measurement of angles (degrees, minutes, seconds). Their numbers were written in cuneiform, or the use of a triangular stylus to make wedges on a clay tablet. A vertical line represented unity and a horizontal wedge mark represented a 10.

For example, within each place value, the number 57 would be represented by 5 horizontal wedges and 7 vertical lines. Expanding the example, a cuneiform number represented in modern form as "3, 4, 57" was equivalent to

$$3(60^2)+4(60^1)+57(60^0)$$
$$=3(3600)+4(60)+57(1)=11097.$$

The Babylonians had neither a symbol for zero as a placeholder nor a symbol to designate the "decimal" point in their sexagesimal fractions. Writing and reading numbers required the Babylonian mathematician to understand the problem's context and the use of a space to represent either an "empty" place value or shift to fractional place values. Thus, the previous number, "3,4,57," possibly was equivalent to:

$$3(60^4)+0(60^3)+4(60^1)+57(60^0)$$
$$\text{or } 3(60^4)+4(60^3)+57(60^2)$$
$$\text{or } 3(60^1)+4(60^0)+57(60^{-1})$$
$$\text{or even } 3(60^{-1})+4(60^{-2})+57(60^{-3}).$$

To avoid ambiguity, modern translations of these numbers would be first "3,0,4,57" or "3,4,57,0,0" or "3,4;57" or "0;3,4,57" respectively, where the semicolon separates whole numbers from fractional numbers. Tablets from the Seleucid period (300 B.C.E.) did include a special symbol that played the double role of a placeholder (zero) and the separator between two sentences.

Babylonian Arithmetic

Using the sexagesimal system, the Babylonians were able to add, subtract, multiply, and divide numbers. Their computations were complemented by the use of extensive tables. Their multiplication tables had products ranging from 1×1 through 59×59, and seeming somewhat unusual, they had access to multiplications tables for "1,20" (or 80), "1,30" (or 90), "1,40" (or 100), "3,20" (or 200), "3,45" (225), and even "44,26,40" (or 160,000). Some of this can be explained by looking at their tables of reciprocals for working with fractions. For example, one table includes the deceptive notation $1 \div 1{,}21 = 44{,}26{,}40$, with the latter value actually being "0;0,44,26,40."

The Babylonians produced extensive tables of squares and cubes, tables of square sides and cube sides (square and cube roots), and sums of squares and cubes. When a table side-value was not available, the Babylonians approximated roots using an interpolation process based on averaging and division; this process was quite fast, producing 26-decimal accuracy in five iterations.

Babylonian Algebra

Though without an algebraic notation, the Babylonians solved numerous types of algebraic equations. Each solution involved the replication of a formulaic prescription represented by a step-by-step list of rules. In effect, their prescriptions invoked algorithms, which were usually specific to a stated problem and not generalized to a class of problems.

For example, consider this Babylonian problem: the area and two-thirds of the side of my square have I added and it is 0;35. In modern notation, their step-by-step solution was: 1, the unit, you take; two-thirds of 1, the unit, is 0;40: Its half is 0;20 and 0;20 you multiply 0;6,40, you add 0;35 to it and 0;41,40 has 0;50 for its square root. 0;20 that you multiplied with itself, from 0;50 you subtract and 0;30 is the side of the square.

In modern mathematics, this same problem would involve solving the quadratic equation:

$$x^2 + \frac{2}{3}x = \frac{35}{60}.$$

The steps in this problem also can be interpreted using geometrical algebra, where the square is "completed" in a manner similar to the derivation of our general quadratic formula.

In their solution of special types of algebraic equations, the Babylonians made extensive use of their tables of the sums of squares and cubes, especially if the equation was of the third or fourth degree. Some of their solutions to algebraic problems were quite sophisticated. For example, one problem involved a system of equations of the form

$$xy = n \text{ and } \frac{mx^2}{y} + \frac{py^2}{x} + q = 0 .$$

Its solution using substitution would normally lead to a single-variable equation involving x^6, but the Babylonians solved it by viewing it as a quadratic equation in x^3.

Babylonian Geometry

Dominated by their work with algebraic ideas, the Babylonians' geometry focused on practical measurements such as the calculation of lengths, areas, and volumes. Again, the Babylonians used prescriptive formulas. For example, to calculate a circle's circumference, they multiplied the diameter by 3, implying their value of π was 3. For the circle's area, they squared the circumference and divided by 12, which is equivalent to our modern formula $A = \pi r^2$ if the correct value of π had been used.

Mathematics historians credit the Babylonians with the division of a circle into 360 degrees. Neugebauer suggests it is related to their Babylonian mile, a measure of long distance equal to about 7 miles. This measure evolved into a time unit, being the time it took to travel this distance. After noting that 12 of these time units equaled a full day or one revolution of the sky, the Babylonians subdivided their mile into 30 equal parts for simplicity, leading to $12 \times 30 = 360$ units in a full circle.

The Babylonians computed areas of right triangles, isosceles triangles, and isosceles trapezoids, as well as the volumes of both rectangular parallelepipeds and some prisms. They had difficulties with certain three-dimensional shapes, being unable to compute correctly the length of the frustum of a pyramid (they claimed it was the product of the altitude by the average of the bases).

The Babylonians did know some general geometric relationships. For example, they knew that perpendiculars dropped from the vertex of an isosceles triangle bisected the base, that corresponding sides of similar triangles were proportional, and that angles inscribed in a semicircle are right angles. The Babylonians used this knowledge to solve difficult geometrical problems, such as their determination of the radius of a circle circumscribing an isosceles triangle.

Evidence suggests that they knew a precursor of the Pythagorean formula. One cuneiform tablet (c. 1700 B.C.E.) includes sexagesimal numbers written along a square's side (30) and diagonal ("42,25,35" and "1; 24, 51, 10").

The latter number is both the product of the other two numbers and a good approximation of the square root of 2 (1.414214). Also, in the Plimpton 322 collection, some of the tablets contain tables of Pythagorean triples ($a^2 + b^2 = c^2$), arranged with increasing acute angle of the associated right triangle.

Signs of Advanced Mathematical Thinking

For the most part, Babylonian mathematics was utilitarian, being tied to solving practical problems. Nonetheless, interpretations of some of the tables on the clay tablets suggest that the Babylonians occasionally explored theoretical aspects of mathematics. Examples include their tables of Pythagorean triples and tables of exponential functions (which perhaps were used to compute compound interest in business transactions). Also, the Louvre tablet (300 B.C.E.) includes two series problems

$$1 + 2 + 2^2 + 2^3 + \cdots + 2^8 + 2^9 = 2^9 + 2^9 - 1$$

and

$$1^2 + 2^2 + 3^2 + 4^2 + \cdots 9^2 + 10^2$$
$$= \left(1\frac{1}{3} + 10\frac{2}{3}\right)(55) = 588$$

but historians do not suggest the Babylonians knew general series formulas such as

$$\sum_{k=0}^{n} r^k = \frac{r^{n+1} - 1}{r - 1}.$$

Specific to number theory, mathematics historians point to the cumbersome nature of the Babylonians' sexagesimal system, making it difficult to explore ideas such as factors, powers, and reciprocals. Some suggest that this is symptomatic of the Babylonian's reasonable choice of 3 for π, rather than the fraction

$$\frac{22}{7}$$

equal to the more complicated repeating expression "3; 8, 34, 17, 8, 34, 17, ..."

Further Reading

Aaboe, Asger. *Episodes From the Early History of Mathematics.* Washington, DC: Mathematical Association of America, 1975.

Friberg, Jöran. *Unexpected Links Between Egyptian and Babylonian Mathematics.* Singapore: World Scientific Publishing, 2005.

Katz, Victor J., ed. *The Mathematics of Egypt, Mesopotamia, China, India, and Islam: A Sourcebook.* Princeton, NJ: Princeton University Press, 2007.

Van der Waerden, Bartel Leendert. *Science Awakening.* Oxford, England: Oxford University Press, 1985.

———. *Geometry and Algebra in Ancient Civilizations.* Berlin, Germany: Springer, 1983.

Jerry Johnson

See Also: Arabic/Islamic Mathematics; Chinese Mathematics; Egyptian Mathematics; Greek Mathematics.

Ballet

Category: Arts, Music, and Entertainment.
Fields of Study: Communication; Geometry; Representations.
Summary: Ballet uses geometry to create captivating moving art.

Ballet can be considered mathematics in motion from basic counting (keeping time with music, and doing *demi-pliés* in childhood dance classes); making lines, angles, and geometric shapes in space via basic positions and choreographed routines of principal dancers and the *corps de ballet*; communicating stories in ballet productions (like the classic *Swan Lake*, or a seasonal favorite *The Nutcracker*); conversing visually among dancers (as in a *pas de deux* with Margot Fonteyn and Rudolf Nureyev); and by representing general emotions, moods, and abstract themes (as in George Balanchine's *Serenade*). Words from the French language may be common in ballet terminology, but concepts from mathematics abound as well. These representations, communications, and geometric creations can all be achieved and evidenced through the dance figures and ballet movements.

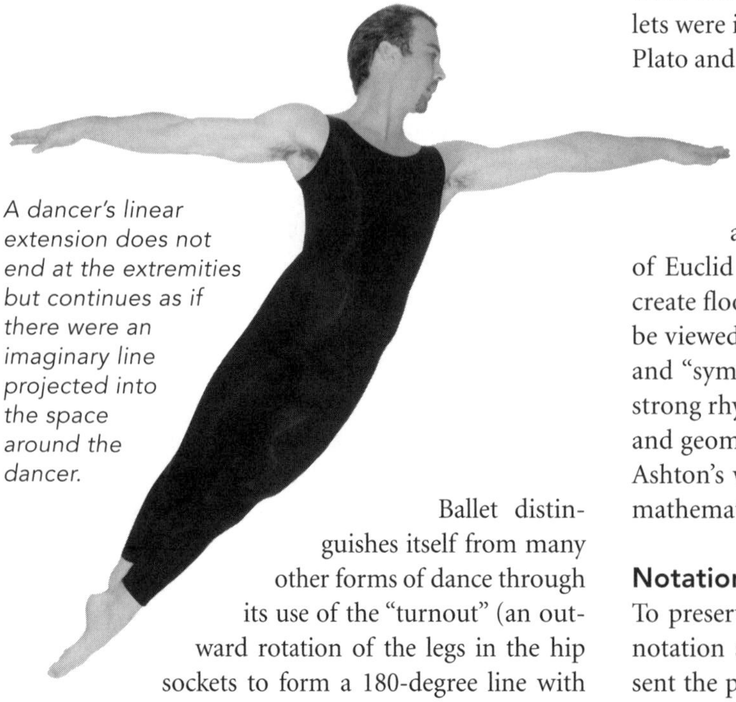

A dancer's linear extension does not end at the extremities but continues as if there were an imaginary line projected into the space around the dancer.

Ballet distinguishes itself from many other forms of dance through its use of the "turnout" (an outward rotation of the legs in the hip sockets to form a 180-degree line with the feet in first position). This turnout gives the dancer a strong base and the ability to move in any direction while allowing a more open body presentation to the audience, yet holding the graceful curves and shapes of the dancer's body to preserve a svelte "line." Other standard positions of the feet, carriage of the arms, or basic movements of the body produce angles such as a 135-degree arabesque, a 90-degree attitude, or a 45-degree *battement tendu.* The *rond de jambe à terre* or *en l'air* utilizes circular movements of the leg to trace semicircles or arcs, on or off the ground. These geometric lines, circles, and angles continue when basic steps become building blocks to more complicated movements. Meanwhile, dancing on the tips of the toes (*en pointe*), another distinctive ballet feature, heightens the dancer's lines in a vertical fashion. The linear extension, from head to toe, fingertip to fingertip, does not end at the extremities but continues as if through an imaginary line into the space around the dancer.

Ballet as Geometry

One of the earliest ballet performances was the sixteenth-century *Le Balet Comique de la Reine* by Balthazar Beaujoyeulx, commissioned by the court of France. During that elaborate production, the dancers performed dozens of geometric figures involving triangles, circles, and squares for their geometric proportions and spatial configurations. These beginning ballets were influenced by the writings of Pythagoras and Plato and represented the cosmic and heavenly significance of numbers and geometry. A twentieth-century choreographer, Frederick Ashton, however, was inspired by mathematics for its sheer beauty in his creation, *Scènes de Ballet.* Working from a book of Euclid theorems, he specifically used geometry to create floor patterns and dance movements that could be viewed from any angle to see the geometric figures and "symmetrical asymmetries." Combined with the strong rhythms and counts of Igor Stravinsky's music, and geometrically patterned costumes and set details, Ashton's work was said to have beautifully combined mathematics and ballet for its visual imagery.

Notation Systems

To preserve these choreographed works of art, dance notation systems were created to symbolically represent the positions, steps, and movements of the dancers. Early seventeenth- and eighteenth-century systems, such as Feuillet notation, recorded mainly floor patterns and feet positions, whereas the twentieth-century notation systems, Labanotation and Benesh Movement Notation (written on vertical and horizontal staffs, respectively), corresponded to the scores of accompanying music. These notation systems detailed the entire body movements from head to toe of every dancer. Even with the advent of video recording, it is these symbolic notations showing graphical representations of the step details that best preserve ballets for future generations.

Further Reading

Cooper, Elizabeth. "Le Balet Comique de la Reine: An Analysis." http://depts.washington.edu/uwdance/dance344reading/bctextp1.htm.

Greskovic, Robert. *Ballet 101: A Complete Guide to Learning and Loving the Ballet.* Milwaukee, WI: Limelight Editions, 2005.

Minden, Eliza Gaynor. *The Ballet Companion.* New York: Fireside, 2005.

Schaffer, Karl, and Erik Stern. "Math Dance Bibliography." http://www.mathdance.org/MathDance-Bibliography.pdf.

Thomas, Rachel. "Scènes de Ballet." http://plus.maths.org/issue24/reviews/ballet/.

Elizabeth A. McMillan-McCartney

See Also: Ballroom Dancing; Contra and Square Dancing; Musical Theater.

Ballroom Dancing

Category: Arts, Music, and Entertainment.
Fields of Study: Communication; Geometry; Representations.
Summary: Ballroom dancing allows students to approach mathematics in a variety of ways.

Ballroom dancing, considered sophisticated for its elegance, is a style of choreographed dance showcasing not only the dancers' technical skill but also their poise and style. Originally danced primarily at balls for the social elite, ballroom dancing has become a competitive sport. Dancing allows students to approach mathematics in a variety of ways, from the basic arithmetic of the beats per minute (bpm) to the geometric spatial relationship with respect to the other dancers. Choreographers Erik Stern and Karl Schaffer have created a dance called a "math dance." The purpose is twofold: to use mathematics to create dance, and to help students learn mathematics concepts through the movements of the dance. Some of the topics explored in math dances are the mathematics of rhythm, polyhedra, symmetry, and dissection puzzles.

History

The phrase "ballroom dancing" derives from the Latin word *ballare* meaning to sweep or to dance. Now considered historical dances, the original forms of ballroom dancing included the minuet and quadrille. Some steps performed in the quadrille, such as the *entrechats* (crossing the legs one in front of the other multiple times) and the *ronds de jambes* (circular movement of the leg while it is extended, toe pointing to the floor), have disappeared from the modern ballroom yet still exist in the ballet world.

In the early 1800s, the waltz made its appearance; the distance between dancing partners was considered scandalous at the time since the waltz required the partners to dance in close proximity. The early 1900s brought the birth of jazz and new dance styles as dancers moved together yet independently of each other. In addition, lively dances such as the Foxtrot, otherwise known as the one-step or two-step, moved away from the traditional placement of feet being turned out and instead called for dancers to have their feet parallel to each other. While many people are unfamiliar with any ballroom dances besides the waltz, competitive ballroom dancing has gained notoriety; it has been showcased on the ABC television show *Dancing with the Stars* and has become an Olympic sport as well.

Beats

Ballroom dancing consists of a series of dance moves, where more complicated dance steps are called "figures" or "dance figures." Each of the formally named dances has a variety of dance moves that can be put together to form a personalized performance. Determining the dance moves to use involves more than merely counting the beats. One can calculate the total number of beats that will occur in a song and then determine how many different dance moves would be necessary. For example, if one hears 12 beats in a five-second segment of the song, it can be calculated that the song has 144 bpm. If the song is exactly two minutes long, one can calculate there are 288 beats to work with for the whole song ($2 \times 144 = 288$). Since each dance move is typically 8 beats, dividing 288 by 8 beats indicates one needs 36 dance moves. The moves can be repeated, using, for example, 9 moves 4 times each or 11 moves 3 times each (the second option gives the dancer three fewer moves than needed, requiring a dramatic flourish to end the dance). The total number of beats combined with the thematic moves of a particular dance and an individual's personal signature steps form a composite whole.

Rhythm

One rhythm option for the American-style Foxtrot consists of *Slow, Quick, Quick*, or half, quarter, quarter in 4/4 time; this approach to the dance gives teachers the opportunity to teach fractions to students using dancing. By creating a dance of successive moves in which two basic steps make one whole move, students will use fractions—adding and subtracting in 4/4 time and introducing the family of fractions

$$\frac{1}{16}, \frac{1}{8}, \frac{1}{4}, \text{ and } \frac{1}{2}.$$

This also can be done in 6/8 time with $\frac{1}{2}, \frac{1}{3}, \frac{1}{6}$, and so on.

Geometry

As the lead dancer gauges the couple's location within the coordinate plane of the dance floor, he or she keeps them spatially equidistant from other couples. In addition to the symmetry involved in the various dance moves on the dance floor, symmetry is considered within each dancer's pose and posture (the form created by the two partners together—symmetrical or asymmetrical). This symmetry can lead to an understanding of angles and curves when various dance poses are examined, and allows students the opportunity to solve problems kinesthetically when they attempt to form a mirror image of their partner while executing the dance moves.

Further Reading

Hackney, Madeleine. "Dancing Classrooms Enhance Math Skills." *Connect* 19, no. 4 (2006).

International Dance Sport Federation. http://www.idsf.net.

National Dance Council of America. http://www.ndca.org.

Watson, Anne. "Dance and Mathematics: Engaging Senses in Learning." *Australian Senior Mathematics Journal* 19, no. 1 (2005).

World Dance Council. "Welcome." http://www.wdcdance.com.

DEBORAH L. GOCHENAUR

See Also: Ballet; Contra and Square Dancing; Geometry of Music; Step Dancing.

Bankruptcy, Business

Category: Business, Economics, and Marketing.
Fields of Study: Data Analysis and Probability; Number and Operations; Problem Solving.
Summary: The value of a business entering bankruptcy is determined by the asset, income, or market approach and creditors are repaid according to their risk.

Bankruptcy of a business occurs when the business is legally declared insolvent (its assets are less than its liabilities). If the debtor files a bankruptcy petition, it is called a voluntary bankruptcy. However, if creditors force the debtor into bankruptcy, then it is called an involuntary bankruptcy. Most bankruptcies are voluntary. In either case, the value of the business needs to be determined for legal purposes. The standard of the value used in the valuation is the fair market value (the value of the price of the firm that a rational buyer is willing to pay to a willing seller in a free market). There are three basic approaches for valuating the business: the asset approach, the income approach, and the market approach. The hierarchy of the creditor in a bankruptcy is determined by the amount of risk the creditor bears: the creditor who bears least amount of risk will have priority to receive payment after liquidation.

Asset Approach

The asset approach determines the value of a company by adjusting its book value of assets to the current market value. It is based on the economic principles of substitution: a rational investor will not pay more for a business asset than the price of a different asset that provides similar utility. There are two methods associated with the asset approach: the adjusted book value method and the replacement cost method.

In the adjusted book value method, the assets and liabilities on the balance sheet are examined item by item by professionals to determine the business's current market value. Once the assets and liabilities have been adjusted to the current market value, the value of the company is calculated as the difference.

In the replacement cost method, the value of each asset and liability on the balance sheet is first determined as the cost to replace it. Then, the value the company is determined as the difference of its assets and liabilities.

The asset approach is not reliable for companies with significant intangible assets because the approach involves professional judgment. It is more suitable for companies that have many tangible assets and few intangible assets.

Income Approach

The income approach determines the fair market value of a firm by discounting its expected cash flows at an appropriate discount rate assuming the firm will continue to operate without liquidation. The discount rate is often chosen to be the firm's weighted average cost of capital (WACC). The procedure is completely analogous to that of determining the net present value of a firm in corporate finance theory. Mathematically, the fair market value under the income approach can be written as

$$\text{FMV} = \frac{E(C)}{\text{WACC}}$$

where FMV is the fair market value, $E(C)$ is the expected cash flows under the assumption that the firm will continue to operate, and WACC is the weighted capital of cost.

In corporate finance theory, WACC is often calculated as the weighted average of the cost of debt of the firm and the cost of equity of the firm

$$\text{WACC} = (1 - T_c) r_D \frac{B}{B+S} + r_E \frac{S}{B+S}$$

where T_c is the corporate tax rate, r_D is the cost of debt, r_E is the cost of equity, B is the market value of the firm's bonds, and S is the market value of the firm's stocks.

WACC takes into consideration the facts of leverage and taxes and thus is the appropriate discount rate used for income approach. The income approach assesses the value of the debtor to the creditors. However, it fails to take account of the value inherent in the flexibility of decision making, which is often valued using a mathematical tool called "decision tree."

Market Approach

The market approach assesses a company's value by comparing it with similar companies in the market. The rationale behind this approach is that the price of the subject company should be very close to the values of the similar companies in the market. There are two methods associated with the market approach: the guideline public company method, and the comparable transaction method. In the guideline public company method, a peer group of public companies with similar sizes, natures, operations, and financial characteristics is first selected. Next, the enterprise value of each company in the group is calculated as

$$\text{EV} = P_S \times N_S + D - C_E$$

where EV is the enterprise value, P_S is the stock price per share, N_S is the number of outstanding shares, D is total debt, and C_E is excess cash.

Then market multiples, such as enterprise value/revenue and enterprise value/earning before interest and tax, will be calculated using the enterprise value. Finally, the value of the subject company is determined by applying the calculated market multiples. For example, if the enterprise value/revenue is used, then the value of the subject company can be calculated as

$$V = \text{EV} \times R$$

where V is the value of the subject company and R is the revenue of the subject company.

In the comparable method, the value of the subject company is determined in a similar fashion as in the guideline public company method. In other words, market multiples are derived, and then they are applied to the subject company to determine its value. However, in the comparable method, public data of comparable transactions are used to calculate the market multiples.

Thus, the comparable method also consists of three steps: selecting a group of comparable transactions, calculating market multiples, and applying the market multiples.

The biggest drawback to the guideline public company method is that it is not applicable for nonpublic companies. The challenge with the comparable method is finding appropriate and reliable comparable transactions.

Paying Creditors

When a company declares bankruptcy, its creditors must be paid, but the creditors receive only some of the money they are owed. For example, if a bankrupt company is ordered to pay 10 cents on the dollar, this means for every dollar the company owes a creditor, it will pay only 10 cents. This is a proportional solution that is easy to arrive at using simple algebra. However, this is not the only payout strategy. There are several mathematical methods that can be used to determine how much money each creditor should receive. In the total equality method, available capital is simply divided equally among debtors, regardless of how much they are owed. A variation, traced back to medieval philosopher Moses Maimonides, proposes giving every debtor as equal a share as possible but never more than they are owed. In modern terms, this is a constrained optimization problem that can be solved using methods such as linear programming. Other decision methods are logically and analytically more complex, like the Shapely value, which considers paying a sequence of creditors their full amounts owed, to the extent of available funds, for all possible orderings. This game-theory approach is named for American mathematician and economist Lloyd Shapely.

Further Reading

Brealey, Richard A., Stewart C. Myers, and Allen Franklin. *Principles of Corporate Finance*. 9th ed. New York: McGraw-Hill, 2008.

Copeland, Thomas E., Fred J. Weston, and Shastri Kuldeep. *Financial Theory and Corporate Policy*. 4th ed. Upper Saddle River, NJ: Pearson Education, 2005.

Newton, Grant W. *Practice and Procedure. Vol. 1, Bankruptcy and Insolvency Accounting*. Hoboken, NJ: Wiley, 2010.

Ratner, Ian, Grant Stein, and John C. Weitnauer. *Business Valuation and Bankruptcy*. Hoboken, NJ: Wiley, 2009.

Liang Hong

See Also: Accounting; Bankruptcy, Personal; Budgeting.

Bankruptcy, Personal

Category: Business, Economics, and Marketing.
Fields of Study: Data Analysis and Probability; Number and Operations; Problem Solving.
Summary: Personal bankruptcy can be caused by exponentially increasing debt, and mathematics is used to calculate payments or to divide assets among creditors.

Personal bankruptcy is a legal proceeding intended to provide relief for the debtor. Personal bankruptcy essentially results from huge debts, which can be caused many factors including unexpected medical bills, huge credit card debts, poorly managed loans, unemployment, and divorce. The fundamental formula that lies behind most large debts is exponential growth.

Legal Procedure

Personal bankruptcy in the United States is usually a court-supervised procedure that provides the debtor with the opportunity for a fresh financial start. The earliest personal bankruptcy law in the United States can be traced to 1800. The most recent personal bankruptcy law passed by the U.S. Congress is the Bankruptcy Code of 1978. Under this law, an individual may file a voluntary petition under either Chapter 7 (liquidation) or Chapter 13 (Reorganization).

If a personal bankruptcy case is filed under Chapter 7, a court-supervised procedure begins. The debtor's assets will be classified as either exempt or nonexempt according to the state law. A trustee will then collect the nonexempt assets of the debtor. The debtor is allowed to keep all the exempt assets provided such an asset is not secured by any property. For example, a mortgage is secured by the house. Thus, debtors can still lose their houses if mortgaged payments fall behind. The debts of the debtor will be wiped out except certain non-dischargeable debts including alimony, child support, student loans, taxes, and any fines resulting from criminal conviction. The record of personal bankruptcy could stay on the debtor's credit history for up to 10 years. In summary, under Chapter 7, the debtor is discharged most of the debts and surrenders all possessions except those necessary for living. However, not everyone is qualified for Chapter 7 bankruptcy. To qualify, the debtor must complete Official Form 22A (Chapter 7) to pass the means test. The personal bankruptcy involves balancing the conflicting interests of the creditors and the debtor. While a qualified debtor can wipe out most debts under Chapter 7, some creditors will not receive any payment. The 2005 Bankruptcy Abuse Prevention and Consumer Protection Act was enacted to prevent the abuse of Chapter 7 and makes it more difficult for a debtor to file under Chapter 7.

If a personal bankruptcy petition is filed under Chapter 13, then the debtor is required to propose a repayment plan that will pay the debts during a specified period of time (typically three to five years). The plan must be reasonable and meet certain requirements. It must be approved by the court. Although the debts of the debtor cannot be written off immediately under Chapter 13, the debtor is protected from debt-collecting actions from the creditors while the repayment plan is in effect. Thus, the Chapter 13 bankruptcy is often chosen by those who have a stable income.

Exponential Growth

Personal bankruptcy results from unmanageably large debts that can be caused by many factors such as medical costs because of under-insurance and uninsured status, compulsive buying habits, loss of job, or irresponsible loans. The fundamental formula that leads to a large debt is the law of exponential growth, which occurs when the growth rate of a quantity is proportional to the current value. In mathematics, exponential functions generally involve the constant *e*. Mathematically equivalent forms with different bases may be used in order to more intuitively correspond to the parameters of a real-life problem, such as interest calculations. The traditional way of calculating interest on a loan is called compound interest, under which the interest earned during each interest measurement period (month, quarter, or year) will automatically be added to the principal to earn additional interest during the next interest measurement period. Mathematically, this can be expressed as

$$B_t = B_0(1+r)^t$$

where B_0 is the principal amount, B_t is the balance of the loan at the end of t-th interest measurement period, and r is the interest effective per interest measurement period.

The loan balance under the compounding interest grows rapidly over a relatively long period, even if the interest rate is not high. For example, consider a person who takes a loan of $10,000 from a bank at a monthly interest rate of 1.5%. The loan balance after one, five, and 10 years will be $11,956, $24,432, and $59,693, respectively.

Mathematical Division of Assets

The ideas of dividing and choosing have existed as long as mankind. The mathematical theory of fair division dates back to World War II, to Polish mathematicians Hugo Steinhaus, Bronisław Knaster, and Stefan Banach. The classic bankruptcy problem in game theory addresses fairness in one way. It involves allocating some amount of resources among two or more individuals who have a claim on them, assuming that any division of the assets is allowable and that there are not enough resources to satisfy all claims. Real-life examples include someone who has declared Chapter 13 bankruptcy and therefore must repay some creditors, or dividing a deceased person's estate among several heirs—especially when the estate cannot satisfy all the deceased's commitments.

Assets may be divided equally (with or without ensuring no claimant receives more than his or her claim), proportionally according to the relative size of the claim, or by other more complex strategies. The cake-cutting problem also tackles the issue of fair allocation but includes more subjective measures of valuation that must be modeled mathematically, and sometimes an asset pool with constraints on the ways in which it may be divided. Cake-cutting problems typically require iterative algorithms to solve.

Further Reading

Anosike, Benji O. *How to Declare Your Personal Bankruptcy Without a Lawyer*. 3rd ed. Newark, NJ: Do-It-Yourself Legal Publishers, 2004.

Haman, Edward A. *The Complete Chapter 7 Personal Bankruptcy Guide*. Naperville, IL: Sphinx, 2007.
Kellison, Stephen G. *Theory of Interest*. 3rd ed. New York: McGraw-Hill, 2009.
Vaaler, Leslie Jane, and James W. Daniel. *Mathematical Interest Theory*. 2nd ed. Washington, DC: MAA, 2009.

LIANG HONG

See Also: Bankruptcy, Business; Budgeting; Credit Cards.

Bar Codes

Category: Business, Economics, and Marketing.
Fields of Study: Algebra; Number and Operations; Representations.
Summary: Bar codes encode numerical data visually for product identification and other purposes.

A bar code is a visual representation of information intended to be decoded by an optical scanner called a bar code reader. The reader illuminates the bar code, thus allowing its light sensor to detect the patterns of dark and light bars. The sequence and width of dark and light bars represents a unique sequence of numbers and letters.

Origins

It took 26 years for the idea of bar codes to be successfully implemented in the retail industry. In 1948, two graduate students at Drexel University, Norman J. Woodland and Bernard Silver, overheard a conversation in which the president of a local supermarket chain in Philadelphia wished to automate the checkout process. At that time, a cashier in a supermarket would have to type into a cash register the price of all items in a purchase—a time-consuming and error-prone task. Woodland and Silver filed a patent application in 1949, obtaining the patent in 1952, for an optical device that would read information automatically. The first prototype was produced by IBM but was impractical because of both its size and the heat generated by the 500-watt light bulb used by the bar code scanner. The patent was sold in 1952 to the Philadelphia Storage Battery Company (Philco), which was also unable to produce a viable prototype, and sold the patent the same year to the Radio Corporation of America (RCA). Bernard Silver died in a 1963 car accident, before the bar code system was implemented in practical settings. The invention of lasers and integrated circuits in the 1960s allowed the manufacture of small, low-energy bar code readers. RCA developed a modern version of bar codes in 1972 in a Kroger store in Cincinnati, but the code was printed in small stripes that were easily erased or blurred by employees who had to attach them manually to each item. Norman J. Woodland was an employee at IBM at the time and led a team that produced bar codes according to a standard known as Universal Product Code (UPC) still in use today. Bar codes are used in nearly all retail products worldwide. The applications of bar codes have also reached far beyond the retail industry; they are now used in such disparate applications as patient identification, airline luggage management, and document management, as well as purchase receipts.

The Mathematics of Bar Codes

The most ubiquitous form of bar codes consists of a visual pattern of long lines (hence the "bar" in "bar code"), which has four well-defined zones (see Figure 1): (1) quiet zone, or empty zone, located in the left and right zones of the code; (2) initial character (right) and final character (left) are standard bars that appear on all bar codes, and indicate where the information begins and ends; (3) variable-length character chain, which contains as many characters as needed to encode the message; and (4) checksum, which is a number that is computed algebraically from the other characters

Figure 1. Zones of a bar code.

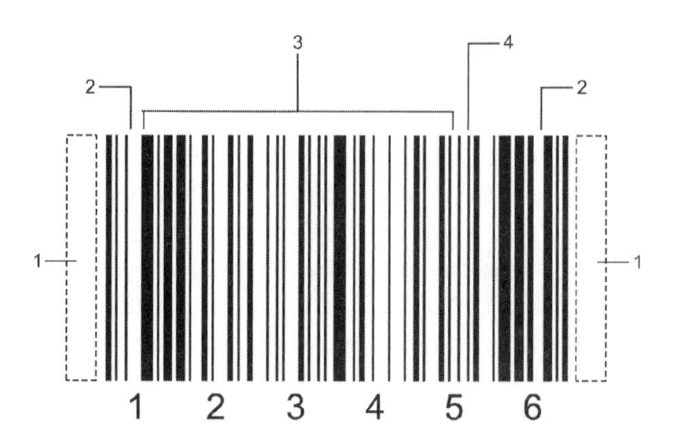

using modular arithmetic, and is used to verify that the characters have been correctly transmitted and interpreted. The digits are either simply added or are weighted. For example, the 10-digit International Standard Book Number (ISBN-10) uses weights based on digit position and modulus 11 arithmetic.

Each digit is encoded by two white and two black bars. The bars have widths of 1 to 4 units, and the total width for each digit is always seven units. Bar code readers are designed to read bar codes irrespective of their size; a magnified bar code encodes the same information as a reduced-size bar code. This property is mathematically known as scale invariance.

Further Reading

Adams, Russ, and Joyce Lane. *The Black & White Solution: Bar Code & the IBM PC.* Dublin, NH: Helmers Publishing, 1987.

Palmer, Roger C. *The Bar Code Book: A Comprehensive Guide to Reading, Printing, Specifying, Evaluating, and Using Bar Code and Other Machine-Readable Symbols.* 5th ed. Victoria, BC: Trafford Publishing, 2007.

Wittman, Todd. "Lost in the Supermarket: Decoding Blurry Bar codes." *SIAM News* 37, no. 7 (September 2004).

Juan B. Gutierrez

See Also: Coding and Encryption; Comparison Shopping; Inventory Models.

Baseball

Category: Games, Sport, and Recreation.
Fields of Study: Data Analysis and Probability; Measurement.
Summary: Baseball is a mathematically rich sport, especially with regard to its array of statistics.

Though America's favorite sport for more than a century, the game of baseball has undergone many changes, many in response to statistics gathered regarding all parts of the game. At first, the statistics were limited to scorecard data but have expanded to include every action and detail of the game. More so, this gathering and analysis of data has expanded beyond the realm of statistical analysis, as mathematics is now used to examine all aspects of baseball—the physical characteristics and performance of its players, the analysis and modeling of each element (hitting, fielding, pitching, strategies), and the combined geometry and physics surrounding the game.

Although some fans object to this intrusion of mathematics into a competitive sport, most accept or even depend on the mathematical aspects as enriching their enjoyment of the game itself. That is, mathematics has become the arbiter in arguments, the stimulus for "hot stove league" discussions, a tool to help identify either patterns of team strengths and weaknesses or optimal strategies, and a decision-making tool for gamblers and fantasy league participants.

Sabermetrics

Bill James, a baseball writer, historian, and statistician, gave authenticity to the use of statistics in analyzing all aspects of baseball through his pioneering mathematics and statistics work. Multiple editions of his *Baseball Abstract* in the 1980s changed not only the play of the game itself but also how it is viewed by fans, and are the predecessor to many modern Web sites dedicated to analysis of the sport. James revolutionized the way mathematics is used to analyze sports to determine why some teams win and others lose. He coined the term "sabermetrics," which is derived from the Society for American Baseball Research acronym SABR, for his analytical and modeling methods. In 2006, *Time* magazine named him one of the most influential people in the world.

Mathematical statistics provide perspectives that explain game occurrences, provide comparative rankings of teams and players, and assist in managerial decision making. The primary example is the simple use of ratios, means, and medians as both descriptive and inferential statistics for a player, position, game, season, or career. Some examples include the following:

- Batting average, slugging percentage, on-base percentage, and batter's run average
- Effect of artificial turf on numbers of ground ball hits or base stealers' performances
- Performance of hitters and pitchers in different environments (outdoor versus dome stadiums; night games versus day games)

- Expected strike zones for umpires, given a pitcher or batter is right- or left-handed

Going beyond these descriptive statistics, the game of baseball can be analyzed using very sophisticated techniques. Some examples include the following:

- Connections between a player's characteristics and training regimens relative to game performance, or even to document the effects of steroid use
- Trend analysis, based on either a player's or team's performance (hitting, pitching, fielding) over the past five, 10, and 15 games
- Importance of pitcher throwing a "first strike"
- Effects of bringing in the infield when the bases are loaded with less than two outs
- Team winning tendencies based on run differential in innings seven, eight, and nine
- Impact of rule changes on pitching and hitting, such as the effects of elevating the pitching mound or changing foul-line distances to outfield fences
- Determination of coaching strategies such as sacrifice bunts, pitch-outs, stealing home, intentional walks, shifts of fielders for certain hitters, or use of pinch hitters and relief pitchers
- Determining the "best" all-time player in a particular position (for example, centerfielder, hitter, relief pitcher, base-stealer)
- Selection of players by professional teams during annual drafts, using both historical data for each player's performance and physical data
- Use of statistical data during contract negotiations between a player and management, or even the release or trading of players based on team needs

Mathematical probabilities, odds, and expected values can help examine the chances of particular events happening within a game or across games:

- Probability that the World Series will go four, five, six, or seven games
- Use of odds to determine personal or professional betting strategies

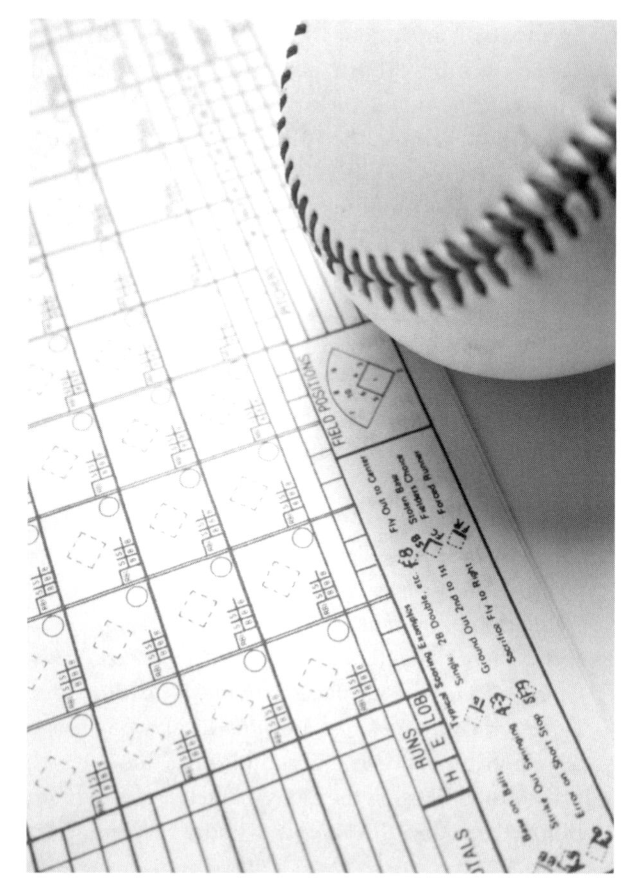

Although some fans object to using mathematics in baseball, some enjoy it as much as the game itself.

- Use of conditional probabilities to determine lineups or use of pinch hitters, reflecting the probability of a batter getting a hit given that the pitcher is right- or left-handed
- Correlations between a team's wins per season and player payrolls, or pitcher salaries and their ERAs
- Probability of a record being broken, either by a team or player, such as Joe DiMaggio's 56-game hitting streak

Though difficult to implement practically, geometry, trigonometry, and calculus can shed light on other important ideas:

- Length of a home run
- Actions of different pitches such as a curve ball, slider, fastball with movement, or forkball

- Determination or alteration of a hitter's batting stance or position in the batter's box
- Use of angles in fielding balls off outfield walls

Game theory also is used as part of the decision-making process within a baseball environment, leading to choices of optimal tactics. Some specific decisions are as follows:

- A manager's choice of batting lineups and pitching moves, relative to the opposing manager's choices
- A manager's calling for shifts of fielders, pitch-outs, or steals at times within a game
- A manager trying to argue, influence, or reverse decisions by umpires
- A manager's use of techniques to motivate specific players
- A team's selection of players during a draft, dependent on the player's apparent abilities, the inferred needs of other teams, and the specific draft round
- Contract negotiations involving players, agents, and team management

Finally, using all of these statistical data and mathematical modeling techniques, one can create realistic simulations of baseball games or end-of-year series, possibly using computer animations.

At the collegiate and professional levels, managers are increasingly using mathematics to remain competitive, even hiring mathematical statisticians as important parts of their staff. However, some authors and fans suggest that the team with the best players and managers will usually win, despite any use of sophisticated mathematics.

Further Reading

Albert, Jim, and Jay Bennett. *Curve Ball: Baseball, Statistics, and the Role of Chance in the Game*. New York: Springer-Verlag, 2001.

Cook, Earnshaw. *Percentage Baseball*. Cambridge, MA: MIT Press, 1966.

Eastway, Rob, and John Haigh. *Beating the Odds: The Hidden Mathematics of Sport*. London: Robson Books, 2007.

Ross, Ken. *Mathematician at the Ballpark: Odds and Probabilities for Baseball Fans*. New York: Pi Press, 2004.

Schell, Michael. *Baseball's All-Time Best Hitters: How Statistics Can Level the Playing Field*. Princeton, NJ: Princeton University Press, 1999.

Schwarz, Alan. *The Numbers Game: Baseball's Lifelong Fascination with Statistics*. New York: St. Martin's Press, 2004.

Jerry Johnson

See Also: Basketball; Football; Hitting a Home Run; Hockey; Soccer.

Basketball

Category: Games, Sport, and Recreation.
Fields of Study: Data Analysis and Probability; Geometry; Measurement.
Summary: Play can be analyzed geometrically and probabilistically to inform strategy or construct simulations.

Basketball is an international sport that can be enjoyed either as a participant or as a spectator, regardless of one's sex or one's age. A growing number of coaches, reporters, and ardent fans are using mathematics to examine all aspects of basketball—the physical aspects and performance of its players, the analysis of each element (shooting, defense, strategies) of the game, and the combined geometry and physics surrounding the game. Perhaps as expected, this mathematical analysis can have opposite effects, either enriching or ruining the sports experience itself.

Basketball was intended to be a dynamic, fair competition between two teams; however, mathematical concepts and techniques can be used in a basketball environment to identify patterns of strengths and weaknesses, suggest optimal strategies for coaches and players, stimulate discussions, and resolve arguments. Statistician Dean Oliver is a well-known contributor to the statistical evaluation of basketball, which is called APBRmetrics. The name comes in part from the Association for Professional Basketball Research (APBR). This methodology is a very similar to the analysis of professional baseball using sabermetrics. Though difficult to implement practically, geometry, trigonometry, and calculus can shed light on these important ideas:

- Given a player's height, the best angle and velocity for shooting a basketball, assuming the intent is to have the basketball's parabolic arc pass through the basket (often called the "Shaq phenomena")
- The connection between the angle of shooting a ball and the event known as an "all-net" basket
- The connection between a player's height and where a player should aim a shot—at the center of basket, the front of the rim, or the back of the rim
- Use of angles in making bounce passes
- Determining defensive positions that maximize centers of gravity
- The connection between a player's position on the court and decisions to bank the basketball off the backboard as the best shot
- Comparison of the merits of shooting a free-throw underhand versus overhand

By gathering and analyzing the myriad of available data provided by a game experience, mathematical probabilities can help examine the chances of particular events happening within a game, including the following:

- The likelihood of a player making 0, 1, or 2 points in a 1-and-1 free throw opportunity

Mathematics can be used to examine the physical aspects and performance of players, the analysis of each element of the game, and the combined geometry and physics surrounding basketball.

- The reality of a player having a "hot-hand," based on his or her making successive shots
- The decision as to which player should be purposely fouled at the end of a close game
- The evaluation of a player's performance in terms of "per-possession efficiency"
- The probability of a record being broken, either by a team or a player

Similarly, the collection and organization of mathematical statistics can provide perspectives that explain game occurrences, provide comparative rankings of teams and players, and assist in future decision making by coaches and team management. The usual sources of statistics are data regarding shooting, rebounding, free throws, turnovers, defensive gains, and time management. Some specific examples include the following:

- The simple use of ratios, means, and medians as descriptive statistics for a player, a position, a game, or a season
- Connections between a player's characteristics and training regimens relative to game performance
- Trend analysis, based on either a player's or a team's performance in specific ways over the past five, 10, and 15 games
- Winning tendencies based on connections to lead changes during a game or knowledge of the team leading at the end of the third quarter
- The impact of rules changes on scoring and defenses within the sport itself, such as the observed effects of expanding either the three-point arc or the free-throw lane
- Determining the "best" all-time player in a particular position (for example, center), at a particular time in a game (for example, last-second shot), or in an era
- The seeding and selection of teams in a bracketed tournament, possibly as part of a betting pool with stated odds
- Selection of players by professional teams during the annual draft, using historical data for each player's performance in conjunction with physical data
- The use of statistical data as part of contract negotiation between players and management
- The release or trading of players based on team needs

The ideas of mathematical game theory have been applied to the decision-making process within a basketball environment, leading to choices of optimal tactics. The specific decisions range considerably:

- A coach's choice of designed offenses and defense strategies, relative to the opposing coach's choices
- A coach's calling of time-outs at opportune times within a game
- A coach trying to influence or reverse decisions by game officials
- A coach's use of techniques to motivate specific players
- A team's selection of players during a draft, dependent on the player's apparent abilities, the inferred needs of other teams, and the specific draft round
- Contract negotiations involving players, agents, and team management

Finally, using all of these available statistical data and mathematical modeling techniques, one can create realistic simulations of basketball events, full games, or even tournament series. At the collegiate and professional levels, coaches are increasingly using mathematics to remain competitive, even hiring mathematical statisticians as important parts of their staffs. Some mathematicians are even found on the court.

Retired San Antonio Spurs player Michael Robinson earned a bachelor's degree in mathematics from the U.S. Naval Academy, and is considered by many to be the best basketball player that school has ever seen. However, there are still some authors and fans who suggest the team with the best players and coaches will usually win, despite the use of sophisticated mathematics.

Further Reading

Bennett, Jay, and James Cochran. *Anthology of Statistics in Sports*. Philadelphia: Society for Industrial and Applied Mathematics, 2005

De Mestre, Neville. *The Mathematics of Projectiles in Sport*. Cambridge, England: Cambridge University Press, 1990.

Friedman, Arthur. *The World of Sports Statistics: How the Fans and Professionals Record, Compile, and Use Information*. New York: Athenaeum, 1978.
Oliver, Dean. *Basketball on Paper: Rules and Tools for Performance Analysis*. Washington, DC: Brassey's, 2004.

JERRY JOHNSON

See Also: Baseball; Football; Hockey; Soccer.

Basketry

Category: Arts, Music, and Entertainment.
Fields of Study: Algebra; Geometry; Number and Operations.
Summary: Basket shapes and patterns are created by varying the weave.

Baskets are woven containers made of plant or artificial strips, such as palm fronds, willow branches, or fabric. People were already making baskets at least 10,000 years ago. Historians conjecture that basketry played a major role in the development of pattern, structure, and number in human cultures. Early humans could have observed birds and animals that wove to first learn the craft. Tracing basket-weaving patterns through cultures assists in creating models of human migration. "Underwater basket weaving," which is a technique in wicker, is a humorous idiom describing academic courses with low education standards or very narrow specializations. Mathematicians study and model patterns found in baskets from around the world, including those created by the Hopi people of the southwestern United States, various African peoples, and Pacific Islanders.

Weaves

There are several types of basket weaves, each with infinitely many possible patterns. Coiled baskets are made with two types of fiber—one thick, and one soft and pliable. The thick cord or vine forms the coil. Flat, pliable strips of materials such as grass or fabric are wound a number of times around the cord, then a number of times around its previous row or several rows in the coil, connecting the rows. Craftspeople can change patterns and shapes of baskets by varying these weave numbers. Wicker is a type of basket weave consisting of relatively stiff fibers of two types. One material, the foundation, is completely rigid, and the other, the weft, is more pliable. The pattern of individual weft fibers going over and under the foundation spokes determines the look of the basket's surface. Such patterns can become very complex. Weft fibers are often soaked to make them soft during weaving.

Twining also requires rigid foundation fibers and pliable weft. Several strands of pliable fiber—usually two—go around a foundation spike on either side, cross or twist in the middle, then go around the next spike. Twining patterns are created by changing the number of wefts or formulas of skipping spikes, and introducing braiding between spikes.

Plaited baskets consist of pliable fibers woven over and under one another, typically at right angles. This weave is very similar to how woven textiles are made, and some historians believe that textiles originated from this type of basket. Formulas—whose variables are the number of fibers that go over and under in each row—determine the pattern.

The physical properties of baskets are determined by the weave, the materials, and the pattern. Wicker baskets can be very sturdy, and wicker has been used in making fences, houses, and furniture like baby cradles.

Basket Patterns

Patterns found in baskets come up in many areas of mathematics. Frieze patterns have translational symmetry along lines, and there are seven types of them, all of which appear in traditional basket making. They are a part of more general wallpaper groups, of which there are seventeen types.

Some mathematics historians observed differences in patterns that involve six-fold symmetry, such as honeycombs, and more complex five-fold symmetry that comes up in basket weaving. For example, the traditional woven Malaysian ball is similar to the modern soccer ball (also known as "football") in that it contains pentagons.

Shapes

Baskets take a variety of three-dimensional shapes, such as cylinders, cubes, and prisms. Properties of weaving often determine the shape. For example, the stiff foundation fibers of twined or wicker baskets are usually straight lines, which only allows so-called ruled surfaces. By definition in analytic geometry, ruled surfaces are generated by straight lines. Cylinders, prisms, and cones are ruled surfaces and can be made by wicker. Spheres cannot be made out of straight lines, but spherical baskets are made by coiling, plaiting, or using bendable foundations in wicker and twining. Mathematicians and mathematical artists who use basket weaving to create striking sculptured models of complex surfaces have to select appropriate weaving techniques for their projects.

Further Reading

Gerdes, Paulus. *African Basketry: A Gallery of Twill-Plaited Designs and Patterns.* NP: Lulu Publishing, 2008.

University of East Anglia. "Basket Weaving May Have Taught Humans to Count." *ScienceDaily* (June 8, 2009). http://www.sciencedaily.com /releases/2009/06/ 090604222534.htm.

Zaslavsky, Claudia. *Multicultural Mathematics: Interdisciplinary Cooperative-Learning Activities.* Portland, ME: Walch Education, 1993.

Maria Droujkova

See Also: Sculpture; Surfaces; Symmetry; Textiles; Transformations.

Bees

Category: Weather, Nature, and Environment.
Fields of Study: Algebra, Geometry; Representations.
Summary: Geometry explains why honeycombs are made of hexagonal cells, while bee movement patterns communicate information visually.

Honeycombs are remarkable for their beauty, precision, and symmetry. The honeycomb corresponds to a mathematical concept known as a "tiling of the plane." That bees use regular hexagons for this tiling (built to a remarkable level of precision) has fascinated human beings throughout history. At the end of the twentieth century, mathematician Thomas Hales rigorously proved a long-standing conjecture that fully justifies to humans what the bees have apparently known all along: the most efficient way to repeatedly enclose a fixed amount of storage space is to use regular hexagons to form the boundaries.

Honeycomb: How to Choose a Cell

Bees use honeycomb cells for storage. It takes work and material (wax) to create the boundary of each cell, so the bees want cells with as little boundary (perimeter) as possible, given that each cell should enclose a certain amount of storage (area). If a bee only needed to make one cell to store honey, it would likely use a shape other than a regular hexagon. For instance, a regular octagon holding the same area has less perimeter; a regular decagon will have less perimeter still. The more sides a polygon has, the smaller the perimeter will be, with the circle having the smallest perimeter-to-area ratio. That a circle is the least-perimeter shape to enclose a given area is a famous problem that goes back to the wonderful tale of Queen Dido of Tyre.

For example, suppose a bee wanted to enclose one square unit of area. The square that accomplishes this has a perimeter of 4. If the bee used an equilateral triangle instead, the necessary perimeter is larger, about 4.56. But the regular hexagon's perimeter is smaller, at

The most efficient way to repeatedly enclose a fixed amount of storage space is to use regular hexagons.

just over 3.72. The pattern of increasing the number of sides leading to a lower perimeter holds for all whole numbers $n > 2$, and every such regular polygon enclosing one unit of area has greater perimeter than a circle holding the same area. The circle that encloses one unit of area has a perimeter of approximately 3.54.

Tilings: Fitting the Cells Together

Bees do not need just one cell; they need many consecutive cells in which to place their honey, and therefore essentially have to create a "tiling" (a pattern involving polygons that will completely cover their work space without overlapping while leaving no space unused). Circular cells simply don't fit together as well because there are gaps between consecutive circles.

Many different kinds of floors and ceilings are tiled—usually with congruent squares or rectangles. Why don't bees use square cells in their honeycomb, rather than hexagons? Or equilateral triangles? It turns out that equilateral triangles, squares, and regular hexagons can all be used to tile the plane, as shown in the figures below. Bees choose hexagons from among these three options since a regular hexagon of unit area uses less perimeter (wax) than does a square or equilateral triangle; the hexagon is a more efficient choice (see Figure 2).

Figure 2.

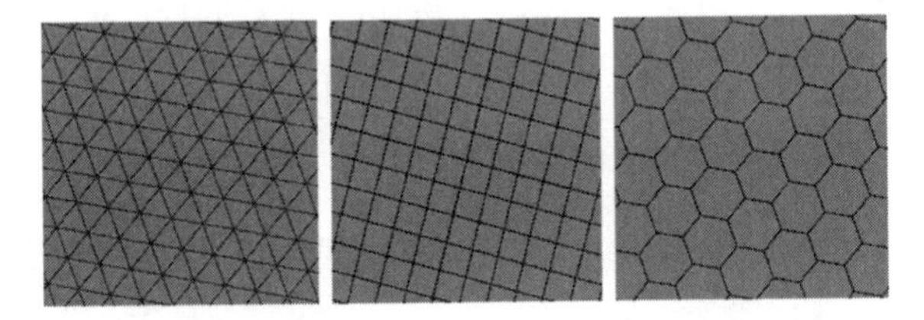

So why not use regular octagons? Here it is not the efficiency of the individual cell that governs the choice but rather the overall packing of them: regular octagons cannot be used to tile the plane.

To understand why triangles, squares, and hexagons tile the plane, but octagons do not, observe that in a regular polygon with n sides, the sum of its interior angles is $180(n-2)$ degrees, and each of its n individual interior angles has the measure

$$\frac{180(n-2)}{n}.$$

For instance, with the square, each interior angle has the measure

$$\frac{180(4-2)}{4} = 90 \text{ degrees.}$$

Four squares arranged at a single vertex fit together perfectly, creating a full 360 degrees around the shared corner. Likewise, six equilateral triangles (each having 60-degree angles) can fit together perfectly for a full 360 degrees, as can three regular hexagons with their 120-degree interior angles.

But for the octagon with $n = 8$, each interior angle has the measure of 135 degrees. Three octagons put together at a shared vertex would have $135 \times 3 = 405$ degrees, which is simply impossible—as would be attempting to only have two octagons meet at a single vertex. Regardless of the number of sides of the regular polygon, the measure of the polygon's interior angle will need to divide evenly into 360 degrees. This forces

$$\frac{2n}{n-2}$$

to be an integer, and the only values of n for which that is true are $n = 3, 4,$ and 6: triangle, square, and hexagon! That the only ways to tile a flat surface using congruent regular polygons are with triangles, squares, or hexagons is a result often taught in high school geometry courses.

Irregular and Non-Polygonal Tilings

Since the time of the ancient Greeks, mathematicians conjectured that among all the ways to tile the plane so that each tile encloses just one unit of area, the way that uses the least perimeter is the tiling that uses all regular hexagons. This conjecture is much harder than it sounds to prove: one must consider irregular polygons (with sides of different lengths), as well as the possibility that the sides of some tiles might be curved. The first possibility is not too difficult to eliminate. For instance, it is straightforward to show that a regular hexagon with all sides of equal length will use less perimeter than any other hexagon to enclose the same area.

But the second possibility—using non-polygonal shapes—proved to be much, much more challenging. In this situation, one must consider the possibility of a shape that bows out on one side and, to fit into a tiling, bows in on another. Obviously, the part that bows out picks up area, while the part that bows in loses area. In 1999, mathematician Thomas Hales proved that any advantage that comes from a side of the tile bowing out is more than cancelled out by the disadvantage that follows from another side having to bow in. Thus, the ideal tile is one that has no bulges: a polygon!

What Professor Hales proved is essentially what the bees knew all along: of all possible tilings, the one using regular hexagons is the most efficient way to enclose cells of the same area.

Other Mathematical Aspects of Bees

Another way that mathematics relates to bees is when mathematicians work with bee researchers to solve problems such as those related to viral disease infection and pollination. Mathematics is also used to model the ways in which bees communicate locations. When a bee finds a source of food, it returns to the hive and performs an elaborate dance that conveys the direction and distance from the hive. Ethologist Karl von Frisch was one of the first to explore the meaning of the honeybee dance, and he won a Nobel Prize for his work. The angle that the bee dances expresses the direction. For example, if a bee dances in a straight line toward the upper part of the hive, then the flowers are located in the direction of the sun. The bee also takes into account the fact that the sun moves; the angle it describes inside the hive changes as the sun does. The duration of the dance and the number of vibrations give the exact distance. Other features of the dance remained unexplained until Barbara Shipman theorized that the honeybee's complex choreography is a projection of a six-dimensional space, and she was able to use this representation to reproduce the entire bee dance in all its parts and variations. To her, this implies that bees can sense the quantum world, although some researchers dispute her conclusions.

Further Reading

Austin, David. "Penrose Tiles Talk Across Miles." http://www.ams.org/featurecolumn/archive/penrose.html.

Frank, Adam. "Quantum Honeybees." *Discover* 18 (1997).

Hirsch, Christian R., et al. *Core-Plus Mathematics: Contemporary Mathematics in Context, Course 1.* 2nd Ed. Columbus, OH: Glencoe/McGraw-Hill, 2008.

Morgan, Frank. "Hales Proves Honeycomb Conjecture." http://www.maa.org/features/mathchat/mathchat_6_17_99.html.

Peterson, Ivars. "The Honeycomb Conjecture: Proving Mathematically That Honeybee Constructors Are on the Right Track." *Science News* 156, no. 4 (July 24, 1999). http://www.sciencenews.org/sn_arc99/7_24_99/bob2.htm.

Serra, Michael. *Discovering Geometry: An Inductive Approach.* 4th ed. Emeryville, CA: Key Curriculum Press, 2008.

University of Guelph. "U of G Using Math to Study Bees." http://www.uoguelph.ca/news/2009/12/u_of_g_using_ma.html.

Matt Boelkins

See Also: Animals; Farming; Polygons.

Betting and Fairness

Category: Games, Sport, and Recreation.
Fields of Study: Algebra; Data Analysis and Probability.

Summary: Mathematics is used to analyze betting and probabilities for games of chance and for investing in the stock market.

A pivotal moment in the early development of probability occurred in 1654, as the French mathematicians Blaise Pascal and Pierre de Fermat exchanged a series of letters. Pascal and Fermat were wrestling with questions involving the fair payoff for a gambler who is forced to quit in the middle of a game. In modern language, they were calculating the "expected value" of the game's payoff (the average payoff under the various possible outcomes, weighted according to the likelihood of those outcomes). A bet is said to be "fair" if the price of placing it is equal to the expected value of the payoff. Betting plays an integral part in our modern society. People place bets in casinos and at sporting events, as well as by buying lottery tickets. They are also placing bets when purchasing insurance or investing in the stock market. Some of these bets are fair, some are unfair, and some cannot be objectively categorized.

The primary problem that Pascal and Fermat solved (each employing a different method) can be used to illustrate some important ideas on fairness. In the problem, two gamblers are playing a game in which a coin is repeatedly tossed. The game is interrupted at a point where 2 more heads are required for Player A to win and 3 more tails are required for Player B to win (whichever occurs first). How should the potential winnings be divided at this stage of the game?

Fermat solved the problem by observing that at most 4 tosses remain in order to identify the winner, and that there are 16 equally likely ways in which 4 tosses could occur:

HHHH, HHHT, HHTH, HTHH, THHH, HHTT, HTHT, HTTH, THHT, THTH, TTHH, HTTT, THTT, TTHT, TTTH, and TTTT.

In 11 of these possibilities (the first 11 items on the list), Player A would win, because 2 heads occur before 3 tails; in the other 5 possibilities, Player B would win, because 3 tails occur first. Therefore, Fermat reasoned that Player A should receive 11/16 of the winnings, and Player B should receive 5/16 of the winnings. In modern language, Player A would win the game with probability 11/16 and Player B would win with probability 5/16; Fermat was calculating the "expected value" of the winnings for each player.

Suppose that up to this point in their game, neither Player A nor B has paid any money for the opportunity to play, but that they are now required to pay a total of $1, altogether, and that this dollar will constitute the winnings. Fermat's solution to the previous problem allows for a fair method of dividing the payment: Player A should pay 11/16 of the dollar and Player B should pay 5/16, so that the payments match the expected winnings. In other words, if the game is being played for a $1 payoff, then the price for a fair bet is 11/16 of a dollar for Player A and 5/16 of a dollar for Player B.

Lotteries and Casinos

State-run lotteries are unfair to the player who purchases a ticket, because some of the revenue goes to the state and is not redistributed to the winner(s). Of course, even if all of the ticket revenue were paid to the winner(s)—so that the bets were fair—a lottery would be unfavorable to almost every player. Nonetheless, lotteries attract large numbers of players because people are willing to pay a small amount for the minuscule chance of winning a fortune.

A similar motivation attracts bettors to casinos, where almost all games are unfair. This casino advantage is known as the "house edge." On average, the house edge at a casino is 2% to 3%, which means that for each dollar that is bet, the house makes a profit of 2 or 3 cents. Over thousands of bets, this adds up to a significant profit. Some games, like slot machines, can have a house edge of up to 15%. Typically (in roulette, slot machines, and craps, for instance), the odds for each bet are slightly in favor of the house. Blackjack is a rare example of a casino game in which a player might be able to place bets that are better than fair from the player's perspective. In blackjack, two initial cards are dealt to each player as well as to the dealer. Certain strict rules dictate whether additional cards are dealt to the dealer, while each player has the choice of whether to receive additional cards. The objective of each player is to hold a total card value closer to 21 than the dealer holds, without going over. Each player knows which cards he or she holds, as well as some of the cards held by the dealer and other players, since some cards are dealt face up. An adept player can also keep track of cards that have been used in previous games following the last shuffle—though casinos often dissuade such card counting by combining several decks and shuffling regularly. By using this information, it is possible for a player to calculate the probability of drawing a particular card and, therefore, the expected value of the payoffs under the options of either receiving an additional card or not; often, one of these expected values is greater than the amount of the bet.

Subjective Probabilities

Early in the twentieth century, mathematicians realized the need to define probability in a rigorous way, if it were to be a formal part of mathematics. In problems involving tossing fair dice or coins, or counting card hands, it was obvious what should constitute the probabilities of the various occurrences, but in many other situations it was unclear. Usually people thought of probabilities as idealized frequencies: if a fair coin is tossed many times, for example, then the fraction of tosses which land heads should be approximately 1/2; so a fair bet for a $1 payoff on heads should cost $0.50. But there is not an obvious analogy for two boxers, for example, about to fight a match. Also, probability was becoming an increasingly important tool for the physical sciences, and mathematical theorems were required. As such, an axiomatic system was necessary. The Russian mathematician A. N. Kolmogorov and the Italian philosopher and mathematician Bruno de Finetti independently provided such a framework in the 1930s. Although different in appearance, their definitions are equivalent in most situations.

De Finetti's concept of a probability stems from gambling: the probability of an event is the price for a $1 payoff bet on that event. These prices may be assigned in whatever way one wants (hence the label "subjective probabilities"), provided certain consistency conditions are met. For example, suppose even money is coming into a betting house on two teams preparing to play a baseball game. This indicates that the bettors collectively value the two teams as equally likely to win the game. Ignoring the house fees, the price for a $1 payoff bet on either team is $0.50, because after the game, the entire pool of money will be redistributed to those who bet on the winning team.

Suppose, however, a particular bettor favors the home team, believing that team to have a 3/4 probability of winning the game. Then this bettor would price a $1 payoff bet on that team at $0.75; for this bettor, the $0.50 price generated by the betting pool is a bargain. From this bettor's perspective, a bet on the home team is better than fair: the price for a $1 payoff bet is $0.50, but the expected value of the payoff is $0.75. Such situations occur beyond sporting events, perhaps most prominently in the stock market. The fact that individuals' valuations often differ from those of the collective public is the driving force behind the trading of stocks. Individuals buy stocks that they believe to be undervalued and sell stocks that they believe to be overvalued. Because they are predicting the future performance of these stocks, they are essentially placing bets that they believe to be better than fair. In 1956, John Larry Kelly, Jr., a physicist who worked at Bell Labs, formulated and described the Kelly criterion. This algorithm for determining an optimal series of investments (or bets) is based on probability and economic utility theory, which tries to mathematically quantify satisfaction. In recent years, the Kelly criterion has been incorporated into many mainstream investment theories and betting strategies.

Further Reading

David, F. N. *Games, Gods and Gambling: A History of Probability and Statistical Ideas*. New York: Dover Publications, 1998.

Devlin, Keith. *The Unfinished Game: Pascal, Fermat, and the Seventeenth-Century Letter That Made the World Modern*. New York: Basic Books, 2008.

Epstein, Richard A. *The Theory of Gambling and Statistical Logic*. 2nd ed. San Diego, CA: Academic Press, 2009.

Packel, Edward. *The Mathematics of Games and Gambling*. 2nd Ed. Washington, DC: The Mathematical Association of America, 2006.

Von Plato, Jan. *Creating Modern Probability*. New York: Cambridge University Press, 1994.

John Beam

See Also: Dice Games; Expected Values; Probability.

Bicycles

Category: Travel and Transportation.
Fields of Study: Algebra, Geometry.
Summary: Bicycle geometry impacts performance, aerodynamics, efficiency, and stability.

The first bicycles of the early nineteenth century were simple designs of wooden frames and metal hoops for wheels. Though these early bicycles were propelled by feet pushing along the ground, soon pedals were added to the front axle allowing the rider to drive the front wheel for locomotion. It was not until the late 1880s when the first chain-driven bicycle was introduced, thereby separating the axles from the primary point of locomotion and overcoming problems with handling, steering, and weight distribution. This explosive decade of development also saw the first pneumatic tires, gearing, and coaster brakes, the latter allowing the rider to brake by pedaling backwards. Another series of innovations a century later was spurred by an explosion in frame design and fabrication techniques including the use of better materials such as aluminum, titanium, and, eventually, carbon fiber.

Bicycles serve as the primary means of transportation in several cultures, especially in southeast Asia. European communities are also known for embracing the bicycle as a legitimate form of transportation.

Mechanics

Bicycles have two in-line wheels and are driven by pedaling. The wheels each spin on axles rotating on bearing surfaces and most commonly support the rims via tension spokes. Pneumatic tires are secured to the outer surface of the rims to provide the primary contact with the ground. The centrally located bottom bracket is the rotating connection point of the pedals. Power is transferred to the rear wheel via a chain. Brakes are usually found on both wheels; most bicycles' brakes squeeze braking pads on the rim surface to create friction and slow the wheel and, as a result, the bicycle. Many newer mountain bicycles use disc brakes for increased stopping power. The rider sits on a saddle atop the bicycle and leans forward on handle bars, which provide support and the ability to steer. Many bicycles, especially mountain bicycles, have shock absorbers built into the front fork to provide cushioning over rough terrain. Some bicycles also feature rear suspension, which allows the rear triangle of the frame to rotate and further absorb the impacts of uneven terrain.

Gears (chain rings on the bottom bracket, a cassette on the rear axle) allow the rider to alter the ratio of pedal rotation to wheel rotation in order to go faster or slower. The gear ratio is determined by the diameter of the chain ring divided by the diameter of the rear cog. Since the number of teeth is proportional to diameter, tooth count is more typically used. For example, a 39-tooth chain ring used with a 15-tooth cog produces a gear ratio of

$$\frac{39}{15} = 2.6$$

that is, one revolution of the pedals produces 2.6 revolutions of the rear wheel. A standard 700C wheel (70 centimeters in diameter) will travel $0.7\pi = 2.2$ meters (7.2 feet) along the road with each revolution. Thus, a single rotation of the pedals produces

$$\frac{39}{15}(0.7)\pi = 5.7 \text{ meters } (18.7 \text{ feet}) \text{ of travel.}$$

Speed and distance traveled can then be calculated based upon the rider's revolutions per minute.

More than 10,000 riders have participated in the Tour de France since it began in 1903. Out of that group, only around 6,000 have been able to complete the grueling race that averages about 2200 miles.

Types of Bicycles

A bike with square wheels demonstrated at Macalester College.

Reflecting their wide versatility, bicycles come in a multitude of different styles. One of the most common is the road bicycle, which is distinguished by thin tires; a drop-style handlebar; and a stiff, light frame. Road bicycles are designed for fast travel over smoother road surfaces. The other most common bicycle is the mountain bicycle, which features wide, knobby tires designed for increased traction in the dirt; flat handlebars for a more upright position; and a wide range of gears, including very low gears for steep climbing. Most mountain bicycles have a front suspension fork and many feature a rear suspension as well.

Cyclocross bikes are closely related to road bikes but have slightly wider tires and lower gears for racing on cyclocross race courses or for exploring gravel roads. Comfort bicycles, commuters, and hybrids are usually compromises between the stiffness of a road bicycle and the comfort of a mountain bicycle; these bicycles' lower prices are often aimed at entry-level riders who are seeking practicality over high performance. Bicycle motocross (BMX) bicycles are single speed (no gears) with smaller, wider tires designed for racing on BMX courses. There are additional niche bicycles for special purposes such as time trialing, track racing, snow riding, and touring. Though most people cannot imagine a bicycle having anything but circular wheels, since that shape travels smoothly on flat roads, mathematicians have modeled as well as built wheels with other shapes, such as squares, three-leaf clovers, star-like shapes, and triangles. They found that a square-wheeled bike will travel smoothly on a road made of inverted catenaries, and each of the other types has at least one solution as well. A differential equation can be used to generally solve the problem of noncircular wheels.

Racing and Performance

Bicycle racing is a popular sport with a surprisingly active history. Near the end of the nineteenth century, bicycle racing was one of the most popular sports, drawing huge crowds of spectators across Europe and the United States. Today, bicycle racing is popular worldwide but has a stronger European following. Why certain cyclists are more successful than others can be analyzed in part using mathematics. Average riding speed, efficiency, and power are all calculated metrics useful for assessing performance. Seven-time Tour de France winner Lance Armstrong has been studied and modeled extensively throughout his career. American cyclist Greg LeMond overcame a 58-second deficit and won the 1989 Tour de France by 8 seconds over French favorite Laurent Fignon, which is generally attributed by most to the innovative aerodynamic handlebars he used in the last stage. Companies now routinely use mathematical modeling for cycling equipment, as well as to test aerodynamics and other essential properties, and teams use optimization strategies to construct bicycles within the sport's guidelines, since seconds can make the difference between victory and second place.

For the average rider as well as for professionals, the geometry of a bicycle plays a large role in its overall performance and stability. For example, the distance between the axles and the angle the front fork makes with respect to the ground are both important, according to bicycle makers. Some mathematicians have explored stability issues. In study released in 2007, researchers investigated and dynamically modeled 25 parameters believed to be important, with the goal of being able to construct bicycles targeted toward riders' specific needs.

Further Reading

Burke, Ed. *High-Tech Cycling.* 2nd ed. Champaign, IL: Human Kinetics, 2003.

Hall, Leon, and Stan Wagon. "Roads and Wheels." *Mathematics Magazine* 65, no. 5 (1992).

Herlihy, David. *Bicycle: The History*. New Haven, CT: Yale University Press. 2006.

McGann, Bill. *The Story of the Tour de France.* Indianapolis, IN: Dog Ear Publishing, 2006.

Peveler, Will. *The Complete Book of Road Cycling and Racing.* New York: McGraw-Hill, 2008.

Matt Kretchmar

See Also: Extreme Sports; Perimeter and Circumference; Pi; Wheel.

Billiards

Category: Games, Sport, and Recreation.
Fields of Study: Geometry; Number and Operations.
Summary: Playing billiards depends on an understanding of spin, momentum, and angles.

Billiards is a cue sport game that involves the use of a rectangular table, billiard balls, and a stick called a "cue." Mathematics and physics are two important components of playing the game well. There are many different games within the cue sports that Americans typically name "billiards." Billiard tables with pockets comprise games that are termed as "pool" or pocket billiards. The rectangular table has two long sides (twice the short side) and two short sides with six pockets—one at each corner, and one midway along the longer two sides of the table. The object of the game is to hit the billiard balls into the pockets using a cue ball (the lone white ball in the set). Gaspard Coriolis, known today for the Coriolis effect, wrote a work on the mathematics and physics of billiards in 1835. He stated that the curved path followed by the cue ball after striking another ball is always parabolic because of top or bottom spin. Further, the maximum side spin on a cue ball is achieved by striking it half a radius off-center with the tip of the cue.

The game of billiards is also a source for interesting mathematical problems, which are connected to dynamical systems, ergodic theory, geometry, physics, and optics. In mathematical billiards, the angle of incidence is the same as the angle of reflection for a point mass on a frictionless domain with a boundary. The dynamics depend on the starting position, angle, and geometry of the boundary and the table. Mathematicians investigate the motion and the path of the ball on a variety of differently shaped flat and curved tables, like triangular or elliptical boundaries or hyperbolic tables. In 1890, mathematician Charles Dodgson, better known as *Alice in Wonderland* author Lewis Carroll, published rules for circular billiards and may have also had a table built. In 2007, mathematician Alex Eskin won the Research Prize from the Clay Mathematics Institute for his work on rational billiards and geometric group theory.

Billiard players can use transformational geometry to try to hit the ball so that it will return to a pocket.

Eight Ball

Eight ball is the pool game most commonly played in the United States, and it involves 16 billiard balls. To begin the game of Eight Ball, the numbered balls are placed in a triangular rack that sets the 8-ball in the middle position of the third row of balls with a single lead ball opposite the cue ball. The cue ball is placed on the midpoint of the line parallel to the short side at one-quarter of the long side known as the head spot. The point of the triangular-shaped racked set of billiard balls is placed on the opposite short end at one-quarter of the length of the long side from the other short side and is known as the foot spot. After one player "breaks" by hitting the cue ball from the head spot into the racked set of balls, the player then hits a set of balls into the pockets. A shot that does not cause a ball of his or her set to go into the pocket results in the next shot going to the other player.

Billiards Geometry and Physics

Shooting the balls into the pockets requires an understanding of angles and momentum, as well as placement of the cue so that the correct spin is achieved to place the cue ball where it can achieve the target ball going into a pocket. Coriolis investigated 90-degree and 30-degree rules of various shots and measured the largest deflection angle the cue ball can experience. Both skill and geometric understanding contribute to successful shots. Some shots require straight shooting; some shots need to be "banked" in by using the table sides. Players can use transformational geometry to approximate where on the table to hit the ball for it to return to a pocket. By measuring the angle from the ball to the side being used to bank off and reflecting the same angle with the cue stick, one can see the most viable spot to aim for so that the path of the caromed ball ends in a pocket. Using the diamonds found on the sides of most tables is one way of measuring these angles, and some systems for pool and billiards play use the diamonds. Using the diamond system for a different billiard game, Three Cushion Billiards is demonstrated on the 1959 Donald Duck Disney cartoon *Donald in Mathmagic Land*. The demonstration shows that it is possible to use subtraction to know where to aim the ball in relation to a diamond to make sure that all three balls are hit.

Further Reading

Alciatore, David. "The Amazing World of Billiards Physics." http://billiards.colostate.edu/physics/Alciatore_SCIAM_article_posted_version.pdf.

Tabachnikov, Serge. *Geometry and Billiards*. Providence, RI: American Mathematical Society, 2005.

Linda Hutchison

See Also: Geometry in Society; Mathematical Modeling; Movies, Mathematics in; Surfaces; Transformations.

Binomial Theorem

Category: History and Development of Curricular Concepts.

Fields of Study: Algebra; Communication; Connections; Number and Operations.

Summary: The binomial theorem is the basis of Pascal's Triangle and is used to solve a variety of problems.

A binomial is an algebraic expression with two terms, like $x+y$. When binomials are multiplied together, they produce higher powers of the individual terms that are called "binomial coefficients." The binomial theorem states that for any real numbers x and y, and whole number n:

$$(x+y)^n = c_0x^n y^0 + c_1x^{n-1}y^1 + c_2x^{n-2}y^2 + \cdots + c_n x^0 y^n$$

where c_k is the binomial coefficient

$$\frac{n!}{k!(n-k)!}$$

and $n!$ is the product of the numbers 1 through n. These coefficients are the entries in what is referred to as Pascal's Triangle, named for mathematician Blaise Pascal. Students typically encounter this theorem in middle school or high school algebra, and in high school or college calculus. It also has uses in other areas of mathematics, such as in combinatorics, where it helps in calculations for certain counting problems.

The binomial theorem is found across ages and cultures. It appears in the ancient world in the work of Greek mathematician Euclid of Alexandria. His formula was for the square $(n=2)$ of a binomial, but it was described geometrically rather than algebraically. There is also evidence that the Hindu scholar Aryabhata knew the theorem for cubes in the sixth century. At least as early as the eleventh century, Chinese mathematicians such as Jia Xian and later Zhu Shijie knew the binomial coefficients in the form of Pascal's Triangle. They used the binomial theorem to find square and cube roots, and evidence suggests they knew of the binomial theorem for large values of n. Around the fifteenth century, the binomial theorem and binomial coefficients to at least the seventh power were found in the writings of Islamic scholars including Omar Khayyam, Abu Bekr ibn Muhammad ibn al-Husayn Al-Karaji, Abu Ali al-Hasan ibn al-Haytham (Alhazen), and Ibn Yahya al-Maghribi al-Samaw'al.

In the sixteenth century, European mathematicians began using the binomial theorem and binomial coefficients. For example, mathematician Michael Stifel's 1544 work *Arithmetica integra* contained the binomial coefficients. Other contributors include François Viète, Blaise Pascal, James Gregory, Sir Isaac Newton, and Niels Abel. John Wallace's seventeenth century book *De Algebra Tractatus* is cited as the first published account of Newton's binomial work. While Pascal was not the first to study the binomial coefficients, he is credited with linking algebraic and combinatorial interpretations of the coefficients.

Pascal's Triangle is a triangular representation of the binomial coefficients, which may be attributed to him because of his 1653 work *Traité du Triangle Arithmétique* in which he compiled and expounded on much of what was known about binomial coefficients. The related Pascal matrix is a symmetric, positive definite matrix with the Pascal triangle represented on its antidiagonals.

Generalizations and Extensions

James Gregory and Sir Isaac Newton generalized the binomial theorem to allow first fractional, and then real powers, which requires replacing the finite sum with an infinite series and extending the definition of the binomial coefficients. When generalizing to an infinite series, another issue that must be considered is convergence, which imposes restrictions on the numbers x and y for which the series converges to the binomial. Newton came to this generalization indirectly while trying to calculate areas under certain curves.

Another way to generalize the binomial theorem is to broaden the types of values that x and y can take. One such generalization allows x, y, and n to be complex numbers. The definition of the binomial coefficients has to be generalized to complex numbers, and certain restrictions on the variables are required for convergence of the resulting infinite series. Alternatively, one can allow x and y to be commuting elements of a Banach algebra, a normed algebra studied in such fields as complex analysis, real analysis, and functional analysis. Banach algebra is named for twentieth century mathematician Stefan Banach.

One more type of generalization considers not just the sum of two numbers x and y, but sums with more terms. Such a sum would be called a multinomial, and the multinomial coefficients would be appropriate generalizations of the binomial coefficients. Pascal's Pyramid or Pascal's Simplex are extensions of Pascal's Triangle for three or more dimensions.

Applications

The binomial theorem gives a quick way of expanding a power of the form $(x+y)^n$, making the formula useful for basic algebraic calculations. The binomial theorem, along with De Moivre's formula, can be used to prove the trigonometric double-angle identities, as well as more general formulas for $\cos(nx)$ and $\sin(nx)$. The mathematical constant e also can be written as the infinite limit of

$$\left(1+\frac{1}{n}\right)^n.$$

Mathematical induction and the binomial theorem, or the multinomial theorem, can be used to prove what is known as "Fermat's little theorem," named for mathematician Pierre de Fermat. This result in number theory states that if p is a prime number and n is an integer not divisible by p, then $n^p - n$ is divisible

Probability Theory

In probability theory, the binomial distribution uses binomial coefficients in the computation of probabilities. A binomial distribution is used to model a situation or process in which a series of independent trials occurs. Each trial may have only one of two possible outcomes, traditionally labeled "success" and "failure." In each trial, the chance of success or failure is constant, such as flipping a fair coin and getting a head, or rolling a fair die and getting a 6.

In this context, the binomial coefficient indicates the number of permutations there may be of a specific number of successes in a given number of trials; for example, the orderings of 2 heads and 8 tails in a series of 10 coin tosses. Mathematicians such as Jacob Bernoulli, Abraham de Moivre, Pierre de Laplace, Simeon Poisson, and Pascal worked on the binomial distribution and extensions, such as the limiting Poisson distribution. Mathematician and statistician Samuel Wilks, as well as others, developed the multinomial distribution to extend the binomial to cases with more than two possible outcomes on each trial.

by p. Fermat's little theorem is itself used in cryptography, providing an indirect application of the binomial theorem. One theorem in graph theory states that a graph with n vertices and adjacency matrix A is connected if and only if all the entries in the matrix $(1+A)^{n-1}$ are positive. This theorem is proved using the binomial theorem, generalized to certain matrices, and some basic graph theory results. Certain colorings of Pascal's Triangle produce fractal figures like Sierpinski's Triangle, named for mathematician Waclaw Sierpinski. In set theory, the regions of a Venn diagram for n distinct sets are in one-to-one correspondence with the binomial coefficients c_k for k ranging from 0 to n. Venn diagrams are named for John Venn.

Further Reading

"The Binomial Theorem." In Math. http://www.intmath.com/series-binomial-theorem/4-binomial-theorem.php.

Chauvenet, William. *Binomial Theorem and Logarithms.* Self-published: Biblio Bazaar, 2008.

Coolridge, J. L. "The Story of the Binomial Theorem." *American Mathematical Monthly* 56, no. 3 (March 1949).

Friedberg, Stephen H. "Applications of the Binomial Theorem." *International Journal of Mathematical Education in Science and Technology* 29, no. 3 (1998).

Fulton, C. M. "Classroom Notes: A Simple Proof of the Binomial Theorem." *American Mathematical Monthly* 59, no. 4 (1952).

Vesta Coufal

See Also: Arabic/Islamic Mathematics; Chinese Mathematics; Cubes and Cube Roots; Permutations and Combinations; Sequences and Series.

Birthday Problem

Category: Friendship, Romance, and Religion.
Fields of Study: Algebra; Data Analysis and Probability; Measurement; Number and Operations.

Summary: The Birthday Problem is a classic example of how probability can reveal counterintuitive truths.

The Birthday Problem is a classic probability problem first presented by mathematician and scientist Richard von Mises in 1939, though the fundamental combinatorial concepts involved can be traced back as far as India in the sixth century B.C.E. Today, it is one of the most-explored problems in classrooms, often introduced as early as the middle grades. The problem asks: Given that there is some number of people (n) in a room, what is the probability that at least two of them share the same birthday? One of the aspects that makes this problem so intriguing is that the answer is much different than people intuitively expect.

Solving the Birthday Problem

The extreme cases of the problem are easy to determine logically. If there are fewer than two people, then it is impossible to have two who share the same birthday, making the probability 0. If there are more people than days in a year, then at least two people must share the same birthday, making the probability 1. Von Mises assumed a fixed 365 days per year, ignoring February 29 as a possible birthday, so the probability is always 1 if there are 366 or more people. More interesting and challenging are the cases for which there are anywhere from 2 to 364 people. For the purposes of modeling and computation, it is assumed that it is equally likely that someone will be born on one day of the year versus another, so there is a

$$\frac{1}{365} \text{ chance}$$

that a person will be born on any particular day.

The Birthday Problem is solved using the mathematical ideas of permutations and combinations, and it is more easily approached if one asks a slightly different but complementary question: What is the probability that everyone in the room has a unique birthday? That is, that no one shares. If there are two people in the room, the first can be born on any of 365 days of the year and the second must be born on any of the remaining 364 days. If there are three people in the room where the first is born on a particular day, the second must be born on one of the remaining 364 days of the year and the third on one of the remaining 363 days. The probability is

$$\frac{365}{365}\times\frac{364}{365}\times\frac{363}{365}.$$

If there are four people, the probability is

$$1\times\frac{364}{365}\times\frac{363}{365}\times\frac{362}{365}.$$

This pattern can be generalized as

$$P_{(\text{no match})}(n)=\frac{{}_{365}P_n}{365^n}.$$

The probability that at least two people in a room of n people share a birthday is 1 minus the probability that there is no match, which can be used to generate the following probabilities.

Number of people in the room	Probability that two of these people share a birthday
2	.00274
10	.117
20	.411
23	.507
30	.706
50	.970
60	.995

When there are 23 people in the room, there is slightly more than a 50% chance that two people will share the same birthday, which answers the original question. The probability of at least one match increases quickly and nonlinearly with the number of people, so that when the number reaches 60 (well below the certainty value of 366 people), there is a 99.5% chance that there will be a match—*almost* certain. For example, as of 2010, there are six pairs of men who share a birthday among the 74 unique winners of the Academy Award for Best Actor. For women, there are three pairs among the 69 unique Best Actress winners.

Applications of the Birthday Problem

Applications of the Birthday Problem exist in many fields. One is called Class Phenotype Probability. Given six characteristics (blood type, RH positive/negative, sex, mid-digital hair positive/negative, earlobes attached/unattached, and PTC taste receptor), it is possible to determine the probability that a particular combination exists and also the probability that two people share the same combination. This possibility is quite valuable in medicine when considering the likelihood of finding matches between donors and recipients. In computer security, a birthday attack is a computationally intensive strategy used to break encrypted digital signatures. A "collision" occurs when different sets of data yield the same cryptographic hash value, which is a function of the input data. The attack repeatedly evaluates a hash-generating function using random inputs until the output creates a collision with the true hash value it seeks to duplicate. On average, $1.2 \times \sqrt{k}$ trials are needed to get a match, where k is the number of possible outputs (for example, a 64-bit hash value has about 1.8×10^{19} outputs). The birthday attack strategy becomes much less efficient as the hash length increases.

There are interesting extensions of the Birthday Problem based on slightly altering the question or assumptions. The first comes from considering the chance that *three* or more people share a birthday (or four, or five, and so forth). The Almost-Birthday Problem expands the problem to finding at least two people whose birthdays are within one day of each other. The Movie Line Problem states that the first person in a line for a movie whose birthday matches someone in front of them wins free tickets, and it seeks to find where someone should stand to have the best chance of winning. The Goldberg Extension computes the expected number of different birthdays in a group, while the Tuesday Birthday Problem is given as, "I have two children, one of whom is a boy born on a Tuesday. What is the probability that my other child is a boy?" Other variations assume unequal distributions of birthdays throughout the year. As with the original problem, solutions usually run contrary to most people's intuition. The ideas provide the basis for many applied investigations, such as the photon behavior modeling done by mathematical physicist Satyendra Nath Bose, after whom the subatomic particle "boson" is named.

Further Reading

Borja, Mario Cortina, and John Haigh. "The Birthday Problem." *Significance* 4, no. 3 (2007).

Goldberg, Samuel. "A Direct Attack on a Birthday Problem." *Mathematics Magazine* 49, no. 5 (1976).

Mosteller, Frederick. *Fifty Challenging Problems in Probability With Solutions*. New York: Dover Publications, 1987.

LIDIA GONZALEZ

See Also: Bar Codes; Coding and Encryption; Permutations and Combinations; Probability; Statistics Education.

Black Holes

Category: Space, Time, and Distance.
Fields of Study: Geometry; Measurement; Number and Operations; Representations.
Summary: Black holes were implied by Einstein's general relativity and have challenged physicists' theories since.

A black hole is a finite region of space during a period of time (called space-time) subject to a singularity caused by a large concentration of mass in its interior. This massive object generates a gravitational field so powerful that atoms are compacted in super-high densities, which in turn increases the gravitational pull. A singularity is created in space because no particle of matter, not even light photons, can escape from that region. Hence the name: a black hole is an invisible region because it does not reflect any light (all light is absorbed). Many aspects of black holes can be described and studied using algebraic and geometric concepts, but the existence of black holes is still under debate. For example, Australian mathematician Stephen Crothers argues that black holes are inconsistent with general relativity and critiques the mathematics used by others to demonstrate their existence. It is believed that black holes originate when stars runs out of gas needed to maintain their temperature, causing a decrease in volume. As volume decreases, the proximity of particles increases the gravitational pull in a positive feedback loop; as particles get closer, the gravitational force keeps increasing. This compaction process continues until a singularity, called the "event horizon," is created. The event horizon is defined as a boundary in space and time beyond which events cannot affect an outside observer. The event horizon separates the black hole region from the rest of the universe and is the boundary of space from which no particle can leave, including light.

The singularity caused by a black hole is considered as a curvature in space-time. This curvature is explored by Albert Einstein's general relativity theory, which predicted the existence of black holes—though Einstein himself did not believe in them. In the 1970s, Stephen Hawking, George Ellis, and Roger Penrose proved several important theorems on the occurrence and geometry of black holes. Previously, in 1963, Roy Kerr had shown that black holes in a space-time have an almost-spherical geometry determined by three parameters: their mass, their total electric charge, and angular momentum.

It is believed that at the center of most galaxies, including the Milky Way, there are supermassive black holes. The existence of black holes is supported by astronomical observations, in particular through the emission of X-rays. Some black hole candidates have been identified experimentally using observations and data. There are different types of black holes, such as rotating black holes and stationary black holes, and these are described by using various metrics in physics and differential geometry.

Origins of Human Awareness of Black Holes

The concept of a body so dense that even light could not escape was described in a paper submitted in 1783 to the Royal Society by an English geologist named John Michell. By then, Isaac Newton's theory of gravitation and the concept of escape velocity were well known. Michell computed that a body with a radius 500 times that of the sun and the same density, would, on its surface, have an escape velocity equal to that of light and would therefore be invisible.

In 1796, the French mathematician Pierre-Simon Laplace explained in the first two editions of his book *Exposition du Système du Monde* the same idea; however, the concept that light was a wave without mass and therefore unaffected by gravitation was prevalent in the nineteenth century, and Laplace discarded the idea in later editions.

In 1915, Einstein developed his general relativity theory, and showed that light was influenced by the gravitational interaction. A few months later, Karl Schwarzschild found a solution to Einstein's equations, where a heavy body would absorb the light. We now know that the Schwarzschild radius is the radius of the event horizon of a black hole that will not turn, but this was not well understood at the time. Schwarzschild himself thought it was just a mathematical solution, not physical. In 1930, Subrahmanyan Chandrasekhar showed that any star with a critical mass (now known as the Chandrasekhar limit) and that does not emit radiation would collapse under its own gravity. However, Arthur Eddington opposed the idea that the star would reach a size zero, implying a naked singularity of matter; instead, the black hole should have something that will inevitably put a stop to collapse, an idea adopted by most scientists.

In 1939, Robert Oppenheimer predicted that a massive star could suffer a gravitational collapse and therefore black holes might be formed in nature. This theory did not receive much attention until the 1960s because after World War II he was more interested in what was happening at the atomic scale.

In 1967, Stephen Hawking and Roger Penrose proved that black holes are solutions to Einstein's equations and that in certain cases the creation of a black hole is the inevitable consequence of a star aging. The black hole idea gained force with the scientific and experimental advances that led to the discovery of pulsars. Soon after, in 1969, John Wheeler coined the term "black hole" during a meeting of cosmologists in New York, to designate what was formerly called "star in gravitational collapse completely."

The Entropy of Black Holes

The mathematical tools used to model black holes use fundamental laws of physics, particularly relativity and thermodynamics. According to initial theories by Stephen Hawking, black holes violate the second law of thermodynamics (the entropy, or disorder, of isolated systems tend to increase over time), which led to speculations about travel in space-time wormholes (tunnels that would allow time travel or fast travel over very long distances). Hawking has recanted his original theory and has admitted that the entropy of the matter is kept inside a black hole. According to Hawking, despite the physical impossibility of escape from a black hole, it may end up evaporating by constant leakage of X-ray energy that escapes the event horizon, called Hawking radiation.

An artist's concept chronicles a time lapse from left to right of an intact sun-like star (left) coming too close to a black hole (right) and its self-gravity becoming overwhelmed by a black hole's gravity.

According to this model, black holes have intrinsic gravitational entropy, which implies that gravity introduces an additional level of unpredictability over the quantum uncertainty. It appears, based on the current theoretical and experimental capacity, as if nature took decisions by chance or, more generally, far from precise laws.

The hypothesis that a black hole contains entropy and, furthermore, it is finite, required to be consistent with such holes emitting thermal radiation, at first seems contradictory. The explanation is that the radiation escapes the black hole in such a way that an external observer knows only the mass, angular momentum, and electric charge. This means that all combinations or configurations of radiation of particles having energy, angular momentum, and electric charge are equally likely. Physicists such as Jacob D. Bekenstein have been linked to black hole entropy and information theory.

Further Reading

Hawking, Stephen. *A Brief History of Time.* New York: Bantam Books, 1998.

———. *The Universe in a Nutshell.* New York: Bantam Books, 2001.

Poisson, Eric. *A Relativist's Toolkit: The Mathematics of Black-Hole Mechanics.* Cambridge, England: Cambridge University Press, 2007.

Sagan, Carl. *Cosmos.* New York: Random House, 2002.

Juan B. Gutierrez

See Also: Astronomy; Einstein, Albert; Temperature.

Blackmun, Harry A.

Category: Government, Politics, and History.
Field of Study: Connections.
Summary: Harry A. Blackmun was a U.S. Supreme Court justice who applied mathematical logic in his judicial career.

Harry A. Blackmun (1908–1999) is best known as the author of the majority opinion in *Roe v. Wade*, the U.S. Supreme Court case that recognized a constitutional right to abortion. Blackmun, however, was perhaps the only Supreme Court justice to hold a degree in mathematics (A.B., Harvard, 1929), and one of the very few who has carefully applied mathematical concepts in judicial opinions.

Blackmun graduated from Harvard Law School in 1932 and served as law clerk to Judge John B. Sanborn of the U.S. Court of Appeals for the Eighth Circuit. His law practice, with a prominent Minneapolis law firm and the Mayo Clinic, focused primarily on tax law and estate planning. He succeeded Judge Sanborn on the Eighth Circuit in 1959 and was appointed to the U.S. Supreme Court by President Richard M. Nixon in 1970 following the Senate's rejection of two previous nominees. Both of Blackmun's judicial appointments were promoted by his childhood friend, Chief Justice Warren E. Burger. So close was this connection that for their first decade on the Supreme Court, the two men were referred to as the "Minnesota Twins," but thereafter they went separate jurisprudential ways.

Blackmun and Mathematical Reasoning

A particularly striking example of Blackmun's use of mathematical reasoning involves his application of the binomial distribution as a method for assessing discrimination claims. In *Castaneda v. Partida* (1977), a Texas county was accused of systematically discriminating against Mexican Americans in the selection of grand jurors. Mexican Americans constituted nearly 80% of the population but only 39% of the grand jurors during the 11-year period at issue. Blackmun's opinion for the Court noted the substantial absolute disparity but used the binomial distribution to explain the unlikelihood that such a disparity would have arisen by chance. Eschewing mathematical symbols, he verbally explained the calculations involved and the formula for making them. The difference between the observed and the expected number of Mexican-American grand jurors during this period was approximately 29 standard deviations, a result that would occur by chance less than once in 10^{149} times. The use of statistical evidence has since become standard practice in discrimination cases.

Justice Blackmun also relied heavily on empirical evidence in *Ballew v. Georgia* (1978), a case that established the minimum constitutionally acceptable size for juries. American juries traditionally had 12 members, but—in a case considered shortly before Blackmun took his seat—the Supreme Court held that six-member juries were permissible. *Ballew* held that a five-person jury was too small to satisfy the requirements of due process. Blackmun's opinion on this issue spoke for all of his colleagues, some of whom wrote separately on other issues. Engaging with a substantial body of statistical and experimental research, he observed that smaller juries are less likely to engage in effective deliberations, take diverse viewpoints seriously, and reflect a fair cross-section of the community than 12-person juries and that smaller juries are more likely to reach inaccurate judgments. Blackmun conceded that the differences between five- and six-person juries might be difficult to discern empirically, but he concluded that a line had to be drawn somewhere to avoid further reductions in jury size. The data actually raised troubling questions about the earlier decision upholding six-person juries, but the Court was unwilling to overrule that precedent.

Finally, Blackmun addressed statistical issues relating to the imposition of the death penalty, both on the Supreme Court and on the court of appeals. On the Eighth Circuit, in *Maxwell v. Bishop* (1968), Blackmun rejected a statistical study purporting to show that African Americans were much more likely than whites to get the death penalty in rape cases. The study was limited in scope, did not relate to the county where the case arose, failed to account for relevant variables, and did not show that racism affected the verdict or sentence in that case. The Supreme Court set aside this decision because of an intervening ruling on a different issue.

Two decades later, in *McCleskey v. Kemp* (1987), Blackmun dissented from a decision that rejected a claim of racial discrimination in a Georgia death penalty case. Blackmun explained that this claim was supported by a multiple regression analysis of every homicide case in that state during the relevant time period. The study, which included 230 variables, found that persons accused of killing whites were 4.3 times as likely to receive a death sentence as persons accused of killing African Americans and that African-American defendants were significantly more likely to be sentenced to death than white defendants. Blackmun emphasized the "sophistication and detail" of this study and concluded that it showed an unacceptable risk that racism had affected the decision to impose the death sentence.

Further Reading

Greenhouse, Linda. *Becoming Justice Blackmun: Harry Blackmun's Supreme Court Journey*. New York: Times Books, 2005.

Yarbrough, Tinsley E. *Harry A. Blackmun: The Outsider Justice*. New York: Oxford University Press, 2008.

JONATHAN L. ENTIN

See Also: Data Analysis and Probability in Society; Number and Operations in Society.

Blackwell, David

Category: Mathematics Culture and Identity.
Fields of Study: Algebra; Data Analysis and Probability; Number and Operations.
Summary: Statistician and game theorist David Blackwell (1919–2010) became one of the most esteemed African-American mathematicians.

David Harold Blackwell was one of the most famous American mathematicians. He was Professor of Statistics at the University of California at Berkley.

Early History

David Blackwell was born on April 24, 1919, in Centralia, Illinois, and was an African American. David was the oldest of four children in his family. His father worked for the Illinois Central Railroad looking after the locomotives, and his mother looked after the family. David attended the integrated elementary school rather than the existing segregated school for African Americans in Centralia. He said, "I had no sense of being discriminated against. My parents protected us from it and I didn't encounter enough of it in the schools to notice it."

Blackwell enjoyed geometry very much. In high school, he applied his mathematical skills to games. His interest in mathematics continued to grow after entering the University of Illinois, in 1935, at age 16. In a course on real analysis, he was especially interested in calculus, and he was excited by Newton's method for solving equations. The course on real analysis turned him on to a career in mathematics. He remarked upon the course, saying, "That's the first time I knew that serious mathematics was for me. It became clear that it was not simply a few things that I liked. The whole subject was just beautiful."

During his study at the university, his father had to borrow money to finance his education. David took jobs such as dishwashing to help earn money and, at the same time, he took courses over the summers and was able to graduate with a B.A. in 1938. After graduating, Blackwell continued to study at the University of Illinois for his master's degree, which he was awarded in 1939, and then for his doctorate, supervised by Joseph Doob. This was awarded in 1941 when Blackwell was only 22 years old. The dissertation title was "Some Properties of Markoff Chains." After that, Blackwell received a one-year appointment as a Rosenwald Postdoctoral Fellow at the Institute for Advanced Study in Princeton. This appointment caused some turmoil, of which he was not fully aware, because he was an African American.

Personal and Professional Life

From 1942 to 1943, he had a post at the Southern University at Baton Rouge, followed by a year as an instructor at Clark College in Atlanta. Blackwell was appointed as an instructor in 1944 at Howard University and, after only three years, he was promoted to full professor and head of the Department of Mathematics.

In 1944 he married Ann Madison, with whom he had eight children.

He left Howard University in 1954 to take up a professorship at the University of California at Berkeley, where he taught students up to his retirement, and, after that, as Professor Emeritus.

During three summers between 1948 and 1950, he worked at the Research and Development Corporation (RAND). At about the time of his arrival and the beginning of his work at Berkeley, Blackwell's interests turned toward statistics, and he became a theoretical statistician. In 1956, he became chairman of the Department of Statistics. He was one of the eponyms of the Rao–Blackwell theorem, a famous theorem in statistical theory. Thanks to him, the areas of game theory and topology were connected as well.

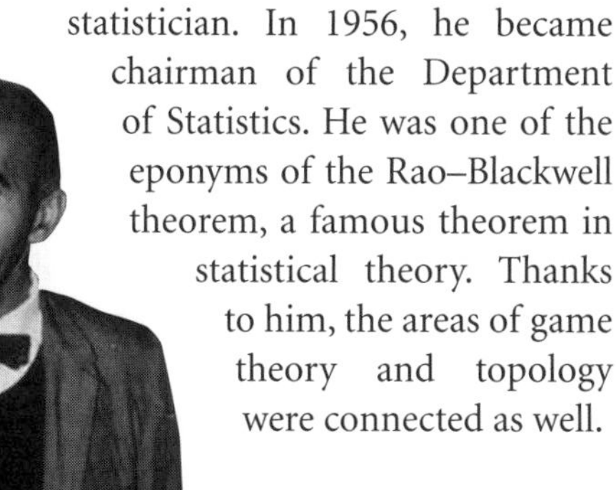

David Blackwell photographed in Seattle in 1967.

Awards and Honors

Blackwell received honorary Doctorate of Science degrees from 12 universities. He was selected to be president of the Institute of Mathematics Statistics, the International Association for Statistics in Physical Sciences, and the Bernoulli Society. He was vice president of the International Statistical Institute, the American Statistical Association, and the American Mathematical Society. He was elected to the National Academy of Science, the first African-American mathematician to do so. He was also elected to the American Academy of Arts and Sciences. He held memberships in numerous professional organizations, including being a life member of National Association of Mathematicians. He was an Honorary Fellow of the Royal Statistical Society.

David Blackwell lived in Berkeley until his death in July 2010.

Further Reading

Degroot, Morris H. "A Conversation With David Blackwell." *Statistical Science* 1, no. 1 (1986).

National Visionary Leadership Project. "David Harold Blackwell: Visionary Videos: NVLP: African American History." http://www.visionaryproject.com/blackwelldavid.

O'Connor, J. J., and E. F. Robertson. "Blackwell Biography." http://www-history.mcs.st-and.ac.uk/Biographies/Blackwell.html.

BILJANA POPOVIC

See Also: Game Theory; Probability; Statistics Education.

Board Games

Category: Games, Sport, and Recreation.
Fields of Study: Algebra; Data Analysis and Probability; Geometry.
Summary: While some games are explicitly mathematical, others are implicitly governed by math.

Humans have been playing games for as long as they have been around. Johan Huizinga was the first to call the attention to the fact that play precedes culture.

Board games, a very organized form of play, are part of human social nature. Human communities may differ in many ways, but they all play games. From the ancient Mancala, practiced for millennia in Africa, to our Monopoly, we find board games in many societies. Besides their cultural relevance—they are studied by anthropologists, historians, and others—board games are characterized by their sets of rules, which show mathematical structures and connections that are at times very surprising.

Game Classifications

Chess and Go come to mind as examples of traditional board games, and Monopoly and Scrabble are examples of proprietary games. The distinction between the two types of games is not always easy to identify. In chess, the movements of the pieces and the other rules are the main considerations. Chess is an abstract game, not considering the fact that it originally emulated a battle between two armies. Chess does have similarities with other games. When playing representational games like Monopoly or Diplomacy, players find themselves focusing on the possibilities and strategic choices, forgetting the particular settings. Accoring to David Parlett, positional games refer to games where pieces are played in a board or any other set of markings, as chess, checkers, and Go, and "theme" games are generally representational and commercial, like Monopoly and Diplomacy.

Board game classification has been inspired in the fact, first noted by H. J. R. Murray, that games are typical of early activities of man—the battle, the siege, the race, the hunt, alignment, arrangement, and counting. Parlett's classification, which evolved from Murray's and others, is as follows. In race games, the board is a linear track where each player tries to be the first to reach a particular cell or remove a set of pieces from the board. Most of the games under this category use dice or other randomizing devices, like Chutes & Ladders, Ludo, and Backgammon, but not all, such as Hare & Tortoise.

Space games, typically two-dimensional and free placing, comprise the alignment games, as Nine Men Morris; connection games, as Hex and Twixt; traversal games, in which a player tries to have one or several pieces cross the board, as Breakthrough, Halma, and Chinese Checkers; configuration games, where players try to achieve certain displays with their pieces, as Agon; restriction games, where the aim is to try to block the adversary, like Pentominoes; and occupation games, in which the winner is the player who achieves more space in the board, as in Go and Othello. Chase games are asymmetrical, one player having several pieces while the other has only one or two. Their goals are also distinct, as in Fox & Geese. "Displace games" include chess and checkers, where a player aims at capturing most of his opponent's pieces (as in checkers) or a particular one (as in chess), and other war games; the family of Mancala games belongs also to this class.

History

The Royal Game of Ur, also known as the Game of Twenty Squares, was found in the south of Iraq in the 1920s and is about 4500 years old. The board shows twenty squares, 12 in a three-by-four rectangular array, six in two rows of three, and two connecting cells. The reverse of the board corresponding to the 12 cells showed a zodiac, illustrating that in the past, the same object could be a board game and a divinatory device. Two cuneiform clay tablets give the exact rules for this game. Each player had seven pieces, which moved across the board according to the toss of three tetrahedral dice.

A similar game is found in Ancient Egypt, Senet or the Game of Thirty Squares. It was a race game as well, but it was more than a simple toy. In funerary monuments that date from 4000 years ago, images are shown of the deceased playing Senet against an invisible adversary. Osiris, which is present but not shown, decides on matters of life after death.

The Royal Game of Ur and Senet can be viewed as the oldest relatives of the modern Backgammon, a game in which the moves are decided by the players upon tossing two cubic dice. The player who better understands the probability laws that rule the dice is most often the winner.

The Chinese game Go is four millennia old. Nowadays, it remains one of the most complex games, despite the simplicity of its rules. Go is played on the intersections of a 19-by-19 grid, and each player fights to control the largest area.

Pure strategy games could also be found in Ancient Greece, like Petteia. This game, and the Roman Ludus Latrunculorum, shared the shape of the board, checkered, and the orthogonal movement of the pieces.

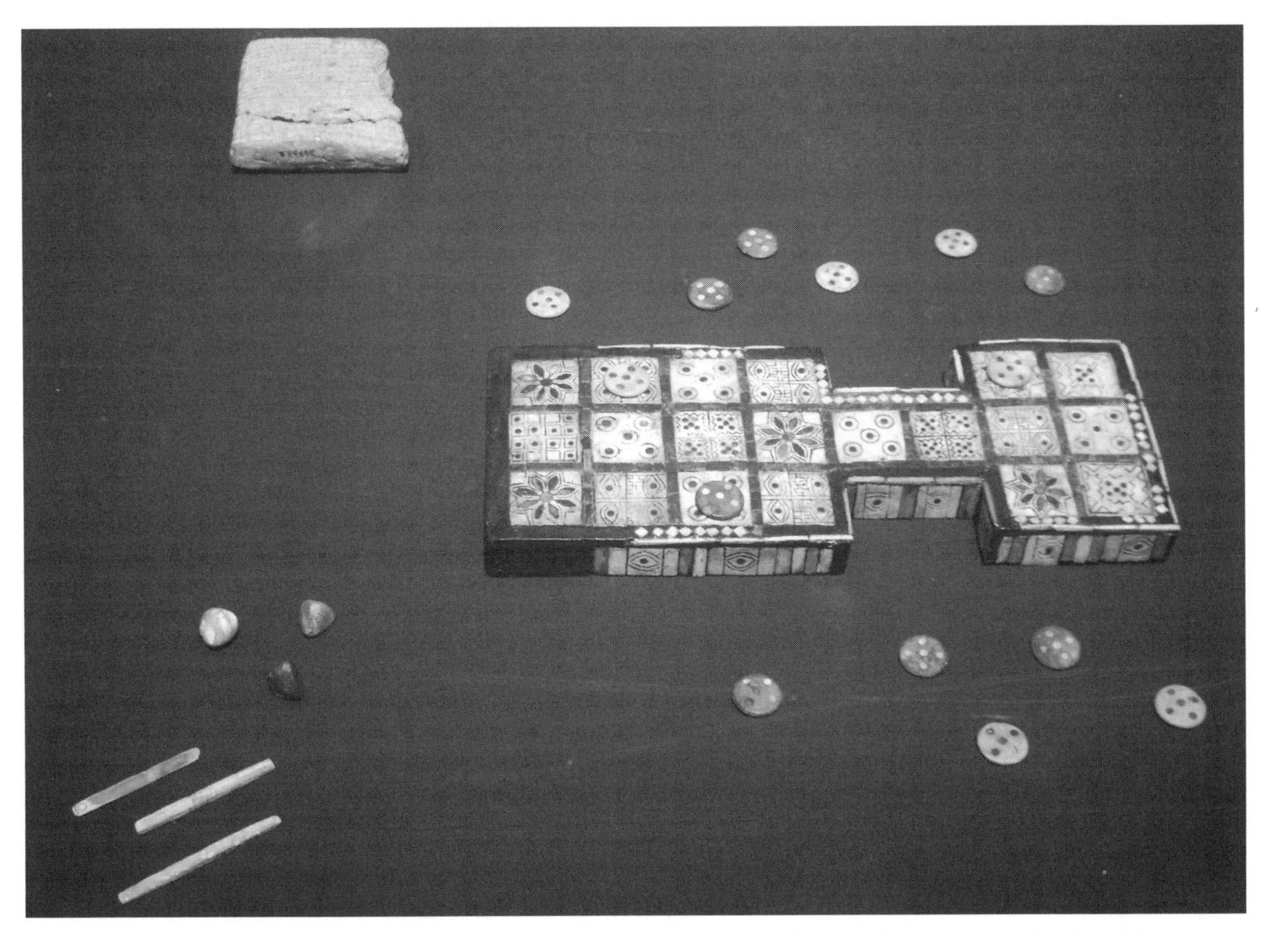

Two ancient copies of the Royal Game of Ur exist; one is exhibited in the British Museum in London (above). The game dates back to Iraq in 2600 B.C.E. and is thought to be oldest board game set ever found.

Chess, which originated in India about 1400 years ago, traveled to the West with the Arabs, and saw its rules evolve in the process. It was originally created as a war game between two armies, and its pieces represented the actors of the battle. However, the abstract shapes that reached Europe gave way to the symbolic representation of the European medieval society.

The Arabs introduced several other games in Europe. One game they introduced, Alquerque, was played on the intersections of a five-by-five lined board. The adaptation of this game to the chessboard originated the game of Checkers.

Board Games and Mathematics

The oldest known pedagogical game is Rithmomachia, also known as Philosopher's Game. It was invented in the eleventh century as a didactical device to teach mathematics. It was practiced wherever Boethius's arithmetic was taught. Pythagorean in nature, this tradition of mathematics dominated teaching at churches and universities for more than 500 years. In an eight-by-16 board, two armies fought each other. Pieces carried numbers and could have one of three shapes: circular, triangular, or square.

The movements depended on the shape of the piece played; the captures depended on the numbers and on arithmetical calculations. Victory was attained by means of a configuration of pieces holding numbers in progression (arithmetic, geometric, harmonic, or combinations of the three). This game spread throughout Europe, and only when the mathematical curriculum at universities changed in the sixteenth century did it vanish. Losing its pedagogical goal turned out to be fatal, as Rithmomachia lacked the qualities to survive as a purely recreational activity. Chinese scholars of the eleventh century also published work on permutations based on the Go board. John H. Conway's twentieth-century research on the game contributed to the invention of surreal numbers and the development of combinatorial game theory.

Ludus Astronomorum was a board game for seven players based on Ptolemaic astrological principles. In the sixteenth century, William Fulke, a professor at Cambridge who had written a manual of the Philosopher's Game, created two other games. One, intended to improve on the astronomy game, was Ouranomachia, the other, created to teach geometry, was Metromachia. Fulke published one book on each.

In the eighteenth century, George Berkeley invented a game to help teach algebra, a subject Berkeley had in very high consideration. The game was Ludus Algebraicus and essentially functioned as a randomizing device to generate algebraic equations.

Charles Dodgson invented a game in the nineteenth century to practice logical deduction and wrote a book about it, *The Game of Logic*, under his pen name, Lewis Carroll.

In Ireland, mathematician William Hamilton created in 1857 the Icosian Game and soon after Traveller's Dodecahedron. This comprised a dodecahedron and a piece of thread that should touch every vertex according to some rules. It was this game that gave rise to the concept of Hamiltonean Graph.

The familiar game of Nim in which a move consists of choosing from one of a pile of beans and reducing its cardinality, was first solved mathematically at the beginning of the previous century. In its normal form, where the winner is the one who takes the last bean, is the paradigm of a class of games studied in Combinatorial Game Theory. The familiar children's game Dots & Boxes was also treated mathematically with the same techniques. Some traditional games, like Konane, can be approached the same way.

The game Hex was invented independently by both Piet Hein and John Nash in the 1940s. It is a connection game played on a diamond-shaped board of hexagonal cells. David Gale noted that a game of Hex can never end in a tie, and that this fact is logically equivalent to a deep theorem in topology.

Abstract games with complete information and no chance devices are also called mathematical games. The mental processes present in their practice and in a typical mathematical activity, like problem solving, are far from disjointed.

Further Reading

Avedon, Elliot M., and Brian Sutton-Smith. *Study of Games*. New York: John Wiley & Sons, 1971.

Berlekamp, Elwyn R., John H. Conway, and Richard K. Guy. *Winning Ways for Your Mathematical Plays*. Natick, MA: Ak Peters, 2001.

Huizinga, Johan. *Homo Ludens*. New York: Routledge, 2008.

Murray, Harold James Ruthven. *A History of Board-Games Other Than Chess*. New York: Hacker Art Books, 1952.

Parlett, David. *Oxford History of Board Games*. New York: Oxford University Press, 1999.

Jorge Nuno Silva

See Also: Dice Games; Mathematical Puzzles; Puzzles; Video Games.

Body Mass Index

Category: Medicine and Health.
Fields of Study: Algebra; Number and Operation.
Summary: Body Mass Index is a statistically useful index of a person's relative weight.

Body mass index (BMI) is an index of the relative weight of a person. In other words, it is an estimate of a person's weight, adjusted for height. The formula for calculating BMI is

$$\text{Weight in kilograms}/(\text{Height in Meters})^2.$$

The equation implies that, holding other factors constant, weight is proportional to height squared, at each level of height. This equation is counterintuitive because of the common assumption that to calculate mass or volume, a cubic function is necessary. In fact, dividing mass by height cubed was historically a popular method, called the "Ponderal index." However, Adolphe Quetelet (1796–1874) observed that, for an "average man," a squared function was a better fit to the data. With increasing age, humans' height increases at a faster rate than width. Taller adults tend to be slimmer—waistlines do not usually grow in proportion to increasing height. Quetelet observed that people do not expand equally in all three dimensions. If they did, then the Ponderal index would still be valid. In reality, he noted that, "weight increases nearly as the square of the height"—particularly between puberty and age 25. Before puberty, the Ponderal index does increase more proportionally to width. Quetelet stressed that there was considerable variance in body shape and size, which are determined by biological, psychological, and social factors. For example, he noted that "young persons who apply themselves to study, and persons in the affluent classes generally, are taller than others."

The Quetelet Index

The popularity of the Quetelet index increased following World War II, when epidemiological evidence began to accumulate that excess weight was a risk factor for premature mortality and morbidity. Historically, and in several cultures today, excess weight (corpulence) was considered healthy and desirable. Given this new evidence, actuaries needed a quick and reliable way to predict who might be most at risk, so that insurance premiums could be loaded against those with excess weight. They creating height-weight charts, based on Quetelet's data, which provided the typical weights expected at various levels of height for the average person—assuming that they were age 25. The index was later termed "body mass index."

Insurance companies, clinicians, and researchers began using BMI as a proxy variable for measuring excess weight at all ages, not simply as an index of relative weight at age 25, as originally intended. The popularity of BMI remains today. For example, the World Health Organization uses BMI in its definition of obesity, where "overweight" is defined as BMI equal to or greater than 25, and obesity is defined as a BMI more than 30. There are no agreed BMI cutoffs for childhood obesity in the same way that there are for adults. The categorization of continuous data (for example, overweight, obese) is controversial because it results in a loss of information. BMI is indeed a risk factor for chronic diseases, despite its usage deviating from the original intended purpose. In clinical settings, BMI is usually supplemented with other information regarding disease risk, such as blood pressure or lifestyle factors including cigarette smoking. Additionally, it may be necessary to take into account whether the person has an ectomorphic, mesomorphic, or endomorphic body type.

The Quetelet index was first formally evaluated by epidemiologists working on data from a large cohort study, called the Framingham Heart Study. They noticed that Quetelet's index was being widely used as an indicator of excess weight, not simply weight adjusted for height as it was originally intended. The epidemiologists wanted to evaluate the validity of this assumption, by comparing different methods for measuring relative weight against three criteria:

1. The proxy should not correlate with height.
2. The proxy should correlate highly with skinfold thickness measurements, since these are valid proxies for the thickness of the subcutaneous fat layers in different parts of the body, in turn.
3. The proxy should be easy to calculate.

After analyzing the data, they concluded that Quetelet's index was indeed the best available measure, and renamed it the Body Mass Index. However, it should be noted that correlations between BMI and skinfold thickness measurements varied considerably, and the highest was 0.8. The researchers noted that if height and weight are the only data available, excess body fat is unlikely to be measured in a satisfactory way. The lower a correlation between a proxy variable and the variable it is intended to measure, the less well that proxy will correlate with health outcomes. It should also be noted that the third criterion (the ease of calculation) is not statistical—the validity of an index or test should be based on how well it performs against a gold standard, not simply because it is easy to use.

Criticisms

Researchers have since argued that valid proxies for excess body fat should take into account its distribution in the body. Excess fat in the abdominal region (visceral fat) is a risk factor for metabolic diseases, regardless of total fat volume in the body. Waist circumference correlates highly with visceral fat, leading some researchers to suggest that waist circumference is a better proxy for excess weight than BMI. Similar alternatives include the ratio of waist circumference to hip circumference (waist-to-hip ratio). Because waist circumference is associated with increased morbidity and mortality risk, holding BMI constant, it provides additional information that is not captured by BMI. Both are considered independent risk factors, such that it may be necessary to measure both BMI and waist circumference. In fact, a consensus statement from Shaping America's Health concluded that waist circumference predicted cardiometabolic outcomes, and should therefore be measured in clinical settings as a matter of routine. However, waist circumference is difficult to measure reliably. BMI remains a useful index for many purposes.

Further Reading

Eknoyan, Garabed. "Adolphe Quetelet (1796–1874): The Average Man and Indices of Obesity." *Nephrology Dialysis Transplantation* 23, no. 1 (2008).

Ferrera, Linda A. *Body Mass Index and Health.* Hauppauge, NY: Nova Biomedical Books, 2005.

Friedlander, Joel. *Body Types.* San Rafael, CA: Marin Bookworks, 2010.

Keys, A., F. Fidanza, M. J. Karvonen, N. Kimura, and H. Taylor. "Indices of Relative Weight and Obesity." *Journal of Chronic Diseases* 25 (July 1972).

National Institutes of Health. "Calculate Your Body Mass Index." http://www.nhlbisupport.com/bmi/.

Quetelet, Adolphe. *A Treatise on Man and the Development of His Faculties.* Edinburgh, UK: W. and R. Chambers, 1842.

Gareth Hagger-Johnson

See Also: Cubes and Cube Roots; Growth Charts; Measurement in Society; Normal Distribution; Squares and Square Roots.

Brain

Category: Medicine and Health.
Fields of study: All.
Summary: The brain is studied through models and through algorithm-dependent medical technology. The neurology of mathematical thought is a vibrant field.

The applications of mathematics to the study and understanding of the brain have been varied and widespread. They include models to predict the start of seizures using dynamical systems; maps of the brain using projective, hyperbolic, or other geometries, as well as graph theory; applications of morphometrics, which is the statistical study of shapes, to schizophrenic brains; and dynamic simulations and visualizations of electrochemical activity in neurons. Other models are used to study how electrical signals propagate along nerve cells and the way in which electrical discharges in nerve cells tend to synchronize and form waves. Medical technology used in brain treatment and studies uses mathematical algorithms; for example, to create and process computer-generated images of brain cells, as well as to measure functions like blood flow, glucose consumption, and electrical activity. Mathematics is also important in modern medical devices that involve nerve fibers within or leading to the brain, such as cochlear implants. How mathematical thought arises in the brain—from arithmetic to abstract thinking—is also of great interest. Mathematics and the Brain was the theme of Mathematics Awareness Month in 2007.

Brain Composition and Structure

Before proceeding with some applications of mathematics in the study of the brain, it is important to have an idea of brain composition and structure. In humans, this complex organ consists of perhaps 100 billion nerve cells (or neurons), with roughly a total of 100 trillion connections between neurons (or synapses). Although some nerve cells do regenerate, and new connections between nerve cells are made, overall these numbers tend to decline after birth. Even with advances in computer processing and storage, the sheer number of neurons and connections hints at the enormous scope of the problem inherent in understanding the brain. By comparison, the nematode *Caenorhabditis elegans* has 959 cells in the entire organism, 302 being nerve cells, which result in over 5000 connections between neurons. Even for something of this vastly smaller scale, the nematode's neural connections were initially mapped after more than 10 years of effort by the mid-1980s, and earned a Nobel Prize for Sydney Brenner—who famously called *C. elegans* "nature's gift to science." Those results have since been updated.

A single neuron generally consists of: a main cell body (or soma); many filamentous dendrites, which are where signals from other neurons are usually received; and a single axon, which typically communicates to the dendrites of other neurons. The electrical voltage across the neuron's cell membrane varies as the concentrations of calcium, sodium, potassium, and chloride ions fluctuate, producing a fluctuating electrical signal. The electrical signal is transferred from one neuron's axon to another's dendrite across a gap known as a "synapse." Some synapses, known as "electrical synapses," involve a direct channel that connects the two cells' cytoplasm and allows for very fast electrical transmission. By contrast, chemical synapses involve molecules known as "neurotransmitters," which mediate signal transmission. The human brain utilizes more than 100 types of neurotransmitters. However, just two of these types arise at the vast majority of synapses; namely, glutamate and gamma aminobutyric acid. In addition to neurons, glial cells serve various support functions for neurons. One important function of special kinds of glial cells—namely, special kinds of oligodendrocytes—is the myelinization of axons. Myelin, a fatty substance, essentially electrically insulates neurons. Because they are pinkish white, and white when stored in formaldehyde, bundles of myelinized axons make up what is known as "white matter" in the brain. On the other hand, "grey matter," as seen on the surface of the cerebral cortex in a typical brain slice image, comprises of the soma, dendrites, and other kinds of glial cells, such as astrocytes. While *C. Elegans* has fewer than 60 glial cells, the human brain likely has at least as many glial cells as neurons, although the ratio varies widely in different brain regions.

Applications of Neural Networks

How neurons collectively convey information is also of much interest to researchers. Interestingly, attempts to model so-called "artificial neural networks" have led to highly useful algorithms used in many areas of mathematics, science, and engineering in their own right, having nothing to do with the study of the brain. Computers are often "trained" with data sets using such neural networks to help process data. Neural networks can be found in software used in fields as varied as financial analysis and fraud detection, robotics, handwriting analysis, and voice recognition. As another example, much mathematics is used in processing and analyzing the enormous amount of neuron image data, and neural network algorithms are now being used to help automate that processing to help computers track neural connectivity.

Brain Mapping and Study

Mathematics also has been used to help in producing accurate maps of the cortex of various parts of the brain. The extensive folding in the human brain in the cerebral cortex, which produces peaks or ridges (or *gyri*) and valleys or furrows (or *sulchi*) makes it difficult to compare two different brain surfaces. A calculus-based geometry is used to find effective maps. As another example of an application, mathematics is used extensively in devices such as cochlear implants, useful to deaf individuals who still have a functioning auditory nerve. In humans, as many as 30,000 individual nerve cells in the inner ear pass through the auditory nerve to the brain. Different sound frequencies innervate different nerve cells; roughly speaking, lower frequencies innervate nerve cells in the basilar membrane closer to the beginning of the cochlea, as opposed to higher frequencies innervating cells further along. But the precise mapping of which cells are affected by which frequencies follows a logarithmic mathematical pattern, as a function of distance in the cochlea. Using various radio signal technologies, external sounds are transmitted to a receiver in the inner ear, which connects to implanted electrodes for nerve innervation. There, mathematics is used in the computational processing to convert the received frequencies into the appropriate electrical innervations, so that only certain nerve cells are stimulated for certain frequencies.

Examples of mathematics applied to the study of the brain abound in the five-year, National Institutes of Health–funded Human Connectome Project. This project, somewhat analogous to the Human Genome Project, was funded in 2010 for approximately $40 million. Mapping all the connections between neurons in the human brain in a meaningful way is the goal of the Connectome project. One component involves constructing connection data from 1200 individuals, including numerous twins. Developing effective ways to collect the data set, as well as analyze the results, involves several areas of mathematics in crucial ways. First, instruments must be able to create high-resolution images of the brain tissue of living humans in a completely noninvasive way. Next, the enormous image data must yield to automated computer analysis that can determine the actual neural connections within the brain. Finally, the connection data set must be amenable to meaningful analysis by researchers interested in understanding normal brain processing as well as diseases. At each stage, mathematics plays a crucial role.

Brain Imaging Technologies

Magnetic resonance imaging (MRI) is commonly used today for noninvasive imaging of the internal structure of the human body; for example, to help determine if knee surgery or back surgery is warranted. In standard MRIs, a powerful magnetic field changes rapidly, and, by doing so, it manipulates the minute magnetic fields produced by protons in water molecules inside the body—a weak signal can be detected externally from the protons being flipped around by the strong magnetic fields. From these weak and indirect measurements, solving the inverse problem using mathematics related to calculus is used to create what appear as two-dimensional slices through the body. In the case of brain studies, the resolution of standard MRIs is adequate to see tumors but is too crude to see individual neurons or even to effectively track bundles of neurons. Since myelin is a fatty substance, water outside neurons will generally not diffuse into axons; rather, this water will tend to diffuse along the length of axons—the water percolates along axons or white matter.

Diffusion tensor imaging (DTI) uses a variation of a standard MRI to determine the diffusion direction, and hence determine bulk nerve fibers. While DTI can produce high-resolution images of nerve fibers, difficulties arise when fibers cross. Water diffusion in this case can now take multiple paths at the crossing points, and it is thus difficult to track nerve fibers at these crossing points. Diffusion spectrum imaging (DSI) involves more mathematics that determines more precisely how water diffuses and is not limited to thinking that water diffuses in only one direction. Roughly speaking, the mathematics is a mixture of calculus and statistical ideas, and it is interesting that two-dimensional ellipses and three-dimensional ellipsoids play a role in the mathematics of DTIs and DSIs. The resulting images of nerve fibers are visually striking.

Not all techniques for imaging neurons rely on such indirect approaches as conventional MRI, or the MRI-based DSI and DTI. Recall that those techniques are used primarily for imaging nerve fibers, not individual neurons. Techniques for higher-resolution imaging of actual neurons are somewhat direct. Jeff Lichtman and others developed the use of genes encoding three proteins that fluoresce in, essentially, the colors red, blue, and green. Genetically modifying mice with these genes, as well as an enzyme that randomly arranges the genes amongst neurons, allows for mice neurons to appear in one of now approximately 150 colors, creating what Lichtman has termed a "brainbow." Another approach relies on a genetically modified version of the rabies virus, which ordinarily is well suited for traveling from neuron to neuron on its journey to the brain. By tagging the modified virus with a fluorescent molecule, one obtains bright images of neurons connecting to just one other neuron to further aid in understanding neural connectivity. All in all, exceptionally striking images are displayed in many places including on the Internet. While the imaging of individual neurons is often more direct and makes less use of mathematics than the MRI approaches to imaging nerve fibers, much mathematics is subsequently used in automating the process of tracking neurons and nerve fibers and, ultimately, the connections found might be described and analyzed by an area of mathematics known as "graph theory."

The approaches to imaging nerve fibers discussed above, such as DTI and DSI, rely on MRI instruments; that is, they rely on indirect methods involving minute variations in very small magnetic fields from protons that are assailed by powerful externally generated magnetic fields. They use indirect information coming from throughout the local environment inside the brain, and mathematics is crucial to inverting the recorded data to recover what is going on at a particular location in the brain. But not all imaging inside the brain focuses just on the connections between individual neurons or bundles of neurons. Other areas of interest include determining which parts of the brain are stimulated at which times by which activities. Here, other inversion processes are used to see what is happening in the brain, including functional MRI (fMRI), positron-emission tomography (PET), and electroencephalograms (EEG). Blood oxygen level dependence (BOLD) uses MRI technology that takes into account the very slight differences in the magnetic fields from water molecules in blood, depending on whether the blood is carrying oxygen. Hemoglobin bound to oxygen is diamagnetic, essentially repelled by a magnetic field. Hemoglobin without bound oxygen is paramagnetic, or attracted to a magnetic field. In either case, it affects the overall magnetic fields from the water molecules. This effect leads to functional MRI (fMRI), which is used to examine oxygenated blood in the brain. The principle is that high neural activity is probably associated with increased blood flow.

PET, another imaging approach, uses an analog of glucose with a radioactive fluorine atom attached. When it decays, it produces a particle known as a positron that is quickly annihilated upon encountering an electron, and two photons stream out in opposite directions. Photomultiplier detectors essentially notice the two photons, and mathematics is used to invert this problem and determine where the annihilation occurred, which presumably is near where the brain was consuming the glucose-like food. A single-photon to PETs, Single Photon Emission Computed Tomography (SPECT), is also utilized.

As the final example of imaging approaches, EEGs focus on using electrical activity recorded on the scalp to see what voltages are created by bulk neurons extending over somewhat larger regions of the brain, as neurons synchronize their electrical signaling. EEGs thus have less spatial resolution than some of the other imaging approaches. Magnetoencephalography (MEG) is a magnetic analog of the EEG in that it is also a noninvasive procedure. Rather than using electrodes attached to a person's scalp for measurements, as in an EEG, very precise superconducting quantum interference devices (SQUIDs) detect weak magnetic fields directly arising

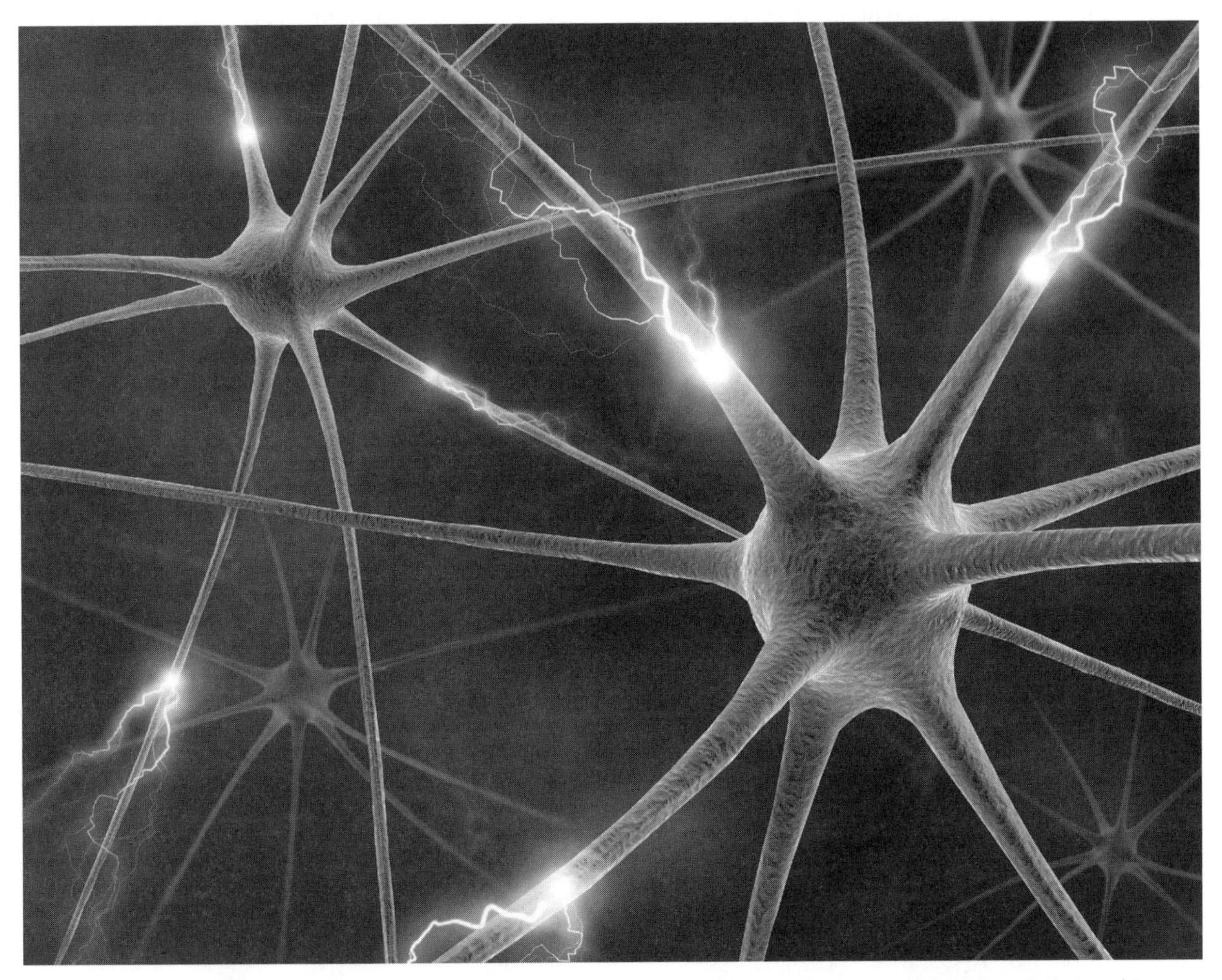

An illustration of hardwired neurons, transferring pulses and generating information. The electrical signal is transferred from one neuron's axon to another's dendrite across a gap known as a synapse.

from electrical brain activity. A mathematical inverse process makes the externally obtained magnetic data usable and converts it to internal electrical activity. MEG typically offers greater resolution, so it can localize the electrical activity more precisely, than EEGs.

Much mathematics is used to model the flow of electrical impulses in the brain. Wave phenomena in the brain arise in varied contexts, from the propagation of signals down a neuron, to collective behavior of many neurons resulting in rhythmic activity. More specifically, the Nernst equation, named for Walther Nernst, and its generalization, the Goldman equation, named for David Goldman, help relate ion concentrations to voltages. How those voltages change in a neuron as it is stimulated by other neurons is modeled by the Hodgkin–Huxley set of equations, which are a calculus-based set of differential equations that resulted in a Nobel Prize for Alan Hodgkin and Andrew Huxley. Next, an area of mathematics known as "dynamical systems" helps model how the firing of individual neurons can naturally become synchronized and produce wave behavior at different frequencies, which are ultimately recorded on EEGs. Normal rhythmic activity is important in activities such as sleeping, breathing, or walking. Abnormal rhythmic activity is manifested in various diseases; for example, forms of schizophrenia, Parkinson's disease, and epilepsy demonstrate deviations from what is considered typical rhythmical behavior.

Neurology of Learning Mathematic

How the brain learns mathematics is another area of interest to researchers. Psychology and other social sciences bring light to bear on this subject but so too does the study of various neural pathologies. As an example, dyscalculia, which has been called a form of "number blindness" (by analogy to color blindness), is a pathology wherein individuals cannot acquire arithmetic skills. For instance, individuals fully capable of language communication who cannot tell if one whole number is larger than another or are unable to do 2-digit computations are considered "number blind." For other examples, there are cases of individuals with increasing difficulties with speech—primarily because of atrophy in the temporal lobes leading to dementia—having highly reduced vocabulary including an inability to name common objects, yet whose arithmetic abilities remained virtually flawless. Similarly, there are instances of autistic individuals essentially unable to speak or understand speech, who nevertheless can perform computations. Infants can notice when the number of objects in a display changes or when a number of objects are hidden behind a screen.

Finally, there are instances of stroke victims who have fully intact language but lack numerical skills, such as not being able to count past 4, or say how many days are in a week. These examples indicate that language is not crucial for arithmetic computations, and, further, language may not be necessary for learning to calculate. Generally speaking, computations seem to be localized to the parietal lobe at the top of the brain, whereas key language areas, such as Broca's (frontal lobe), named for Paul Broca, and Wernicke's (temporal lobe), named for Carl Wernicke, reside elsewhere. However, there are ongoing debates among neuroscientists regarding what the highlighted areas on images mean with regard to brain functionality.

A related issue is how mathematical thinking beyond the level of simple arithmetic evolved in humans, including its relationship with the development of language and increasingly abstract reasoning. There are different and intriguing hypotheses regarding why language evolved roughly 200,000 years ago, whereas various forms of numerical and algorithmic abstraction evolved within the past few thousand years.

Further Reading

Bookstein, Fred. "Morphometrics." *Math Horizons* 3 (February 1996).

Joint Policy Board for Mathematics. "Mathematics Awareness Month, April 2007: Mathematics and the Brain." http://www.mathaware.org/mam/07/announcement.html.

Martindale, Diane. "Road Map for the Mind: Old Mathematical Theorems Unfold the Human Brain." *Scientific American* 285, no. 2 (August 2001).

Schoonover, Carl. *Portraits of the Mind: Visualizing the Brain from Antiquity to the 21st Century*. New York: Abrams, 2010.

Sousa, David. *How the Brain Learns Mathematics*. Thousand Oaks, CA: Corwin Press, 2008.

Rick Kreminski
Dedicated to Wanda Kreminski (1925–2010)

See Also: EEG/EKG; Medical Imaging; Nervous System; Optical Illusions; Visualization.

Bridges

Category: Architecture and Engineering.
Fields of Study: Algebra; Geometry.
Summary: Bridges are subject to various complex forces, the distribution of which are determined by their structures.

Bridges are structures built to span a gap or a physical obstacle such as a road or body of water. The many forces acting on bridges make different designs variously suited to different conditions, uses, and building materials. The earliest manmade bridges emulated naturally occurring bridges, like fallen trees that spanned rivers, and were improved upon by lashing logs into place, cutting planks to form a more even travel surface, and eventually building bridges out of stone. The mathematics of bridges was not well understood and most improvements were achieved through trial and error, one of the most significant being the advent of the arch bridge, introduced in Greece in 1300 B.C.E. and used extensively by the Romans. Arch bridges use arch-shaped abutments, sometimes in a series, to distribute much of the bridge's load into horizontal thrust the abutments can restrain—not only a major improvement over earlier designs, but a design well-suited to the simple building materials of the time as stone is strong in compression but weakly resists tension. As applied mathematics became more sophisticated, bridges were often objects of study.

Most bridges are built for functional purposes, but some of them are works of art, like the Golden Gate Bridge or the London Bridge. Mathematicians have long worked on various aspects related to the design and construction of bridges. For instance, Charles Hutton worked on equilibrium principles and Claude-Louis Navier developed a theory for suspension bridges. Applied mathematician P. Joseph McKenna analyzed bridge oscillations and differential equation models of the collapse of the Tacoma Narrows Bridge. The configuration of bridges in Konigsberg served as the subject of mathematical study for Leonhard Euler and is sometimes noted as the beginning of graph theory.

The Tacoma Narrows Bridge

The Mathematics of Bridges

A bridge has to support various forms of forces: tension, compression, bending, torsion, and shear. It has to carry its own weight (or "dead load"), the weight of the traffic for which it was intended (or "dynamic load"), and it should resist various natural forces, such as wind or earthquakes. The Tacoma Narrows Bridge is often presented in engineering, physics, or mathematics classes as an application of oscillation problems or differential equations. It was a 1.1 mile (1.9 kilometer) long suspension bridge and collapsed in 1940—four months after being opened—because a 35–46 mile per hour wind produced an oscillation, which ultimately broke the entire construction.

Types of Bridges

There are various types of bridges. Beam bridges consist of a horizontal beam with two supports called "piers" at the ends. Arch bridges are one of the oldest types of bridges and distribute the load of the bridge outward along the curve of the arch to the supports at the ends. Suspension bridges are light and strong and can span longer distances than any other type of bridge, but they are expensive to build. Large bundles of cables suspend the roadway from one end of the bridge to the other. Early Asian suspension bridges were suspended with bamboo cables. Cable-stayed bridges look like suspension bridges, but their cables are secured to towers that bear the load of the bridge. They cost less and their construction is faster than suspension bridges, since they need fewer cables and builders can use pre-cast concrete sections. Movable bridges can be occasionally levered for making way for ships or other kinds of traffic. Double-decked bridges have two levels and are used for multiple forms of traffic—subway, pedestrian, automobile, or bicycle.

The Seven Bridges of Konigsberg

Mathematician Leonhard Euler posed the problem of the seven bridges of Konigsberg in a 1736 paper. The town of Konigsberg contained an island with two branches of a river flowing around it. There were seven bridges spanning the river, and the question was whether a person could start at some point and follow a path that would cross each bridge exactly once and return to the starting point. Euler proved that there was no such path.

Famous Bridges

Millau Bridge, France, is 1125 feet high—higher than the Eiffel Tower. Hangzhou Bay Bridge, China, is 22 miles long. The Rolling Bridge, England, is 39 feet long and rolls itself up until the two ends meet, using a hydraulic press. Tower Bridge, England, is a landmark of London and opens in the center, allowing ships to sail through. Ponte Vecchio, Italy, is considered by some to be the oldest stone arch bridge in Europe. Lake Pontchartrain Causeway, Louisiana, is 24 miles long. Vasco da Gama Bridge, Portugal, is 10.5 miles long. Confederation Bridge, Canada, is 8 miles long. Golden Gate Bridge, California, is one of the most famous symbols of San Francisco. Evergreen Point Floating Bridge, Washington, is a 1.5-mile-long floating bridge.

Further Reading

Blockley, David. *Bridges: The Science and Art of the World's Most Inspiring Structures*. New York: Oxford University Press, 2010.

Hopkins, Brian, and Robin Wilson. "The Truth About Konigsberg." *College Mathematics Journal* 35, no. 3 (2004).

Humphreys, Lisa, and Ray Shammas. "Finding Unpredictable Behavior in a Simple Ordinary Differential Equation." *The College Mathematics Journal* 31, no. 5 (2000).

Peterson, Ivars. "MathTrek: Rock-and-Roll Bridge." *Science News*, January 30, 1999. http://www.sciencenews.org/pages/sn_arc99/1_30_99/mathland.htm.

Picon, A. "Navier and the Introduction of Suspension Bridges in France." *Construction History* 4(1988).

Simone Gyorfi

See Also: Engineering Design; Graphs; Highways; Levers; Tunnels.

Budgeting

Category: Business, Economics, and Marketing.
Fields of Study: Measurement; Number and Operations.
Summary: Creating a viable budget requires mathematical analysis.

The word "budget" originally meant a small pouch. By the end of the sixteenth century, people used the word to refer to the contents as well as the bag. The connection with finance dates back to at least 1733. In general, a budget is a balanced plan for spending and saving that includes expected incomes and expenditures. Individuals or families use budgets to manage earnings; pay bills; save for events like retirement, college, or vacations; and to plan for large purchases like a car or a home.

Businesses manage revenues and expenses for materials, taxes, advertising, and payroll using budgets. They may also have smaller budgets for individual projects. City, state, and national governments use budgets to distribute incomes from taxes and other sources among expenditures like infrastructure, social programs, national defense, and debt. Mathematicians play a large role in developing mathematically sound budgets at all levels, especially accountants and actuaries. In the past, budgeting in classroom settings was confined largely to home economics classes, but now budgeting activities are often used to teach various mathematical principles in context.

Some budgets are created using known amounts. Other times, the values are forecasts of income or expenditures based on data or mathematical models. Budgets themselves can also be used for modeling and production. For example, a static budget is a fixed budget created before any input and output values are

The 10-10-80 principle is to give 10% to charity, save 10%, and live on the remaining 80% of your earnings.

known, while a flexible budget can be adjusted based on information about actual activity. A metric called "flexible budget variance" compares flexible budgets to actual results to determine the effects of economic variables on business operations. Sales volume variance compares flexible budgets to static budgets to determine the effect a company's activity had on its operations. Budget accuracy ratios also quantify differences between various budgets or actual production. These can be used to create more accurate future budgets and to plan operations. Budgeting concepts can also apply to resources other than money. Lisa Sullivan, a senior budget analyst working for the U.S. Department of the Navy, regularly uses algebra, statistics, mathematical modeling, and operations research to explore resource allocation problems that affect budgeting. She often works on unique mathematical problems that do not occur in private industry, such as determining the optimal number of Navy surgeons needed in wartime.

Budgeting Basics

Creating a spending plan can be complicated; however, the easier the plan, the more likely it is to be followed. One of the simplest budgets used is the 10-10-80 principle. John D. Rockefeller, the first person in the world to amass a fortune of $1 billion and the wealthiest American ever when adjusted for inflation, is reported to have used the 10-10-80 principle. The crux of the principle is simple: give 10% of your earnings to charity, save 10% of your earnings, and live on the remaining 80%.

Anytime you receive income (for example, paycheck, gift, or prize), first multiply that amount by 10%. Finding 10% of an amount is a relatively easy process: move the decimal point to the left one place value. For example, if you received earnings of $342.57, multiplying by 10% would yield $34.257 (rounded to $34.26). Based on the 10-10-80 principle, you should first give $34.26 away to charity. Many people donate this money to religious institutions or charities such as the Red Cross or the United Way. One argument for giving, besides being altruistic, is to show ourselves that we have control of our money. By freely and willingly giving some of it to others instead of tightly holding onto it, we gain confidence that we have enough and can therefore live on what we are given.

The next 10%, or $34.26 in this example, is given to yourself into some sort of savings vehicle like a savings account or a money market fund. Ideally, this money is never needed as it becomes part of your long-term savings. This money may go toward retirement or an emergency fund in case of job loss or major disaster. Many people are tempted to use this long-term savings for expenses like taking vacations, buying a car, or replacing an appliance. However, these foreseeable expenses should be budgeted as part of the remaining 80%.

Once you have given 20% of your income away (10% to others and 10% to yourself), the remaining 80% can be used for living expenses (including short-term savings). How that 80% is spent can vary depending on many factors including how many people are being supported (for example, you do not need to buy as much food for a single adult as you do for a family of five). Usually the biggest expenditures are for housing and transportation.

Combined, these two categories should not account for more than 50%, or half, of your income. Of course, the less you spend on these the more you have to spend on other areas. Housing, by itself, should account for less than 35% of your income. In the previous example, 35% of $342.57 is $119.91. Set aside this $119.91 to cover any housing expenses you have.

Housing expenses include not only the obvious rent or mortgage but also utilities (heat, electric, plumbing, sanitation, telephone, and Internet), insurance, property taxes, and property maintenance (property maintenance is usually about 5% of the property value each year).

If housing and transportation together should be 50% (or less) of your income, then 15% should be used on transportation. In the example, 15% of $342.57 is $51.39. This amount becomes earmarked to cover all transportation expenses. These expenses include car payments, insurance, license, gasoline, parking, and maintenance (car maintenance is usually about 10% of your transportation costs).

If you spend 50% of your income on housing and transportation, this leaves a mere 30% for everything else. If you have been spending more than you earn, you probably have credit card debt or other personal debt. Ideally, your debts (not including housing or transportation debts) do not account for more than 5% to 10% of your income. What remains should be used to pay for food, life insurance, medical insurance, medical and dental expenses, clothing, entertainment, short-term savings (for vacations and replacement costs), and other miscellaneous spending.

Further Reading

Johnson, Kay. *The Mathematics of Budgeting: Mathematics for Everyday Living*. Erie, PA: Meridian Creative Group, 1999.

Joshi, Mark. *The Concepts and Practice of Mathematical Finance*. 2nd ed. Cambridge, England: Cambridge University Press, 2008.

Shim, Jae. *The Art of Mathematics in Business: Analyzing Facts and Figures for Smart Business Decisions*. Sterling, VA: Global Professional, 2009.

Chad T. Lower

See Also: Comparison Shopping; Coupons and Rebates; Home Buying; Money.

Burns, Ursula

Category: Business, Economics, and Marketing.
Field of Study: Connections.
Summary: The CEO of the Xerox Corporation and *Forbes'* 20th most powerful woman in the world, Ursula Burns (1958–) is an accomplished mathematics education advocate.

Ursula Burns was the first African-American woman to be named chief executive officer (CEO) (2009) and ultimately chairman (in 2010) of a Fortune 100 company. Burns has a bachelor of science degree in mechanical engineering from Polytechnic Institute of NYU and a master's of science degree in mechanical engineering from Columbia University. In addition to her work at the Xerox Corporation, she has been passionate about mathematics, science, and engineering education.

Career

Burns has dedicated her entire professional life to the Xerox Corporation. She began her career with the company in 1980 as a mechanical engineering summer intern. After completing her master's degree, she joined the company as a full-time employee. In her early career, she worked in product development and planning, and later in manufacturing and supply chain operations. She noted, "This company was my family." In 2007, she was appointed president of the company, and in 2009, at the age of 50, she succeeded another female CEO, Ann Mulcahy, to become the first African-American woman named CEO of a Fortune 100 company. She advises students, "Find something that you love to do, and find a place that you really like to do it in.…I'm a mechanical engineer by training, and I loved it. I still do.…I got to work on these great problems."

Commitment to Mathematics and Science

In addition to her business successes, Burns advocates for stronger educational efforts in science, technology, engineering, and mathematics. In a 2010 interview with *Fortune* magazine, she reflected, "If you get kids when they're young from just about any background, you can create people who are capable of utilizing science, technology, math, and engineering to solve problems. If you look at the list of the top nations and try to find out where we are in reading, math, and any science, it is stunning. I don't look at the list anymore because it's an embarrassment. We are the best nation in the world. We created the Internet and little iPods and copying and printing machines and MRI devices and artificial hearts. That's all science and engineering. Who's going to create those things?" Her concern is supported by data from sources such as the Trends in International Mathematics and Science Study (TIMSS).

In response to her strong dedication to mathematics and science learning, in 2009, President Barack Obama appointed Mrs. Burns a member of Educate to Innovate, a White House Initiative on Science, Technology, Engineering, and Mathematics Education (STEM) to help lead a national program to improve student learning in these fields. This committee was also charged with creating public-private partnerships to foster innovation and creativity in the STEM fields.

This program was expanded in 2010 to include Change the Equation, a CEO-led initiative. Funded in part by the Bill and Melinda Gates Foundation and the Carnegie Corporation of New York, the group provides financial support to assist high school students in passing advanced placement tests in science and mathematics, as well as promoting the professional development of STEM teachers. Burns has been quoted as saying, "If we inspire young people today, we secure our ability to innovate tomorrow...[because] [i]nnovation is central to our nation's overall growth."

Accomplishments

In 2010, Burns was named the 20th most powerful woman in the world by *Forbes* magazine. The award, in part, was based on her commitment to mathematics and science education. Similarly, Fortune magazine has ranked her as number nine on its list of the 50 most powerful women for 2010. In addition to her position at Xerox, she is on the board of directors for several organizations including the American Express Corporation, FIRST Robotics, the National Academy Foundation, MIT, and the University of Rochester. Burns credits her success in part to the lessons in life she learned from her mother. She is often quoted as saying, "Don't ever do anything that won't make your mom proud." Clearly, her accomplishments would please any mother.

Further Reading

Bryant, Adam. "Xerox's New Chief Tries to Redefine Its Culture." *New York Times*, February 20, 2010. http://nytimes.com/2010/02/21/business/21xerox.html.

Colvin, Geoff. "Ursula Burns Interview." *Fortune* 161, no. 6 (2010).

Ferrin, Lindsay. "A Long Way From Delancy." *Rochester Woman*, January 15, 2010. http://www.rochesterwomanmag.com/JanArticles/RWM_JanCoverStory.htm.

"50 Most Powerful Women in Business 2010." *Fortune* (September 30, 2010). http://money.cnn.com/magazines/fortune/mostpowerfulwomen/2010/index.html.

"The World's 100 Most Powerful Women." *Forbes*. http://www.forbes.com/wealth/power-women?boxes=Homepagelighttop.

Konnie G. Kustron

See Also: Careers; Curriculum, K–12; Fax Machines.

Bus Scheduling

Category: Travel and Transportation.
Fields of Study: Algebra; Data Analysis and Probability; Measurement; Number and Operations.
Summary: Mathematical modeling can be used study and create optimal bus schedules.

Public transportation systems, like buses, are the primary mode of transportation for millions of people worldwide. Many people advocate for the increased use of buses to alleviate problems such as pollution and roadway congestion. Most public bus systems use fixed routes and schedules that specify the times and places at which the bus will stop so that people can plan their travel. However, most bus riders have had an occasion when their bus arrived late or have seen several buses arrive in quick succession. At peak times, buses may also be too full to admit new riders. Operations research is a subdiscipline of mathematics that focuses on these sorts of scheduling problems and mathematicians in a wide variety of areas work on related theories, problems, and applications.

Since buses usually travel several circuits in the same closed loops, and since there may be several buses following the same path, scheduling buses is similar to the problem of people waiting in a line or queue at the grocery store or the movies. Queuing theory uses mathematical techniques and concepts such as Markov chains, boundary models, series and cycles, numerical methods, simulation, and stochastic modeling to optimize scheduling. These problems can be challenging because of the need to quantify human behavior. Of particular interest to some mathematicians is the amount of slack that must be allowed in the schedule to allow buses to complete their routes in a timely and efficient fashion while accounting for natural variability and unexpected events. A related phenomenon is "bunching," which happens when buses traveling the same route get too close together. Both result in delays, lack of reliability, and customer dissatisfaction. In 2006, engineers Maged Dessouky, Jiamin Zhao, and T. S. Bukkapatnam published a mathematical model that created curves to correlate average delay times and slack time ratios with passenger waiting times. The curves were used to estimate optimal slack as a function of total round-trip travel time. They found an exact solution for the simplified case of a single bus on a closed loop with a known distribution of travel delays, with approximate extensions for more buses. In contrast, physicists Petr Seba and Milan Krbalek studied unscheduled, privately owned buses in Mexico. Passengers waited at known stops, and the drivers competed for passengers rather than assigning specific pickup times. While this system may appear to be chaotic, it has been shown in some studies to be more efficient than scheduled stops, and it can be modeled with a mathematical concept known as random matrices. Theoretical physicists have used these matrices since the 1970s to model complex quantum systems. They also have applications such as describing the distribution of prime numbers, and the possible arrangements of shuffled playing cards.

Queues

Both the problems of scheduling and queuing have commonality and are studied under the title "Queues" or "Queuing Theory." To think of the simplest problem is to consider a single customer service counter where the server takes a random amount of time serving each customer, and customers come one by one to the counter. A customer arriving at the counter is served straight away if the counter is idle when the customer arrives. If a customer is being served when additional customers arrive, then these new customers have to wait in a queue for their turn to be served. This method by which a queue forms leads to several interesting questions. How does one model the arrival pattern of customers? What is the expected time of service for different customers in the queue? What is the expected length of the queue as a function of time? How many customers will be served in a day given a model for the random service times?

This queuing problem can be translated to scheduling a bus to run in a city. Various specific questions arise in this scenario. When should the bus start? Which route should it take so that the service is available to the maximum number of commuters? How much time should the bus wait in intermediate bus stops? How often should the bus repeatedly go in the same route? Here, the objectives may be to maximize the utility of people who commute using the bus, minimize the fuel costs for running the bus, and optimize the use of the available buses. The problem of finding the optimal routes is called a "routing problem" or "bus scheduling problem." Given information about the number of buses available, the layout of the city, and the number of commuters who are likely to use the bus facility in the city, the scheduling problem can be posed as an optimization problem. To find out the number of commuters who may use the facility, one can perform a pilot study to ascertain the views of the people who may be interested in using a bus for their transport.

Modeling bus schedules is necessary to predict the arrival time of a bus at a particular station. Stochastic modeling must be employed since many random factors are involved like possible delay in the starting station because of commuter rush, and unexpected hurdles in the route because of weather. Modeling also helps in avoiding the clustering of buses at some points in a route. Another application of modeling is to track the buses and monitor the speed of buses on the routes. Once the bus scheduling is completed, service reliability has to be studied so as to make adjustments in the bus scheduling for improving the service. Efficient bus scheduling also helps in increasing the profitability of running bus service.

Scheduling Factors and Models

Bus scheduling involves a lot of random factors. Some of the factors are the number of people who will use the service, the amount of time the bus takes to cover a particular route, the delay caused by traffic jams, the number of commuters getting on and off at a particular bus station, the monthly income generated by the bus service along a particular route, and the maintenance costs for the bus. This necessitates stochastic modeling for the bus schedules. Models can be proposed based on historical data, pilot studies, and experiments. One of the important parameters considered in bus scheduling is the waiting time of commuters at a particular station. The objective of scheduling should be to minimize the waiting times of commuters at several points along a route, and for this, it is necessary to provide the most accurate bus schedules possible so that commuters get the maximum benefit. Queuing theory addresses most of these problems discussed and is a good source for solutions to problems in bus scheduling. Data mining techniques can also be used to look at patterns of commuter behavior across routes and this may be helpful in improved bus scheduling.

Further Reading

Eastaway, Rob, Jeremy Wyndham, and Tim Rice. *Why Do Buses Come in Threes? The Hidden Mathematics of Everyday Life*. Hoboken, NJ: Wiley, 2000.

Gross, Donald, John Shortle, James Thompson, and Carl Harris. *Fundamentals of Queueing Theory*. 4th ed. Hoboken, NJ: Wiley, 2008.

Ravi Sreenivasan

See Also: Animation and CGI; Data Mining; Probability.

C

Calculators in Classrooms

Category: History and Development of Curricular Concepts.
Fields of Study: Communication; Connections; Number and Operations.
Summary: Calculators can be used in classrooms to augment rather than replace learning mathematical calculations.

Calculators have a long history. They can be traced back to the ninth century when the original compact calculator, the abacus, was developed in China. Nowadays, calculators are small (often handheld), electronic, digital, and inexpensive devices to perform various operations of mathematics. There are many kinds of calculators. Simple calculators just perform the basic operations of arithmetic.

Advanced calculators include scientific calculators and graphing calculators. Scientific calculators can work on complex operations such as trigonometric, logarithmic, and statistical calculations; some are even able to perform computer algebra. Graphing calculators usually have similar abilities to scientific calculators; however, they can be used to graph functions defined on the real line or higher dimensional Euclidean space. Since the advent of handheld calculators in the 1970s, the daily lives of people have been changed profoundly. Almost each business office and every high school student has at least one calculator. People can solve tedious computations in a few seconds by calculator, which was impossible before 1970.

Calculators in Primary Mathematics Classrooms

The availability of inexpensive calculators in primary classrooms has been increasing rapidly. However, the debate on their effectiveness in student learning and their role in mathematics instruction continues. Whether calculators should be used in primary classroom remains a controversial issue. On one hand, some people worry that calculators may hinder students' learning and obstruct the development of basic arithmetic operations such as addition, subtraction, multiplication, and division. On the other hand, research has shown that calculators can promote problem solving in students with a strong foundation in basic skills.

In primary classrooms, the use of calculators aims at facilitating the learning of mathematics rather than replacing mental arithmetic and written calculation. Pupils can use calculators to explore number patterns, construct concepts, and check different methods and results of problem solving. With the help of calculators, children can strengthen their abilities in mental arithmetic and estimation, as well as judge the sensibleness of the results of calculation. For instance, pupils may be asked to estimate the sum of $9 + 99 + 999$ and explain how they get the answer. One method of estimating the sum is the calculation $10 + 100 + 1000 - 3$.

After they have done the estimate mentally, they can check their estimation by calculator. Depending on their abilities, pupils may be asked to estimate the sum of more complicated operations such as $999 + 9999 + 99{,}999 + 999{,}999 + 9{,}999{,}999$ and then check the answer by calculator. This kind of activity facilitates the development of inquiry mind and higher order thinking in children. When pupils are allowed to use calculators to check the answers they have come up with by themselves, they have immediate feedback, have more time for solving additional problems, and make fewer errors. Calculators help pupils concentrate on thinking rather than on computation.

Pupils cannot benefit much if they are requested to compute traditional calculations such as $2 + 7$ or 3×5 by using a calculator. However, when they are asked to explore what calculations would give an answer of 10 with the aid of calculator, the effect can be very positive. Pupils may find patterns such as $1+9,\ 2+8,\ 3+7, \ldots$; $11-1,\ 12-2,\ 13-3, \ldots$; and $10 \div 1,\ 20 \div 2,\ 30 \div 3, \ldots$; and so on. Such open-ended tasks provide opportunities for children to explore basic arithmetic operations, natural numbers, fractions, and decimals. Through these exploratory activities, children can develop number sense and strengthen inquiry mind by making and testing conjectures. Calculators help children quickly detect and correct their misunderstandings experientially.

There are many good calculator activities that will enrich the learning experience for pupils. Different activities may be suitable for different classrooms; however, the focus should be on the mathematics rather than the calculator.

Calculators in Secondary Mathematics Classrooms

The use of scientific and graphing calculators in secondary school causes much less controversy than the use of simpler calculators in elementary schools. In fact, many countries allow their secondary students to bring in approved calculators for their university-entrance examinations.

Over the past 10 years, many innovative methods of teaching secondary mathematics have been developed with the advancement of handheld calculators and the needs of society. Many of the ideas require only basic calculators, but scientific and graphing calculators open up more possibilities—particularly for the learning of complicated functions, shapes, and graphs.

A graphing calculator typically refers to a class of handheld calculators that are capable of plotting graphs, solving systems of equations, and performing numerous other tasks with variables. For instance, graphing calculators allow students to explore the effect of varying the coefficients in the quadratic equation $y = ax^2 + bx + c$ by plotting graphs for different set of values of a, b and c in seconds. Plotting quadratic graphs by paper and pencil would consume a lot of time and effort, which would slow down the learning pace and reduce learning interest in the topic.

Graphing calculators save students from laborious work and provide opportunities to facilitate independent learning and give scope for open-ended exploration. If students go further in their investigation, they may generalize the conditions under which only one solution is obtained for the quadratic equation $0 = ax^2 + bx + c$.

The power of calculators advances rapidly. Some people worry about the use of symbolic calculators that can perform symbolic computations. They have argued that the use of symbolic calculators can cause core mathematical skills to wither, or that such use can prevent understanding of advanced concepts. It is not unusual that students use a symbolic calculator to find

$$_{x}\lim_{\infty}\left(\frac{1}{1+x}\right)$$

without realizing the mathematical principle or skills involved.

Concerns on Usage

From time to time, calculators are accused of making children lazy and replacing the need for them to use or remember number facts. They provide a means for getting answers without understanding mathematical processes. Some people worry that the extensive use of calculators in mathematics instruction interferes with students' mastery of basic mathematical skills and the understanding they need for more advanced mathematics.

In reality, the calculator is a tool that, if used in the right way, can support and encourage children's mathematical thinking. It is not calculators themselves that matter but when and how they are used that is important. To avoid overemphasis on the use of calculators, students should be guided to recognize the functions and limitations of calculators, so as to strengthen their abilities in exploring and solving mathematical problems. For instance, in a classroom activity, all pupils are given the same set of calculation questions, such as $789 + 0$, 25×4, 17×8, and 299×10. Pupils work in pairs; one is requested to find the answers by mental computation while the other uses a calculator.

At the end, they have to record the time needed and the number of correct answers. Pupils have to discuss and identify which calculations can be easily done mentally and which cannot. This activity can facilitate pupils' communication in mathematics and understanding that mental calculation sometimes is more powerful than the calculator.

Research also suggests that inadequate guidance in the use of calculating tools can restrict the kind of mathematical thinking that students engage in. Therefore, it is important that schools implement a balanced program that develops students' understanding of the appropriate use of the calculator.

Further Reading

Baxter Hastings, Nancy, ed. *Workshop Precalculus: Discovery With Graphing Calculators*. Emeryville, CA: Key College Publishing, 2002.

Guin, D., K. Ruthven, and L. Trouche, eds. *Didactical Challenge of Symbolic Calculators: Turning a Computational Device Into a Mathematical Instrument*. New York: Springer Science+Business Media, 2005.

Rising, Gerald R. *Inside Your Calculator: From Simple Programs to Significant Insights*. Hoboken, NJ: Wiley Interscience, 2007.

Sparrow, Len and Paul Swan. *Learning Math With Calculators: Activities for Grades 3–8*. Sausalito, CA: Math Solutions Publications, 2001.

Eddie Leung

See Also: Calculators in Society; Software, Mathematics.

Calculators in Society

Category: Communication and Computers.
Fields of Study: Algebra; Measurement; Number and Operations.
Summary: Advancements in calculator technology have profoundly changed society and mathematics education.

In the decades since the invention of a truly handheld calculator, these devices have evolved from four-function curiosities costing hundreds of dollars to sophisticated machines capable of performing a wide range of mathematical and statistical functions at the same cost as that "four-banger" from the early 1970s. The effect on society has been considerable, as the laborious arithmetic involved in routine scientific or financial calculations can be done by nearly anyone with minimal effort and accuracy that was unthinkable in the 1950s. A variety of technological advances and a new market for calculating power during the 1970s led to the "calculator wars" among a variety of manufacturers, and frequent major advances in the power of a calculator were marketed to a willing society. These powerful calculators have changed the school mathematics curriculum in a variety of ways and brought a new focus to the Advanced Placement (AP) calculus exams.

Early History of Calculators

In their 1951 textbook *Mathematics of Investment*, Paul Rider and Carl Fischer made occasional reference to the ability of "computing machines" to facilitate involved calculations in financial mathematics. However, since such machines were by no means common in the 1950s, the book includes 123 pages of numerical tables, roughly one-third of the book's total length. These references were essential to actuarial calculations for many years, and their analogous tables of values of trigonometric functions, exponentials, and logarithms were a staple of mathematics textbooks for a comparable time period.

The rapid rise of low-cost electronic calculators—a generation beyond the electric computing machines to which Rider and Fischer referred—reduced those tables to a mere historical curiosity within a generation. In 1958, Texas Instruments (TI) engineer Jack Kilby invented the integrated circuit, which became known as the "calculator on a chip," that revolutionized the world of calculating devices. Large electromechanical desktop calculators soon gave way to more compact electronic machines, which culminated with the development of the Cal-Tech handheld calculator in 1965 at Texas Instruments. The Cal-Tech was a simple four-function calculator that used a paper tape for output. With a new standard for what was possible, the rush to advance calculating machines, both handheld and desktop, was on.

Engineers at Hewlett-Packard (HP) merged the old with the new in 1968 with the development of the HP-9100A, the first fully electronic desktop calculator. The 9100A was considerably larger than the Cal-Tech but was much more versatile. Its function set included all of the functions found on a modern scientific calculator—trigonometric functions, logarithms, reciprocals, and others—and it was fully programmable. On viewing the 9100A, company founder Bill Hewlett included among his words of praise for the developers the challenge that the world needed a similar machine that would fit into a shirt pocket.

In 1972, Texas Instruments introduced the Datamath, a four-function calculator released under the TI name. This was a departure for the company, which until then had confined its calculator work to manufacturing parts for other companies' machines. Indeed, the Cal-Tech was built primarily to show other manufacturers what the company's parts could do, not as an eventual consumer product. In that same year, Hewlett-Packard engineers developed the HP-35, a fully scientific calculator that could fit into a shirt-pocket. With these two companies at the forefront of a rapidly advancing technology, and with many other manufacturers in close competition, the "calculator wars" began. The rapid evolution of affordable competing calculators from a variety of manufacturers went on throughout the 1970s and into the early 1980s.

A major innovation was TI's introduction of the TI-30 scientific calculator, which sold for under $30 beginning in 1976. The full scientific function set of the TI-30 on a low-priced machine was a huge advance over the $395 price tag of the original HP-35, and the TI-30 was regarded for many years as the best-selling calculator of all time.

HP introduced the first handheld programmable calculator, the HP-65, in 1974 (fewer than two years after its first scientific calculator), and followed it up in 1977 with the HP-67. TI countered with the SR-52 in 1975, which was succeeded by the TI-58 and TI-59 in 1977. Each of these milestone calculators allowed the user to specify a sequence of steps into a special memory. These steps could then be repeatedly executed as many times as desired. The HP models and the 52 and 59 provided the option of recording programs onto small magnetic cards for permanent storage, while the 58 and 59 came equipped with a slot for read-only memory (ROM) cartridges with space for dozens of specialized prewritten programs that were stored on the chip and could be run as needed without the need for repeated keying.

Special-Purpose Calculators

Special-purpose calculators are preprogrammed with functions and formulas that are specific to a particular profession or interest. Among the earliest were calculators designed for financial mathematics, with keys and

Hewlett-Packard developed the world's first pocket scientific calculator, the HP-35, in 1972.

routines for solving the time value of money problems and automating interest calculations—and here was where Rider and Fischer's prediction was exceeded. These business calculators were considerably more sophisticated than could have been imagined in 1951.

By far the most successful business calculator is Hewlett-Packard's HP-12C, which was introduced in 1981 and is still in production 30 years later. In most senses, the 12C is the industry standard financial calculator, and it has been the key to HP's successful focus on the business calculator market. In 2003, the 12C got a facelift—and a faster processor—as the HP 12C Platinum Edition.

Unit conversion calculators inspired by the push in the 1970s to introduce the metric system in the United States live on in a variety of construction calculators, many of which have been produced by a small company, Calculated Industries (CI). CI was founded in the 1978, and its first product was a real estate calculator dubbed "The Loan Arranger." Future financial calculators from CI would expand in capability to accommodate more sophisticated calculations, and a separate line of CI financial calculators is specific to Canadian interest calculations. Later product lines from CI included the Construction Master and Measure Master lines—which were specialized for the building industry. CI also produces a series of electrical engineering calculators and a pair of professional plumbing calculators.

CI also manufactures special-purpose calculators for a variety of niche markets. Do-it-yourselfers can find the calculations they need preprogrammed into the ProjectCalc series. Several of these have been rebranded by Sears under the Craftsman line. The KitchenCalc Pro is preset to convert cooking measurements and includes a built-in timer. The Quilters' FabriCalc is one of the company's most successful hobbyist calculators and automates the considerable mathematics involved in quilting. Most recently, the Mr. Gasket Hot Rod Calc was developed to serve performance automotive enthusiasts with a collection of functions for use in assessing an automobile's performance.

Calculators in the Classroom

In 1976, Texas Instruments released the Abstract Linking Electronically (ABLE) calculator system, which represented the first attempt to manufacture a calculator specifically designed for elementary school classrooms beginning in the earliest grades. The ABLE system consisted of a standard four-function calculator with six interchangeable faceplates. These faceplates blocked access to some of the calculator's functions and could be switched out to allow a richer selection of options as a child's mathematical sophistication grew.

There was then, and continues to be, considerable tension over the question of calculator use in school mathematics. The conflict is generated by the ability of inexpensive calculators to automate routine arithmetic problems, which had led one side of the debate to suggest that there is no need to require computational automaticity, such as memorizing multiplication tables, which a calculator can handle. These advocates then assert that calculators free up room in the curriculum for what are called "higher-order" mathematical thinking skills. Those opposed to this view assert that higher-order skills are not useful without a sound foundation based on mastery of routine calculations. Sensible middle ground exists between these two viewpoints, and a variety of combinations of these approaches are advocated in textbooks and available to teachers.

In 2000, TI expanded the Explorer line to include the TI-15 Explorer calculator, which was designed for use in grades 3–6. This calculator contains specialized keys for computations like place value calculations and fraction operations without cluttering the keyboard with higher-level computations, like trigonometric functions, that are not studied in elementary school. Additionally, the TI-15 Explorer includes two keys that can be programmed to repeat simple operations, a randomized arithmetic tutor, and tools for exploring inequalities. A simpler companion calculator, the TI-10, was introduced in 2002 and is aimed at kindergarten through third grade classrooms.

At higher grade levels, one effect was far less controversial. With the advent of inexpensive powerful scientific calculators, there was no longer a need for extensive tables of functions in precalculus textbooks.

Graphing Calculators

In 1985, Casio introduced the first graphing calculator, the fx-7000G. In addition to serving as a fully functional scientific calculator, the fx-7000G had a large (1.4-by-2-inch) LCD screen on which graphs of functions could be displayed. This allowed students to work with functions from both numerical and graphical perspectives, and set the stage for a revolution in mathematics teaching. Graphing calculators soon came to be seen as one of the primary components of this shift in teaching and learning.

Hewlett-Packard advanced handheld capacity further with the HP-28C, introduced in 1987. In addition to numerical and graphical approaches to functions, the 28C was able to perform symbolic algebra and calculus, working with variables directly without the need for numbers. Texas Instruments released its first graphing calculator, the TI-81, in 1990, and the TI-85 soon after. The TI-82, 83, 84+, 86, and 89 have extended this successful product line, while the TI-80 and 73 have reached downward into middle schools.

As graphing calculators and computer algebra systems, such as Derive and Mathematica, competed for space in calculus classrooms around the world, it became clear that standardized testing would have to accommodate these new devices. Beginning in 1995, the Advanced Placement calculus exams have required the use of a graphing calculator on part of the exam, one that can plot graphs of functions, solve equations numerically, compute numerical derivatives, and evaluate definite integrals numerically. The College Board, which administers the AP exams, draws the line at calculators with a typewriter-style QWERTY keyboard, such as the TI-92 (introduced in 1996) and Voyage 200 (introduced in 2002) from Texas Instruments. The concern here is for the security of the tests, as the typewriter keyboard and text-processing capability are thought to make it too easy to collect confidential test questions and remove them from the testing site.

The Future of Calculators

It is unclear what new ground remains to be broken in future calculators. Three-dimensional graphing is available on a variety of TI and HP machines, but the size of the screen and the challenge from computer algebra systems, such as Mathematica, have limited the reach of this feature. Calculating power is finding its way into a variety of other handheld devices. Just as many people no longer wear watches because they can get the time from their cell phones, calculator applications for cell phone platforms may render the cell phone an attractive alternative to a specialized calculator. While there are cost and durability issues to be considered in this comparison, CI has recognized this alternate platform by marketing its Construction Master Pro software for the iPhone.

Further Reading

Ball, Guy, and Bruce Flamm, *The Complete Collector's Guide to Pocket Calculators*. Tustin, CA: Wilson/Barnett Publishing, 1997.

Hicks, David G. "The Museum of HP Calculators." http://www.hpmuseum.org.

Sippl, Charles J., and Roger J. Sippl. *Programmable Calculators*. Champaign, IL: Matrix Publishers, 1978.

Woerner, Joerg. "Datamath Calculator Museum." http://www.datamath.org.

Mark Bollman

See Also: Calculators in Classrooms; Calculus and Calculus Education.

Calculus and Calculus Education

Category: History and Development of Curricular Concepts.

Fields of Study: Calculus; Communication; Connections.

Summary: Once reserved for upper-level majors, the study of derivatives and integrals of functions has been mainstreamed by its applications.

Calculus, which takes its name from the Latin word for "pebble," is one of the most important branches of mathematics and one of the cornerstones of mathematics education. In ancient history, pebbles were used for counting, and "calculus" initially referred to that. The word now represents the method of calculation linked often with the study of change attempting to investigate motions and rates of change. From its mathematical development to the philosophy of calculus education, calculus has been fraught with rigorous debate and change.

Its appreciation and deeper understanding is a fundamental requirement in order to proceed toward a more advanced mathematical education and be involved with topics such as mathematical analysis. The topic finds its use and application in a vast number of different applied disciplines, such as biology, engineering, physics, population dynamics, statistics, and, in general, any scientific area that involves the study of instantaneous change.

Calculus education has a rich and varied history. Takakazu Seki is remembered as an influential teacher who passed his form of calculus on to his students. However, during the seventeenth century, secrecy surrounded rival schools in Japan, so it is difficult to determine his exact contributions. Successful calculus textbooks date back to at least Maria Gaetana Agnesi in the eighteenth century. She wrote *Analytical Institutions*, probably as a textbook for her brothers. She mastered many languages, which were useful when she integrated the knowledge of the time. She also added her own examples and expositions. Her book was widely translated and used all over the world, making the concepts of calculus more accessible.

Calculus education underwent many changes in the twentieth century. Early on, calculus was often an upper-division college course in North America while it was a pre-college course in France. U.S. President John F. Kennedy's race to the moon impacted calculus education in the United States. More engineering students were recruited, and as a result, calculus shifted earlier in the college curriculum. Another change was an emphasis on set theory in such texts as Tom Apostol's *Calculus*. Beginning in the latter half of the twentieth century, high schools offered AP Calculus. The shift of calculus to lower-level students also occurred in other countries, such as in Japan.

However, students who did not have the aptitude to succeed in competitive programs were filtered out in lower-level college courses, and educators debated this problem internationally. A calculus reform movement in the United States originated in the late 1980s, epitomized by the slogan "Calculus should be a pump, not a filter." Teachers debated the roles of lectures, technology, and rigor. With minimal theoretical support for the choice of teaching strategies, mathematicians relied on empirical studies to determine what would help calculus students succeed. Educators tested many different approaches, such as those emphasizing active learning, graphing calculators, computer algebra software, historical sources, writing, humanistic perspectives, real-life applications, distance education, or calculus as a laboratory science. New teaching approaches were met with widespread acceptance on some campuses and rejection and backlash on others. Some schools reported a decline in the number of students failing calculus. In the early twenty-first century, mathematicians continue to discuss and refine the calculus course.

Calculus—A Journey Through History

Even though counting as a process appears from the very first stages of humanity and its traces are lost in history of various civilizations, calculus was officially introduced as a realization of the deeper need to set rules and construct generally approved techniques that would assist toward quantification of any kind of change in time or space.

It could also perform modeling of systems that continuously evolve, and hence aid the interpretation and deduction of the consequences of the existence of such systems. Basic ideas of calculus involve limits, continuity, derivatives, and integrals.

Archimedes is one of the main scholars of ancient history who is linked with the ideas of calculus (c. second century B.C.E.). However, two important scholars of the seventeenth century made significant contributions to the introduction and the establishment of calculus as a quantitative language. Isaac Newton (1642–1727) and Gottfried Wilhelm Leibniz (1646–1716) are both recognized as fathers of modern calculus. Even though they worked independently and were influenced by two different areas—Newton by physics and Leibniz by geometry—they both reached into discovering the same fundamental ideas of calculus.

Differential Calculus

If x is a variable that changes with time (for example, $x = x(t)$ is a function of time t) then one denotes

$$\dot{x} = \frac{dx}{dt}$$

the first derivative of the function x, which represents the rate of change of $x(t)$ with respect to time t. Newton used the notation $\dot{x}$ while Leibniz used

$$\frac{dx}{dt}.$$

In case of a moving object in one dimension, $x(t)$ represents the position of the object and $\dot{x}(t)$ its velocity.

The term "function" was first introduced by Leibniz and is one of the fundamental terms of mathematics. In practice, a quantity y is defined as a function of another quantity t if there is a rule (method or process) in which a unique y will be assigned to any t. Leonhard Euler (1707–1783) introduced the notation $y = f(t)$ to identify a function f.

"Method of fluxions" is the term Newton used for his set of techniques to study the continuous flow of change. The process of finding a formula for the function x, given the formula for the function of x, is known as "differentiation" and the methods used for this belong to the field of differential calculus.

Rate of Change

There is a particular interest in studying the change of a quantity and by extension the rate of change of a quantity as another quantity changes in a very small amount. As Newton and Leibniz were developing calculus, they both used "infinitesimals" in order to emphasize the idea of such a small quantity that is not zero and that cannot be measured ("infinitesimal calculus" or "calculus of infinitesimals"). Hence, the infinitesimal number dx was considered to be different from zero and less than any positive real number.

Their approach raised criticism among other well-known scholars such as George Berkeley (1685–1783), and the idea of using infinitesimals became gradually unpopular. The introduction of "limits" from Augustin Louis Cauchy (1789–1857), Karl Theodor Wilhelm Weierstrass (1815–1897), and Georg Friedrich Bernhard Riemann (1826–1866) led in a better realization of the fundamental ideas of calculus and reestablished the topic within a more sound framework. However, nonstandard analysis (Abraham Robinson, 1960) and smooth infinitesimal analysis as introduced in the twentieth century have brought back into use the idea of infinitesimals.

Limit

The definition of limit is a cornerstone for advanced mathematics and especially for mathematical analysis. Limit is what distinguishes calculus from other areas of mathematics, such as algebra, geometry, and number theory. Even though mathematics has a history of more than 3000 years, limits were treated as a special area of mathematics only from 1823 C.E. when the concept was published for the first time in Cauchy's book *Résumé of Lessons of Infinitesimal Calculus*. The first appearance of the term belongs to the Greek mathematician and philosopher Zeno of Elea (495–435 C.E.). However, the definition that was finally accepted and used by the mathematical community is the (ε, δ) definition as stated by Weierstrass.

Weierstrass Definition

Assume that L is a real number and that $f(x)$ is defined in an open interval where x_0 belongs. Then the limit of $f(x)$ as x tends to x_0 is equal to L and is denoted with

$$\lim_{x \to x_0} f(x) = L$$

if the following is true: for any real number ε, there exists another real number δ such that for all x in the interval $x_0 - \delta \le x \le x_0 + \delta$ the value $f(x)$ of f lies within the range from $L - \varepsilon$ to $L + \varepsilon$.

In terms of infinitesimals, the limit is defined as follows: L is the limit of $f(x)$ as x tends to x_0 and is denoted as

$$\lim_{x \to x_0} f(x) = L$$

if the following is true: for any infinitesimal number dx, the value of $f(a + dx)$ is finite, and the standard part of $f(a + dx)$ equals to L.

Equation of Tangent

The term "derivative" as introduced from Newton and Leibniz signified a new era in mathematics. The term assisted mathematicians in finally solving rigorously the problem of constructing a unique tangent passing from a point of a curve. Historically, mathematicians since Archimedes' period were constantly trying to solve the prob lem of a unique tangent on a point of a curve. Ancient Greeks believed first that the tangent at a point of a circle should be the line that passes from the particular point and is vertical to the radius of the circle. Archimedes devoted a significant part of one of his books to this specific problem, which is known as "Archimedean spiral." However, it was because of the introduction of the first derivative that the researchers could actually provide the equation for the tangent line of a curve $C: y = f(x)$ at a point $(x_0, f(x_0))$ as

$$y = f(x_0) + f'(x_0)(x - x_0) \quad \text{where}$$

$$f'(x_0) = \lim_{x \to x_0} \frac{f(x) - f(x_0)}{x - x_0}$$

is the rate of change of the function at that point

$$(x_0, f(x_0))$$

corresponds and defines the slope of the line tangent to the curve C at point

$$(x_0, f(x_0))$$

if and only if the limit exists.

Higher Order Derivatives

Thinking of the example of a moving object in time, it can be easily identified that there is a need for estimating the acceleration of the object. Acceleration is the rate at which an object changes its velocity. Therefore, acceleration in mathematical terminology is nothing else but the derivative of the derivative of x denoted

$$\ddot{x} = \frac{d^2x}{dt^2}$$

and called "second derivative." Since the first derivative provides information on the rate of change of a function, the second derivative refers to the rate of change of the rate of change. In general, a higher order derivative is denoted as

$$\frac{d^n x}{dt^n}.$$

In a more geometric framework, the first and second derivatives can be used to determine the concavity; in other words, the way that the slopes of tangent lines of a curve $y = f(x)$ change as x changes in an interval (a, b).

If f is a differentiable function and f' is increasing on (a, b), then f is concave up on (a, b). The slopes of the tangent lines of the graph of f increase as x increases over (a, b); a concave up graph looks like a right-side up bowl.

If f' is decreasing on(a, b), then f is concave down on (a, b). The slopes of the tangent lines of the graph of f decrease as x increases over (a, b); a concave down graph looks like an upside-down bowl.

Points where the concavity changes are known as inflection points of f. Given that a function is increasing throughout an interval, if its first derivative is positive throughout the interval and vice versa, and given that f' is differentiable, then the following can be obtained: If

$$(f')'(x) > 0$$

for all x in (a, b), then f' is increasing on (a, b) and therefore f is concave up on (a, b). If

$$(f')'(x) < 0$$

for all x in (a, b), then f' is decreasing on (a, b) and therefore f is concave down on (a, b). A natural application of this concept is to find the maximum or the minimum of a function in a case in which the function is concave down or concave up throughout the whole domain respectively. This can be used further to solve problems where an optimal solution is requested.

According to Hans Hahn (1879–1934), the fundamental problem of differentiation can be expressed by two problems: (1) if the path of a moving object is known, estimate its velocity, and (2) given the existence of a curve, estimate its slope. Therefore, the inverse of these problems are (1) if the velocity of a moving object at every instance is known, estimate its path, and (2) if the slope of a curve is known, find the curve.

Integral Calculus

Generally, the process to find a formula for a function of x given the formula for the derivative of the function of x is known as "integration" and the methods used to find the formula belong to the field of integral calculus.

Historically, integral calculus was motivated by the geometric problem of estimating the area of a region in xy-plane bounded by the graph of f, the x-axis, and the vertical lines $x = a$ and $x = b$. The solution of this problem came as a realization for the need of integral calculus and is linked with

$$\int_a^b f(x)\,dx$$

which is known as the "definite integral."

It is not known exactly for how long the aforementioned problem troubled the mathematical world. In

1858, Alexander Henry Rind (1833–1863), an Egyptologist from Scotland, discovered parts of a handwritten papyrus document that is considered to have been written in 1650 B.C.E. *The Rind Papyrus*, as it is known today, consisted of 85 problems by the Egyptian scribe Ahmes, who claimed that he had copied these problems from an older document. Problem number 50 indicates that before 4000 C.E., Egyptians knew how to compute the area of a circle by using the formula Area = 3.16 × radius^2.

Eminent interest toward computations of areas of regions bounded by different kinds of curves is also seen in ancient Greece. Archimedes, whom several scholars consider as the "father of integral calculus" because of his method to estimate that the area bounded from the parabola $y = x^2$ and the rectangular lines $x = 1$ and $y = 0$ would be equal to 1/3. His method, which is known as the "method of exhaustion," was an attempt to approximate the area of a curve by inscribing first in it a sequence of polygons and computing afterward their area, which must converge to the area of the containing curve. However, this method was first developed by Eudoxus; Archimedes just applied this method in order to establish the said area. This method was later generalized in what is known now as "integral calculus."

The fundamental problem of integration focuses on finding the actual function (or, equivalently, its indefinite integral) if the derivative of the function is known.

Assume that function f exists. If there is a function $F: y = F(x)$ such that

$$F'(x) = \frac{dF}{dx} = f(x)$$

then F is called the indefinite integral or antiderivative of f and it is denoted as

$$F(x) = \int f(x)\,dx = \int f = I$$

where I stands for the first letter of the word "integral."

Cauchy was most probably the first mathematician who provided a rigorous definition for the integral by using the limit of a sum. Riemann, later on, influenced by the theory of trigonometric series of the form

$$\sum \left(a_n \cos(nx) + b_n \sin(nx)\right)$$

continued Cauchy's work and defined the integral in a similar way, with the only difference that he studied the whole family of functions that can be integrated—functions for which the integral exists. During Cauchy and Riemann's period, mathematicians were mainly concerned with integrating bounded functions. However, the need for integrating functions that cannot be bounded was soon apparent. Carl Gustav Axel Harnack (1883) and Charles De La Vallée-Poussin (1894) were among the first mathematicians to be occupied with such a problem. However, Henri Léon Lebesgue (1875–1941) is the one who, with his Ph.D. thesis titled "Integral, Length and Area" published in 1902, brought integral calculus into a new level. He defined the Lebesgue integral, which is a generalization of the Riemann integral, and defined a new measure known today as the Lebesgue Measure, which extends the idea of length from intervals to a large class of sets.

Other important scholars whose names are tightly linked with the development of modern calculus are Frigyes Riesz (1880–1956), Johann Radon (1887–1956), Kazimierz Kuratowski (1896–1980), and Constantin Caratheodori (1873–1950). They succeeded in generalizing and extending even further Lebesgue's work.

The symbol $\int$, which is used for integration, is a big S (the first letter of the German word *summe*, meaning "sum") and was used for the first time by Leibniz. There are several theories regarding the origin of the symbol. F. M. Turrell has supported the theory that almost every botanist knows that if an apple is peeled by hand, and, with the help of a knife, starting from the stem and continuing in circles around the central axis without cutting off the apple skin until the opposite end is reached, then a spiral is produced that creates an extended S once placed on the top of a horizontal surface with the inner part of the skin facing upward. This observation, according to Turrell, could possibly explain the symbol of integration. Finally, the Greek letter Σ is strongly linked with $\int$ as Euler used it to denote a sum.

The fundamental theorem of calculus asserts that differentiation and integration are inverse problems. If a function f is continuous on the interval $[a,b]$ and if F is a function whose derivative is f on the interval (a,b), then

$$\int_a^b f(x)\,dx = F(b) - F(a).$$

This realization has proved to be a very useful technique to estimate definite integrals in an algebraic way. Isaac Barrow, Newton, Leibniz, and Cauchy worked on the concepts and early proofs, and Riemann and Vito Volterra explored what conditions on functions were necessary in the theorem. Lebesgue's definition of integrals avoided some of the previous problems.

Probability theory and statistics are disciplines that use calculus. A valuable application is to determine the probability of a continuous random variable from an assumed density function and define the average of the variable and a range of variation around it. The basic method used to approach the underpinning problems is to find the area under the corresponding curves (compute an integral).

For the study of joint distributions of several random variables (multivariate distributions), students and researchers need to be familiar with the fundamental ideas of multidimensional calculus. Optimization in statistics is another area where calculus is significant; when, for instance, there is a demand to find an estimator of an unknown parameter that satisfies an optimality criterion, such as minimum variance.

Other Types of Calculus

Other calculi that are linked strongly with the undergraduate and postgraduate curriculum, indicating the broadness of the topic, are vector calculus and calculus of variations.

- Stochastic calculus is tightly linked with financial calculus. It is mostly found in higher levels of mathematical education as it requires knowledge of measure theory, functional analysis, and theory of stochastic processes.
- Malliavin calculus or stochastic calculus of variations was initiated by Norbert Wiener (1894–1964) in an attempt to provide a probabilistic proof of Hörmander's "sum of squares" theorem. It is an infinite-dimensional differential calculus on a Gaussian space with features that can be applied in a wide variety of advanced topics of stochastic analysis. Its development has enormously facilitated the study of stochastic differential equations where the solution is not adapted to the Brownian filtration.
- Quantum and quantum stochastic calculus, which have gained the interest of quantum mechanics specialists, use infinitesimals rather than limits.
- π-calculus and λ-calculus offer a simpler syntax, which is highly appreciated by those in computing, offering an easier development of the theory of programming languages: network calculus and operational calculus.

Calculus, with its all variations, can be characterized as the mathematical language that unifies science by linking different disciplines together; this is why it plays a central role in the mathematical curriculum with students exposed to its basic ideas from the high school level.

The appreciation of the influence of calculus upon the vast majority of disciplines promotes a simultaneous intuitive approach by providing sufficient examples that illustrate the applicability of the topic. Modern technology in the form of computers and graphical calculators provides the tools that can assist not only in applying the mathematical techniques but also in a smooth transmission of the scientific ideas and basic mathematical concepts.

Further Reading

Bardi, Jason. *The Calculus Wars: Newton, Leibniz, and the Greatest Mathematical Clash of All Time.* New York: Basic Books, 2007

Friedler, Louis. "Calculus in the U.S.: 1940–2004." *Studies in College Mathematics* 8, no. 3 (2005) http://gargoyle.arcadia.edu/mathcs/friedler/Calculus1940-2004.pdf.

Ganter, S. L. *Changing Calculus.* Washington, DC: Mathematical Association of America Notes #56, 2001.

Khuri, André I. *Advanced Calculus With Applications in Statistics.* Hoboken, NJ: Wiley, 1993.

Rassias, John M. *Geometry, Analysis and Mechanics.* Singapore: World Scientific Publishing, 1994.

Rudin, W. *Principles of Mathematical Analysis.* New York: McGraw-Hill,1953.

Zill, D. G. *Calculus With Analytic Geometry.* Boston: Prindle, Weber & Schmidt, 1985.

Matina J. Rassias

See Also: Archimedes; Calculus in Society; Curriculum, College; Curves; Functions; Software, Mathematics.

Calculus in Society

Category: School and Society.
Fields of Study: Calculus; Connections.
Summary: Since its introduction in the seventeenth century, calculus has been applied to more and more practical endeavors, from engineering and manufacturing to finance.

Since its initial development in the seventeenth century, calculus has emerged as a principal tool for solving problems in the physical sciences, engineering, and technologies. Applications of calculus have expanded to architecture, aeronautics, life sciences, statistics, economics, commerce, and medicine. Contemporary society is impacted continually by the applications of calculus. Many bridges, high-rise buildings, airlines, ships, televisions, cellular phones, cars, computers, and numerous other amenities of life were designed using calculus.

Since the 1970s, calculus in conjunction with computer technology has resulted in the emergence of new areas of study such as dynamical systems and chaos theory. Such vast applications have established the study of calculus as essential in preparation for numerous careers. Indeed, calculus is considered among the greatest achievements of humankind, making it worthy of study in its own right in a society that places rational thought and innovation in highest esteem. Recent curricular and pedagogical reforms in calculus have made it more academically accessible to the school population.

What Is Calculus?

Calculus originated from studying the physical motions of the universe, such as the movement of planets in the solar system and physical forces on Earth. It involves both algebra and geometry, in combination with the concepts of infinity and limits. In contrast to algebra and geometry, which focus on properties of static structures, calculus centers on objects in motion. There are two principal forms of calculus, differential calculus and integral calculus, which are inversely related. At its most basic level, differential calculus is used in determining instantaneous rates of change of a dependent variable with respect to one or more independent variables; integral calculus is used for computing areas and volumes of nonstandard shapes.

Who Invented Calculus?

In the late seventeenth century, Isaac Newton (1646–1727) of England and Gottfried Wilhelm Leibniz (1646–1716) of Germany independently invented calculus. Isaac Newton began his development of calculus in 1666 but did not arrange for its publication. He presented his derivations of calculus in his book, *The Method of Fluxions*, written in 1671. This book remained unpublished until 1736, nine years after his death. Gottfried Leibniz began his work in calculus in 1674. His first paper on the subject was published in 1684, 50 years earlier than Newton's publication. Because of these circumstances and fueled by the eighteenth-century nationalism of England and Germany, a bitter controversy erupted over who first invented calculus. Was it Isaac Newton or Gottfried Leibniz?

Investigators found that Leibniz had made a brief visit to London in 1676. Supporters of Newton argued that during that trip, Leibniz may have gained access to some of Newton's unpublished work on the subject from mutual acquaintances within the mathematics community. However, these two prominent and outstanding mathematicians used their own unique derivations and symbolic notations for calculus, with Newton developing differential calculus first and Leibniz developing integral calculus first. For many decades, the calculus feud divided British mathematicians and continental mathematicians, and it remains a historical mystery into the twenty-first century. It was an unusual controversy in that it erupted rather late in the development of calculus and was ignited by the respective followers of Newton and Leibniz. In the twenty-first century, the general consensus is that both Newton and Leibniz invented calculus, simultaneously and independently.

Isaac Newton (1646–1727): The Man

Isaac Newton was revered in England during his lifetime and is recognized as one of the foremost mathematicians and physicists of all time. In addition to his invention of calculus, Newton is famous for designing and building the first reflecting telescope, formulating the laws of motion, and discovering the white light spectrum. He held many prestigious positions, including Fellow of Trinity College, Lucasian Professor of Mathematics, Member of Parliament for the University of Cambridge, Master of the Royal Mint of England, and many others. Even though Newton was extremely productive and admired universally for his work, on a personal level he was humble, cautious of others, and angered by criticism. His modest nature is embodied in his famous statement, "If I have seen farther than others, it is because I stood on the shoulders of giants." His works in mathematics and physics were recognized throughout Europe when he was honored as Fellow of the Royal Society of London in 1672. He subsequently served as the Society's president from 1705 until his death. In 1705, Newton was knighted in Cambridge by Queen Anne of England for his contributions to the Royal Mint. In 1727, Newton's name was immortalized in English history by his burial in London's Westminster Abbey and by the accompanying monument honoring his contributions to mathematics and science.

Gottfried Leibniz (1646–1716): The Man

Gottfried Leibniz is recognized as one of Germany's greatest scholars of philosophy, history, and mathematics. He was the son of a philosophy professor and a leader in the philosophy of metaphysics. His optimism is reflected in his words, "We live in the best of all possible worlds." On a personal level, Leibniz was considered likeable, friendly, and somewhat boisterous. Professionally, Leibniz was employed by a succession of German princes in the capacities of diplomat and librarian. He planned and founded several academies throughout Europe. For his knowledge of law, he was appointed Councilor of Justice for the Germanic regions of Brandenburg and Hanover. Similarly, Russian Tsar Peter the Great appointed Leibniz as Court Councilor of Justice for the Habsburgs. For his work in mathematics (derivations in calculus and invention of the binary number system), in 1673, Leibniz was appointed Fellow of the Royal Society of London, a society honoring outstanding mathematicians and scientists throughout Europe. By 1706, however, Leibniz's stellar reputation had begun to disintegrate. Accusations of plagiarism regarding the invention of calculus were unrelenting until Leibniz's death in 1716. In contrast to Newton, the only mourner at Leibniz's funeral was his secretary. Eventually, more than a century after his death, Leibniz's outstanding contributions to mathematics were recognized in Germany when a statue was erected in his honor at Leipzig, one of Germany's major centers of learning and culture.

Interestingly, it is Leibniz's symbolic notations for calculus, namely dy/dx and $\int y\, dx$, that have stood the test of time. These notations are most prevalent in calculus classrooms in the twenty-first century because of their consistency with the operations of differential equations and dimensional analysis. The most significant contribution to mathematics by Newton and Leibniz was their derivations of the Fundamental Theorem of Calculus, a theorem that unites both differential and integral calculus.

Building on Newton's and Leibniz's Work

Following the invention of calculus, additional contributions to calculus were made by John Wallis (1616–1703), Michel Rolle (1652–1719), Jacob Bernoulli (1654–1705), Guillaume de l'Hôpital (1661–1704), Brook Taylor (1685–1731), Colin Maclaurin (1698–1746), Joseph-Louis Lagrange (1736–1813), Bernard Bolzano (1781–1848), Augustin-Louis Cauchy (1789–1857), Karl Weinerstrasse (1815–1897), and Bernhard Riemann (1826–1866).

The Power of Calculus

The power of calculus in contemporary society rests primarily in its applications in the physical sciences, engineering, optimization theory, economics, geometrical measurement, probability, and mathematical modeling.

The following is a sampling of basic applications using the two major branches of calculus.

Applications of Differential Calculus

- *Environmental science*: An oil tanker runs aground and begins to leak oil into the ocean and surrounding land areas, resulting in potentially devastating consequences. Differential calculus can be used to supply information essential for assessing the leakage and resolving the problem. For example, the rate and volume at which the oil is leaking can by determined using calculus.
- *Business and economics*: Important applications of calculus in business and economics involve marginal analysis (known as the first derivative). Marginal costs, revenues, and profits represent rates of change that result from a unit increase in product production. This information is valuable in developing production levels and pricing strategies for maximizing profits.

- *Medicine*: Calculus can be used for evaluating the effectiveness of medications and dosage levels. For example, calculus can be used in determining the time required for a specific drug in a patient's bloodstream to reach its maximum concentration and effectiveness.
- *Biology and chemistry*: Assessments of chemical treatments for reducing concentration levels of biological contaminants (such as insects or bacteria) can be determined by calculus. For instance, calculus can be used in measuring the concentration levels, effectiveness, and time necessary for a chemical treatment supplied to a body of water to reduce its bacterial count to desired minimal levels.
- *Physics (velocity and acceleration)*: For moving objects (such as rolling balls or hot-air balloons), their maximum velocities, accelerations, and elevations can be determined using calculus.
- *Politics*: The number of years required in a city for the rate of increase in its voting population to reach its maximum can be determined using calculus.
- *Manufacturing*: The design of containers, meeting specific constraints, can be determined using calculus. For example, calculus will supply the dimensions of a container that will maximize its volume or minimize its surface area.

Applications of Integral Calculus

- *Inverse of differential calculus*: In mathematics, most operations have inverse operations. In calculus, the inverse of differentiation is integration. Therefore, a fundamental application of integral calculus is to find functions that produce the answers to a problem in differential calculus.
- *Measurement, area, and volume*: Integral calculus can be used to find (1) the areas between the graphs of functions over specified intervals, (2) the surface areas of three-dimensional objects, and (3) the volumes of three-dimensional objects.
- *Centroids*: The centroid (or center of mass) of an object can be found using integral calculus. For two- and three-dimensional objects, the centroid is the balancing point of the object. Calculus can be used to locate the position of the centroid on the object.
- *Fluid pressure*: Integral calculus is essential in the design of ships, dams, submarines, and other submerged objects. It is used in determining the fluid pressure on the submerged object at various depths from the water's surface. This information is essential in the design of submerged objects so they will not collapse.
- *Physics (work)*: When a constant force is applied to an object that moves in the direction of the force, the work produced by the force is found by multiplying the force by the distance moved by the object. However, when the applied force is not constant or is variable, calculus can be used in determining the work produced by the variable force (for example, the variable force needed to pull a metal spring, or the force exerted by expanding gases on the piston in an engine).

The aforementioned applications are examples of the most elementary applications of calculus. In the technological world of the twenty-first century, applications of calculus continue to evolve. The consequences of calculus are ubiquitous in contemporary society and impact every walk of life.

Recommendations for Mathematics Curriculum Reform

In 1983, following a harsh report from the National Commission on Excellence in Education, U.S. society began to question seriously the effectiveness of its educational systems. The report, titled *A Nation at Risk: the Imperative for Educational Reform*, was commissioned by U.S. President Ronald Reagan. The report cited U.S. students for their poor academic performance in every subject area at every grade level and their underachievement on national and international scales. The Commission warned the United States that its education system was "being eroded by a rising tide of mediocrity." In the years that followed, the Commission's explicit call for educational reform in U.S. schools served to generate numerous curricular reform efforts at the pre-college and college levels.

In response to this call for reform, in 1987, the Mathematical Association of America (MAA) and the National Research Council (NRC) co-sponsored a conference held in Washington, D.C., titled Calculus for a New Century. The conference was attended by more than 600 college and pre-college calculus teachers. The conference focused on the nature and need for calculus reform in college and pre-college institutions throughout the nation. During that conference, the phrase "Calculus should be a pump, not a filter in the pipeline of American education" became a national mantra for calculus reform.

National educational assessments conducted in 1989 further supported initiatives for calculus reform. During the 1980s, approximately 300,000 U.S. college students were enrolled annually in science-based calculus courses. Of that number, only 140,000 students earned grades of D or higher. Thus, more than 50% of U.S. college students were failing the calculus courses required for their majors, which included mathematics, all of the natural and physical sciences, and computer science. These bleak statistics served to motivate concerned calculus teachers to examine the traditional calculus curriculum, as well as their own teaching methodologies, with the intention of increasing course enrollments, student achievement, and enthusiasm for the subject.

Their efforts resulted in major calculus reform initiatives as early as 1989. The first set of recommendations for reform in school mathematics (grades prekindergarten–12) came from the National Council of Teachers of Mathematics (NCTM). These recommendations were delineated in NCTM's publication, *Principles and Standards for School Mathematics* (also known as NCTM Standards).

Four overarching standards (called Process Standards) were identified for improving mathematics instruction at all levels. These standards identified problem solving, reasoning and proof, connections, and communications as the four primary foci for mathematics instruction. During the 1990s, most U.S. states adopted this document as their curriculum framework for school mathematics. Decisions regarding the mathematics curriculum, textbook selections, and instructional strategies were revised in accordance with the recommendations of the NCTM Standards. Interestingly, the same document served to inspire pedagogical reform in mathematics at the college level, especially in calculus.

Traditional Calculus Versus Reformed Calculus

Until 1990, the calculus curriculum had remained basically the same for decades. The traditional calculus curriculum reflected formal mathematical language, mathematical rigor, and symbolic precision. Computations with limits, mathematical proofs, and elaborate mathematical computations were common practice in calculus classrooms. Students took careful notes, asked clarifying questions, and completed voluminous amounts of homework in preparation for test questions similar to those completed for homework. Instruction was teacher-centered and delivered through a lecture approach. Relevant applications were seldom considered, and graphing calculators and computers were rarely used in calculus instruction, and students were not allowed to use them for computations, graphing, or problem solving. Mathematics educators attributed the dismal performance of the majority of students in the nation's calculus classes to this traditional calculus curriculum. Consequently, by the mid-1990s, calculus reform movements had been initiated in many of the colleges and pre-college classrooms throughout the nation.

Calculus reform efforts at the college level in the 1990s often applied the pedagogical recommendations found in NCTM Standards. These pedagogical recommendations were also reflected in the revised Advanced Placement Calculus (AP Calculus) and International Baccalaureate Calculus (IB Calculus) courses offered in the nation's high schools. A measure of the subsequent success of the calculus reform movement at the pre-college level can be seen in the dramatic increase in numbers of students who took these courses from the 1980s into the twenty-first century. Specifically, the National Center for Education Statistics reported that the percentage of students completing calculus in high school had risen from 6% to 14% in the years from 1982 to 2004. The number of students completing calculus in high school continues to grow exponentially, at an estimated rate of 6.5% per year.

Several reform calculus curricula originated in the 1990s, and continue into the twenty-first century. The following examples are prominent reform calculus projects: Calculus, Concepts, Computers and Cooperative Learning (C^4L) conducted at Purdue University; the Calculus Consortium at Harvard (CCH) conducted at Harvard University; and Calculus and Mathematica (C&M) conducted at the University of Illinois at Urbana-Champaign and at Ohio State University.

While these three reform calculus projects differ from each other in significant ways, they share the following characteristics:

- They use graphing calculators, computers, and computer algebra systems (CAS) extensively for instruction, exploration, and visual representations. Supporters argue that technology serves to alleviate the huge burden of algebraic computation so characteristic of traditional calculus. The rationale for this reform is that technology facilitates instructional processes that focus on the principles of calculus rather than on computational procedures. Moreover, the graphical and visual representations provided by these technologies offer alternative modalities for learning that accommodate students' different learning styles. The curricula for CCH and C^4L focus heavily on graphing calculators, whereas the curriculum for C&M relies heavily on the computer software, Mathematica.
- The teacher serves as a facilitator of learning rather than the main conveyor of knowledge. While the teacher continues to initiate instruction and answer questions, mathematical situations are often explored by groups of students, using cooperative learning strategies. Using the principles of constructivist learning, students are guided to discover mathematical properties for themselves in a laboratory setting.
- A major focus is placed on real applications from multiple disciplines. The intention is to raise students' interest in the subject and motivate them with relevant applications.
- Mathematical rigor and formal language are de-emphasized. The abstractions of mathematical proof and rigor are postponed for several semesters to provide sufficient time for students to gain practical and intuitive knowledge of the subject.
- Assessment focuses heavily on students' writing, explanations of problem solutions, and open-ended projects. Sometimes students' explanations are valued as highly as the accuracy of their answers.

Whereas all of the above instructional practices have shown varying degrees of success in reform calculus classrooms, some areas of concern have been identified by those involved in the projects. Specifically:

- Focusing heavily on relevant applications sometimes results in the omission of important calculus content that cannot always be motivated by applications.
- The use of everyday language sometimes results in imprecise and incorrect mathematical definitions.
- Overuse of technologies for computation and graphing can weaken the development of students' quantitative reasoning and computational skills in calculus.
- Real-world problems are sometimes too complex and frustrating to students because of the extraneous and irrelevant information they usually contain.
- Short-answer problems for assessment are often easier for students than describing their problem-solving procedures in writing.
- Constructivist approaches are often too time consuming, allowing insufficient time for covering the entire calculus curriculum during class time.

Resolution of these concerns will surely be addressed in future curriculum revisions, and changes or modifications will be made accordingly. However, these accommodations are consistent with the historical evolution of calculus, which is the study of change and systems in perpetual motion.

Summary

In the past, calculus was taught in ways that made it accessible to only a small proportion of the population. However, recent curricular and pedagogical reforms in calculus, both at the college and pre-college levels, have served to increase student success, include twenty-first-century-technologies, and triple course enrollments. Statistics indicate that calculus enrollments will continue to increase exponentially. These findings suggest that calculus instruction in the United States is responding positively to the academic needs of society.

Indeed, by combining the power of technology with calculus, new areas of mathematics are emerging (for example, fractals, dynamical systems, and chaos theory). These new branches of mathematics have allowed humans to mimic nature's designs of mountain ranges, oceans, and plant growth patterns—which once were considered random acts of nature. In conclusion, calculus as a subject is still growing, and its applications are continually expanding to meet the needs of a dynamic, diverse, and technologically driven society.

Further Reading

Barnett, Raymond, Michael Ziegler, and Karl Byleen. *Calculus for Business, Economics, Life Science, and Social Science*. Upper Saddle River, NJ: Prentice-Hall, 2005.

Bressoud, David M. "AP Calculus: What We Know." June 2009. http://www.maa.org/columns/launchings/launchings_06_09.html#Q1.

Calinger, Ronald. *A Contextual History of Mathematics*. Upper Saddle River, NJ: Prentice-Hall, 1999.

Calter, Paul, and Michael Calter. *Technical Mathematics with Calculus*. 4th ed. Hoboken, NJ: Wiley, 2007.

Dubinsky, Edward. "Calculus, Concepts, Computers and Cooperative Learning." May 2004. http://www.pnc.edu/Faculty/kschwing/C4L.html.

Ferrini-Mundy, Joan, and K. Graham. "An Overview of the Calculus Curriculum Reform Effort: Issues for Learning, Teaching, and Curriculum Development." *The American Mathematical Monthly* 98, no. 7 (1991).

Gleaso, Andrew M., and Deborah H. Hallett. *The Calculus Consortium Based at Harvard University*. Spring 1992. http://www.wiley.com/college/cch/Newsletters/issue1.pdf.

International Baccalaureate Organization. *Diploma Programme Mathematics HL*. Wales, UK: Peterson House, 2006.

Johnson, K. "Harvard Calculus at Oklahoma State University." *The American Mathematical Monthly* 102, no. 9 (1995).

Murphy, Lisa. "Reviewing Reformed Calculus." http://ramanujan.math.trinity.edu/tumath/research/studpapers/s45.pdf.

Rogawski, Jon. *Single Variable Calculus*. New York: W. H. Freeman, 2008.

Silverberg, J. "Does Calculus Reform Work?" *MAA Notes* 49 (1999).

Steen, Lynn A. *On the Shoulders of Giants: New Approaches to Numeracy*. Washington, DC: National Academy Press, 1990.

Tucker, Thomas, ed. *Priming the Calculus Pump: Innovations and Resources*. Washington, DC: Mathematical Association of America, 1990.

Sharon Whitton

See Also: Algebra and Algebra Education; Archimedes; Calculus and Calculus Education; Function Rate of Change; Functions; Geometry and Geometry Education.

Cameras

See *Digital Cameras*

Calendars

Category: Space, Time, and Distance.
Fields of Study: Measurement; Number and Operations; Representations.
Summary: Various calendars use different methods of resolving the need for "leap" days, months, or years.

Even the earliest human beings must have noticed the astronomical cycles: the alternation of day and night, the pattern of the changes in the moon's shape and position, and the cycle of the seasons through the solar year. It must have been frightening every autumn as the days became shorter, causing concern that the night might become permanent. This led to celebrations of light in many areas as the days began to lengthen again. Once the repetitions of the patterns were recognized, people could count them to keep track of time. Longer cycles helped avoid difficulties in keeping track of large numbers—once approximately 30 days had been counted, people could, instead, start counting "moons."

This same technique of grouping also occurred in the development of counting systems in general—leading to place-value structures in numeration systems.

The problem was that the shorter cycles did not fit evenly into the longer cycles. Trying to fit the awkward-length cycles together actually led to some mathematical developments: two different cycles would come together at the least common multiple of the lengths of their cycles; modular arithmetic and linear congruences were methods of handling leftover periods beyond the regular cycle periods.

A 1412–1416 illumination depicting the month of March with the constellations of the zodiac on top.

The Julian and Gregorian Calendars

The Romans developed the Julian calendar (named for Julius Caesar), recognizing that the exact number of 365 days in one year was slightly too short and would soon throw the calendar off the actual cycle of the solar year. They found a remedy by assuming the solar year to be 365.25 days. To handle the one-quarter day, they added one full day every four years—the day that we call "leap-year day" on February 29 of years whose number is a multiple of four. This gives $3(365)+366=1461$ days in four years, or an average of 365.25 days per year as desired. However, the actual solar year is 365.2422 days long (to four decimal places), about 11 minutes less than the Romans' value. Even in a human lifetime, this is negligible. Over centuries, however, the extra time builds up so that by the 1500s, the calendar was 10 days off from the solar cycle (for example, the vernal equinox seemed to be coming too late).

In 1582, Pope Gregory XIII assembled a group of scholars who devised a new system to fit better. It kept the Roman pattern except that century years (1600, 1700), which should have been leap years in the Roman calendar, would not have a February 29 unless they were multiples of 400. For example, 1900 was not a leap, year but 2000 was. In the full 400-year cycle, there are (400×365) regular days + 97 leap-year days = 146,097 days, making an average of 365.2425 days per year. This cycle is only .0003 days (about 26 seconds) too much; in 10,000 years, we would gain three extra days. This system was called the Gregorian calendar. Since the longer Julian calendar had fallen behind the solar year by about 10 days, the changeover to the Gregorian required jumping 10 days.

Various countries in Europe changed at different times, with each switch causing local controversy as people felt they were being "cheated" out of the skipped days. The effects of the change are noticed in history. When Isaac Newton was born, the calendar said it was December 25, 1642; but later England changed the calendar, so some historians today give Newton's birthday as January 4, 1643. The Russians did not change their calendar until after the 1917 October Revolution, which happened in November by the Gregorian calendar.

The Lunar Calendar

The other incongruity of calendar systems is that the moon cycle of 29.53 days does not fit neatly in the 365.2422 days of the year. Twelve moon periods is 11 days shorter than a year, and 13 "moons" is 18 days too long. It is interesting to note that of the three major religious groups of the Middle East—the Christians, the Muslims, and the Jews—each chose a different way to handle "moons/months." The Christians (actually, originally, the Romans) ignored the moon cycle and simply created months of 30 and 31 (and 28 or 29) days. The Muslims considered their year to be 12 moon cycles and ignored the solar year. This means that dates in the Muslim calendar are shifted back approximately 11 days each year from the solar calendar, and Muslim festivals move backward through the seasons.

People in the Jewish faith chose to keep both the solar and lunar cycles. After 12 lunar months, a new year begins—as in the Muslim calendar—11 days "too early." However, after the calendar slips for two or three years—falling behind the solar calendar by 22 or 33 days—an extra month is inserted to compensate for the loss. There is a 19-year pattern of the insertion of extra months, which keeps the year aligned with the solar year. Interestingly, the traditional east Asian calendar follows a pattern very similar to the Jewish calendar.

The Mayan Calendar

The Mayans of Central America had a very complex pattern of cycles leading to a 260-day year for religious purposes, and a regular solar year that was used for farming and other climate-related activities. Their base-20 numeration system, which should have had place-value columns of 1-20-400-8000, was adjusted to 1-20-360 to fit into the 365+ days of the year. They were also notable for developing massive cycles of years lasting several millennia, including one ending in late 2012 of the Gregorian calendar.

Further Reading

Aslaksen, Helmer. "The Mathematics of the Chinese Calendar." http://www.math.nus.edu.sg/aslaksen/calendar/chinese.shtml.

Crescent Moon Visibility and the Islamic Calendar. http://aa.usno.navy.mil/faq/docs/islamic.php.

Duncan, David Ewing. *Calendar: Humanity's Epic Struggle to Determine a True and Accurate Year*. New York: Harper Perennial, 2001.

Rich, Tracey R. "Judaism 101: Jewish Calendar." http://www.jewfaq.org/calendar.htm.

Richards, E. G. *Mapping Time: The Calendar and Its History*. New York: Oxford University Press, 2000.

Stray, Geoff. *The Mayan and Other Ancient Calendars.* New York: Walker & Company, 2007.

Lawrence H. Shirley

See Also: Astronomy; Incan and Mayan Mathematics; Measuring Time.

Canals

Category: Architecture and Engineering.
Fields of Study: Algebra; Geometry; Number and Operations; Problem Solving.
Summary: Modern canal design, particularly the challenges of a lock system, depends on partial differential equations and other mathematics.

Canals are human-made channels for water, including both waterways big enough to be traversed by ship (built for transportation), and aqueducts (built for water supply and irrigation). The building of canals was critical to the formation of many ancient civilizations, which needed to manipulate water access in order to enable an early urban lifestyle. Many ancient mathematics texts address such large-scale ancient engineering projects.

A number of the surviving Babylonian tablets dealing with geometry were composed for canal projects: they calculated the number of workers necessary to build the canal in a given number of days, the dimensions of the canal, and the total wage expenses so that the ruler for whom they were built would know how much the project would cost. Mathematical problems related to the construction of canals can also be found in the fifth chapter of *The Jiuzhang Suanshu* (*Nine Chapters on the Mathematical Art*), one of the earliest surviving ancient Chinese mathematics texts. Mathematicians and engineers have long investigated canals.

For instance, Jacopo Riccati worked on hydraulics and constructed dikes in Venice, and Barnabé Brisson employed descriptive geometry in the design and construction of ship canals. Mathematicians like George Green and Joseph Boussinesq analyzed and modeled wave motion in canals. John Russell tested and studied steam-powered canal transportation and wave creation for the Union Canal Company. Mikhail Lavrentev created a theoretical foundation for large projects on the Volga, Dnieper, and Don rivers. Mathematics theories and techniques are critical when engineers, mathematicians, and software programmers model the

The Gatun (above) and Miraflores Locks of the Panama Canal can be viewed from a webcam.

Major Canals

Significant canals include the Erie Canal in the United States, the Suez Canal in Egypt, the Panama Canal in Panama, and the Grand Canal in China, each of which was constructed as a major operation for the sake of hastening trade and transport. Judge Benjamin Wright, who some call the father of American civil engineering, was appointed the chief engineer of the Erie Canal. Astronomer and mathematician Guo Shoujing (also known as Kuo Shou-ching) was the head of the Water Works Bureau in the thirteenth century. He made improvements to control the water level in existing canals and built new ones.

The Suez Canal was imagined long before it was completed, and the Egyptians were masters of large-scale engineering projects. Napoleon Bonaparte, during the French invasion of Egypt, reportedly discovered ruins of an ancient canal, which inspired him to order a preliminary survey exploring the possibility of a north–south canal joining the Mediterranean and the Red Sea (the ancient canal had been east–west and was intended to link the Red Sea and the Nile). The project was abandoned—possibly because of the belief that the Red Sea was higher than the Mediterranean—and so the canal remained unbuilt for 70 years.

changing flow rates and levels of a canal. They rely on mathematics like the Saint-Venant equations (partial differential equations that are named after mechanic and mathematician Jean Claude Saint-Venant).

The simplest canals are merely trenches through which water runs, usually lined with some kind of construction material. Canals need to be level in order to be navigable (a ship cannot move "uphill"). When the land itself is not level, a lock system must be used. Locks are systems for raising and lowering boats from one stretch of water to a stretch of water at a different level. The most common type of canal lock—used in ancient China, and most likely in the ancient West, and still common today—is the pound lock, which consists of a watertight chamber with gates at either end to control the water level in the chamber.

Engineer Chiao Wei-Yo is credited with the design of the lock system, which he used on the Grand Canal in the tenth century. In the pound lock system, a ship enters the chamber (the "pound") from one length of canal; water is raised or lowered to bring the ship to the level of the next length of canal; and the ship exits the chamber. The necessity of locks added much complexity, time, and room for error to the construction of canals, which would have been sufficient to discourage Napoleon's aims. In 2010, the Panama Canal commemorated its one-millionth transit, and engineers plan to expand the canal by adding more locks. It has been referred to as one of the seven wonders of the industrial world.

Further Reading

Bernstein, Peter. *Wedding of the Waters: The Erie Canal and the Making of a Great Nation*. New York: W. W. Norton, 2006.

Karabell, Zachary. *Parting the Desert: The Creation of the Suez Canal*. New York: Alfred A. Knopf, 2003.

Montañés, Jose. "Mathematical Models in Canals." In *Hydraulic Canals: Design, Construction, Regulation and Maintenance*. New York: Taylor & Francis, 2006.

Parker, Matthew. *Panama Fever: The Epic Story of One of the Greatest Human Achievements of All Time*. New York: Doubleday, 2007.

Bill Kte'pi

See Also: Floods; Tides and Waves; Tunnels; Water Distribution.

Carbon Dating

Category: Weather, Nature, and Environment.
Fields of Study: Algebra; Data Analysis and Probability.
Summary: Exponential and logarithmic functions are used in carbon dating—a method of determining the age of plant and animal fossils.

As is demonstrated throughout this encyclopedia, mathematics provides explanations for many interesting physical phenomena, and enables humankind to better understand its surrounding world. One of our ongoing intellectual projects is simply to make sense of the world we inhabit, based on the evidence that surrounds us. As anthropologists, archaeologists, and geologists have worked to determine the age of the earth and to track the evolution of species, radioactive isotopes have played a prominent role in efforts to create a timeline that charts a wide range of historical developments. In particular, carbon-14 dating has provided a fundamental test enabling scientists to accurately date certain plant and animal fossils that are approximately 60,000 years old or less. Willard Libby was one of the first to research radiocarbon dating, and he won a Nobel Prize in chemistry. Carbon dating is not an exact science, and statistical methods are used to enhance the reliability of the methods.

The Mathematics of Carbon Dating

Left alone, a radioactive quantity will decay at a rate proportionate to the amount of the quantity present at a given time. More specifically, a radioactive chemical element (such as uranium) is one that is unstable; as it decays, it emits energy and its fundamental makeup changes as the mass of the element is changed to an element of a different type. Because such an element is losing mass at a rate proportionate to the available mass at time t, an exponential function may be used to model the amount of the isotope that is present.

Letting $M(t)$ represent the mass of the element at time t, it turns out that $M(t) = M_0 e^{-kt}$, where M_0 is the mass at initial measurement (at time $t = 0$), and k is a constant that is connected to the rate at which the element decays. Furthermore, k is tied to the isotope's half-life (the amount of time it takes for 50% of the mass present to decay). In the given model, if h represents the half-life, then when $t = h$, it follows that

$$M(h) = \frac{M_0}{2}.$$

That is, the equation

$$\frac{M_0}{2} = M_0 e^{-kh}$$

must hold. Dividing both sides by M_0, yields

$$\frac{1}{2} = e^{-kh}$$

and using the natural logarithm function, one may solve for k and thus rewrite the most recent equation as

$$-kh = \ln\left(\frac{1}{2}\right).$$

This can be rewritten as

$$k = \frac{-\ln\left(\frac{1}{2}\right)}{h}.$$

A property of the natural logarithm is that

$$-\ln\frac{1}{2} = \ln(2)$$

so that in slightly simpler terms,

$$k = \frac{\ln(2)}{h}.$$

Therefore, the model for radioactive decay of an element having half-life h is

$$M(t) = M_0 e^{-(\ln(2)/h)t}.$$

With this background in place, one is now ready to understand how carbon dating works.

All living things contain carbon, and the preponderance of the carbon present in plants and animals is its stable isotope, carbon-12. At the same time, every living being takes in radioactive carbon-14, and this carbon-14 becomes part of our organic makeup. While carbon-14 is constantly decaying simply by doing the normal things that come with being alive, each living organism continuously replenishes its supply of carbon-14 in such a way that the ratio of carbon-12 to carbon-14 in its body is constant.

When no longer living, a plant or animal lacks the ability to ingest carbon-14, and thus the ratio of carbon-12 to carbon-14 starts to change, and this ratio changes at the rate that carbon-14 decays. Chemists have long known that carbon-14 has a half-life of approximately $h = 5700$ years, and this knowledge, together with the exponential model

$$M(t) = M_0 e^{-(\ln(2)/h)t}$$

enables people to determine the age of certain fossils. Consider, for example, the situation where a bone is found that contains 40% of the carbon-14 it would be expected to have in a living animal. With less than half the original amount present, but more than 25%, it can be determined that the bone is somewhere between one and two half-lives old; that is, the animal lived between 5700 and 11,400 years ago.

Through our understanding of exponential functions and logarithms, this estimate can be made much more precise.

Specifically, let $t = 0$ be the year the animal died. The present year t satisfies the equation $M(t) = 0.4M_0$, since 40% of the initial amount of carbon-14 remains. From the model, it is known that t must be the solution to the equation

$$0.4\,M_0 = M_0 e^{-(\ln(2)/5700)t}.$$

First, divide both sides by M_0 to get $0.4 = e^{-(\ln(2)/5700)t}$ and then, taking the natural logarithm of both sides ofthe equation, it follows that

$$\ln(0.4) = \frac{-\ln(2)}{5700}t.$$

Thus, solving for t yields

$$t = \frac{-5700\ln(0.4)}{\ln(2)} \approx 7500 \approx \text{years}$$

and the skeletal remains have been dated according to their carbon content.

Limitations of Carbon Dating

Carbon dating does have some reasonable limitations. One of these involves the complications of measuring only trace amounts of carbon-14, and emphasizes the behavior of functions that model exponential decay. For each half-life that passes, half of the most recent quantity of the element remains. That is, after one half-life,

$\frac{M_0}{2}$ remains; after two, half of that amount,

or $\frac{M_0}{4}$ is left;

after three, $\frac{M_0}{8}$ is present.

The quantity rapidly diminishes from there. For instance, after 10 half-lives have elapsed, there is

$$\frac{M_0}{2^{10}} = \frac{M_0}{1024} \text{ or approximately}$$

$0.0009766M_0$ left. Because each living organism only contains trace amounts of carbon-14 to begin with (of all carbon atoms, only about one-trillionth are carbon-14), after 10 half-lives elapse, the remaining amount of carbon-14 is so small that it is not only difficult to measure accurately, but it is difficult to ensure that the measured carbon-14 actually remains from the organism of interest and was not somehow contributed from another source. Ten half-lives is approximately 60,000 years, so any organism deemed older than that needs to be dated in another manner, typically using other radioactive isotopes that have considerably longer half-lives.

Finally, because radiocarbon dating depends on naturally occurring radioactive decay, its accuracy depends on such decay not being accelerated by unnatural causes. In the 1940s, the Manhattan Project resulted in humankind's development of synthetic nuclear energy and weapons; subsequent nuclear testing and accidents have released radiation into the atmosphere that makes the accuracy of carbon-14 dating more suspect for organisms that die after 1940.

New Developments

The exponential model $M(t) = M_0 e^{-kt}$ of radioactive isotope decay has enabled humans to better understand our surrounding world, and to know with confidence key information about the history of the existence of plant and animal life on Earth. Even today, there are new developments in the science of radiocarbon dating as experts work to understand how subtle changes in Earth's magnetic field and solar activity affect the amounts of carbon-14 present in the atmosphere. In addition to continuing to help analyze fossil histories, carbon-14 dating may prove an important tool in ongoing research in climate change.

The Accelerator Mass Spectrometry method of dating directly measures the number of carbon atoms rather than their radioactivity, which allows for the dating of small samples. Other methods under development include nondestructive carbon dating, which eliminates the need for samples. A group of Russian mathematicians have proposed a new chronology of history based on other methods for dating; however, many have dismissed their work as pseudoscience. Physicist Claus Rolfs explores methods to accelerate radioactive decay in the hope of reducing the amount of radioactive material.

Further Reading

"Archaeological 'Time Machine' Greatly Improves Accuracy of Early Radiocarbon Dating." *Science News Daily* (February 11, 2010) http://www.sciencedaily.com/releases/2010/02/100211111549.htm.

Brain, Marshall. "How Carbon Dating Works." http://www.howstuffworks.com/carbon-14.htm.

Comap. *For All Practical Purposes: Mathematical Literacy in Today's World.* 7th ed. New York: W. H. Freeman, 2006.

Connally, E. et al. *Functions Modeling Change: A Preparation for Calculus.* Hoboken, NJ: Wiley, 2007.

Matt Boelkins

See Also: Algebra and Algebra Education; Calculus and Calculus Education; Exponentials and Logarithms; Functions.

Carbon Footprint

Category: Weather, Nature, and Environment.
Fields of Study: Algebra; Data Analysis and Probability; Measurement; Representations.

Summary: A carbon footprint is a mathematical calculation of a person's or a community's total emission of greenhouse gases per year.

Carbon footprint is intended to be a measure of the ecological impact of people or events. It is a calculation of total emission of greenhouse gases, typically carbon dioxide, and is often stated in units of tons per year. There is no universal mathematical method or agreed-upon set of variables that are used to calculate carbon footprint, though scientists and mathematicians estimate carbon footprints for individuals, companies, and nations. Many calculators are available on the Internet that take into account factors like the number of miles a person drives or flies, whether or not he or she uses energy efficient light bulbs, whether he or she shops for food at local stores, and what sort of technology he or she uses for electrical power. Some variables are direct, such as the carbon dioxide released by a person driving a car, while others are indirect and focus on the entire life cycle of products, such as the fuel used to produce the vegetables that a person buys at the grocery store and disposal of packaging waste.

The notion of a carbon footprint is being considered in a wide range of areas, including the construction of low-impact homes, offices, and other buildings. The design must take into account not only the future impact of the building in terms of carbon emissions, but carbon-related production costs for the materials, labor, and energy used to build it. Mathematical modeling and optimization helps engineers and architects create efficient, useful, and sometimes even beautiful structures while reducing the overall carbon footprint. Mathematicians are also involved in the design of technology that is more energy efficient, as well as methods that allow individuals and businesses to convert to electronic documents and transactions rather than using paper. These methods include using improved communication technology, faster computer networks, improved methods for digital file sharing and online collaboration, and security protocols for digital signatures and financial transactions. Manufacturers are increasingly being urged and even required to examine their practices, since manufacturing processes produce both greenhouse gasses from factory smokestacks and waste heat. Mathematicians and scientists are working on ways to recycle much of this heat for power generation. One proposed device combines a loop heat pipe, which is a passive system for moving heat from a source to another system, often over long distances, with a Tesla turbine. Patented by scientist and inventor Nikola Tesla, a Tesla turbine is driven by the boundary layer effect rather than fluid passing over blades as in conventional turbines. It is sometimes called a Prandtl layer turbine after Ludwig Prandtl, a scientist who worked extensively in developing the mathematics of aerodynamics and is credited with identifying the boundary layer.

These are in turn related to the Navier–Stokes equations describing the motion of fluid substances, named for mathematicians Claude-Louis Navier and George Stokes. The Navier–Stokes equations are also of interest to pure mathematics, since many of their mathematical properties remain unproven at the beginning of the twenty-first century.

Carbon Footprints of People

A calculation of the carbon footprints of different aspects of people's lives, and then the aggregate for a year, is always an estimate. For example, different towns use different methods for generating electricity. Entering data for an electric bill allows for a rough estimate of the household's carbon footprint, but not exact numbers, which would depend on the electricity generating methods. Houses contribute to carbon footprints through their building costs, heating and cooling, water filtration, repair, and maintenance—all of which use products with carbon footprints.

Travel is another major contributor to peoples' carbon footprints. Daily commutes and longer trips with any motorized transportation contribute to carbon dioxide emissions. When computing carbon footprints, fuel production and storage costs have to be taken into consideration.

The food that people eat contributes to the carbon footprint if it is transported by motorized vehicles before being eaten. The movement of locavores (people who eat locally grown foods) aims to minimize the carbon footprint of food. Also, different farming practices may contribute more or less to the carbon footprint of food.

The objects people use contribute to their carbon footprints. Recycling and reusing reduces the need for landfills, waste processing, and waste removal, all of which have carbon footprints. There are individuals and communities who avoid waste entirely; several countries, such as Japan, have plans to mandate zero-waste practices within the next few decades.

Economy and Policy

There are two main strategies for addressing carbon footprints. The first strategy is to lower the carbon footprint by modifying individual behaviors, such as traveling by bike, eating locally, and recycling. The second strategy is to perform activities with negative carbon footprints, such as planting trees, to match carbon footprints of other activities.

Some companies incorporate activities that offset the carbon footprint of their main production into their business plans, either lowering their profit margins or passing the cost to their customers. There are economic laws and proposals that attempt to integrate carbon footprint considerations into the economy, usually through taxes on use of fuel, energy, or emissions. Carbon dioxide emissions, in economic terms, are a negative externality (a negative effect on a party not directly involved in the economic transaction). Money collected through carbon taxes is generally used to offset the cost to the environment.

Emissions trading is another mathematics-rich area of dealing with carbon footprints economically. Governments can sell emission permits to the highest-bidding companies, matching their carbon footprints, and capping the total emission permits sold. This method allows prices of permits to fluctuate with demand, in contrast with carbon taxes in which prices are fixed and the quantities of emissions can change. Economists model the resulting behaviors, and advise policymakers based on the models' outcomes.

Marginal Abatement Cost Curve

"Marginal cost" is an economic term that means the change of cost that happens when one more unit of product is made, or unit of service performed. For physical objects, the curve is often U-shaped. The first units produced are very costly because their cost production involves setting up the necessary infrastructure. As more units are produced, and the infrastructure is reused, the price goes down until the quantities of production reach such levels that the logistic difficulties drive the price per additional units higher again.

A marginal abatement curve shows the cost of reducing emissions by one more unit. These curves are usually graphed in percents. For example, such a curve can be a straight line, with the cost of eliminating the first few percent of emission being zero or even negative. This happens because it can be done by changing practices within existing economic infrastructures, such as cheap smart switches into the residential sector's lighting grids. Additional lowering of the carbon footprint, however, requires deeper and costlier changes to the way of life. For example, there are relatively high costs involved in switching to wind and solar power, or switching to the use of crop rotations that do not require high-carbon fertilizers.

Country by Country

The average carbon footprint of citizens varies by country. For example, in late 2000s, the average annual carbon footprint of a U.S. citizen was about 30 metric tons per year, and a Japanese citizen about 10 metric tons per year. However, these calculations are extremely complicated because of global trade. For example, many developed countries "export" or "outsource" their carbon emissions to developing countries. Products imported from developing countries account for anywhere from a tenth to a half of the carbon footprints of developed nations.

International calculations indicate a strong correlation between the average carbon footprint of a country's citizen and the average per capita consumption. The higher the consumption rates, the higher the average carbon footprint.

The categories used for calculation for countries are similar to those used for individuals and include construction, shelter, food, clothing, manufactured products, services, transportation, and trade. The ratios of these items to one another in the carbon footprints vary by country. For example, the greatest item in the U.S. carbon footprint is shelter (25%), with mobility

Carbon footprints are calculated to include travel, fuel production, transportation, and storage. In Canada, mobility is the highest contributor to the national carbon footprint.

being second (21%). In contrast, Canada's greatest item affecting carbon footprint is mobility (30%), and its second greatest is shared between shelter and service (18% each).

Further Reading

Berners-Lee, Mike. *How Bad Are Bananas? The Carbon Footprint of Everything*. Vancouver, BC: Greystone Books, 2011.

Goleman, Daniel. *Ecological Intelligence: How Knowing the Hidden Impacts of What We Buy Can Change Everything*. New York: Broadway Books, 2009.

Maria Droujkova

See Also: City Planning; Climate Change; Electricity; Energy; Farming; Fuel Consumption; Green Design; Green Mathematics; Recycling; Traffic; Wind and Wind Power.

Careers

Category: Mathematics Culture and Identity.
Fields of Study: All.
Summary: There are a wide variety of careers in many disciplines available to those with a mathematics background.

"What can one do with a mathematics degree other than teaching?" It is a question asked by many aspiring mathematicians. In fact, a more accurate question to ask should be "What can't one do with a math degree?" Actually, the study of mathematics extends far beyond mere number crunching and doing fast mental arithmetic in grocery stores. The fact is that studying mathematics can prepare one for numerous careers.

In general, companies believe that studying mathematics develops analytical skills and the ability to work in a problem-solving environment. These are the skills and experiences that are essential assets to one's success in the workplace. Precisely, mathematics is often the quintessential element to fluently communicate with people of various backgrounds. It is the ability to efficiently process a manifold of information and deliver the technical details to a general audience that makes mathematicians valuable. Having a mathematics background not only helps people broaden their pool of career options, it also helps to land some of the best jobs available.

According to an article published in the *Wall Street Journal* on January 26, 2009, a "mathematician" is considered to be the best occupation in the United States. This ranking was determined based on five criteria inherent to every job: environment, income, employment outlook, physical demands, and stress. In fact, five out of the six "best jobs" in terms of low stress, high compensation, autonomy, and hiring demand in the *Job Related Almanac* by Les Krantz are all mathematics related: (1) mathematician, (2) actuary, (3) statistician, (4) biologist, (5) software engineer, and (6) computer systems analyst. In this entry, a collection of possible career opportunities appropriate for someone with a mathematics background is provided, and a list of resources is given on how to find a job with different levels of academic degrees. The lists are by no means exhaustive and should only be used as a reference.

Analytical Thinking

Why is mathematics a required subject in school curricula at all levels? Why is mathematics so essential for the proper functioning of everyday tasks in society? Why do most people who excel in their field credit their success to their formal training in mathematics? One possible reason is that a proper training in mathematics provides people with abilities to think and solve problems critically in novel settings.

A Web site sponsored by the Department of Mathematics at Brigham Young University provides a list of possible career options for someone with a background in mathematics. Some of the more common professions include actuary, architect, chemical engineer, college professor, computer scientist, cryptanalyst, economist, mechanical engineer, quantitative financial market analyst, and statistician; some less well-known career options include air traffic controller, animator, astronaut, epidemiologist, geologist, hydrologist, lawyer, market research analyst, composer, physician, technical writer, and urban planner. Certainly, a fixed set of mathematics curriculum will not prepare one for all the jobs listed here. What will be consistent is gaining the ability to solve problems analytically and critically.

Not many people know that the San Antonio Spurs Basketball Hall of Famer David Robinson had a B.S.

in mathematics from the U.S. Naval Academy. Even Michael Jordan toyed with the idea of being a mathematics major in his early college years. It is perhaps not surprising that one of the world's most influential bankers and financiers, J. P. Morgan, majored in mathematics, but not many would think that mathematics would find its way into entertainment. For example, American actress Danica McKellar, who had a leading role in a television comedy-drama *The Wonder Years*, is a well-known mathematics author and education advocate. The popular television drama, *Numb3rs* featured a mathematician who helped his brother in the FBI to solve crimes with his mathematical genius. A few popular movies that successfully portray mathematicians in society include *Good Will Hunting* (1997), *A Beautiful Mind* (2001), and *Proof* (2005).

Although it is seemingly impossible to categorize every branch of mathematics in society, career options available for those who study under common branches of mathematics include the areas of applied mathematics, actuarial mathematics, financial mathematics, and other emerging fields.

Applied Mathematics

Applied mathematicians often solve problems that originate in physics, chemistry, geology, biology, or various disciplines of engineering. Mathematics is used to model physical phenomena, to answer questions derived from observations, to learn characteristics of large quantities, and to make predictions and improvements for future events. A representative mathematical training includes coursework in numerical analysis and methods, computer programming, computer languages, applied and experimental statistics, and probability theory, as well as a few courses in another field of interest.

Often, a typical applied or computational mathematics problem is interdisciplinary in nature and derived from realistic demands in industry. People who wish to gain a general sense of what these types of problems entail are encouraged to attend mathematics-in-industry workshops that are available in Europe and some parts of the United States and Asia. Mathematics in Industry and International Study Groups maintains a Web site that provides updated information for future study groups and meetings. The Society for Industrial and Applied Mathematics (SIAM) maintains a list of example organizations, corporations, and research institutions that hire mathematicians and computational scientists with an applied mathematics training. These organizations, corporations, and research institutions include the following:

- Aerospace and transportation equipment manufacturers such as Aerospace Corp., Boeing, Ford Motor Co., General Motors, Lockheed Martin, and United Technologies
- Chemical and pharmaceutical manufacturers such as DuPont, GlaxoSmithKline, Kodak, Merck & Co., Pfizer, and Wyeth
- Communications service providers such as Clear Channel Communications, Qwest Communications and Verizon
- Electronics and computer manufacturers such as Bell Laboratories, Alcatel-Lucent, Hewlett-Packard, Honeywell, IBM Corporation, Motorola, Philips Research, and SGI
- Energy systems firms such as Lockheed-Martin Energy Research Corporation and the Schatz Energy Research Center (SERC)
- Engineering research organizations such as AT&T Laboratories—Research, Exxon Research and Engineering, and NEC Laboratories America
- Federally funded contractors such as the Mitre Corporation and RAND
- Medical device companies such as Baxter Healthcare, Boston Scientific, and Medtronic
- U.S. government agencies such as the Institute for Defense Analyses, NASA's Institute for Computer Applications in Science and Engineering, National Institute of Standards and Technology, Naval Surface Warfare Center, Supercomputing Research Center, and the U.S. Department of Energy
- U.S. government labs and research offices such as the Air Force Office of Scientific Research, Lawrence Berkeley National Laboratory, Los Alamos National Laboratory, Oak Ridge National Laboratory, Pacific Northwest National Laboratory, and Sandia National Laboratories
- Producers of petroleum and petroleum products such as Amoco, Exxon Research and Engineering, and Petróleo Brasileiro S/A, Petrobras

Actuarial Mathematics

An actuary is a risk management professional who helps design insurance plans by recommending premium rates and making sure companies are designating enough funds to pay out on claims. Actuaries may also help create new investment tools for financial institutions. The main type of mathematics an actuary uses on a daily basis is applied statistics, which involves arithmetic, basic algebra, and practical applications such as using numbers and math to generate tables and graphs. Actuaries should also have a general understanding of business, economics, and corporate finance, all of which have mathematical components.

Most actuaries have at least a four-year degree in mathematics, business, economics, statistics, or, in some cases, a specific degree in actuarial science. As computer modeling replaces traditional graphs and tables, computer and programming skills have become increasingly important as well. The last step to becoming a licensed actuary is to get certified by passing a series of exams sponsored by either the Society of Actuaries or the Casualty Actuarial Society. The list of possible job choices for someone with an actuarial background is relatively small compared to that of the applied mathematicians. These include the following:

- Consulting firms such as Daniel H. Wagner Associates, Deloitte Touche Tohmatsu, Ernst & Young, Hewitt, McKinsey & Company, and KPMG
- Banks or related financial institutions such as AIG, ING, Capital Management, Chase Manhattan Bank, CitiGroup, Fidelity Investments, Goldman Sachs & Co, HSBC, JP Morgan Securities, Lehman Brothers, Mercer Investment Consulting, Merrill Lynch, Morgan Stanley & Co, Standard and Poor's, TD Ameritrade, and Wachovia Securities
- Brokers such as Acordia, Benfield, Cooper Gay, Heath Lambert, HLF Group, March & McLennan, and Willis Group
- Actuarial software development companies such as Actuarial Resources Corp. (ARCVal, HealthVAL, STAR, UltraVAL, CARVM), BLAZE SSI Corp., EMB America, Integrated Actuarial Services (Total Solution, RAAPID), TAG, and WySTAR Global Retirement Solutions (DBVAL, DCVAL, OPEVS)
- Miscellaneous jobs in large companies and government agencies such as ACTEX Publications, Casualty Actuarial Society, Coca-Cola, Ford Motor Co., International Actuarial Association, National Association of Insurance Commissioners, and the Society of Actuaries
- Insurance companies including both property and liability insurance, and life and health insurance such as AFLAC, AAA of CA, Allstate, Blue Cross and Blue Shield, Safeco, Sun Life, Universal Care, and WellPoint

Financial Mathematics

Financial mathematics is the development of mathematical tools and computational models used in the financial industry and on Wall Street. People in this profession are referred to as "quantitative analysts," or "quants." As new quantitative techniques have transformed the financial industry, banks, insurance companies, investment and securities firms, energy companies and utilities, multinationals, government regulatory institutions, and other industries have all come to rely on applied mathematics and computational science.

Sophisticated mathematics models and the computational methods and skills needed to implement them are used to support investment decisions, to develop and price new securities, and to manage risk, as well as for portfolio selection, management, and optimization. For example, modern hedge funds depend on these sophisticated techniques, as do pricing of bonds and commodity futures. Typically, someone who is interested in working in financial service and investment firms such Citibank, Moody's Corporation, Morgan Stanley, or Prudential will need to have a solid background in mathematical modeling, numerical and computational mathematics, applied statistics, business, economics, and finance.

Emerging Fields

Biomathematics and Bioinformatics. This emerging field can be thought of as a computer science/mathematics/biology hybrid that integrates mathematics and computer technology in the study of biological sciences. Broadly speaking, bioinformatics is the recording, annotation, storage, analysis, and searching/retrieval of nucleic acid sequence (genes, RNAs, and DNAs), protein sequence, and structural information. Mathematicians in this area contribute to the development of new algorithms with which to detect patterns and assess relationships among members of large data sets.

Computer Visions and Computer Graphics. Mathematicians in the field of computer vision work on developing theoretical machine learning algorithms to extract meaningful information from images. The images take on various forms such as waveforms from voice recorders or three-dimensional images from a magnetic resonance imaging (MRI) device. Its example applications include (1) artificial intelligence and controlling processes (for example, industrial robots and autonomous vehicles), (2) pattern recognition and verification (for example, public surveillance and biometric identification), (3) modeling and processing (for example, medical image analysis and terrain modeling), and (4) communication (for example, brain-computer interface for people with disability). Mathematicians in the field of computer graphics develop ways to represent and manipulate image data to be used by computers. The most well-known applications under this category are the video game and computer animation industries, where various transformation matrices and interpolation techniques are used to create smooth and believable subjects in successive frames. Companies such as Pixar and DreamWorks hire mathematicians in their research divisions to come up with innovative ways to enhance visual effects to be more aligned to reality. Other companies that hire mathematicians with backgrounds in computer vision and graphics include Siemens, Hewlett-Packard (HP), Honeywell, Flash Foto, GeoEye, Nokia, Microsoft, Apple Inc., Amazon.com, and Google.

Operations Research. This is a highly interdisciplinary branch of applied mathematics that uses methods such as mathematical modeling and optimization to solve problems that require a complex decision-making process. Mathematical areas such as game theory and graph theory have become useful tools in solving problems under the umbrella of operations research (OR). Examples of disciplines that use OR are financial engineering, environmental engineering, manufacturing and service sciences, policy-making and public sector work, revenue management, and transportation. Almost all companies hire operations research analysts to use mathematics and computers to develop software and other tools that managers can use to make decisions such as how many people to hire and retain in order to maximize productivity and minimize costs.

It is worth reemphasizing that having a mathematics degree or a mathematics-related degree increases one's chance of securing a position in nearly any company. Even areas that are traditionally viewed as pure mathematics such as combinatorics, number theory, topology, algebraic and differential geometry, analysis, and algebra often turn out to have real-world applications; for example, number theory in cryptography, Fourier analysis in speech recognition, and differential geometry in face recognition. Some additional career choices are as follows:

- Nonprofit organizations such as the American Institute of Mathematics (AIM), and SIAM
- Publishers and online products such as Birkhauser, Springer, and Elsevier Science
- University-based research organizations such as the Institute for Advanced Study, the Institute for Mathematics and Its Applications (IMA) and the Mathematical Sciences Research Institute (MSRI)
- Government agencies such as the National Security Agency (NSA) and the U.S. Department of Defense (DoD)
- Teaching at academic institutions. To teach at the high school level, one needs a bachelor's degree in mathematics and a teaching credential; to teach at the community college level, one needs a Master of Science or Master of Art degree in mathematics; to teach at the college level, one needs a Ph.D. in mathematics, mathematics education, applied mathematics, or statistics

Online Mathematics Jobs Listings

The American Mathematical Society (AMS) has an extensive set of resources to help someone in the market for academic positions and is the premier source for information on careers in mathematics. This includes a list of job postings organized by country and state. It has useful features such as an e-mail service that notifies applicants of all new job listings and an online storage of curriculum vitae (academic resume) and transcripts that can be used repeatedly for different applications. In addition, it allows one to register for the job fairs at the annual AMS meetings and has a list of graduate programs for students.

The Math-Jobs Web site lists international and national job openings for mathematicians in both industry and academics.

The Mathematical Association of America (MAA) has a comprehensive set of resources for students, faculties, professional mathematicians, and all who are interested in the mathematical sciences. In particular, MAA Math Classifieds helps people to find career in the diverse field of mathematics.

The *Chronicle of Higher Education* has academic and nonacademic job advertisements. Use the searchable index to find mathematics jobs.

The Mathematical Sciences Career Information by AMS-SIAM has information on nonacademic jobs, profiles of mathematicians in industry, job search tips, and links to many online job-posting services.

Further Reading

Lambert, Stephen, and Ruth J. DeCotis. *Great Jobs for Math Majors*. Chicago: VGM Career Horizons, 1999.

Sterrett, Andrew. *101 Careers in Mathematics*. 2nd ed. Washington, DC: The Mathematical Association of America, 2003.

Tyler, Marya Washington. *On-the-Job Math Mysteries: Real-Life Math From Exciting Careers*. Waco, TX: Prufrock Press, 2008.

JEN-MEI CHANG

See Also: Accounting; Mathematics, Applied; Problem Solving in Society; Professional Associations.

Caribbean America

Category: Mathematics Around the World.
Fields of Study: All.
Summary: The diverse islands of Caribbean America have produced notable mathematicians.

The Arawaks, Caribs, and other pre-Columbian peoples lived in the area of the Caribbean Sea before Spanish, French, English, or Scottish sea traders settled there. Linguists explore the different languages that were spoken in the Caribbean, traces of which can be found in the twenty-first century. Along with these languages, there were possibly different numerical systems. Sea merchants needed bookkeepers and accountants to keep track of their business, and although Port Royal and other places in the seventeenth-century Caribbean were notorious for piracy and lawlessness, there were also many counting houses and legitimate business operations.

The development of schools and universities led to more mathematical opportunities. According to the United Nations, the Caribbean America region encompasses Anguilla, Antigua and Barbuda, Aruba, the Bahamas, Barbados, the British Virgin Islands, Cayman Islands, Cuba, Dominica, the Dominican Republic, Grenada, Guadeloupe, Haiti, Jamaica, Martinique, Montserrat, Netherlands Antilles, Puerto Rico, Saint-Barthélemy, Saint Kitts and Nevis, Saint Lucia, Saint Martin (French part), Saint Vincent and the Grenadines, Trinidad and Tobago, Turks and Caicos Islands, and the U.S. Virgin Islands. By the end of the twentieth century, there were numerous Caribbean mathematicians, and *The Caribbean Journal of Mathematical and Computing Sciences* has published volumes of research articles.

Mathematicians in the Caribbean, and around the world, have also worked on mathematics history and research that is specifically related to the Caribbean area, like C. Allen Butler, who investigated optimal search techniques for smugglers in the Caribbean. Mathematicians have also discussed the high numbers of university graduates who have left the Caribbean, and they have created educational initiatives and mathematical texts designed for Caribbean children. The Caribbean and Central America areas combine for a joint Mathematical Olympiad. The most well-known mathematician in the region is perhaps Keith Michell from Grenada who, after completing his doctoral thesis from the American University, was a professor at Howard University, and then returned to Grenada, becoming prime minister in 1995, a position he held until 2008.

Barbados

On the island of Barbados, although education was an important facet of colonial life from the late nineteenth century on, few students were able to continue with mathematics. One exception was Merville O'Neale Campbell, who had become fascinated with mathematics at an early age and won a scholarship to study at Cambridge University in England. He then went to teach at the Gold Coast (now Ghana), completing his doctoral thesis, "Classification of Countable Torsion-

Free Abelian Groups," from the University of London, and is noted as the first Barbadian to have a Ph.D. in mathematics. His daughter, Lucy Jean Campbell, also completed her doctoral thesis in mathematics, and specializes in geophysical fluid dynamics, nonlinear waves, and a variety of numerical and analytical methods at Carleton University in Ottawa. Other prominent Barbadian mathematicians include Charles C. Cadogan, who has edited the *Caribbean Journal of Mathematical and Computing Sciences* and has contributed papers in journals around the world; and Hugh G. R. Millington, who completed his doctorate, "Cylinder Measures," from the University of British Columbia and then worked at the University of the West Indies, Cave Hill, Barbados.

British Caribbean

Well-known mathematicians from the British Caribbean include those from Jamaica. Earl Brown, who was the head of the Department of Science & Mathematics at University of Technology (Jamaica) from 1997 to 2000, completed his doctoral thesis at the Massachusetts Institute of Technology. Joshua Leslie completed his doctoral thesis from the Sorbonne in Paris, and was the chair of the Mathematics Department at Howard University; and Kweku-Muata Agyei Osei-Bryson from Kingston completed his doctoral thesis, "Multiobjective and Large-Scale Linear Programming," at the University of Maryland—College Park in 1988, and from 1993 until 1997 was the Faculty Fellow (Information Systems) for the U.S. Army, The Pentagon. Other prominent mathematicians from Jamaica, or whose ancestors were from Jamaica, include Garth A. Baker, Charles Gladstone Costley, Leighton Henry, Fern Hunt, Lancelot F. James, Clement McCalla, Bernard Mair, Claude Packer, Paul Peart, Donald St. P. Richards, and Karl Robinson.

Elsewhere in the British Caribbean, there have also been a number of mathematicians who held senior positions in the region and in the United States including Ron Buckmire from Grenada, who has specialized in computational aerodynamics; Edward Farrell from Trinidad, who has published extensively on polynomials; and Velmer Headley from Barbados, who has concentrated on the study of differential equations.

Cuba

One notable Cuban mathematician is Argelia Velez-Rodriguez, who was born in Havana and won her first mathematics competition when she was 9. She was the first Afro-Cuban to complete a doctorate from the University of Havana but left Cuba two years later to live in the United States. Since the 1959 Revolution, there has been an increased emphasis on the education system in Cuba, and Cuban students have long shown a high aptitude for mathematics.

French Caribbean

French Caribbean mathematicians include those from Haiti, with a desperately poor education system, and Guadeloupe. Louis Beaugris completed his doctoral thesis, "Some Results Related to the Generators of Cyclic Codes Over Zm," at the University of Iowa. Serge A. Bernard completed his doctoral thesis, "A Multivariate EWMA Approach to Monitor Process Dispersion," at the University of Maryland—College Park; and Jean-Michelet Jean-Michel completed his doctorate at Brown University. Alex Meril from Guadeloupe completed his thesis at the University of Bordeaux and worked at the University of Guadeloupe.

Further Reading

Nieto Said, José, and Rafael Sánchez Lamoneda. "Ten Years of the Mathematical Olympiad of Central America and the Caribbean." *World Federation of National Mathematics Competitions* 22, no. 1 (2009).

University of the West Indies. "Caribbean Journal of Mathematical and Computing Sciences." http://www.cavehill.uwi.edu/fpas/cmp/journal/cjmcs.htm.

Williams, Scott. "Mathematics Today in the Caribbean." http://www.math.buffalo.edu/mad/Caribbean/Caribbean.html.

Justin Corfield

See Also: Central America; North America; South America.

Carpentry

Category: Architecture and Engineering.
Fields of Study: Algebra; Geometry; Measurement.
Summary: Precise measurement is the foundation of the building trades.

While the word "carpentry" originally comes from the Latin root for chariot maker, today, the term refers to a number of trades that use wood for the construction of buildings and other articles. As there is a wide range of activities involved in carpentry tasks, carpenters must possess many different manual and intellectual skills to function in the profession.

Types of Carpenters

Carpenters who work on houses often fall into one of two broad categories: framing carpenters who work on the rough frame of a building, and finish carpenters who complete trim, stairs, railings, shelves, and other detail work. However, in practice, many carpenters end up doing some of each type of work, and carpenters who specialize in remodeling may not only do framing and finish carpentry but also tasks that are not strictly carpentry at all, such as plumbing, wiring, sheetrock finishing, and painting. There are also carpenters who specialize more narrowly, such as cabinet makers or carpenters who work on the specialized joinery between large posts and beams required in timber frame and log cabin construction.

Carpenters need to be able to read blueprints, and measure, cut, fasten, and finish a variety of materials.

Tasks of the Carpenter

Carpentry requires a variety of skills, including reading blueprints, measuring, cutting, fastening, and finishing. In addition, a carpenter must have knowledge of materials, including a variety of wood products and fasteners; and tools, including measuring devices, saws, drills, hammers, planes, and sanders. Carpenters who work on their own or as subcontractors on larger jobs must also have skills in cost-estimation and billing.

Consider, for instance, a carpenter who has been hired to add a covered deck onto a house. This carpenter might begin by working with the homeowner to determine the size and shape of the deck, possibly using a Computer Assisted Design (CAD) program to generate three-dimensional representations of how the finished project will look. After deciding on a design, the carpenter will need to use structural engineering tables to assess structural issues related to the design, such as the dimensions required for posts, the placement and size of cross-bracing, and the sizes of timbers that will be needed to span the distance between posts. From the calculations, the carpenter will then generate a price estimate, based on a materials list and an estimate of labor. The actual construction will include pouring concrete footers for the posts, measuring and cutting posts and joists with a circular saw, fastening materials to one another and to the house, screwing decking materials to the framing, framing a roof, installing roofing materials, constructing a railing, and building and finishing a set of stairs from the yard to the deck.

A Carpenter's Calculations

In the process of creating a simple covered deck, this carpenter will be making many measurements, calculations, and decisions regarding:

Layout: The initial position of the deck must be laid out so it is square to the house. To do this, the carpenter will construct a set of batter boards that are set outside the corners of the proposed deck and allow strings to be pulled to mark the edges of the deck. Employing the rule that the diagonals in a rectangle are equal to one another, the carpenter adjusts the strings to bring the corners to 90 degrees. Corner square may also be established and checked using the Pythagorean theorem.

Footers: Each post will be anchored to a concrete footer that will prevent it from moving or sinking into the ground. The bottom of the holes for these footers must be dug below the freeze level for the geographic area where the deck is being built so that the footers will not be heaved out of place by the freezing and resulting expansion of the soil. By consulting the building code, the carpenter will determine the appropriate area for the footer in square feet; multiplying by the height will give the cubic feet. If this is a large project, where the concrete will be delivered, the carpenter will have to convert cubic feet to cubic yards, as this is the unit in which concrete is ordered.

Raising the Posts: After pouring the footers, the carpenter will raise the posts for the deck being built. Since these posts will also support the roof in this example, they must be cut carefully to take into account any variation in the height of the footers. This measurement will be done by using a transit, a laser level, or a water level to assess the difference in the height of the footers. The carpenter will then add or subtract length to the height of each post to compensate. Once the posts are cut, they can be raised into position, ensuring each is plumb (perfectly vertical) using a level.

Joists and Decking: The sizing for all the wooden parts of the deck is determined by calculating how long a distance must be spanned and the weight the span will carry. The timber that is parallel to the house and runs between the posts must be sized to be strong enough to carry all the weight between each pair of posts; the longer the span between posts, the larger this timber must be. Similarly, the floor joists that butt into this timber will need to be large enough to carry the weight over their length, and the decking will be sized so that it does not sag between the joists.

Fasteners: In our example, the deck will be fastened to the building using bolts, and held together using nails, while the decking itself will be screwed on. The carpenter has many fasteners to choose from with many different finishes. Each type of fastener has special characteristics that make it useful for certain tasks. Nails are typically sold by the pound and come in sizes from large 20d framing nails (often called 20 penny nails) to small 6d finish nails. Screws are also sold by the pound but are sized by length and by a number that can be converted, using a chart, to their diameter. Bolts are sold by diameter and length; as is the case with all fasteners, there are many different types among them, lag bolts, carriage bolts, and through bolts.

The Roof: The roof over the deck will be set at an angle so water runs off it and away from the house. The pitch of a roof is typically measured in "rise over run," with the denominator of this fraction always given as 12. Thus, a roof that goes up four feet over a run of 12 feet is said to be a "4:12 roof." The carpenter will use a special tool called a "speed square" that allows the direct conversion of roof pitch to angles and mark rafters for cutting.

Stairs: While stairs can be constructed to be more or less steep, a carpenter must keep in mind a basic mathematical relationship between tread length and riser height that will make a set of stairs comfortable to ascend. It turns out that because of the characteristic of the human gait, the steeper a stair, the less wide each tread should be. The formula that carpenters use is that for each stair, twice the rise plus the run should equal 24–26 inches.

Of course, once the carpenter is done with the project, there are still numerous other tasks to complete, including building railings and benches, as well as finishing and waterproofing the surfaces. If the homeowner were to want an outdoor grill area, with built-in cabinets, the carpenter would have a whole new set of challenges worthy of a cabinet maker and finish carpenter.

Further Reading

Gerhart, James. *Mastering Math for the Buildings Trades.* New York: McGraw-Hill, 2000.

Webster, Alfred P. *Mathematics for Carpentry and the Construction Trades.* 2nd ed. Upper Saddle River, NJ: Prentice Hall, 2001.

Jeff Goodman

See Also: Geometry in Society; Measurement, Systems of; Pythagorean Theorem.

Castillo-Chávez, Carlos

Category: Mathematics Culture and Identity.
Fields of Study: Algebra; Connections; Data Analysis and Probability; Measurement.

Summary: Carlos Castillo-Chávez works in the field of mathematical epidemiology, which deals with the spread, treatment, and eradication of diseases.

Carlos Castillo-Chávez (1952–) is a Mexican-American applied mathematician, eminent in the field of mathematical epidemiology. His research and writing has advanced human understanding of the mechanisms by which diseases spread and by which they can be contained. The specific diseases that he has worked with extensively include human immunodeficiency virus and acquired immune deficiency syndrome (HIV/AIDS), tuberculosis, influenza, and many others.

He grew up in Mexico, where he excelled academically. Motivated in part by the Tlatelolco massacre in which hundreds of Mexican students were killed, he emigrated to Wisconsin in 1974. In 1984, he was awarded a Ph.D. in mathematics from the University of Wisconsin—Madison. He spent 18 years as a professor at Cornell University before coming to Arizona State University, where he is both professor of mathematical biology and executive director of the Mathematical and Theoretical Biology Institute, as well as the Institute for Strengthening the Understanding of Mathematics. He is considered an important voice of the mathematical biology community and has served on many influential committees and panels, including the National Science Foundation, the National Institutes of Health, the Society for Industrial and Applied Mathematics, and the American Mathematical Society.

Mathematics and Biology

In the past decades, mathematics and biology have enjoyed an increasingly symbiotic relationship. Mathematical biology is a wide area of applied mathematics, focusing principally on modeling. A "model" of a biological process or phenomenon is a mathematical system that obeys certain rules and properties abstracted from what we know (or suspect) about the biology in the real world. Two primary mathematical tools for mathematical biology are the study of dynamical systems and differential equations, since we are often most interested in how certain quantities change in response to other quantities.

A distinguishing feature of mathematical biology is the mutual feedback between the mathematicians and the scientists involved. A model based on today's understanding of a certain disease (or of the action of neurons, or of cellular growth) may make certain predictions, suggesting that certain experiments be performed. The results of these experiments can improve, correct, and refine scientists' understanding of the underlying biology. Mathematicians can then incorporate this new knowledge into more sophisticated, more accurate models.

Carlos Castillo-Chávez is a leader in the area of mathematical epidemiology, the branch of mathematical biology dealing with the spread, treatment, and eradication of diseases. Mathematical epidemiologists can use mathematical modeling techniques to predict how certain diseases might affect the population. More sophisticated models can incorporate the effects of various proposed treatment and control options. Properly applied, these techniques can enable epidemiologists to effectively predict the effects of methods of prevention, allowing for a more effective allocation of resources in responding to disease threats.

Minorities in Mathematics and Science

Carlos Castillo-Chávez is an outspoken advocate of minorities, women, and other underrepresented groups in mathematics and the sciences. He has expressed the belief that people from different backgrounds may bring different perspectives to mathematics and science, leading them to directions of research that may have gone unnoticed or uninvestigated. Since mathematics and the sciences are driven by the questions that participants pursue, asking a richer set of questions leads to a fuller body of knowledge; supporting students from underrepresented groups minorities is therefore a matter both of social justice and of enhancing the discipline.

Dr. Castillo-Chávez has supported these beliefs with his actions at all stages of his career. As a Ph.D. student in Milwaukee, he spent his summers teaching mathematics to Latino students in the area. He has served as a mentor to numerous female and minority students, helping and encouraging them at all stages of education. He is also an active member of the Society for the Advancement of Chicanos and Native Americans in the Sciences (SACNAS); during his time at Cornell, he was the founding president of a northeast chapter of SACNAS and was instrumental in initiating a special summer program intended to provide Latino, Chicano, and Native American students with mentorship, encouragement, and training in the sciences. Carlos

Castillo-Chávez is highly acclaimed for his work in this regard.

Further Reading

Blower, Sally, and Carlos Castillo-Chávez, eds. *Mathematical Approaches for Emerging and Reemerging Infectious Diseases: An Introduction*. New York: Springer, 2002.

Brauer, Fred, and Carlos Castillo-Chávez. *Mathematical Models in Population Biology*. New York: Springer, 2001.

Castillo-Chávez, Carlos, ed. *Bioterrorism: Mathematical Modeling Applications in Homeland Security*. Philadelphia: SIAM, 2003.

Castillo-Chávez, Carlos. "SACNAS Biography: Dr. Carlos Castillo-Chávez—Mathematical Biologist." http://www.sacnas.org/biography/Biography.asp?mem=35&type=2.

Michael "Cap" Khoury

See Also: Disease Survival Rates; Diseases, Tracking Infectious; Mathematics, Applied; Viruses.

Castles

Category: Architecture and Engineering.
Fields of Study: Algebra; Geometry; Measurement.
Summary: Mathematics has been used to both construct and study castles.

Castle are fortified structures, used as residences by European nobles in the Middle Ages. Early castles were often made of wood, but with the development of better attack methods, castle builders switched to stone as the main building material. With the extensive use of artillery, residential castles became indefensible. They were replaced by purely military forts (not used for administrative and residential purposes) and decorative residences resembling castles (not used during wars). The geometry of a castle was often dictated by defense considerations. Architect Benjamin Bramer fortified castles and published a work on the calculation of sines.

The Alhambra, a fourteenth-century palace and fortress, is well known for its mathematical tiles. In the early twenty-first century, the American Institute of Mathematics proposed a headquarters in California that would be modeled after the Alhambra, popularly referred to as a "math castle."

Castles are frequently found in fantasy and horror literature. One common image is that of Dracula's castle. Dracula author Bram Stoker earned a degree in mathematics. Some mathematics teachers use castles like Cinderella's castle or sand castles to explore concepts such as ratios, fractions, volume, statistics, and geometric shapes. Scientists, including physicist Mario Scheel, explore the physical properties of sand-like material, and researchers in experimental archaeology model and design castles.

Geometry of Castle Defense

Both the layouts of castles and the shapes of their parts were dictated by defense needs. For example, concentric

The Alhambra, in Granada, Spain, is known for its Moorish use of mathematical symmetry groups.

castles consisted of several concentric walls. The barbican (the outer wall) had relatively many entrances, while the inner wall had few, making the attacking army crowd between walls and, thus, become vulnerable to defenders.

Keeps and towers were mostly round to allow for a larger arc of shooting coverage from each arrowslit. In addition, the isoperimetric theorem states that for a given area, the circle has the least perimeter among all shapes, thus minimizing the amount vulnerable walls (not to mention reducing the costs of building materials). Each corner introduced blind spots where enemies could avoid arrows, and circles have no corners. Also, corners are more vulnerable for mining.

Cylindrical towers led to the invention of spiral staircases. Most castle staircases were built so attackers would ascend clockwise, making the central shaft of the staircase interfere with their right hands—often the hand that held the sword.

Stonemasons building castles used simple tools, such as compasses, dividers, and straightedges. Their manuals included descriptions for creating a variety of shapes with these tools. For example, pointed and rounded arches, including Tudor, lancet, and horseshoe arches, could be traced with compasses and straightedges.

Shooting from high towers allowed for better view, and also used gravity to add acceleration to arrows and other projectiles. When glass windows were installed in circular towers, they were made by blowing glass inside a cylinder, cutting it, and then connecting multiple pieces with lead to match the curvature of the castle wall.

Castle builders used terrain geometry to support defense. In addition to the height advantage of the castle walls and towers, castles were frequently situated on hills (either natural or artificial) or on earthen mounds called "mottes." Defensive ditches around castles, called "moats," prevented siege towers from coming close. When moats were filled with water, they could also make digging tunnels for mining the walls more difficult.

The construction of moats led to the invention of drawbridges and the mechanisms of raising and lowering them. The drawbridge mechanisms involved levers and pulleys.

Logistics and Finance

Building a large castle was a major financial undertaking spanning many years, and occasionally bankrupted the ruler attempting it, such as King Edward I. Supplying the castle, especially with enough supplies to withstand lengthy sieges, presented another organizational problem. A siege was a common method of castle attack in which the attackers surround the castle grounds and waited for the defendants to starve. The siege process could sometimes last for months or even years.

Experimental archaeology is a new field of study that combines archaeological research, computer modeling, and actual building. Observations in building experiments allow for conclusive results of how models can be made to work. For example, Project Gueledon is a real-size castle built recently to help give people a deeper understanding of how castles were constructed in medieval times. The researchers used building methods and materials similar to those used by thirteenth-century castle builders, with a team of 50 workers from various professions.

Further Reading

Holden, Constance. "A Castle Fit for a Mathematician." *Science* 314, no. 6 (October 2006).

Whitney, Elspeth. *Medieval Science and Technology*. Westport, CT: Greenwood Press, 2004.

Maria Droujkova

See Also: Archery; Artillery; Geometry in Society; Middle Ages; Tunnels.

Caves and Caverns

Category: Weather, Nature, and Environment.
Fields of Study: Algebra; Geometry; Representations.
Summary: Several metrics are used to describe caves while mathematical measurements can detect them.

Caves are underground spaces large enough for a human to enter. The science of studying caves is called speleology and the practice of exploring caves is spelunking. Caves can be formed through a variety of ways, such as solutional caves (made by rocks dissolving in acids in water) or littoral caves (made by waves pounding cliffs). They are also categorized by the passage patterns, such as angular networks or ramiform caves. Mathematical techniques are used to model and understand the structures and ages of caves and caverns. For instance, the topology of the cave highlights the number of tunnels and how they are connected while the geometry shows accurate distances, curvatures, and steepnesses. Statistical methods as well as fractal concepts of self-similarity have been used to estimate the number of entranceless caves. Archaeology has revealed that caves are among the oldest known human habitations.

Some researchers analyze ancient cave paintings for mathematical, astronomical, or geographical interpretations. Mathematical objects and mathematicians have also been connected to caves and caverns. The Lebombo bone was discovered in a Swaziland cave in the 1970s. It dates to approximately 35,000 B.C.E and is thought to be the oldest known mathematical artifact. The bone holds 29 tally notches and it has been compared to calendar sticks that are still in use in Namibia. In France, numerous mathematicians trained at École des Mines including Henri Poincare, who was employed as a mine engineer and was eventually promoted to inspector general.

Visitors today can enter Pythagoras' cave in Samos, where he apparently lived and worked on mathematics. In *The Republic*, Plato imagines chained prisoners in a cave who can only see shadows of the movement behind them. Similar metaphors continue to be explored in order to explain higher dimensional realities and other concepts in mathematics, physics, and philosophy, including investigations of quantum caves.

Cave Measurement and Records

There are several metrics used to measure caves, including total length of passages, depth from the highest entrance to the lowest point; total volume, or height, depth, length, area; and volume of individual passages, shafts, and rooms. The deepest cave is 2191 meters meters (7188 feet) deep and the greatest total length cave is 591 kilometers (67 miles) long. These numbers are updated as more parts of caves are explored and new caves are discovered.

Geophysical Detection of Caves

The mapping of hidden caves and smaller karst formations is done for scientific and recreational explorations, as well as to ensure the stability of constructions, such as houses and bridges. Geophysical detection methods use contrasts in a physical property, such as electric resistance or density, between different parts of the underground medium. To detect variations, scientists measure microscopic changes in gravity caused by empty spaces, or transmit electromagnetic waves into the ground and measure their reflections. Another method is to transmit an electric current and measure changes in ground resistance. Seismic tomography depends on collecting massive amounts of data from inducing stress through boring holes, but it can be very accurate. All these methods depend on mathematical models of changes in physical properties between different surfaces.

All geophysical techniques require contrasts of some physical property (density, electrical resistivity, magnetic susceptibility, seismic velocity) between subsurface structures.

Cave Patterns

The geometry of a cave depends on many geological factors, such as the structures dominant in the rock and the sources of water for solution caves. Spongework caves consisting of large, connecting chambers formed in porous rocks. If the rock also fractures easily, large chambers will be interspersed with long passages formed by fracturing in a pattern called "ramiform" (branchlike). Nonporous rock that fractures will produce a distinct pattern called rectilinear branchwork, with straight passages at angles to one another. Lava tubes are round in cross-section, long, and relatively even; they are formed by a lava flow that develops a hard crust.

Cave Meteorology and the Geothermal Gradient

Heat in caves comes from water or air entering the cave, or from overlying and underlying rock. Overlying rock does not transmit the surface heat well. For example, a difference of 30 degrees Celsius between day and night on the surface translates into 0.5 degrees Celsius difference one meter (3.28 feet) deep into limestone. Seasonal fluctuations penetrate deeper but still become negligible at depths of 10 or so meters (32.8 feet).

In most parts of the world, the temperature increases by about 25 degrees Celsius for every kilometer of depth, because of the molten interior of Earth, the rate called "geothermal gradient." As one goes deeper into a cave that starts at a sea level, the temperature first drops because of insulation from the surface but then increases because of the geothermal gradient. In areas of high volcanic activity near the surface, caves can be very hot, or even contain molten lava. Some of the deepest caves in the world are cold, because their entrances are high in the mountains.

Further Reading

Curle, Rane. "Entranceless and Fractal Caves Revisited." In: Palmer, A. N., M. V. Palmer, and I. D. Sasowsky, eds. *Karst Modeling, Special Publication* 5, Charlottesville, VA: Karst Water Institute, 1999.

Maurin, K. "Plato's Cave Parable and the Development of Modern Mathematics." *Rendiconti del Seminario Matematico, Università e Politecnico di Torino* 40, no. 1 (1982).

O'Connor, J. J., and E. F. Robertson. "Mactutor History of Mathematics Archive: Poincaré—Inspector of Mines." http://www-history.mcs.st-and.ac.uk/HistTopics/Poincare_mines.html.

Palmer, Arthur. *Cave Geology*. Trenton, NJ: Cave Books, 2007.

Maria Droujkova

See Also: Geothermal Energy; Measurements, Area; Measurements, Volume; Stalactites and Stalagmites; Temperature; Tides and Waves; Tunnels.

Cell Phone Networks

Category: Communication and Computers.
Fields of Study: Algebra; Geometry; Measurement.
Summary: Mathematics is involved in the design of the cell network and the assignment of calls to frequencies, as well as in data compression and error compression.

Cell phones have grown from a novelty, to a luxury, to a virtual necessity since the 1990s, with the number of cell phone subscribers in the United States growing from about 91,000 in 1985 to 276 million in 2009. Part of the reason that cell phones have become so reliable, cheap, and secure has to do with mathematics. Mathematics is involved in the design of the cell network and the assignment of calls to frequencies (or channels), as well as data compression and error compression that allow a large number of clear calls to be carried over a small bandwidth. The concept of a tree from graph theory can be used to understand cell phone networks, which are challenging because of the large amount of data and links. Mathematicians like Vincent Blondel analyze millions of users and months of communication.

Cellular Radio Networks

Cell phones work by communicating via radio signal with a nearby cell phone tower. In a cellular radio network, the type of system used for cell phone coverage, the land area to be supplied with coverage is divided into regular shaped regions (or "cells"), each of which has a corresponding radio base station or cell tower. Phones within a particular cell connect via radio signal to the tower for that cell, which then connects to the public telephone network through a switch. The range of a tower may be about one-half mile in urban areas up to about five miles in flat rural areas.

Because of this relatively short transmission range, cell phones and towers can use low power transmitters. In addition to allowing phones to be small and use smaller batteries, the low power also means the radio frequencies can be reused by towers not too far away from each other without any interference between the transmissions. This function allows cell phone networks to carry a larger number of calls in a smaller bandwidth. Typically, cell companies will divide their coverage area into regularly shaped cells or regions with each one covered by a single tower. In fairly flat areas, these regions are usually hexagonal in shape—an idea developed by Bell Labs engineers W. Rae Young and Douglas Ring in the middle of the twentieth century.

The frequencies used by a particular tower for transmissions in its region cannot be used by any of the six regions with which it shares a boundary. The Four Color Theorem from graph theory indicates that only four frequencies are needed to ensure that regions that share a boundary do not use the same frequency. However, companies usually want to further buffer the distance between reuse of the same frequency, so they divide the frequencies up into seven bundles and use a different one on each of the six cells sharing a boundary with a given cell.

A cell phone tower disguised as an evergreen tree. Cell signals are sent through towers via radio signals.

Cell Phone Channels

During the twentieth century, there were many discussions among professionals at the Federal Communications Commission regarding the possibility of opening up frequencies for phone use. Cellular networks began to appear around the world. For instance, Japan offered a 1G system in 1979, and, in 1983, AT&T and Ameritech tested a commercial cellular system in Chicago. Much of the advancement in cell network technology has been focused on the frequency band within a cell, which must be divided up to carry several calls at the same time. In first-generation cell technology, calls were transmitted in analog, which allowed only one call per frequency. Typically, a cell phone carrier was assigned 832 radio frequencies to use in a city. Each call was full duplex, meaning that it used two frequencies: one to transmit and one to receive.

Thus, typically there were 390 voice channels with the remaining 42 radio frequencies used for control channels that were used to locate and communicate with phones but not to carry calls. If the 395 voice channels were divided into seven frequency bundles, that made 56 voice channels per region. So if more than 56 calls were in progress in a given region at a given time, then one of the calls would be disconnected or dropped. Fortunately, first-generation technology is no longer in use. With second-generation (2G) cell technology, calls were no longer analog signals but were converted to a digital (0 and 1) format. This shift is similar to the change from cassette tapes to compact discs in the recording industry.

The greatest advantage to digital technology is that it allows for sophisticated data compression techniques to be used without losing acceptable call quality. Data compression allows for between three and 10 digital calls to be carried in the bandwidth necessary for a single analog call. Further advancements in compression have allowed for even newer third-generation (3G) technology. 3G networks have much faster transmission speeds and allow the use of smartphones that can transmit data fast enough to surf the Internet, send and receive e-mail, and even instant message with a cell phone. Newer 4G technology adds even more speed and capacity to cell phone networks.

Further Reading

Agar, Jon. *Constant Touch: A Global History of the Mobile Phone.* Cambridge, UK: Icon Publishers, 2003.
Brain, Marshall, Jeff Tyson, and Julia Layton. "How Cell Phones Work." *HowStuffWorks.* November 14, 2000. http://electronics.howstuffworks.com/cell-phone.htm.

MARK GINN

See Also: Coding and Encryption; Digital Storage; Telephones; Wireless Communication.

Census

Category: Government, Politics, and History.
Fields of Study: Data Analysis and Probability; Measurement; Representations.
Summary: Conducting a valid and reliable census depends on mathematical and statistical methods.

The term "census" comes from the Latin word *censere*, meaning "to assess." A census is a systematic collection of data about an entire population of interest. Usually the population is people but historically it has also been done for land, livestock, and trade goods. Sometimes a census is a one-time event, or it may be repeated periodically, like the decennial census in the United States.

There are also two primary philosophies of data collection that can affect the outcome of a census: *de jure* and *de facto*. *De jure* counts people at their usual place of residence, while *de facto* counts people where they are on the day of the census.

For example, one biblical account of the birth of Jesus involves a census in which individuals were required to return to their town of origin rather than being counted where they lived, as opposed to the U.S. census, which is centered about people's permanent residences. Archaeological records indicate that many ancient civilizations conducted censuses, the purpose of which was often taxation or military recruitment. The constitutionally stated purpose of the U.S. census is to determine each state's congressional representation, though it has grown to include additional descriptive and predictive activities. The U.S. Census Bureau is one of the largest employers of mathematicians and statisticians, who not only collect and analyze data but also lead the way in developing new data collection and analysis methods.

Statisticians work internationally as well. For example, in 1949, British statistician Frank Yates was appointed to the United Nations Commission on Statistical Sampling and published *Sampling Methods for Censuses and Surveys*, which is widely acknowledged to have been influential in establishing sound principles and technical terminology. Overall, mathematical and statistical procedures improve the quality, reliability, and representation of census data, and the methods used by census-takers are constantly evolving.

Census Controversies

While the aim of collecting census data is to provide complete record of data on a population, there can be many difficulties in obtaining such comprehensive data. Past problems have included members of the population objecting to the potentially intrusive nature of such a full-scale inquiry, which has the potential to be misused, and difficulties reaching the entire population.

This second problem was especially problematic in the 1990 decennial census and spurred a great deal of developmental activity with regard to statistical survey methods. Even further in the past, there was a heated debate among the U.S. Founding Fathers about how to account for slaves in the U.S. census, as these counts had the potential to dramatically shift the balance of representative power between the northern and southern colonies. Even now, evolving social constructs and definitions of significant demographic variables, like race, can be a controversial topic.

The History of the Census

The practice of completing a census for an entire population occurred in many ancient civilizations. Records suggest that the Babylonians conducted a census in about 3800 B.C.E., and that Egyptians did so in the second millennium B.C.E. Male Roman citizens had to register for a census every five years and declare both family and property. Elected censors oversaw and coordinated the census process. The censors would then summon every tribe in the country to appear before them so they could record the relevant details. In ancient Rome, the census recorded the names of the family members, along with details of any property or land they owned. This provided the leaders of the country at the time the ability to tax their citizens according to the amount they owned. William the Conqueror carried out a census in Britain in 1086 C.E. for taxation purposes. This census took years to complete and attempted to compile a comprehensive list of all land and property in Norman Britain. Such a comprehensive exercise was previously unheard of in Europe, and it preceded an early example of a modern census by nearly 600 years.

Instructed by King Louis XIV in 1666, Jean Talon, a French colonial administrator, conducted a census in order to expand the colony in New France, North America. Talon used the *de jure* method and visited many of the colonial settlers personally, compiling data on settlers' names, age, sex, and occupation. The aim of this census was to help the colony settle by using the statistics to decide how best to develop agriculture, trade, and manufacturing industry. In all, Talon managed to compile details of 3215 inhabitants and paved the way for the development of a number of further censuses in the New World. In Britain, a 1798 paper written by demographer Thomas Malthus discussed the possibility that not knowing the population size and growth rate (demographics), of a country could lead to food shortages and overuse of other resources, resulting in famine and disease as the population is unable to sustain itself. These revolutionary modeling ideas led the British government to pass through parliament the Census Act of 1800. The first modern British census took place in 1801; the process has been repeated decennially since then, except for in 1941 during World War II.

The Modern-Day Census—Data Collection

The U.S. census is required by constitutional law to take place every 10 years and involves sending forms to every residence in the United States and Puerto Rico. The data are then analyzed in order to determine how each state is represented in the U.S. House of Representatives and to provide the correct resource allocation for the current population, that is, how much of the federal fund is given to hospitals, schools, and other public services. Individual responses to the U.S. census are kept confidential for 72 years.

A similar process is used in the United Kingdom, although the census details are kept confidential for 100 years. A key difference in the census forms in the United Kingdom (UK) and the United States is that the U.S. form has just 10 questions and is two pages long. The UK census form for 2011 contained 43 questions in a 32-page booklet.

In Canada, a national census is taken every five years. Each household receives a census questionnaire, to either be filled out online or returned in the post.

Practical Problems With Census Taking

A number of problems can arise when attempting to take a census of an entire population. For the data to be useful, the characteristics of the whole population need to be reflected. This requirement means that any nonresponse could jeopardize the quality of the data. Nonresponses can happen, for example, when an address list is not comprehensive, or people fail to fill in their forms fully and return them. There are a number of measures used to prevent this, including following up with nonrespondents in a face-to-face interview, and setting fines for nonrespondents.

A statistical technique called "imputation" was used by the U.S. Census Bureau in its 2000 census to create data using the nearest neighbor "hot deck" method. Where a gap in the counting (for example, an entire household's data was missing) was identified, data from the geographically closest neighbor were used instead. Where a household had not completed every question fully, the missing data were imputed from a nearest neighbor record where the households are of the same size. Where a respondent within a household gave incomplete data, the characteristics were imputed from the characteristics of other household members. This technique enabled the U.S. Census Bureau to produce a more complete set of data on the U.S. population.

In order to overcome the obstacle of an incomplete address list, a number of different address lists can be combined to get a more complete list—thus ensuring a wider population is reached, and improving the reliability of the data. Alternatively, another statistical method, called "sampling," can be used to estimate features of the population. A forward-thinking statistical sampling plan was proposed by many mathematicians and statisticians after the 1990 census turned out to be particularly problematic in terms of issues such as undercoverage of certain subpopulations. The U.S. Supreme Court refused to permit sampling to completely substitute for counting. These mathematical methods are used, however, for other types of estimation and to gauge how much undercoverage or other biases might exist.

Analysis of Census Data

A number of mathematical and statistical techniques can be used to draw the most descriptive and predictive information possible from raw census data. For example, to identify resource need, data can be ranked in such a way to identify areas where there are more children, thus enabling the government to plan where to locate schools. Alternatively, areas with a high percentage of elderly people could be identified and provided with more social care. Since the 1990s, census data have become a major resource for both amateur and professional genealogists now that older records are being digitized. Census data are also used to find ways to make future collections efforts better.

Edna Lee Paisano grew up on a Nez Perce Indian Reservation in Sweetwater, Idaho. Talented in both mathematics and science, she attended the University of Washington and earned a graduate degree in social work, studying statistics in the process. In 1976, she was hired by the U.S. Census Bureau to work on issues regarding Americans and Alaskan Natives, and was the Bureau's first full-time Native American employee.

Using data from both the 1980 census and a survey she developed, Paisano discovered that Native Americans in some locations were undercounted. This was a serious issue, as allocation of federal funds to tribal units is based on census figures. She used statistical methods to improve the accuracy of the census and encouraged others in the Native American community to become educated in mathematics-related fields such as computer science, demography, and statistics. The 1990 census showed a 38% increase in U.S. residents counted as American Indians.

Further Reading

Aly, Gotz, and Karl Heinz Roth. *Nazi Census: Identification and Control in the Third Reich.* Philadelphia: Temple University Press, 2004.

Anderson, Margo. *The American Census: A Social History.* New Haven, CT: Yale University Press, 1990.

Kertzner, D. *Census and Identity: The Politics of Race, Ethnicity, and Language in National Censuses.* New York: Cambridge University Press, 2002.

Wright, Tommy, and Joyce Farmer. *A Bibliography of Selected Statistical Methods and Development Related to Census 2000.* Washington, DC: U.S. Bureau of the Census, 2000.

AMY EVERTON

See Also: Congressional Representation; Data Analysis and Probability in Society; Probability; Sample Surveys.

Central America

Category: Mathematics Around the World.
Fields of Study: All.
Summary: Mesoamericans were sophisticated mathematicians, and mathematics continues to be important in the area.

Central America is defined as the southern part of the North American continent, reaching from Mexico to Panama. The portion of the region in which corn, beans, and squash were reliable crops during the pre-Columbian era is referred to as "Mesoamerica," reaching from the mountains of Mexico to Guatemala and down the Pacific coast into Nicaragua. Teotihuacan, Olmec, Maya, and Aztec were among the many cultures sharing the same prehistoric land and cultural legacy.

The development of the area and its perspective on mathematics were shaped in part by the origins of civilization isolated from the other large centers of civilization in the Eastern Hemisphere. Spanish colonization in the sixteenth century brought the first introduction to European cultures. Efforts at spreading Christianity resulted in the loss of much of their rich, ancient heritage. The area had gained independence by the mid-nineteenth century, variously structured as separate nations and unified groups. Struggles to achieve stability continue into the twenty-first century in many parts of the region. Education and mathematics are highly valued as keys to further progress.

Ancient Mesoamerica

Without the benefit of influence from other cultures, the ancient Mesoamericans built large city-states sometimes supporting several hundred thousand people, and extensive empires, with no domesticated large mammals and with no use of the wheel, other than in children's toys. They mastered basic arithmetic, with a concept of zero evident a millennium before European civilizations. They shared a counting system based on 20 rather than 10. Numeral representations included dots for units, bars for five, and a circle or seashell for zero. Ancient ruins show evidence of meticulous accounting of trade and personal lives. The construction of imposing pyramids and other structures aligned to astronomical features and adorned with harmonic geometric design reveal an advanced level of engineering, architecture, and astronomy to rival that found in Europe at the same time. From as early as 2000 B.C.E., the people of the area had sophisticated calendars, which were used in tandem to mark time reflecting both human and solar cycles. Ethnomathematicians continue to study ancient and modern Central America, and many teachers use Mesoamerican mathematics concepts as the basis of lesson plans and assignments.

Modern Central America

Central America is defined by the United Nations to include the modern countries of Belize, Costa Rica, El Salvador, Guatemala, Honduras, Mexico, Nicaragua, and Panama. The countries share ethnic, economic, and geological features. The peoples are primarily Spanish, Amerindian, or Mestizo (a mixture of the two). The climate ranges from mountainous to tropical coastline. While significant portions of the population are centered in large urban areas, much of the population of the region is located in small villages, sometimes isolated by rugged terrain.

Education

Central Americans are continuously improving their education systems, and efforts at reform often include careful inclusion of children from both urban and rural areas with the use of radio, television, and computer technologies. Teacher salaries and the contrasts of management of schools by local or federal administrators are recent areas of research. United Nations data report high participation in formal schooling. Private schools usually are more prestigious than public schools in most areas.

As calendars held power in ancient Mesoamerica, knowledge of mathematics is held to be essential for the people in modern Central America. High school graduates receive extensive content instruction in mathematics and science but historically with little emphasis on mathematics applications. Teachers are encouraged to teach mathematics in context rather than as an isolated, esoteric discipline both for the better understanding and for the application of learning to solve problems and promote progress. Recent research in mathematics from the region includes a diverse range of areas like topology, noncommutative geometry, and applied mathematics.

Mathematics researchers gather for conferences, research seminars, educational forums, and social events. For example, the Sociedad Matemática Mexicana (Mexican Mathematical Society) was founded in 1943. The society's goals include encouraging mathematical research, including cooperation with related scientific disciplines; improving mathematics education at primary, secondary, and college levels; and providing various forums for discussion and dissemination, including journals and conferences.

Further Reading

Evans, Susan. *Ancient Mexico & Central America: Archaeology and Culture History.* London: Thames & Hudson, 2008.

Jimenez, Emanuel, and Sawada, Yasuyuki. "Do Community-Managed Schools Work? An Evaluation of El Salvador's EDUCO Program." *World Bank Review* 13, no. 3 (1999).

Sociedad Matemática Mexicana. http://smm.org.mx/smm.

Valero, Paolo. "Deliberative Mathematics Education for Social Democratization in Latin America." *Mathematics Teaching and Democratic Education, Part 2. ZDM The International Journal of Mathematics Education* 31, no. 1 (1999).

Vegas, Emiliana, and Llana Umansky. "Improving Teaching and Learning Through Effective Incentives: Lessons From Educational Reform in Latin America," *The World Bank*, February, 2005. http://info.world

bank.org/etools/docs/library/242822/day5Improving%20teaching%20and%20learning_Final.pdf.

Judith E. Beauford

See Also: Calendars; Castillo-Chávez, Carlos; Curricula, International; Incan and Mayan Mathematics.

Cerf, Vinton

Category: Communication and Computers.
Fields of Study: Connections.
Summary: Computer Scientist Vinton Cerf helped create the Internet and continues to be a leader in Internet innovation.

Vinton Cerf playing a game on the Computer History Museum's PDP-1 computer from 1959.

Vinton Gray Cerf is an American computer scientist and is one of the creators of the Internet. He worked on Internet architecture and the design of TCP/IP protocols in the 1960s and 1970s, eventually moving from academia, to government, and to corporations like MCI and Yahoo. He continues to work in advancing Internet applications and policies, such as laws regarding "net neutrality." He has won many prestigious awards in conjunction with collaborator Robert E. Kahn, including the U.S. National Medal of Technology, the Association for Computing Machinery's Alan M. Turing Award (sometimes called the "Nobel Prize of Computer Science"), and the Presidential Medal of Freedom, which is the highest civilian award given in the United States. In December 1994, he was also listed as one of *People* magazine's "25 Most Intriguing People." Cerf and his wife, Sigrid, have been married since 1966, and he has spoken of her support regarding his education and career. They have two sons, and, at times, his family's needs have influenced where he has decided to work. Since 2005, he has been a vice president at Google, and continues to be a leader in Internet innovation.

Reflecting on his own education, Cerf traced his interest in mathematics to primary school. He cited his fifth grade mathematics teacher as being an influential force. When Cerf complained of boredom with the standard curriculum, the teacher introduced him to more advanced mathematics. Cerf said, "I fell in love with algebra. It was wonderful.…Frankly, I liked the word problems the best because they were like little mystery stories.…I still love word problems. To this day, give me an algebra word problem, and I'll have a great old time with it." Outside the classroom, Cerf also enjoyed the camaraderie and challenge of his high school math club and mathematics competitions led by a young teacher named Florence Reese. Of the experience, he noted positively, "It would be weeks and weeks of just working problems, and then the morning of the event we'd all get up and have a big steak and egg breakfast at 7:00 in the morning.…You didn't want to dull your brain with a lunch of any kind." He went on to earn a B.S. in mathematics from Stanford University, then a M.S. and a Ph.D. in computer science from the University of California, Los Angeles, along with multiple honorary doctoral degrees from universities around the world. Regarding his change of field from undergraduate to graduate school, Cerf said, "I had already figured out that I wasn't going to be a world-class mathematician. I sort of broke my pick on Riemannian geometry...." At the same time, he credited his education in geometry with developing his thinking skills, saying, "...I enjoyed the reasoning part of it, which is probably one of the reasons why I've enjoyed being a programmer, because you have to go through the same line of thinking."

As a graduate student, Cerf was part of Professor Leonard Kleinrock's data packet networking group when they conducted the first connection tests and demonstrations of the Advanced Research Projects Agency Network (ARPANet), one predecessor to today's Internet. ARPANet was created as a joint project between MIT and the Defense Department's Defense Advanced Research Projects Agency (DARPA). After earning his doctorate, Cerf returned to Stanford as an assistant professor from 1972 to 1976, where he continued to work on packet networking and worked with Robert E. Kahn—who was instrumental in ARPANet's hardware design—to develop the TCP/IP protocol for the Department of Defense. Various protocols had to be developed to enable computers to communicate with one another. TCP/IP was a suite of two such protocols: the Transmission Control Protocol, used to exchange data; and the Internet Protocol, which handles routing and addressing. The early version of TCP/IP was introduced in Cerf and Kahn's 1974 paper "A Protocol for Packet Network Interconnection," published by the Institute of Electrical and Electronic Engineers (IEEE). In the twenty-first century, it continues to be the protocol that most Internet applications rely on, including e-mail, file transfer, and the World Wide Web.

Cerf left Stanford in 1976, to work for DARPA directly until 1982, roughly the dawn of the personal computer era, when he was hired as vice president of MCI Digital Information Services (which has since been acquired by Verizon Communications). Cerf oversaw the development of MCI Mail, the first commercial e-mail service, which was officially in service from 1983 to 2003. Messages over MCI Mail were sent over any standard telephone landline with the use of a modem and could be delivered to any other MCI Mail user, a telex, or an MCI Mail print site—an important option in days when access to a personal computer was often limited. Eventually, messages could be sent to any e-mail user regardless of his or her service, as well as to FAX dispatchers. He also led teams at MCI that developed Internet solutions for data, voice, and video transmissions. As the Internet became more widespread, he continued to be an advocate for its use and development. For example, from 1999 to 2000, he served on the board of the Internet Corporation for Assigned Names and Numbers, and some attribute the group's survival to Cerf's business prowess, technical knowledge, and ability to work with players at all levels of Internet governance. He has consulted with NASA's Jet Propulsion Laboratory to develop an Internet standard for planet-to-planet communication and testified before the U.S. Senate in favor of "net neutrality" as that has become an increasing concern of the twenty-first century.

In 2005, Google hired Cerf as vice president and "Chief Internet Evangelist," which has given him a prominent platform from which to address issues from environmentalism, to artificial intelligence, to the imminent transformation of the television industry's delivery model. When asked about the process of innovation and where innovators like him get their ideas, Cerf said, "Part of it is being willing to think literally, out of the box....The people I find most creative are also the ones who really know a lot about what they're doing. They either know a lot of physics, or a lot of math." In addition, he noted that "depth of understanding" means not only knowing the terms of a formula but being able to convey the intuitive meaning of the mathematics.

Further Reading

Abbate, Janet. *Inventing the Internet*. Cambridge, MA: MIT Press, 2000.

Morrow, Daniel. Computerworld Honors Program International Archives: Vinton G. Cerf Oral History." http://www.cwhonors.org/archives/histories/cerf.pdf.

BILL KTE'PI

See Also: Data Mining; Internet; Personal Computers.

Cheerleading

Category: Games, Sport, and Recreation.
Fields of Study: Geometry; Number and Operations.
Summary: Cheerleading demonstrates and depends on an understanding of gravity and other forces.

Cheerleading is an activity that can be considered both recreation and a competitive sport, depending on the context. It typically consists of choreographed routines that require energy, discipline, and stamina, and may include chants, dance, tumbling, and other physical stunts. Cheerleaders make what they do look easy when, in reality, the underlying mathematics, such as symmetry, sequences, and physics, helps them to conquer gravity and fly. In 2008, the show *Time Warp* on the Discovery Channel analyzed the physics of cheerleading and gymnastics using slow-motion cameras.

History

In 1898, University of Minnesota football student Jack "Johnny" Campbell became the first person to lead football fans in cheers, using a megaphone, which had been invented by Thomas Edison in 1878, in order to spur his school's football team to victory. This cheering gave rise to organized cheerleading. Women joined the sport in the 1920s, bringing an opportunity to add gymnastics and throws to the cheerleading repertoire. Showmanship and pom poms were later added to the sport. The Dallas Cowboys cheerleaders' skimpy outfits in the 1970s changed the outward appearance of cheerleaders, while the 1980s brought the pursuit of more technical stunt sequences. In the new millennium, the *Bring It On* movies highlight the sport's challenges as well as its technical aspects. Although college squads are currently about 50% male, youth cheerleading is predominantly female. Cheerleaders are now found all around the world.

Focusing on symmetry helps to maintain an equal distribution of weight when acting as a base.

The Physics of Cheerleading

Cheerleaders are focused on center of mass and axes of rotation in order to maintain balance and complete pivots, jumps, and flips. Focusing on symmetry not only helps both their formations and individual poses have a more appealing look but also keeps them focused on maintaining an equal distribution of weight when they act as "bases" for a "climber" or "flyer. "

Cheerleaders need a firm grasp of gravity and the physics involved in their work, including Newton's Third Law, which states that for every action there is an equal and opposite reaction. For example, in a "full extension," the climber pushes off the two bases' shoulders and pulls up with his or her own shoulders to bear some the weight. The two bases move into a "chest prep" with their knees locked, their arms extended and locked, holding the climber's feet at chest level; the climber is now referred to as a flyer. The back person, or "spotter," will often be used as an additional holder to both hold some of the flyer's weight as well as to solidify the overall hold.

As the bases bend their knees, preparing to exert upward force in order to toss the flyer, each base's arms hold half the flyer's weight—uneven distribution of weight is seen when the bases' hips are uneven, exhibiting a loss of symmetry. The bases will extend their knees, letting go of the flyer's feet, to give the flyer upward force; the flyer lands exerting greater force on the way down, so the bases bend their knees and lock hands to cushion the catch. If the bases have not evenly distributed the weight, or have exerted unequal amounts of force, the flyer will not go straight up and the bases will need to move to catch the flyer.

In preparing to execute a flip, the cheerleader bends his or her knees to exert the upward force. To execute, the cheerleader needs to stay tight, keep the axis of rotation steady, point the feet, and land lightly, snapping together to a final pose to stop his or her momentum. The cheerleader's angular speed can change by changing the distance of mass to the axis of rotation; the cheerleader gets momentum from the push off as well as from reducing the distance from mass to axis of rotation by tucking the body in as he or she rises from the ground.

Further Reading

Lesko, Nancy. "We're Leading America: The Changing Organization and Form of High School Cheerleading." *Theory and Research in Social Education* 16, no. 4 (1988).

Pennington, Bill. "As Cheerleaders Soar Higher, So Does the Danger." *New York Times* (March 31, 2007). http://www.nytimes.com/2007/03/31/sports/31cheerleader.html?_r=1&ref=cheerleaders.

Physics of Cheerleading. http://thephysicsofcheerleading.homestead.com.

Deborah L. Gochenaur

See Also: Gymnastics; Knots; Skating, Figure.

Chemotherapy

Category: Medicine and Health.
Fields of Study: Algebra; Measurement.
Summary: Mathematical modeling has improved chemotherapy protocols and saved patients' lives.

Chemotherapy is the use of chemical drugs to kill cancerous cells in the body. Although cancerous cells are the target of chemotherapy, traditional chemotherapies do not distinguish between "good" and "bad" cells. Hence, chemotherapy often results in side effects, such as hair loss and toxicity damage to body organs. Because of these chemotherapy side effects, chemotherapy protocols attempt to kill as much of the tumor as possible while incurring as little damage to the patient as can be managed. Thus, chemotherapy regimens are managed according to different variables, including how much drug is given in a treatment, how frequently treatments are given, and the total number of treatments given. Historically, chemotherapy protocols were designed only through experimental data from clinical trials and practice. However, such experiments can be costly or even pose ethical dilemmas. Chemotherapy variables are quantitative—each lends itself to a mathematical understanding and description that can be used to model and simulate treatment experiments, adding to the information gained in clinical settings.

Mathematics in Cancer Chemotherapy News

Mathematics is becoming an increasingly powerful tool in cancer chemotherapy treatments, especially in the dosing and management of chemotherapy protocols. For example, in 2004, Dr. Larry Norton received the American Society of Clinical Oncology's David A. Karnofsky Award, which is given for an outstanding contribution to progression in cancer treatment. Norton's award is notable because of his quantitative contribution to the field of chemotherapy dosing. The National Cancer Institute has a Center for Bioinformatics that addresses the issue of systematically studying the vast amounts of data associated with cancer growth and treatment response.

Cancer Geometry and Treatment

Cancer cells appear visibly different in shape and structure than normal healthy cells. This fact helps practitioners identify unhealthy cells. Quantitative measurements are associated with the geometry and the complexity of cancer cells. These measurements are related to fractal geometry. Tumor fractal dimensions reflect more complex structures generally because of the arrangement of blood vessels in the tumor. Abnormal blood vessel arrangements inhibit the tumor's uptake of therapeutic drugs. This understanding has led to the use of anti-angiogenic drugs that inhibit the production of new blood vessels and lower the measurement of the tumor's complexity. These drugs can now be used in concert with other cancer treatments in order to create a more effective cancer-fighting regimen.

Cancer Growth and Chemotherapy Treatment

Historically, it was believed that cancer cells grew in an exponential manner over the entire period of a tumor's growth. In exponential growth, the doubling time of a population is constant. This belief affected the way that chemotherapy was delivered, since chemotherapy works by attacking rapidly dividing cells. If a tumor's growth rate were constant, there would be no difference in how many cells were killed during any chemotherapy treatment, regardless of the size of the tumor. This is the "log-kill" model of tumor growth.

However, in the mid-twentieth century, it was experimentally discovered that many tumors exhibited a different kind of population growth: "Gompertzian growth," named for Benjamin Gompertz. When populations grow in a Gompertzian fashion, they grow very rapidly at first—when the population is small. As the population size increases, the growth rate of the population slows. Thus, many tumors would have a smaller doubling time when smaller, and a larger doubling time when larger. Because chemotherapy attacks the most rapidly dividing cells, smaller tumors would be more susceptible to chemotherapy treatments. Thus, if a tumor has been reduced in size by one chemotherapy treatment, it would be better to give a second chemotherapy treatment as soon as possible without costing the patient in terms of healthy cell function. This Norton–Simon hypothesis, named for Larry Normal and Richard Simon, has led to a change in the frequency of standard chemotherapy regimens—the time between treatments was decreased in order to take advantage of the more rapid growth rate in the smaller tumor that had resulted from the previous treatment. This change in treatment timing has increased the survival time of patients undergoing chemotherapy treatments.

Looking Ahead

Although the Norton–Simon hypothesis is a prominent example of how mathematics has helped improve cancer chemotherapy treatments, there are ongoing studies by mathematicians to further improve treatment of cancer. Using a field of mathematics known as optimal control, some mathematicians study how to make chemotherapy treatments as ideal as possible. Although practitioners can make (and have made) use of the Norton–Simon hypothesis, the increase of chemotherapy treatments for a patient, while better, is not necessarily best. Using optimal control theory on mathematical models of cancer and cancer treatment, researchers can investigate the best timing and dosing strategies for chemotherapy based on the variables mentioned above. This work may even lead to determining cancer treatment plans based on a particular individual or a particular kind of cancer in the future.

Further Reading

Dildine, James. "Cancer and Mathematics." http://mste.illinois.edu/dildine/cancer/cancer.html.

Laird, Anna. "Dynamics of Tumour Growth: Comparison of Growth Rates and Extrapolation of Growth Curve to One Cell." *British Journal of Cancer* 19, no. 2 (1965).

Martin, R., and K. Teo. *Optimal Control of Drug Administration in Cancer Chemotherapy*. Singapore: World Scientific Publishing, 1966.

National Cancer Institute (NCI). *NCI Cancer Bulletin* 3, no. 18 (2006).

Schmidt, Charles. "The Gompertzian View: Norton Honored for Role in Establishing Cancer Treatment Approach." *Journal of the National Cancer Institute* 96, no. 20 (2004).

Piccart-Gebhart, Martine. "Mathematics and Oncology: A Match for Life?" *Journal of Clinical Oncology* 21, no. 8 (2003).

Angela Gallegos

See Also: Mathematical Modeling; Mathematics Research, Interdisciplinary; Medical Simulations.

Chinese Mathematics

Category: Government, Politics, and History.
Fields of Study: Algebra; Connections; Geometry; Representations.
Summary: Chinese mathematicians have a long history of investigation and discovery, sometimes predating similar findings in other cultures.

Chinese mathematics has a very long history, and its development is quite independent of other civilizations before the thirteenth century. Roughly speaking, it has four periods of developments before the middle of the Qing dynasty, namely

- *The early development period*: from ancient times to the Qin dynasty (2700–200 B.C.E.)

- *The foundation period*: from the Han dynasty to the Tang dynasty (200 B.C.E–1000 C.E.)
- *The golden period*: from the Sung dynasty to the Yuan dynasty (1000–1367)
- *The east meets west period:* from the Ming dynasty to the middle of the Qing dynasty (1367–1840)

Early Chinese mathematics is problem based and is motivated by various practical problems, including astronomy, trade, land measurement, architecture, and taxation.

The Early Development Period

It is written in *Yi Jing* (*I-ching* or *Book of Changes*) that, "In early antiquity, knotted cords were used to govern with. Later, our saints replaced them with written characters and tallies." In other words, the ancient Chinese used knotted cords to record numbers. Later, written symbols and tallies were used instead. In the Shang dynasty (1600–1050 B.C.E.), numerals were invented and inscribed on oracle-bones or tortoiseshells for recording numbers. It was a decimal system and was widely used at the time.

In the Zhou dynasty (1050–256 B.C.E.), mathematics was one of the *Six Arts* (*Liu Yi*), which were taught by teachers at schools. The other five arts were rites, music, archery, charioteering, and calligraphy. From this dynasty onward, the ideas of *Taichi*, *Ying Yang*, *Trigrams,* and *Hexagrams* largely influenced the developments of sciences, mathematics, philosophy, arts, architecture, and many other areas in Chinese culture. For example, *Luoshu* (3×3 magic square) is closely related to the eight *Trigrams*. It has both ceremonial and metaphysical importance, which plays a significant role in Chinese philosophy for several thousands of years.

From the Kingdoms of Spring and Autumn (720–480 B.C.E.) to the period of Warring States (480–221 B.C.E.), the Chinese used counting rods to do calculations. Numbers were expressed by nine symbols, and blanks were used to denote zeros. The numeration system was already a decimal place-valued system.

The first definitive work on geometry in ancient China was the *Mo Jing*, which was compiled after the death of Mozi (470–390 B.C.E.). Many basic concepts of geometry can be found in this book. For example, the *Mo Jing* defines a point to be the smallest unit that cannot be divided, and points on a circle to be equidistant from the center. The book also mentions the definitions of endpoints, straight lines, parallel lines, diameter, and radius.

In the Qin dynasty, the famous Great Wall and many huge statues, tombs, temples, and shrines were built, which required sophisticated skills and mathematical knowledge for calculating proportions, areas, and volumes. Unfortunately, not much is known about the actual mathematical development in the Qin dynasty now, because of the burning of books and burying of scholars ordered by Emperor Qin Shi Huang.

The Foundation Period

In 1984, a Chinese mathematics text called *Suan Shu Shu*, completed at about 200 B.C.E., was discovered in a tomb at Zhangjiashan of the Hubei Province. It is about 7000 characters in length, and is written on 190 bamboo strips. Its content is mainly concerned with basic arithmetic, proportions, and formulas of areas and volumes. The next complete surviving text is the *Zhou Bi Suan Jing*, written between 100 B.C.E. and 100 C.E. Although it is a book on astronomy, it contains a clear description of the *Gougu Theorem* (the Chinese version of the Pythagorean theorem), which is very useful in solving problems in surveying and astronomy. This work is perhaps the earliest recorded proof of the Pythagorean theorem.

After the book burning in 212 B.C.E., the Han dynasty (202 B.C.E.–220 C.E.) began to edit and compile the mathematical works lost in the Qin dynasty. The most important one is the *Nine Chapters on the Mathematical Art* (*Jiuzhang Suan Shu*), completed at around 179 C.E. Although the editor is unknown now, this book had a great impact on the mathematical developments in China and its neighboring countries, such as Japan and Korea. It contains a collection of 246 mathematical problems on agriculture, engineering, surveying, partnerships, ratio and proportion, excess and deficit (the method of double false positions), simultaneous linear equations, and right-angled triangles.

The general method of solutions is provided, but no proof is given in the Greek sense. Most of the methods are of computational nature, and they can be applied to solve problems algorithmically. For instance, square roots, or cubic roots, can be found in a finite number of steps by using a procedure called *Kai Fang Shu*. For skillful users of this method, the answers can be computed efficiently by manipulating the counting rods.

For circular measurements, the approximated value of π is taken as 3. Some problems are expressed in terms of a system of linear equations and then solved by algebraic techniques. For instance, a problem in Chapter Eight leads to the system

$$3x + 2y + z = 39$$

$$2x + 3y + z = 34$$

$$x + 2y + 3z = 26$$

which can be solved by a method like the matrix approach described in modern textbooks.

In the third century, Liu Hui wrote his famous commentary on the *Nine Chapters on the Mathematical Art.* He also wrote a book called *The Sea Island Mathematical Manual* (*Haidao Suan Jing*), to demonstrate how to apply the Gougu Theorem. He was the first Chinese mathematician to deduce that the value of π lies between 3.1410 and 3.1427, by repeatedly doubling the number of sides of a regular polygon inscribed in a circle. It is called the method of dissection of a circle. Liu Hui also discovered the Cavalieri's Principle and used it to find the volume of a cylinder. About two centuries later, Zu Chongzhi (430–501) and his son, Zu Geng, found that the value of π lies between 3.1415926 and 3.1415927, based on the pioneer works of Liu Hui. He also obtained the remarkable rational approximation 355/113 for π, which is correct to six decimal places. Working with Zu Geng, he successfully applied the Cavalieri's Principle to deduce the correct formula for the volume of the sphere by computing the volume of a special solid called *Mouhe Fanggai* (the double vault) as proposed earlier by Liu Hui.

Unfortunately, his own work called *Zhui Shu* was discarded from the syllabus of mathematics in the Song dynasty and was finally lost in the literature. Many believed that *Zhui Shu* probably describes the method of interpolation and the major mathematical contributions by Zu Chongzhi and Zu Geng.

At the beginning of the Tang dynasty, Wang Xiaotong (580–640) wrote the *Jigu Suanjing* (*Continuation of Ancient Mathematics*), a text with only 20 problems that illustrate how to solve cubic equations. His method was a first step toward the *Tian Yuan Shu* (the method of coefficient array), which was then further developed by other mathematicians in the Sung and Yuan dynasties.

In the sixth century, mathematics was a subject being included in the civil service examinations. Li Chunfeng (602–670) was appointed by the Chinese emperor as the chief editor for a collection of mathematical treatises for both teachers and students. The collection is called the *Ten Classics* or the *Ten Computational Canons*, which include the *Zhou Bi Suan Jing*, the *Jiuzhang Suan Shu*, the *Haidao Suan Jing*, the *Sunzi Suan Jing*, the *Wucao Suan Jing*, the *Wujing Suan Shu*, the *Shushu Jiyi*, the *Xiahou Yang Suan Jing*, the *Zhang Qiujian Suan Jing*, and the *Jigu Suan Jing*. The book *Zhui Shu* by Zu Chongzhi had been included in the *Ten Classics* at the beginning, but it was later replaced by the *Shushu Jiyi* because of it being lost in the Sung dynasty.

The Golden Period

No significant advances in mathematics were made between the tenth century and the eleventh century. However, Jia Xian (1023–1050) improved the methods for finding square roots and cube roots, and also extended them to compute the numerical solutions of polynomial equations by means of the *Jia Xian Triangle* (the Chinese version of the Pascal Triangle).

The golden period of mathematical development in China occurs in the twelfth and the thirteenth centuries, which is called the "Renaissance of Chinese mathematics" by some authors. Four outstanding mathematicians appeared in the Sung dynasty and the Yuan dynasty, namely Yang Hui (1238–1298), Qin Jiushao (1202–1261), Li Zhi (also called Li Yeh, 1192–1279), and Zhu Shijie (1260–1320). Yang Hui, Qin Jiushao, and Zhu Shijie all used the Horner–Ruffini method to solve quadratic, cubic, and quartic equations. Li Zhi, on the other hand, revolutionized the method for solving problems on inscribing a circle inside a triangle, which could be formulated as algebraic equations, and solved by using the Pythagorean theorem. Another mathematician, Guo Shoujing (1231–1316), worked on spherical trigonometry for astronomical calculations. Therefore, much of the modern mathematics in the West had already been studied by Chinese mathematicians in this period.

Qin Jiushao (1202–1261) invented the symbol for "zero" in Chinese mathematics. Before this invention, blank spaces were used to denote zeros. Qin Jiushao also studied indeterminate problems and generalized the method of Sunzi to become the now-called "Chinese Remainder Theorem." He wrote the *Shushu Jiu-*

zhang (*Mathematical Treatise in Nine Sections*), which marks the highest point in indeterminate analysis in ancient China.

Yang Hui was an expert in designing magic squares. He discovered elegant methods for constructing magic squares with an order greater than three. Some of the orders are as high as 10. He was also the first in China to give the earliest clear presentation of the *Jia Xian Triangle* in his book *Xiangjie Jiuzhang Suanfa*.

The famous work of Li Zhi is the *Sea Mirror of the Circle Measurements* (*Ce Yuan Hai Jing*). It is a collection of some 170 problems. He used *Tian Yuan Shu* or the Method of Coefficient Array to solve polynomial equations of degree as high as six. He also wrote the book *Yi Gu Yan Duan* (New Steps in Computation) in 1259, which is an elementary book related to solution of geometric problems by using algebra.

The most important text in the thirteenth century is the *Precious Mirror of the Four Elements* (*Si Yuan Yujian*), written by Zhu Shijie in 1303. This book marks the peak of the development of algebra in China. The unknowns that appeared in equations are called the four elements, namely heaven, Earth, man, and matter. This book describes how to solve algebraic equations of degrees as high as 14. The method is the same as the Horner–Ruffini method. Zhu Shijie also used the matrix methods to solve systems of equations. He was also an expert in summation of series. Many formulas on summation of series can be found in the *Precious Mirror of the Four Elements*. He also wrote an elementary mathematics text called the *Introduction to Computational Studies* (*Suanxue Qimeng*) in 1299, which had a significant impact on the development of Japanese mathematics later.

The East Meets West Period

In the Ming dynasty, not much original mathematics work emerged in China. Even the famous work *Suanfa Tongzong* (*General Source of Computational Methods*) by Cheng Dawei (1533–1606) was an arithmetic book for the abacus only. Its style and content were still influenced very much by the *Nine Chapters on the Mathematical Art*. It was only when the Italian Jesuit Matteo Ricci (1552–1610) came to China in 1581 that the development of mathematics in China was influenced by the West from this time onwards. For instance, Xu Guangqi (1562–1633) and Matteo Ricci translated a number of Western books on sciences and mathematics into Chinese, including the famous *Euclid's Elements*, the influence of the Western culture on China became more apparent.

However, the Chinese mathematicians also did an excellent job in editing and recording their traditional mathematics and science works in the early Qing dynasty, so that much of them can come down to us now. For example, Mei Juecheng (1681–1763) edited the famous mathematical encyclopedia *Shuli Jingyun* in 1723, and Ruan Yuan (1764–1849) edited the *Chouren Zhuan* (*Biographies of Astronomers and Mathematicians*) in 1799. Both of these works are very valuable and useful references for historians to study the mathematical developments in China before the middle of the Qing dynasty.

Achievements in Chinese Mathematics

After the decline of Greek mathematics in the sixth century, Western Europe was undergoing the period of Dark Ages. On the other hand, many of the achievements of Chinese mathematics predated the same achievements before and shortly after the Renaissance. For instance, before the fifteenth century, China was able to (1) adopt a decimal placed-value numeral system, (2) acknowledge and use negative numbers, (3) obtain precise approximations for π, (4) discover and use the Horner–Ruffini method to solve algebraic equations, (5) discover the *Jia Xian Triangle*, (6) adopt a matrix approach to solve systems of linear equations, (7) discover the Chinese Remainder Theorem, (8) discover the method of double false position, and (9) handle summation of series with higher order. It was only after the fourteenth century that the development of Chinese mathematics began to decline and lag behind the Western mathematics in the Ming and Qing eras. However, it is worthy to note that the traditional Chinese mathematics still can find its contribution in mechanized geometry theorem proving in the twentieth century, because of its algorithmic characteristics.

Further Reading

Cullen, Christopher. *Astronomy and Mathematics in Ancient China: The Zhou Bi Suan Jing*. Cambridge, England: Cambridge University Press, 1996.

Martzloff, J. C. *A History of Chinese Mathematics*. Corrected ed. New York: Springer-Verlag, 2006.

Shen, Kangshen, John N. Crossley, and W. C. Lun Anthony. *The Nine Chapters on the Mathematical*

Art: Companion and Commentary. Oxford, England: Oxford University Press, 1999.

Swetz, Frank. *Legacy of the Luoshu: The 4000 Year Search for the Meaning of the Magic Square of Order Three.* Wellesly, MA: A K Peters, 2008.

Yiu-Kwong Man

See Also: Asia, Eastern; Binomial Theorem; Greek Mathematics; Pi; Pythagorean Theorem.

Circumference

See *Perimeter and Circumference*

City Planning

Category: Architecture and Engineering.
Fields of Study: Geometry; Number and Operations; Problem Solving.
Summary: Mathematics is used to model optimal city designs and reduce problems of traffic congestion, sanitation, and water distribution.

Also called "urban planning" and "town planning," city planning is a discipline that focuses on the various economic, environmental, historical, physical, political, and social characteristics of the urban environment and their harmonious organization. It encompasses a variety of projects, processes, and goals that involve multiple disciplines and fields of expertise, such as physical design, and quantitative and qualitative research, as well as analysis, forecasting, strategic planning, negotiation, and public mediation. Since the late nineteenth century—and especially during the second half of the twentieth century—the profession has increased its reliance on statistics and mathematics.

Early History

The early origins of urban planning can be traced in the physical design and purposeful spatial organization of some ancient cities in Mesopotamia, Egypt, the Mediterranean Basin, South and Central America, the Yellow River Basin, and along the Indus Valley. Many of these settlements present a hierarchical system of paved streets, often following a rectilinear grid, with water supply and drainage systems. The Middle Ages was not a propitious era for urban planning. It became popular again during the Italian Renaissance with the design of ideal cities. Influenced by the belief that a perfect form was the image of a perfect society, designers opted for radial or centrally planned cities frequently uniting the perfect geometric figures of the square and circle into a star-shape layout. In the seventeenth century, the rise of nation states and absolutism was conducive to the development of the monumental baroque city with its straight and endless avenues, unbroken horizontal rooflines, and repetition of uniform elements, which glorified the ruling power. Simultaneously, the advances of warfare techniques led to the disappearance of the old city walls and the adoption of new complicated systems of fortification with considerable outworks and bastions in spearhead forms.

The Industrial Era

The modern origins of city planning have their roots in the industrial city of the mid- and late nineteenth century. In both Europe and the United States, rapid technical progress, tremendous industrial development, and massive displacements of rural population to urban areas created considerable problems that threatened to disrupt the existing social order. The dreadful conditions experienced by masses of people living in abject poverty and misery in overcrowded slums sprawling around wealthier districts became a source of concern for the general public health. In 1854, Dr. John Snow—the father of modern epidemiology—identified the source of a cholera outbreak in London by studying the patterns of the disease and using statistics and a spot map illustrating the clustered death cases of cholera around the Broad Street pump. The fears of major epidemics resulted in the rise of a social movement for urban reform and planning, which first focused on water supply and sanitation improvement, and later on housing provision.

In the 1880s, the basic lack of information regarding the extent and distribution of poverty in London led English philanthropist Charles Booth to develop a comprehensive and scientific social survey investigating the incidence of pauperism first in East London, and

Lithograph by Currier and Ives of the 1893 World's Columbian Exposition in Chicago. The "White City" in the exhibition was the first large-scale planning execution of the City Beautiful movement in the United States.

later in the entire city. His quantitative statistical analyses and qualitative research presented in 17 volumes with accompanying colored maps indicating the levels of poverty and wealth by street received considerable attention. They were also influential in demonstrating the importance of social surveys for public policy, demographics, and sociology as well as in improving census data collection.

Similar problems affected Paris and, after visiting London, Napoléon III placed considerable emphasis on urban planning to modernize the medieval capital into the capital of light. The large-scale restructuring program under the direction of Baron Haussmann affected not only the center of Paris but also its surrounding suburbs. At the time, it was the largest urban renewal project ever implemented. The plan created a network of large, easily accessible avenues and boulevards with radiating vistas terminated by prestigious public edifices and monuments. In addition to the building of 71 miles of new roads, the layout of 400 miles of pavement, and the doubling of the number of trees lining the streets, the city's infrastructure was entirely renovated. The construction of more than 340 miles of sewers and hundreds of miles of aqueducts increased the water supply by 400%. Haussmann also created two major urban parks and two large natural preserves on the periphery. This urban metamorphosis influenced the design of numerous cities worldwide and in particular the "White City" of the World's Columbian Exposition of 1893 in Chicago, which was the first large-scale project of the City Beautiful movement in the United States. The aim of expanding civic consciousness and raising the standards of civic design culminated in the publication of the famous Plan of Chicago in 1909, which coincided with the first university course in city planning at Harvard and the first National Conference on City Planning. In 1917, the American City Planning Institute was founded.

The Twentieth Century

Nevertheless, with the growth of the automobile as a favorite mode of transportation, it was the "Garden City" concept invented by Ebenezer Howard that became the leading model for the development of U.S. suburban residential communities. As the old city centers became increasingly congested, transportation planning became increasingly important to ensure an efficient balance between land-use activities and the potential communications between them. Transportation planners regularly collect data, which they analyze and process, to forecast future traffic using various techniques such as land-use ratio methods, multiple regression models, category analysis, growth-factor methods, synthetic models, modal split analysis, diversion curves, and geographic information systems (GIS).

After the U.S. Department of Commerce published "A Standard State Zoning Enabling Act" and "A Standard City Planning Enabling Act" in the 1920s and the U.S. Supreme Court upheld the constitutionality of zoning in 1926, most U.S. cities established planning departments to adopt master plans and zoning regulations that allowed them to control land-use development, protect property values, and segregate uses. Cities also started implementing subdivision controls and regulations. These new tools contributed to the belief in part of the planning community of the possible rational and scientific management of cities. On the other hand, idealists such as Frank Lloyd Wright and Lewis Mumford criticized the new pragmatic and technological approach, preferring a philosophy of city development for humanistic and social ends as epitomized in the design of Radburn, New Jersey. Over time, zoning regulations revealed some drawbacks. They often increased traffic congestion, and sometimes prevented the construction of affordable homes. Some courts struck them down as exclusionary.

City planning in the post–World War II era was dramatically affected by four significant federally funded programs: public housing, urban renewal, home mortgage insurance, and highway building. The miserable failure of urban renewal—and the urban crisis of the 1960s that ensued—required new approaches to urban planning. During the second half of the twentieth century, city planning became increasingly defined as a cyclical process attempting to balance conflicting social, economic, environmental, and aesthetic demands while implementing selected objectives and goals. Therefore, regular monitoring became necessary to test, evaluate, and review the strategies and policies adopted on a continuous basis. City planners regularly use a wide range of models ranging from basic descriptive statistics to more complex mathematical models that allow them to understand the nature of various urban components and forecast the consequence of change.

Because of the tremendous complexity of urban systems, models can provide only a simplified representation of the studied phenomena. Consequently, there is considerable attention and controversy regarding the choice of variables, and their level of aggregation and categorization, as well as the handling of time, specification, and calibration. Although deterministic models are the dominant type of predictive models used by urban planners, there has been some attempt at developing stochastic models. Urban planners are also concerned with the accuracy, validity, and constancy of the models they use. Most models tend to be topic specific, focusing, for example, on population, housing, employment, shopping, transport, or recreation, but integrated forecasting systems have become more common as there has been an increasing recognition of the interdependence of the various subfields of a city.

Further Reading

Field, Brian, and Bryan MacGregor. *Forecasting Techniques for Urban and Regional Planning.* Cheltenham, England: Nelson Thornes, 2000.

Freestone, Robert. *Urban Planning in a Changing World.* New York: Routledge, 2000.

Hall, Peter G. *Cities of Tomorrow: Intellectual History of Urban Planning and Design in the Twentieth Century.* 3rd ed. Malden, MA: Blackwell Publishing, 2004.

Kostof, Spiro. *The City Shaped: Urban Patterns and Meanings, Through History.* Boston: Bulfinch Press, 1993.

Krueckeberg, Donald A. *Introduction to Planning History in the United States.* New Brunswick, NJ: Center for Urban Policy Research, Rutgers University, 1983.

Lynch, Kevin. *Good City Form.* Cambridge, MA: MIT Press, 1984.

Moughtin, Cliff, et al. *Urban Design: Method and Techniques.* 2nd ed. Burlington, MA: Architectural Press, 2003.

Catherine C. Galley
Carl R. Seaquist

See Also: Engineering Design; Highways; Water Distribution; Wright, Frank Lloyd.

Civil War, U.S.

Category: Government, Politics, and History.
Fields of Study: Algebra; Geometry; Measurement; Number and Operations; Problem Solving.
Summary: The U.S. Civil War saw numerous advances in firearms, cryptography, and strategy.

The U.S. Civil War, also sometimes known as the War Between the States, was a conflict fought from 1861 to 1865 between 11 southern U.S. states that seceded from the Union to form the Confederate States of America and the remaining United States. Precipitating causes of the war centered on economic issues and states' rights versus federal power, often symbolized by the central dividing issue of slavery. More than 600,000 men on both sides died, which is greater than the combined U.S. losses in all subsequent wars and military conflicts through the beginning of the twenty-first century, though World War II exceeds this count if the metric is combat deaths versus deaths from all causes. Some also refer to the Civil War as the first "industrialized war" because of the extensive use of the telegraph, railroads, and mass-manufactured goods and weaponry.

Mathematics was instrumental in this war in many ways. Introduction of the Spencer repeating rifle has been cited by many as the turning point toward the eventual Union victory. Ciphers and code-breaking efforts were important in communicating military strategies and plans. The U.S. Army Signal Corps, founded in 1860, used both telegraphy and line-of-sight methods, such as the wig-wag signaling system in which the left, right, or upward positions of a single flag represented the numbers 1–3 and specific number combinations corresponded to letters. The ranks of leaders on both sides were filled with mathematically educated graduates of schools like the Unites States Military Academy at West Point. Mathematics education and research were also impacted by the war.

Weaponry

Changes in small arms and artillery that were occurring in Europe and the United States at this time had a tremendous impact on the war. Many different types of smoothbore or rifled artillery were used during the Civil War. One way in which they were distinguished was by their bore size, which was the diameter of their barrels, usually expressed in inches. Another differentiating feature was the weight of the projectiles they fired, in pounds. Different classes of weapons also had different trajectories. Cannons known as "guns" had relatively flat trajectories. Mortar rounds followed a steeply arcing path. Howitzers fell between the other two because the possible angles of inclination and powder charges could be varied more than the other two types.

The most common artillery piece was the Napoleon, a howitzer named after Napoleon Bonaparte, an avid student of mathematics who had revolutionized infantry and artillery warfare. Artillery ammunition included solid shot or balls, grape, canister, shell, and chain shot. Canister and grape shot could be particularly devastating to humans, since they disintegrated into smaller, scattering projectiles along a number of trajectories when fired. At the start of the war, both sides relied primarily on muzzle-loading rifled muskets such as the 58 Springfield, though some smoothbore muskets were also in use. Rifled weapons had greater range—nearly half a mile—versus the 100-yard range for smoothbores, which affected infantry tactics. The breech-loading Spencer repeating rifle was a major innovation, and was considered the most advanced weapon of its time. It used metal cartridge ammunition and it could hold seven cartridges at a time, which greatly increased rate of fire and accuracy, though the Union was initially concerned about the corresponding increase in the demand for ammunition.

Revolvers also replaced muzzle-loading pistols, with similar effects. Minié ball ammunition, named for French military officer Claude Minié, was used extensively in the Civil War. Previously, rifles had been difficult to load because the bullets fit tightly in the bore of the weapon, which was necessary for them to be propelled effectively by explosive powder charges. Despite being called a "ball," the Minié projectile was conical. It was smaller in diameter than older ammunition, and also had grooves that allowed it to fall smoothly and quickly into the barrel of the rifle. A hollow indentation extending from the base caused it to expand to the size of the gun's bore when fired, optimizing the combination of loading time and accuracy.

Cryptography

One problem addressed by mathematical problem-solving approaches was the terrible problem of "hacking." Increasingly, communications were being relayed by the newest technology—the telegraph and Morse code. It was relatively easy for someone to climb the telegraph pole, connect a telegraph key to the wires, and intercept messages that were being relayed back and forth from the front lines to the base camp. Messages needed to be coded so that interceptors could not interpret them, which was not a new problem. The problem of encoding military messages can be dated to at least Julius Caesar and earlier to the Spartan military.

The governors of Ohio, Indiana, and Illinois were close enough to the Confederate border that they felt the need to have their messages encoded. Governor William Dennison of Ohio asked Anton Stager to prepare a cipher (a code) that could be used by these three governors. Stager adapted a transposition coding system that had been used in Great Britain years earlier. Words were rearranged into a grid. The first word of the message was the "key" to indicate how many columns were to be formed and in what order they were to be read. Instructions for these codes were printed on cards about the size of a standard index card, which were the precursors of codebooks. Included on the cards were the route, the keys, the code words, and words used to check the cipher.

This system underwent a number of modifications, and Stager's route cipher was eventually adopted as the Union's official cipher. Increasingly sophisticated ciphers were created during the war and, as a result, the instructions could no longer fit on cards. Some of the resulting codebooks were 48 pages long. The messages were intercepted by the Confederacy and sent to Richmond, Virginia, the capital of the south. By twenty-first-century coding standards, they appear to have been relatively easy codes to break, but evidence suggests that the south did not have either sufficient manpower or mathematical knowledge to decode them.

By contrast, the Union forces had a team to work in breaking the southern codes, including many times President Abraham Lincoln. The codes used by the south at the beginning of the war were not standardized, resulting in many messages that were unreadable. The Confederacy eventually settled on a code from 1587, the Vigenère cipher, named for Blaise de Vigenère—although others before him, such as Giovan

A 12-pdr. howitzer gun captured by Butterfield's Brigade of the 12th Maine Infantry in May 1862.

Bellaso, are also noted as having invented it. This code consisted of a tableau of staggered alphabets. The ease of this code was that the code did not have to change if a coded message was captured, only the code phrase. The problem with the Vigenère code came in errors in transmission over the telegraph. Even though the code was harder to break than the Union cipher, it was more difficult to implement because a missed letter would result in an incoherent message. For instance, General Edmund Smith reportedly spent 12 hours trying to decode a message from General Joseph Johnston during the Vicksburg campaign. The message requested reinforcements, but Smith was unable to read it. He eventually sent a courier, but it was too late for reinforcements; Johnston's army was already cut off. Revisions to the code to avoid this problem in the future made deciphering easier as well.

Mathematically Educated Leaders

Many of the military leaders for both the north and the south were graduates of the U.S. Military Academy at West Point, which was the United States' only engineering school for an extended period of time. The Civil War was fought with 359 generals who graduated from West Point. They served on both sides, 217 for the Union and 142 for the Confederacy. This list of elite officers and leaders includes many well-known officers. Ulysses S. Grant had plans to return to West Point to teach mathematics, and these plans were changed by the outbreak of the Mexican War. Robert E. Lee graduated second in his class in 1829 and served as an assistant professor for mathematics for his first two years at West Point. Edmund Kirby Smith (1845) taught mathematics at the University of the South after the war, where he joined another West Point graduate, Josiah Gorgas. Other well-known graduates who served both sides during the Civil War included Confederate President Jefferson Davis, Braxton Bragg (1837), John Bell Hood (1853), Thomas Jonathan "Stonewall" Jackson (1846), Albert Sidney Johnston (1826), James Longstreet (1842), George E. Pickett (1846), J. E. B. Stuart (1854), William Tecumseh Sherman (1840), George G. Meade (1835), George McClellan (1846), Joseph Hooker (1837), Abner Doubleday (1842), George Armstrong Custer (1861), and Don Carlos Buell (1841).

All these men were mathematically educated, which was unique for that point in U.S. history and likely played a role in many aspects of the war. For example, the maps and messages of the military in the Civil War show the influence of what is now referred to as "descriptive geometry," which was created by Gaspard Monge and was incorporated into the curriculum after engineer Claudius Crozet brought it to the Military Academy. Mathematics textbooks used by most of the leaders on both sides of the Civil War include those written by mathematicians Charles Davies and Albert Church, some of which were adaptations of earlier French works. This education of the leaders of both sides of the conflict may have had a great deal to do with the long length of the conflict and allows historians the opportunity to study other differences in the two sides.

It was not just the military leaders during the Civil War who made use of mathematics. Lincoln purportedly had a great reverence for Euclid of Alexandria and geometry. Some historians assert that he kept a copy of Euclid's *Elements* in his saddlebag and studied it by lamplight to develop his logic and reasoning skills. Phrasing in Lincoln's well-known 1863 address at the Gettysburg battlefield has sometimes been compared to Euclid.

Education

While mathematics undoubtedly influenced the course and outcome of the war, the Civil War also affected mathematics education and research. Antebellum college curriculum in schools such as The Citadel or Harvard consisted of classes in mathematics that were filled with "practical applications," such as mercantile transactions, navigation, surveying, civil engineering, mechanics, architecture, fortifications, gunnery, optics, astronomy, geography, history, and "the concerns of Government." These topics were all expanded in one of the common textbooks of the day, *An Introduction to Algebra*, by Jeremiah Day. Geometry and trigonometry were also commonly taught, and analytic geometry, conic sections, and calculus were often optional classes.

The problems discussed and worked in these classes, both in surveying and in navigation, were carefully chosen and adapted to make them easily done but not extremely realistic. Thus, the navigators and the surveyors being prepared for the Army were ultimately ill prepared to handle the realistic situations of making measurements under fire or in harsh seas. According to the work of Andrew Fiss, the Union army regulations required that surveyors plot the best course for the army to take. The topographical engineers worked so slowly that many of the generals took to asking local citizens for the best directions. By two years into the war the topographical engineers were incorporated into the Army Corps of Engineers. Likewise, the Navy found that the U.S. Naval Academy, founded near the middle of the nineteenth century, could prepare navigators better than a mathematics department. However, the academy was negatively impacted by the temporarily relocation from Maryland to Rhode Island during the war.

After the war, many universities started offering higher level mathematics courses, and some increasingly focused on research. Harvard and John Hopkins University graduated doctorates in mathematics within the next decade.

Further Reading

Antonucci, Michael. "Code Crackers: Cryptanalysis in the Civil War." *Civil War Times Illustrated* 34, no. 3 (1995). http://www.eiaonline.com/history/codecrackers.htm.

Benac, T. J. "The Department of Mathematics." United States Naval Academy. http://www.usna.edu/MathDept/website/mathdept_history.pdf.

Fiss, Andrew. "The Effects of the Civil War on College-Level Math Education." Talk given at the meeting of the History and Pedagogy of Mathematics—Americas Section. March 13, 2010, Washington, DC. http://www.hpm-americas.org/wp-content/uploads/2010/04/Fiss_Civil_War.pdf.

Parshall, Karen Hunger, and David E Rowe. *The Emergence of the American Mathematical Research Community, 1876–1900: J. J. Sylvester, Felix Klein, and E. H. Moore.* Providence, RI: American Mathematical Society, 1994.

Plum, William Rattle. *The Military Telegraph During the Civil War in the United States.* Reprinted in 2 Volumes. Charleston, SC: Nabu Press, 2010 (1882).

Rickey, V. Frederick, and Amy E. Shell. "The Mathematics Curriculum at West Point in the Nineteenth Century." http://www.math.usma.edu/people/Rickey/papers/WP19thCentury/WP19thCentCurr.htm.

Rickey, V. Frederick. "201 Years of Mathematics at West Point." In *West Point: Two Centuries and Beyond*, Lance A. Betros, ed. Abilene, TX: McWhiney Foundation Press, 2004 (1974).

Sauerberg, James. "Route Ciphers in the Civil War." http://www.mathaware.org/mam/06/Sauerberg_route-essay.html.

David C. Royster

See Also: Artillery; Coding and Encryption; Firearms; Revolutionary War, U.S.; Strategy and Tactics.

Climate Change

Category: Weather, Nature, and Environment.
Fields of Study: Algebra; Calculus; Data Analysis and Probability; Problem Solving; Representations.
Summary: Mathematicians and scientists use sophisticated models to track and predict global climate change.

The term "climate change" refers to the changing distribution of weather patterns. Climate is considered to be the average of 30 years of weather. In other words, climate is the distribution from which weather is drawn. Global warming refers to the change in climate in such a way that warmer weather is increasingly likely. In fact, it is not just the warming itself that is of concern but also the rate of change of the warming process since ecological systems typically cannot adapt to a rapidly changing climate. According to the *2007 Synthesis Report* by the Intergovernmental Panel on Climate Change (IPCC), "Warming of the climate system is unequivocal, as is now evident from observations of increases in global average air and ocean temperatures, widespread melting of snow and ice and rising global average sea level." The main cause of changing climate is the increasing atmospheric concentrations of greenhouse gases (carbon dioxide, methane, and nitrous oxide), which effectively act as a blanket over the atmosphere.

The IPCC report noted, "There is very high confidence that the net effect of human activities since 1750 has been one of warming." The evidence for the warming of the climate includes more than the measurement of global average temperatures, as physical evidence such as glacier melt also exists. Most predictions of global warming are based on data models, and mathematics is used extensively to measure and quantify atmospheric carbon dioxide and aerosols, which are believed to add to the problem. The National Oceanic and Atmospheric Administration (NOAA) and the U.S. National Aeronautics and Space Administration (NASA) are two large federal agencies that are involved in the collection, analysis, and dissemination of climate data. They employ a diverse range of mathematicians, statisticians, scientists, and others, and they have partnerships with many academic institutions, government agencies, and businesses around the world. Researchers do agree, however, that the current and future consequences of climate change disproportionately impact the world's poor.

Climate as a Distribution

In order to understand global warming, it is important to understand that the term refers to a distribution. It is easy to dismiss the notion of global warming on a cold winter's day, a mild summer day, or any day where the weather is cooler than expected. In fact, an unusually hot or cold event is not evidence for or against global warming. To aid in understanding climate as a distribution, consider a set of 20 cards numbered 1–20. One card will be drawn at a time with replacement. The value 10.5 is the average of these cards; a card selected below 10.5 will represent a below-average temperature for the day, and one selected above 10.5 will represent an above-average for the day. Further, the farther away the value of a card is from 10.5 will represent a larger deviation from average temperature. This example represents a stable climate. Some days are colder than average, and others are hotter than average. But, over time, roughly an equal number of colder days and hotter days occur. Moreover, if the value of the cards were unknown, basic statistical sampling ideas could be used to estimate the average.

To represent a changing climate, start with the same set of cards and consider values below or above 10.5 as a colder or hotter day. But this time, every time a card is drawn and replaced, the next higher card will be added to the set. For example, after the first card is drawn, a 21 will be added to the set, then a 22 will be added after the second draw, then a 23 after the third, and so on. At first, this change would be barely noticed if at all since the cards drawn will be roughly equal above and below 10.5. After some time, however, one would start to question the assumption that 10.5 is the average. In this case, if the values of the cards were unknown, basic statistical sampling techniques could not be used to estimate the average since the average is in fact changing. In this example, if 10.5 is taken as the average, then values below 10.5 still occur but are becoming less likely. In other words, record lows can still occur—and will still happen—even though climate is warming.

To complicate this example further, consider this same experiment being performed simultaneously by 2000 people to represent different locations around Earth. When the set of cards have values from 1 to 100, one individual would have only a 10% chance of drawing a card below 10.5, but it is expected that approximately 200 of the 2000 experiments will draw a card below 10.5. In other words, even though climate is warming, there will still be places that have colder than average days.

Figure 1. Global temperature anomalies for December 2009.

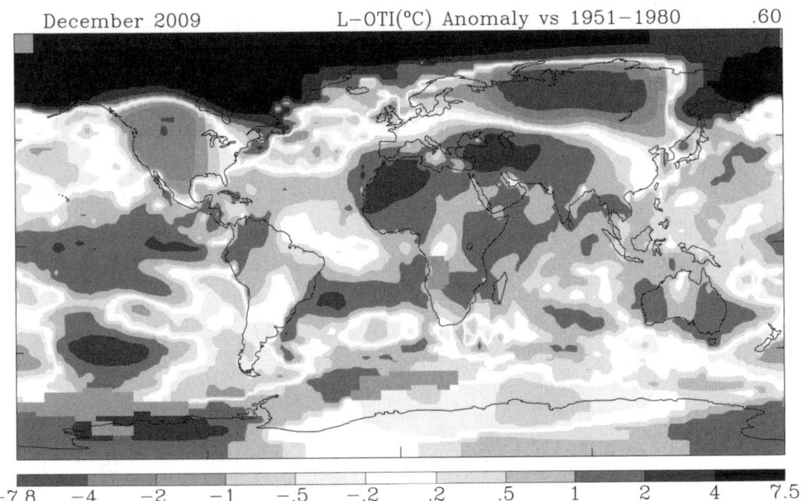

In terms of actual weather, consider Figure 1, which provides the average monthly temperature anomalies in degrees Celsius for December 2009 compared to the average from 1951 to 1980. The month of December was slightly colder for most of the United States. The overall average for the world was 0.60 degrees Celsius higher than the baseline years. One month, or even one year or a few years, of above average temperatures does not provide conclusive evidence for or against global warming as these abnormalities could be explained as normal variations in weather.

Evidence of Warming

Calculating global mean temperatures each year provides one form of evidence for global warming. For example, Figure 2 displays mean global temperature anomalies dating to 1850. Even though the overall trend is upward, variation from one year to the next can go in either direction. Gerald Meehl, who has a Ph.D. in climate dynamics and works at the National Center for Atmospheric Research (NCAR), collected information from 1800 weather stations across the United States that have been operating since 1950. He and his colleagues looked at the ratio of record highs to record lows and grouped the ratios by decade. From the 1950s through the 2000s, the ratio of record highs to lows was 1.09:1, 0.77:1, 0.78:1, 1.14:1, 1.36:1, 2.04:1. From the 1950s through the 1980s, the ratios might be considered to be in the range of normal variation for a stable climate. On the other hand, by the 2000s, it certainly appears that the observations no longer represent normal variation, and that the climate distribution is getting warmer.

In Figure 2, the baseline is the average from 1961 to 1990, with regression lines for different time periods from 1850 to 2009, 1910 to 2009, 1960 to 2009, and 1985 to 2009. The data are the HadCRUT3 data set provided by the University of East Anglia Climatic Research Unit.

Figure 2. Global mean temperature anomolies.

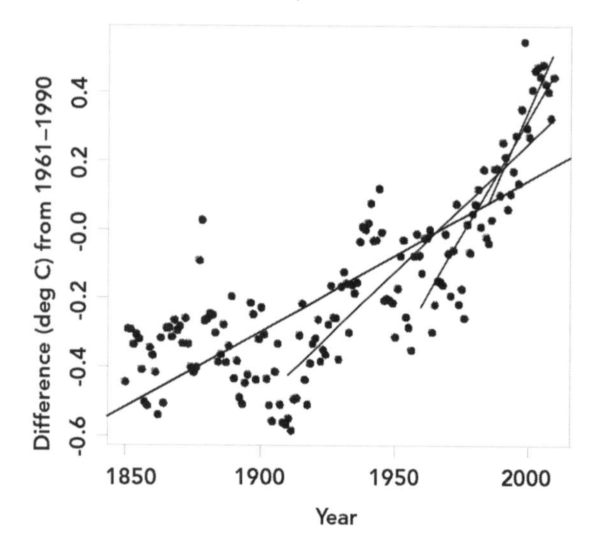

Beyond data, a warming climate should present physical evidence in the form of melting ice. Figure 3 is one of a number of glacier image pairs, which are pictures of glaciers taken from the same vantage but 40–100 years apart. The change is striking. Where there was once ice, there is now ocean water with the glacier retreating about seven miles. In the foreground, thick vegetation exists where there was once rock. This change is because of microclimate changes since the ice is no longer cooling that area. Along with melting glaciers, Arctic sea ice is decreasing rapidly and permafrost is melting. In fact, the entire village of Newtok, Alaska, must be relocated because the loss of permafrost has allowed the banks of the Ninglick River to erode.

Melting ice is just one source of evidence of a changing climate. During most of the twentieth century, sea level was rising at a rate of 0.07 inches per year, but by the 1990s that rate increased to 0.12 inches per year. In 2006, the National Arbor Day Foundation updated its plant hardiness zone maps, and most of the zones shifted northward. In other words, many plants can now be grown where they could not before because of their cold hardiness. There have already been observed shifts in species ranges, a northward shift, as well as shifts in phenology (seasonal biological timing) toward events such as early blooming. In fact, many species have seasonal behavior that is occurring 15–20 days earlier than the behavior occurred in the mid-twentieth century.

The general trend of warming is only part of the story. If the planet warmed a degree or two over millions of years, then ecological processes could adapt and societies could migrate. Figure 2 has least squares regression lines calculated over the time periods of 1850–2009, 1910–2009, 1960–2009, and 1985–2009. The four regression lines are as follows:

$$y_{159}(t) = 0.0041t - 8.67281$$

$$y_{100}(t) = 0.00750t - 14.75315$$

$$y_{50}(t) = 0.01364t - 26.96187$$

$$y_{25}(t) = 0.01801t - 34.67615.$$

In each case, the slope of the line, with units of degrees Celsius per year, is increasing as the time periods are shortened toward more recent years. More importantly, the 95% confidence intervals for the slopes are (0.00387, 0.00495), (0.00657, 0.00844), (0.01151, 0.01577), (0.01254, 0.02347), respectively. The first three intervals do not overlap, and so the slopes of the lines are significantly different. This provides evidence not only for overall warming but also that the rate of warming is increasing. Some species, trees for example, will simply not be able to adjust their ranges quickly enough to adapt to the warming climate.

Climate Science

Climate models, which incorporate mathematical topics such as dynamical systems, statistics, differential equations, and applied probability, are used to predict future global average temperature.

Mathematician Ka-Kit Tung, in his book *Topics in Mathematical Modeling,* provides a simple climate model. The model is

$$R\frac{\partial}{\partial t}T = Qs(y)(1-\alpha(y)) - I(y) + D(y).$$

The left-hand side represents change in temperature. There are three basic terms on the right-hand side that contribute to temperature change. The first term has incoming solar radiation at the top of Earth's atmosphere,

$$Qs(y),$$

where the $s(y)$ term distributes the radiation differently depending on the latitude $y = \sin\theta$ with θ representing latitude. The term also takes into consideration how much radiation is absorbed

$$(1-\alpha(y))$$

where $\alpha(y)$ is the fraction reflected or albedo. The next term,

$$I(y)$$

represents outward radiation, and the last term,

$$D(y)$$

represents heat transportation from warmer latitudes to colder latitudes. In Tung's textbook, this simplified model is analyzed to gain understanding of possible locations in ice lines.

The more complex computer simulations that model climate are built with assumptions related to population growth and societal choices, such as energy use or technological change. These assumptions are then used to predict how greenhouse gases will increase. The effect of increased greenhouse gases in trapping heat is well understood, and in terms of the simple climate model above, the increase in greenhouse gases decreases outward radiation. Beyond that, the increase in carbon levels itself is a problem as oceans work to absorb some of this carbon in the form of carbonic acid. The increase in carbonic acid in the oceans increases the acidity levels, which damages coral, crustaceans, sea urchins, and mollusks.

For each scenario, many different models are considered, and the predictions are averaged to produce the graph on the left side of Figure 4. The three higher curves illustrate the average warming. On the right side of the graph is a range based on the various models. A distribution has been created, and based on the graph, one could say that, by 2100, global mean temperatures will increase between approximately 1.5 degrees Celsius and 3.5 degrees Celsius, but the distribution around the three scenarios presented is from approximately 1 degree Celsius to 6 degrees Celsius. The right side of Figure 4 presents the predicted temperature changes as a distribution across Earth, and it is

Figure 3. Images of the Muir glacier taken from the same vantage on August 13, 1941 (left), and August 31, 2004 (right).

Figure 4. Climate model predictions of future average global temperature and distributional changes of temperature.

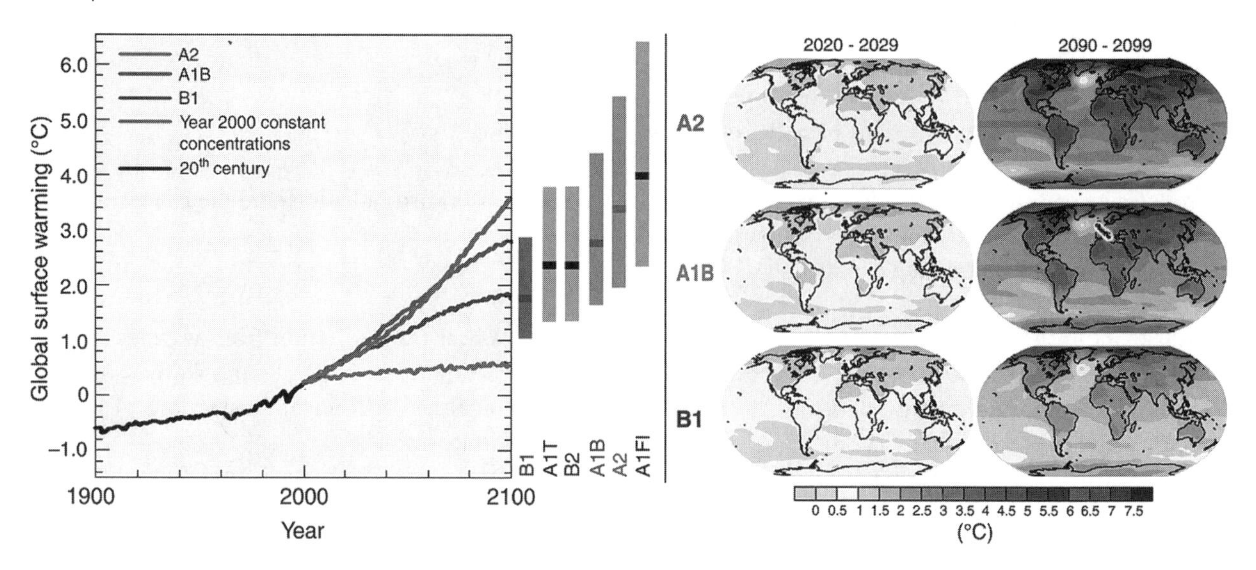

predicted that the Arctic region will warm more than the equatorial region.

A key complication in climate modeling is the existence of feedback loops. A feedback loop is created when a change in one factor causes a change in a second factor that then either reinforces or diminishes the change in the first factor. While each scenario sets out greenhouse gas levels, the models must then attempt to take into account how warming may, in fact, increase warming or decrease warming. For example, one positive feedback loop involves melting ice. As ice melts, the Earth's albedo (reflectivity) changes so that less solar radiation is reflected out to space. In the climate model above, the $\alpha(y)$ term is decreased so that more solar radiation is absorbed. In other words, as the planet warms, ice melts. However, there are now fewer reflective white surfaces and more dark surfaces, which will then absorb even more solar radiation and increase the planet's warming. Another potential positive feedback loop arises from melting permafrost. As the permafrost melts, partially decomposed organic matter will decompose more fully and release carbon into the atmosphere. Even more uncertainty arises with the effect of clouds. Low clouds tend to cool by reflecting more energy than they trap, while the reverse is true for high clouds. As surface temperature increases, there is increased evaporation from the oceans, creating more water vapor and hence clouds. But the type of clouds that arise will depend on whether this is a positive or negative feedback loop.

Of course, to many people, an increase of a few degrees Celsius does not seem to be drastic enough to impact life on Earth significantly. But consider that during the twentieth century, global average temperatures increased by less than 1 degree Celsius. Nevertheless, there has already been observed disappearing glaciers, loss of Arctic sea ice, changing species habitat and phenology, and a new plant hardiness map. In fact, a difference of approximately 0.2 degrees Celsius was the difference between the Medieval Warm Period (c. 950–1250) and the cooling period (c. 1400–1700). The warm period led to the Norse migrating to Greenland and bountiful harvests and population increases in Europe. This period was followed by a cooling period that led to the collapse of the Norse Greenland society and starvation in Europe.

Impacts of Climate Change

The general consensus in the scientific community in 2010 is that warming has occurred and will continue to take place even with changes. Debate continues on precisely how much warming will occur and the exact nature of the ramifications. The questions are by how much, and what should people expect to happen? Species ranges are already changing, and, in some cases, species ranges are disappearing as appears to be the case for polar bears. Unfortunately, the speed of warming will lead to some species not being able to change their range quickly enough, resulting in extinction. The changing phenology is already causing ecological disruption. Some plants are blooming earlier, but the species that feed on them are not arriving earlier, leading to decreased food supply.

As Figure 3 shows, ice is melting and more of that is expected. The loss of Arctic ice will decrease polar bear populations. The melting glaciers of the Tibetan Plateau are of particular importance. These glaciers are responsible for supplying water to about 2 billion people, and data suggests that the Tibetan Plateau is warming twice as fast as the global average. Once these glaciers are gone, so is the water supply. The melting of glaciers and land ice, along with the thermal expansion of water, will raise sea levels. One example is Bangladesh, which faces severe threats from sea level increases since millions of people live along a coastline that may be underwater in the future.

There are additional predictions as of 2010, based on models and scientific expertise. An increase of 2 degrees Celsius from pre-industrial levels would lead to a fall in agricultural yields in the developed world, a 97% loss of coral reefs, and 16% of global ecosystems transformed. With an increase of 3 degrees Celsius, few ecosystems could adapt and an additional 25–40 million people would be displaced from the coasts because of sea level rise. If global average temperatures rose to 4 degrees Celsius above pre-industrial levels, entire regions would be out of agricultural production, including Australia.

Climate Change and Societies

The joint science academies' statement on sustainability, energy efficiency, and climate protection issued in 2007 by the G8 nations and Brazil, China, India, Mexico, and South Africa, said that, "Many of the world's poorest people, who lack the resources to respond to the impacts of climate change, are likely to suffer the most." The warming of the planet will have some advantages and disadvantages, although there will be more disadvantages. Some warmer climate species will have expanded ranges and be able to thrive, while arctic species may lose their entire ecosystem. Some countries will be impacted more than others, and the wealthier countries will have a better ability to adapt. The examples that have been given here of societies that already have been or will likely be impacted are all examples of poorer societies.

The people of Newtok, Alaska, are poor; in 2010, Bangladesh ranked 183 in the world in terms of GDP per capita; and there is considerable poverty in regions in and around the Tibetan Plateau. Part of the tragedy is that these are not the people who are largely responsible for increasing greenhouse gases. China and the United States are the largest emitters of carbon dioxide, but on a per capita basis, the United States far exceeds China. In general, it is the industrialized nations that contribute the most to greenhouse gases. Figure 4 provides different models for future climate change, and these are primarily based on the models that predict future greenhouse gas emissions, and it is the more industrialized nations that have the resources to make reductions in these emissions.

Further Reading

Fleming, James. *Fixing the Sky: The Checkered History of Weather and Climate Control.* New York: Columbia University Press, 2010.

Intergovernmental Panel on Climate Change. "Climate Change 2007." http://www.ipcc.ch/pdf/assessment-report/ar4/syr/ar4_syr_spm.pdf.

Mackenzie, Dana. "Society for Industrial and Applied Mathematics: Mathematicians Confront Climate Change." http://www.siam.org/news/news.php?id=1131.

The National Academies. "Ecological Impacts of Climate Change." http://dels-old.nas.edu/climatechange/ecological-impacts.shtml.

The National Academies. "Understanding and Responding to Climate Change, 2008 Edition." http://dels-old.nas.edu/dels/rpt_briefs/climate_change_2008_final.pdf.

National Oceanic and Atmospheric Administration. "Climatic Data Center." http://www.ncdc.noaa.gov/oa/climate/globalwarming.html.

Neelin, David. *Climate Change and Climate Modeling.* New York: Cambridge University Press, 2011.

Oreskes, Naomi, and Erik Conway. *Merchants of Doubt: How a Handful of Scientists Obscured the Truth on Issues From Tobacco Smoke to Global Warming.* New York: Bloomsbury Press, 2010.

U.S. Global Change Research Program. "Global Climate Change Impacts in the United States." http://down

loads.globalchange.gov/usimpacts/pdfs/climate-impacts-report.pdf.
Weart, Spencer R. *The Discovery of Global Warming: Revised and Expanded Editions*. Cambridge, MA: Harvard University Press, 2008.

THOMAS J. PFAFF

See Also: Coral Reefs; Forecasting; Function Rate of Change; Mathematical Modeling; Weather Forecasting; Weather Scales.

Climbing

Category: Games, Sport, and Recreation.
Fields of Study: Algebra; Data Analysis and Probability; Geometry; Problem Solving.
Summary: Effective climbing relies on mathematical principles, and there are connections between climbing and mathematical problem solving.

Climbing is the use of the human body and assisting equipment to ascend or descend steep surfaces. Climbing can be done professionally, such as for construction or in the military, for exercise or competition, or for performance—in the case of parkour. There are different styles of climbing depending on the object, such as bouldering, ice, tree, and rope climbing. If the weight of the climber is supported by equipment, it is called aid climbing; when the weight is supported only by the climber's muscles, it is called free climbing. Mathematics plays a role in successful climbing and in analyzing various aspects of the discipline. Mathematician Skip Garibaldi said, "Climbing has a lot of puzzles that have to be solved. It's not just strength or skill."

Anthropometry in Climbing

Anthropometry is the mathematical study of body measurements in order to understand human variability. For example, studies show that elite climbers, on average, tend to have small stature, low body mass, and a high handgrip-to-mass ratio compared to the population as a whole. Compared to nonclimber athletes with similar physical conditioning, they are frequently linear, with narrow shoulders relative to hips. Ape index is the ratio of a climber's arm span to height. In adults, it is usually close to one, as illustrated in Leonardo da Vinci's "Vitruvian Man." An ape index greater than one is reputedly advantageous for climbing, and some researchers have found ape index to be a statistically significant predictor of climbing success.

Fall Factor and Impact

Fall factor quantifies how hurtful a fall may be to a roped climber. Mathematicians such as Dan Curtis have derived the fall factor (F_{max}) using differential equations. It is a function of the ratio of the total distance the climber falls (D_T) to the length of the unstretched rope (L) between the climber and belayer or anchor at the rope's other end. It is also a function of the climber's mass (m), the elasticity or "stretchiness" of the rope (k), and gravity (g). Algebraically, it is represented as

$$F_{max} = \sqrt{2mgk \frac{D_T}{L}}.$$

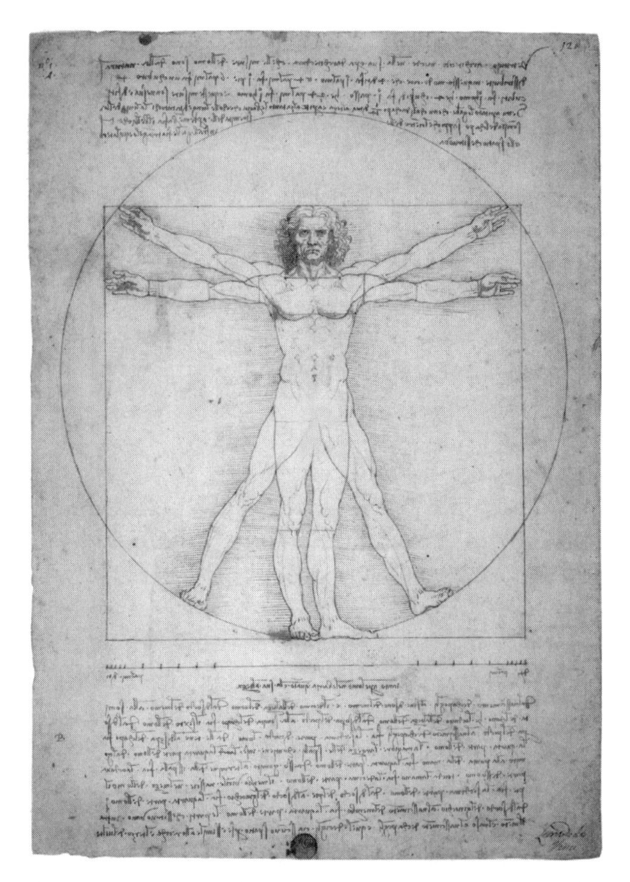

The arm span to height ratio for an adult is generally near one, as shown in da Vinci's "Vitruvian Man."

Climbing ropes must pass a statistically designed drop test to be certified for sale and use. Other critical safety equipment is also designed using mathematics. One example is the curve of cams used in the "friend" devices that secure ropes to crevices in rock walls, which may be optimized using systems differential equations, sometimes with polar coordinates. The devices themselves are an application of logarithmic spirals.

Climbing Theories and Modeling

Many people have drawn parallels between climbing mountains and solving mathematical problems, especially great challenges like summiting Mount Everest and solving a problem like the Riemann hypothesis, first proposed by mathematician Bernhard Riemann. Analyses have shown that Everest climbers engage in multistep problem solving with altitude changes, rates, percentages, conversions, approximations, and division of large numbers. Mathematician-climber John Gill said that problems in both mathematics and climbing are often solved by "quantum jumps of intuition." Patterns found in the natural features of some popular climbing locations can very mathematical. The Navajo Sandstone formation includes rounded domes and saddle shapes with remarkably precise-looking contour lines.

At the same time, the geometric diversity and complexities of climbing surfaces and the variety of techniques used by climbers have made developing a single theory of optimal climbing strategy difficult. However, several methods are used to quantify characteristics of different climbs and probabilistic models can be used to make decisions. Competitive climbers assign climbing grades to climbing routes, using objective and subjective criteria, to describe their difficulty. Other systems assess the technical difficulty of required moves, the stamina necessary, exposure to the elements, or the frequency of difficult moves. Mathematician Alan Tucker demonstrated using graph theory that the classic Parallel Climbers mathematical puzzle has a solution for any mountain range.

Further Reading

Curtis, Dan. "Taking a Whipper: The Fall-Factor Concept in Rock-Climbing." *The College Mathematics Journal* 36, no. 2 (2005).

Garlick, S. *Flakes, Jugs, and Splitters: A Rock Climber's Guide to Geology.* Kingwood, TX: Falcon, 2009.

Tucker, Alan. "The Parallel Climbers Puzzle." *Math Horizons* 3 (November 1995).

Maria Droujkova

See Also: Algebra in Society; Elevation; Extreme Sports; Puzzles.

Clocks

Category: Space, Time, and Distance.
Fields of Study: Measurement; Number and Operations; Representations.
Summary: Clocks are devices for timekeeping and are used for a variety of mathematical calculations, including finding one's longitude.

The term "clock" in a generic sense is applicable to a broad range of devices for timekeeping usually concerning fractions of the natural unit of time—the day. Modern clocks operate through various physical processes. It does not matter what kind of periodic signals a clock produces—ringing a bell, firing a cannon, flashing a light, moving a hand, displaying a number, or generating electric impulses. Mathematics has been fundamental both in the design of clocks and in the measurement of their accuracy. Modular arithmetic, an algebraic concept involving cycles, is sometimes informally known as "clock arithmetic." In the realm of biology, mathematicians have also worked on theories related to the operation of humans' internal biological clocks and bacterial genetic clocks.

History of Clocks

In everyday English language, watches and other timepieces that can be carried individually sometimes continue to be distinguished from clocks. Via Dutch, Northern French, and Medieval Latin, the word "clock" is derived from the Celtic *clagan* and *clocca* meaning "bell." Those old clocks had a striking mechanism for announcing intervals of time acoustically. The history of clocks is much deeper, however. It started in early prehistoric times with sundials (often a vertical post or pillar on horizontal ground exposed to the sun or a post parallel to the Earth's axis) that were the first and

Galileo Galilei, with his design for a pendulum clock around 1641, drawn by Vincenzo Viviani in 1659.

oldest scientific instruments of archaic humankind. They worked only in the daytime. In the terminology of ancient Greece, such a device was called a *gnomon*, and the entire branch of science on sundials is *gnomonics*. Famous Egyptian obelisks—now reerected in some European capitals—were also sundials.

Timekeeping devices of different types were called *horologium* by the Romans. In its corrupted forms, this term later on entered many languages of the world. A noticeable step in the history of timekeeping was the invention of a "water clock" (the specific Greek name is the *clepsydra*). Water clocks could be used at night. Some of the water clocks in China and the Near East were quite large. Another type of simple clock was the "sandglass."

The modern era of clock-art started with the invention of weight-driven mechanical clocks (sometimes known as "chimes"). The inventor of such a novelty is unknown. Because daily prayer and work schedules in medieval times were strictly regulated, religious institutions required clocks, and it is certain that the earliest medieval European clockmakers were Christian clerics. Mechanical clocks were designed en masse in the thirteenth century in Western Europe. They were bulky and appeared on cathedral towers in many countries. Some of them have survived up to now and are among the great artifacts of the medieval epoch.

After the invention of tower clocks, efforts were made to design smaller pieces for tabletops and personal "pocket" clocks (watches) for individuals. Peter Henlein (c. 1480–1542), a locksmith from Nuremberg, Germany, is often credited as the forerunner of the first portable timekeeper, but this claim is disputed. His drum-shaped *Taschenuhr* was too big for a pocket. The first individual clocks were usually worn on the neck or beneath the knee. Timepieces of this type were often known as "Nuremberg eggs." The earliest clocks are very expensive now and are subjects for collectors.

Clocks for Navigation

A great chapter in clock-making began in conjunction with the rapid development of seafaring after the European discovery of the Americas. In order to determine one's position at sea, it is necessary to calculate two geographical coordinates: latitude and longitude. Latitude is easily computed directly from trivial astronomical considerations (the latitude of a locale is equal to the altitude of the celestial pole). As for longitude, it is equal to the difference between local time and the time of a prime meridian chosen specifically for cartographic purposes; navigators used different prime meridians in different countries in different epochs. To discover one's longitude, an observer must know the time at the prime meridian, which requires the art of "transporting" accurate time.

The search for accurate and convenient timekeeping became one of the most impressive scientific and technological challenges of the seventeenth century. Numerous mathematical and astronomical methods were proposed, such as observations of the moon. However, the computations would have been difficult for the typical sailor and the mathematical methods were not yet well-developed enough to provide an accurate prediction. This problem was among the foci of scientific activities of Galileo Galilei of Italy (1564–

1642), who discovered the key property of pendulums that makes them useful for timekeeping: isochronism, which means that the period of swing of a pendulum is approximately the same for different sized swings. Galileo developed the idea for a pendulum clock in 1637, but did not have enough time to complete the design.

Dutch scholar Christian Huygens (1629–1695) successfully built a pendulum clock in 1656 and patented it the following year. Its design incorporated concepts derived from mathematical work on cycloids. The introduction of the pendulum—the first harmonic oscillator for timekeeping—increased the accuracy of clocks enormously, from about 15 minutes per day to 15 seconds per day. In addition to building a clock, Huygens investigated the properties of synchronization of identical pendulum clocks. Researchers have been interested in the subject of synchronization of clocks and oscillators since that time.

The design of the first marine chronometer was performed by the self-educated English carpenter and clockmaker John Harrison (1693–1776). This device dramatically revolutionized and extended the possibility of safe long-distance sea travel. At the time, the problem was considered so intractable that the British Parliament offered a prize of 20,000 British pounds sterling (comparable to about $4.72 million in modern currency) for the solution. Sailors and astronomers continued to be the principal consumers of accurate timekeeping. Precise clocks became essential equipment for each and every astronomical observatory.

Modern Clocks

The problem of "transportation" of accurate time to determine longitudes lost its actuality with the invention of the telegraph and, later on, with utilization of radio signals. But with the advancement of the twentieth century, new scientific and applied challenges demanded increasingly accurate time reckoning. As a result, new clocks were created based on newly discovered physical principles that were operationalized using mathematics. The crucial step in this direction was the invention of so-called quartz clocks. A quartz crystal has the unusual property of piezoelectricity—when stimulated with voltage and pressure, it oscillates at a constant frequency.

The vibration of a quartz crystal regulates the clock very precisely. Quartz crystal clocks were designed in 1927 by two engineers at Bell Telephone Laboratories: the Canadian-born telecommunications engineer Warren Marrison (1896–1980) and an electrical engineer from the Massachusetts Institute of Technology (MIT), Joseph Warren Horton (1889–1967). Since the 1970s, quartz clocks have become the most widely used timekeeping technology. Atomic clocks followed quartz clocks toward the end of the century. The U.S. National Bureau of Standards (now the National Institute of Standards and Technology) based the time standard of the land on quartz clocks between the 1930s and the 1960s. Eventually, it changed to atomic clocks, the best of which are accurate to 5×10^{-15} seconds per day. Researchers are now developing optical clocks that can be up to 100 times more accurate than the best atomic clocks. Further, satellite-based global

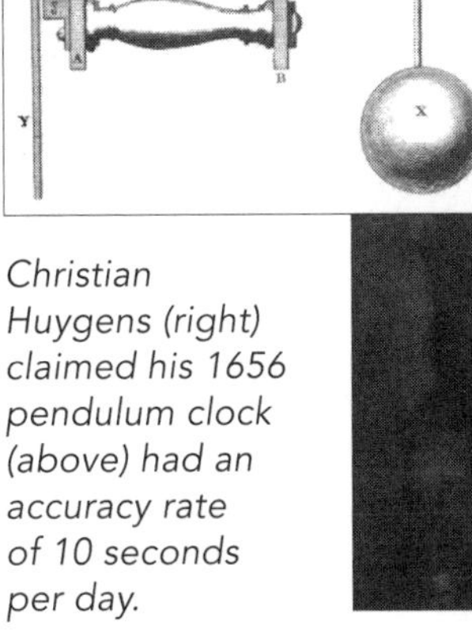

Christian Huygens (right) claimed his 1656 pendulum clock (above) had an accuracy rate of 10 seconds per day.

positioning systems are now a primary source of time for some scientists and people in everyday life. This system provides almost unlimited transportation of time using variety of mobile devices in space and on Earth.

Today, the reckoning and keeping of precise and super-precise time continues to be requisite for numerous scientific and applied problems. Astronomers are still important users of this data. It is important, for instance, in cosmic navigation, in the measurement of variations of the rotation of Earth, and in the implementation of a particular technology into everyday life, such as radio interferometry with a hyperlong base. Every developed country now has a specialized national service for addressing questions regarding precise timekeeping and time reckoning. For a long time in the Paris Observatory, there was the Bureau International de l'Heure (The International Time Bureau), which played an important role in the research of timekeeping. In 1987, the responsibilities of the Bureau were taken over by the International Bureau of Weights and Measures (BIPM) and the International Earth Rotation and Reference Systems Service (IERS).

Further Reading

Bruton, Eric. *The History of Clocks and Watches.* New York: Time Warner Books, 2003.

Collier, J. L. *Clocks.* New York: Benchmark Books, 2004.

Landes, David S. *Revolution in Time: Clocks and the Making of the Modern World.* Cambridge, MA: Belknap Press of Harvard University Press, 2000.

Sobel, Dava. *Longitude: The True Story of a Lone Genius Who Solved the Greatest Scientific Problem of His Time.* New York: Walker, 1995.

Uresova, Libuse. *European Clocks. An Illustrated History of Clocks and Watches.* London: Peerage Books, 1986.

Alexander A. Gurshtein

See Also: Calendars; GPS; Measuring Time.

Closed-Box Collecting

Category: Arts, Music, and Entertainment.
Fields of Study: Data Analysis and Probability; Number and Operations.

Summary: Collecting objects in sets is a popular pastime that can require a great deal of effort and such collections have inspired mathematical investigations.

A set of closed-box collectibles is a set of similar objects that are sold interchangeably. The objects might be anything; for example, cards, figurines, or trinkets. The term "closed-box" means that the consumer purchases each item without knowing exactly which thing the consumer will get—the package will contain a random item from the collection. Because some collectibles may be more rare or more valuable than others and because individual preferences vary, side markets for these collectibles may emerge, with an identified item selling for many multiples of the price of a random one. Baseball cards, collectible card games, and other trading cards give a familiar example of closed-box collectibles. Toy prizes in cereal boxes and Kinder Surprise eggs are other examples. This problem is one of the classics of probability theory. It has many extensions and can be solved by many methods, including combinatorics and generating functions. It is also known as the Coupon Collector Problem.

Promotional Contests

There have been many contests based on closed-box collecting used for promotional purposes by various businesses and products. Two well-known examples include McDonald's annual Monopoly game and Subway's Scrabble game. In these cases, certain purchases come with one or more random game pieces, which will be one of a large number of types. The game pieces come in various groups; a complete group of collectibles can be exchanged for a contest prize.

From the perspective of the business running such a promotion, the contest design creates certain mathematical problems. What proportion of the game pieces should be manufactured of each type? A main goal is to minimize unpredictability. If too many grand prizes are collected, the company may have to pay out a substantial amount of money; this might be too great a risk to tolerate even if it is very rare. On the other hand, if too few major prizes are awarded, the public may become dissatisfied, negating the public relations goals of the promotion. The usual solution for the significant prizes is to make one type of piece in each group extremely rare, manufacturing only as many as they intend to pay out prizes. The other types can then be made relatively common without risk. This system generally has the effect of maintaining public interest by giving a large number of people the "feeling" of getting closer to winning a big prize as they accumulate common tokens in the group, without risking a huge payout.

Expectations in Closed-Box Collecting

Suppose that a consumer is interested in one particular collectible from a set, and the consumer decides to purchase collectibles one at a time until getting the desired one. Assume each collectible purchased will be the desired kind with probability p, independently of the others. (In real life, this assumption will not be strictly valid, but the discrepancy is negligible if the number of collectibles purchased by an individual is small compared to the total number in existence.) The chance that it is not the kind desired is then $1-p$, and this scenario is modeled by a geometric random variable. The probability of getting the desired item on the first try is p, on the second try is $p(1-p)$, on the third try is $p(1-p)^2$, and so on. Then the expectation is

$$\sum_{n-1}^{\infty} p(1-p)^{n-1}.$$

Standard techniques of basic analysis now show that the expected number of purchases needed is $1/p$.

It should be emphasized that this is the expectation in the sense of probability theory and that there are some common misconceptions about what it means. If the probability of getting the desired item is 1/100, this does not mean that 100 is the most likely number of purchases, nor that the 100th item is any more likely to be the desired type than any other. It means that on average—in the long run—it will take 100 tries to get the desired item. This also means, for example, that when rolling a fair die, it will take an average of six tries to roll a 1, squaring well with intuition.

Another important issue in understanding the dynamics of closed-box collecting is the expected number of purchases to collect a complete set. Suppose that there is a set of 100 collectibles, each item purchased being equally likely to be any of the hundred types. If a consumer purchases collectibles one at a time until obtaining a complete set, how many purchases will be made? It will take one purchase to get one item. With one item type, each purchase will add to the collection with probability 0.99, so the consumer expects to purchase 100/99 more items to get the second item. With two item types, each purchase will add to the collection with probability 0.98, so the consumer expects to purchase 100/98 more items to get the third item. This process continues until the consumer has all the items but one; then each purchase will complete the collection with probability .01, so the consumer expects to purchase 100/1 more items to get the last type, completing the collection. This process indicates a total of

$$\frac{100}{100} + \frac{100}{99} + \frac{100}{98} + \frac{100}{97} + \cdots + \frac{100}{1}$$

purchases, about 519. In general, if there are n types, then the expectation is

$$n\left(\frac{1}{1} + \frac{1}{2} + \frac{1}{3} + \cdots \frac{1}{n}\right).$$

For large n a good approximation is

$$n\ln(n) + \gamma n + \frac{1}{2} + o(1) \text{ purchases}$$

to collect all n objects. The constant $\gamma \approx 0.5772156649$ is the Euler–Mascheroni constant, named for Leonhard Euler and Lorenzo Mascheroni, while $o(1)$ is a constant used in computer science meaning a function that converges to zero for very large inputs, such that the value is effectively zero for very large n.

This illustration gives insight into why it seems harder and harder to make further progress in collecting, the further you get. In the example of collecting a complete set of 100 collectibles, with each purchase equally likely to be any of the hundred types, the expected purchases needed is about 519. Suppose now that one has accumulated a collection of 50 different items; is that really halfway to a complete collection? By a similar analysis, the expected number of additional purchases to collect the remaining 50 items is about 450. So there is a meaningful sense in which 450/519 of the collecting task is still undone; a more accurate description of the progress is that the collection is 13.3% completed. In the sense of expectation, one is not really halfway through collecting 100 items until obtaining the 93rd item. While the assumption that all types are equally likely does not usually hold in practice (some types are rarer, some more common), the qualitative conclusion applies in general, unless a few of the items are extremely rare.

Further Reading

Myers, Amy, and Herbert Wilf. "Some New Aspects of the Coupon-Collector's Problem." http://arxiv.org/abs/math/0304229v1.

Ross, Sheldon. *A First Course in Probability*. 8th ed. Upper Saddle River, NJ: Prentice Hall, 2010.

MICHAEL "CAP" KHOURY

See Also: Coupons and Rebates; Expected Values; Probability.

Clouds

Category: Weather, Nature, and Environment.
Fields of Study: Data Analysis and Probability; Geometry.
Summary: The formation and behavior of clouds can be mathematically modeled and studied.

Mathematics has been called "the science of patterns." In clouds and the atmosphere, generally there is no end to the patterns that may be observed, quantified, and more clearly understood using mathematics. Mathematicians have long modeled the behavior and structure of clouds.

Applied mathematicians continue to develop ways to detect clouds and quantify motion, composition, density, top altitude, and the distance between clouds, among other characteristics. In 1999, the U.S. National Aeronautics and Space Administration (NASA) launched the Multi-Angle Imaging SpectroRadiometer to measure environmental and climate data from nine different angles, including cloud data.

The U.S. National Oceanic and Atmospheric Administration (NOAA) is one of the largest organizations specializing in the study of the environment. In 2010, a NOAA team led by physicist Graham Feingold reported its findings that clouds form synchronous patterns, meaning that individual clouds in a group respond to signals from other clouds, an effect also observed in chirping crickets or flashing fireflies. This research has implications for interpreting climate change data. There are also mathematical objects such as point clouds that are of interest in geometry, imaging, and efficient distribution mining. Fractal clouds are appreciated for their mathematical properties and their artistic qualities.

Water in the Air

Air is composed primarily of nitrogen (78%) and oxygen (21%). Argon comprises nearly 1%, leaving little room for the remaining gasses, including carbon dioxide, ozone, and neon. This recitation, however, is for dry air. Water vapor, the invisible gas from which clouds are constructed, can account for 0% to 4% of any given parcel of air. In order to form a cloud, water vapor must change phase to either liquid water droplets or ice crystals.

The Transformation of Water into Clouds

The amount of water vapor that can be held in a parcel of air is determined primarily by the temperature of the air; warm air can hold more and cold air less. The amount of water vapor held in a parcel of air is identified by the mixing ratio:

$$w = \frac{\text{grams of water vapor in a parcel}}{\text{kgs of dry air in the same parcel}}.$$

The amount of water vapor a parcel of air *can* hold is called the "saturation mixing ratio":

$$w_1 = \frac{\text{grams of water vapor in a saturated parcel}}{\text{kgs of dry air in the same parcel}}.$$

Relative humidity is a measure of how much vapor a parcel of air is holding compared to how much it could possibly hold and is expressed algebraically as

$$\text{RH} = 100\frac{w}{w_1}.$$

The dew point is the temperature at which a parcel of air becomes saturated. At this point, the saturation mixing ratio and the actual mixing ratio are equal to one another, and the relative humidity is therefore 100%. A further drop in temperature should produce condensation as water changes phase from vapor to liquid cloud droplets or solid ice crystals—a cloud is born.

The Unstable Atmosphere

Clouds are often the result of lifting in the atmosphere. When a parcel of air rises, it generally cools, and this cooling produces condensation. The way in which the lifting is accomplished can lead to dramatic differences in the appearance of the cloud. When whole layers of air are gently lifted in an atmosphere that is stable, stratus clouds are formed, whereas the more dramatic vertical structure of a cumulus cloud comes from runaway convection, a self-perpetuating process that can build clouds more than 12 kilometers (km) or 40,000 feet tall.

What is a stable atmosphere? Temperatures generally decrease with height. The rate of change is, of course, variable but it is referred to as the "lapse rate" (Γ) of the atmosphere. A parcel of air, distinct from the air that surrounds it, may be forced to rise or descend and will cool or warm as a result. Pressure generally decreases with height, and a parcel that rises into a zone of lower pressure will expand, doing work on the environment and therefore cooling. The rate at which a parcel of air cools as a result of this sort of ascension is known as the "dry adiabatic lapse rate" (Γ_d) which is approximately 10 degrees Celsius per km. When the dew point is reached in the parcel and condensation occurs, latent heat is released as a result of the phase change and the parcel is warmed.

The result is a lower lapse rate, the saturated adiabatic lapse rate (Γ_s). The saturated lapse rate depends on the amount of moisture being condensed but 6 degrees Celsius per km may be used as a rough estimate.

Now if $\Gamma < \Gamma_d$, the atmosphere is stable because unsaturated air that is made to rise will cool at approximately 10 degrees Celsius per km and will find itself in air that is increasingly warmer than itself. The greater the difference $\Gamma_d - \Gamma$, the greater the force restoring the parcel to its previous altitude. The force may be quantified as

$$\frac{g(\Gamma_d - \Gamma)\delta z}{T}$$

where g is the gravitational constant, T is temperature, and δz is a small upward displacement of the parcel from its equilibrium level. Consider the implications of a temperature inversion in which temperature actually increases with height and Γ is a negative quantity. Now consider a situation in which the atmosphere cools strongly with height, that is $\Gamma > \Gamma_d$. Then, the restoring force becomes negative. Air that rises becomes warmer than its surroundings and so continues to rise. This leads to the runaway convection that builds the towering cumulonimbus clouds that can produce thunderstorms, lightning, and hail.

Further Reading

Adam, John. *Mathematics in Nature: Modeling Patterns in the Natural World.* Princeton, NJ: Princeton University Press, 2003.

Feingold, Graham, et al. "Precipitation-Generated Oscillations in Open Cellular Cloud Fields." *Nature* 466, no. 12 (August 2010).

Wallace, John M., and Peter V. Hobbs. *Atmospheric Science.* Burlington, MA: Academic Press, 2006.

Mark Roddy

See Also: Energy; Forecasting; Hurricanes and Tornadoes; Wind and Wind Power.

Clubs and Honor Societies

Category: Mathematics Culture and Identity.
Fields of Study: Communication; Connections.
Summary: Various clubs and honor societies add a social dimension to the enjoyment of doing mathematics and can provide networking and scholarship opportunities for talented students.

Mathematics clubs are often designed to provide a fun atmosphere outside of the classroom environment in order to promote mathematics and create a sense of community and camaraderie. Clubs exists for students of all ages as well as adults. Many undergraduate mathematics clubs are affiliated with organizations like the Mathematical Association of America, the Society for Industrial and Applied Mathematics, or the Association for Women in Mathematics. These clubs are open to all students, regardless of gender, race, color, religion, age, national origin, sexual orientation, or disability. However, other student extracurricular activities like semester or summer programs or camps may use a variety of selection criteria for membership.

Mathematical honor societies like Pi Mu Epsilon often consider both mathematical GPA and overall scholarship in their selection of candidates. Participants in mathematics clubs, programs, and honor societies seem to be attracted to the social aspects and related food opportunities; pizza or donuts are often a component of such activities. Researchers have investigated the impact of participation on achievement. There is some evidence that participation may be correlated with an increase in retention, positive attitudes about mathematics, and higher grade point averages.

Activities and Participation

People in mathematics clubs enjoy a wide range of activities together. Clubs may participate in mathematical contests, homecoming activities, mathematical murder mysteries, π-day celebrations, mathematics *Jeopardy*, or recreational mathematics. Some clubs bring in outside speakers, work on problems together, or write and perform mathematical plays or poetry. Other clubs perform service activities, like volunteering to be tutors. Faculty and teacher advisors run or help student members organize clubs at all levels, from primary school through graduate school, although some clubs may be completely run by student members. Adults also organize mathematical clubs for themselves. For instance, in Los Angeles, California, Math-Club's catchphrase is "Be there and be square." Listed members are employed in a wide variety of careers and include professional Hollywood writers, actors, designers, journalists, and musicians.

Clubs may be funded from schools and private donations or they may raise funds from activities like the sale of mathematical T-shirts with slogans like "Know your limits—don't drink and derive," "Math club is as sweet as π," or "Nerds now, rich later." In fact, some journalists have noticed that members of mathematics clubs may enjoy embracing their status as "nerds" or "intellectuals." This may be connected to the same occurrence in popular culture, where nerds are sometimes hip. Clubs are often open to anyone who wishes to join. Specific clubs also exist for members with more specialized interests, like for prospective mathematics teachers or mathematical knitters.

Honor Societies

Members of honor societies are recognized for their successful pursuit of mathematical understanding. The most well-known mathematics honor society in the United States may be Pi Mu Epsilon. As of April 2010, there were 343 chapters. The organization promotes student scholarly activity through a student research journal as well as grants for contests, conferences, and speakers.

Another college society is Kappa Mu Epsilon, which listed 144 chapters and more than 75,000 members in 35 states as of March 2009. The organization focuses on the power and beauty of mathematics and the connections between mathematics and society through a journal and regional conventions. Mu Alpha Theta is an honor society for high schools and two-year colleges, which listed more than 75,000 members in more than 1,500 schools as of October 2010. There are also mathematics honor societies for the homeschool community as well as for some states, schools, and colleges.

Other Programs

There are many other opportunities for students to engage in club-like or honor society–like activities. The Budapest Semesters in Mathematics (BSM) study-abroad program holds courses in English and is seen by many as a prestigious program for students bound for graduate school. In the summer, students may participate in a variety of mathematics camps, workshops, or research programs, such as the U.S. Space & Rocket Center's mathematics camp, Clarkson University's Roller Coaster Camp, or Research Experiences for Undergraduates. Some programs charge money for such activities, and others are funded by grants.

Further Reading

Cohen, Moshe. "How to Start…and Maintain Your School's Math Club." *Math Horizons* 14 (September 2006).

Mathematical Association of America. "MAA Student Webpage." http://www.maa.org/students.

Tanton, James. *Solve This: Math Activities for Students and Clubs*. Washington, DC: The Mathematical Association of America, 2001.

Sarah J. Greenwald
Jill E. Thomley

See Also: Careers; Competitions and Contests; Professional Associations; Social Networks; Succeeding in Mathematics.

Cochlear Implants

Category: Medicine and Health.
Fields of Study: Algebra; Data Analysis and Probability.
Summary: Cochlear implants use signal processing and algorithms to transmit electrical impulses to the brain to simulate hearing.

A cochlear implant is an electrical device that can help provide a representation of sound to a deaf or severely hard-of-hearing person. Unlike a hearing aid, a cochlear implant does not amplify sound; instead, it directly stimulates the auditory nerve, which sends these signals to the brain, where they can be interpreted as sound. Development of the cochlear implant relied in part on discoveries by French mathematician Joseph Fourier (1768–1830), whose studies in heat transfer led to the development of mathematics that can also be used to describe sound. Fourier analysis allows mathematicians to describe complex wave patterns, including the pressure waves that produce sound, as the combination of a number of component waves. Cochlear implants also draw on the discovery by the Italian physicist Alessandro Volta (1745–1827) that electrical current could be used to stimulate the auditory system and produce the sensation of sound. Practical work on cochlear implants dates back to the mid-twentieth century, and cochlear transplants were first approved by the U.S. Food and Drug Administration in 1984 for adults and in 1990 for children age 2 years and older (a limit since lowered to 12 months for one type of implant).

Sound Waves and Hearing

In a person with normal hearing, sound waves are collected by the outer (visible) ear (pinna) and sent down the ear canal to the eardrum (tympanic membrane). Movement of the eardrum is amplified by three small bones in the middle ear, commonly referred to as the "hammer," the "anvil," and the "stirrup," before being passed on to the cochlea in the inner ear. In the cochlea, this information is converted into electrical impulses by the hair cells of the organ of corti, and these impulses are sent on to the brain, where they are interpreted as sound. The cochlea has a spiral shape (sometimes likened to that of a snail shell) and scientists have recently discovered that the shape itself is significant in the cochlea's function. The spiral shape produces a "whispering gallery" effect as the energy of the frequency waves accumulate against the outer edge of the chamber, increasing humans' ability to detect low-frequency sounds.

Signal Processing

Most sound, including speech, is complex, meaning that it consists of multiple sound waves with different frequencies. In a person with normal hearing, the ear acts as a kind of Fourier analyzer, which decomposes sound into components. A cochlear implant attempts to mimic this activity, translating sound waves into electrical impulses and transmitting them directly to the brain. Two basic signal-processing strategies have been used in designing cochlear implants: filter bank strategies, which use Fast Fourier Transforms to divide sound into different frequency bands and represent

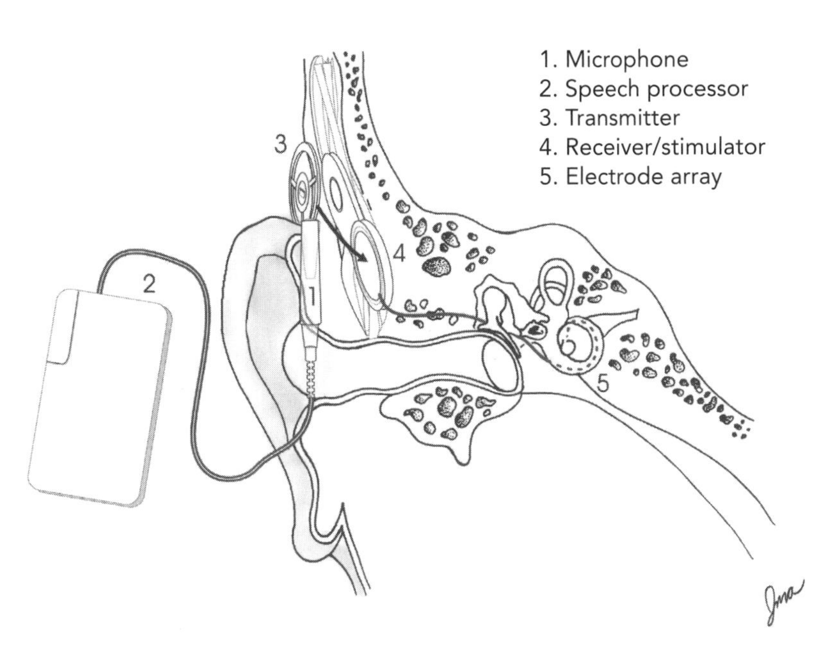

This illustration from the U.S. National Institutes of Health details the components of a cochlear implant.

this information as an analog or pulsatile waveform; and feature-extraction strategies, which use algorithms to recognize and emphasize the spectral features of different speech sounds.

A cochlear implant somewhat simulates normal hearing rather than restoring it, and individuals who receive an implant require special training in order to learn to recognize the signals as sound. In addition, cochlear implants are not advisable for every type of hearing loss, and a number of factors should be considered by the individual and his or her physician before committing to an implant. These factors include current age, age at which the person became deaf, how long the person has been deaf, the availability of support (including financing) to see him or her through the training period, and the health and structure of the individual's cochlea.

Although cochlear implants are growing in popularity and being used for younger and younger children, they are also controversial for several reasons, some of which were discussed in the 2000 documentary film *Sound and Fury*. One is based on the cost of the operation and follow-up therapy necessary to help the recipient learn to process the electrical impulses as sound. Another is that the surgery requires destroying whatever hearing may remain in the ear where the implant will be placed; for this reason, it is common to have the implant in one ear only. In addition, the surgery is done on children as young as 1 year in order to take advantage of peak language learning periods, so parents must make this decision for their children. Finally, many members of the deaf community feel that they should not be regarded as being defective or handicapped, that they can function successfully in the world using sign language and lip reading. They fear that widespread adoption of cochlear implants will ultimately destroy a distinctive and flourishing deaf culture.

Further Reading

Aronson, Josh. *Sound and Fury*. Aronson Film Associates, Inc. and Public Policy Productions, 2000. Filmstrip.

Loizou, Philipos. "Mimicking the Human Ear: An Overview of Signal-Processing Strategies for Converting Sound Into Electrical Signals in Cochlear Implants." *IEEE Signal Processing Magazine* (September 1998).

Manoussaki, Daphne, E. Dimitriadis, and R. S. Chadwick. "The Cochlea's Graded Curvature Effect on Low Frequency Waves." *Physical Review Letters* 96 (March 3, 2006).

National Institute of Deafness and Other Communication Disorders, National Institutes of Health. "Cochlear Implants." http://www.nidcd.nih.gov/health/hearing/coch.htm.

Niparko, John K., ed. *Cochlear Implants: Principles and Practices*. Philadelphia: Wolters Kluwer Health/Lippincott Williams & Wilkins, 2009.

Sarah Boslaugh

See Also: Brain; Functions; Harmonics; Mathematical Modeling; Nervous System.

Cocktail Party Problem

Category: Friendship, Romance, and Religion.
Fields of Study: Data Analysis and Probability; Measurement.
Summary: A metaphorical cocktail party is the setting for a source separation problem and other challenges.

The eponymous Cocktail Party Problem is a source separation problem in digital signal processing, wherein digital systems have difficulty separating out one signal among many—the metaphorical conversation in a noisy cocktail party, which is comparatively easily handled by the human brain. More broadly, distinguishing signal from noise is a data analysis challenge with many specific applications. The metaphor of the cocktail party also lends itself to a number of other problems in combinatorics, graph theory, probability, and functional analysis.

Conversations and Background Noise

With all the noise at a party, it can be difficult to focus on one conversation, although many people are able to do so. Telecommunication professor Colin Cherry conducted experiments in this area, and he is considered by some to be a pioneer in cognitive science. Many people can even recognize the sound of their name from across a noisy room. However, this is not as easy when heard on a recording. One cocktail party problem arises from concerns about separating each individual's voice characteristics in a recording from the other voices and background noise. People in surveillance and intelligence are inherently interested in such a problem, and scientists and engineers have worked on solutions since at least the 1950s. One common method is mathematical signal processing. Mathematicians and engineers digitize a signal using a Fourier transform, named for Joseph Fourier. They process it using a variety of methods to remove noise and other extraneous information, and then reconstruct the signal using the inverse transform.

While the process may result in an improved recording of one person's voice, early twenty-first-century technology and methods do not provide perfect separation, so the recording still includes at least some distracting background noise. However, engineers have conjectured that the signal should be able to be reconstructed without the noisy phase. Mathematicians Radu Balan, Peter Casazza, and Dan Edidin made progress on the problem in 2006, when they showed—using a neural net—that it is mathematically possible to retain the voice characteristics without the noise. Scientists continue to work on developing algorithms for practical use. Casazza made another fundamental mathematical discovery during his work on the cocktail party problem. He and his wife, Janet Tremain, also a mathematician, showed that the Kadison–Singer problem, named for mathematicians Richard Kadison and Isadore "Iz" Singer, is equivalent to other unsolved problems in areas of pure and applied mathematics and engineering, such as operator theory, harmonic analysis, and signal processing.

Mathematicians also investigate other party problems, like the probability that when people at a party are chosen to be partners for a card game—like bridge—no randomly chosen partners will contain spouses or members of the same family. The solutions require finding specific combinations or permutations of the guests. Under certain constraints, the maximal probability for some problems may be bounded at less than certainty as the number of people at the party grows. There are also connections between this question and the card game War, as well as with a related set of problems that focus on orders and arrangements of guests around a single dinner table or in various groupings, with applications in areas like queuing theory and assignment problems. The classic dining philosophers problem is yet another variation that has applications in resource sharing and task allocation in computer science.

Another party problem asks how many people must be present at a party in order to ensure that there will be a group of three people who share the characteristic of being acquaintances or strangers. There is no guarantee that three people will all know each other or will be strangers in parties of five or less people since counterexamples exist. The Java game HEXI, named so because the game is played on the vertices of a hexagon, is modeled on this question. The six vertices are connected by edges and each player takes a turn coloring an edge his or her color. One color represents acquaintances, and the other represents strangers. The goal of the game is to avoid making a triangle of the same color. Mathematicians model this question using graph theory, and show that in any group of at least six people, it is possible to find a group of three people satisfying one of the mutually exclusive relationships. Hence HEXI will always have a loser. Instead of people at a party or vertices of a polygon, one could explore other objects like nations embroiled in a conflict, sequences of randomly generated numbers, or stars. Mathematicians investigate problems like these concerning the existence of regular patterns in sets of objects in Ramsey theory, named for Frank Ramsey.

Further Reading

Albertson, Michael. "People Who Know People." *Mathematics Magazine* 67, no. 4 (1994).

Brodie, Marc. "Avoiding Your Spouse at a Party Leads to War." *Mathematics Magazine* 75, no. 3 (2002).

Casazza, Peter, and Janet Tremain. "The Kadison–Singer Problem in Mathematics and Engineering." *Proceedings of the National Academy of Sciences* 103, no. 7 (2006).

"Mathematicians Solve the 'Cocktail Party Problem.'" PHYSorg.com. August 22, 2006. http://www.physorg.com/news75477497.html.

Sarah J. Greenwald
Jill E. Thomley

See Also: Daubechies, Ingrid; Graham, Fan Chung; Intelligence and Counterintelligence; Probability; Wireless Communication.

Coding and Encryption

Category: Communication and Computers.
Fields of Study: Algebra; Data Analysis and Probability; Number and Operations; Representations.
Summary: Mathematical algorithms are used in modern encryption and decryption.

Human beings have a propensity to preserve and share secret information. Cryptography, from the Greek *kryptos* (hidden) and *graphein* (to write), is the art and science of coding and decoding messages containing secret information. Encryption is the algorithmic process that converts plain-text into cipher-text (looks like a collection of unintelligible symbols), while decryption is the reverse process that converts the cipher-text back to the original plain-text. A cipher algorithm and its associated key control both directions of the sequence, with the code's security level directly related to the algorithm's complexity. The two fundamental types of cryptography are symmetric (or secret keys) or asymmetric (or public-key), with multiple variations. Claude Shannon, an American mathematician and electronic engineer, is known as the father of information theory and cryptography. Some claim that his master's thesis, which demonstrates that electrical applications of Boolean algebra can construct and resolve any logical numerical relationship, is the most important master's thesis of all time.

Around 2000 B.C.E., Egyptian scribes included nonstandard hieroglyphs in carved inscriptions. During war campaigns, Julius Caesar sent coded information to Roman generals. Paul Revere's signal from a Boston bell tower in 1775 is even a simple example of a coded message. Success of the Allies in both World Wars depended on their breaking of the German's Enigma code. With the world-wide need for more sophisticated coding algorithms to transmit secure messages for military forces, businesses, and governments, people began capitalizing on the combined powers of mathematics, computer technology, and engineering.

The simplest examples of ciphers involve either transpositions or substitutions. In 450 B.C.E., the Spartans used transposition ciphers when they wound a narrow belt spirally around a thick staff and wrote a plain-text (or message) along the length of the rod. Once unwound, the belt appeared to be a meaningless sequence of symbols. To decipher the cipher-text, the receiver wound the belt around a similar staff. Variations of transposition ciphers are the route cipher and the Cardan grill.

Julius Caesar used substitution ciphers, where each letter of the plain-text is replaced by some other letter or symbol, using a substitution dictionary. For example, suppose:

Original Alphabet:
A B C D E F G H I J K L M N O P Q R S T U V W X Y Z
Key Dictionary:
K L M N O P Q R S T U V W X Y Z A B C D E F G H I J

where the key dictionary is made by starting with "code" letter K and then writing the alphabet as if on a loop. To encode the plain-text, "The World Is Round," each letter is substituted by its companion letter, producing the cipher-text "CRO FXAUN SB AXDWN." To disguise word lengths and to add complexity, the cipher-text was sometimes blocked into fixed-length groups of letters such as "CROF XAUN SBAX DWN." To decipher the cipher-text, one needed to know only the "code" letter. Though simple and initially confusing, substitution ciphers now are easily broken using frequency patterns of letters and words. Variations of the substitution cipher involve the suppression of letter frequencies, syllabic substitutions, or polyalphabetic substitutions such as the Vigenère or Beaufort ciphers.

The Playfair Square cipher used by Great Britain in World War I is a substitution cipher, but its encryption of letter pairs in place of single letters is more powerful yet easy to use. The cipher-key is a 5 × 5 table initiated by a key word, such as "mathematics."

M	A	T	H	E
I	C	S	B	D
F	G	K	L	N
O	P	Q	R	U
V	W	X	Y	Z

The table is built by moving left to right and from top to bottom (or other visual pattern as in a spiral) by first filling in the table's cells with the keyword's letters—avoiding duplicate letters. Then, the subsequent cells are filled with the remaining letters of the alphabet, using the "I" to represent the "J" to reduce the alphabet to 25 letters (instead of 26). Both the coder and the decoder need to know the both the keyword and the conventions used to construct the common cipher-key.

The coder first breaks the plain-text into two-letter pairs and uses the cipher-key via a system of rules:

- If double letters occur in the plain-text, insert an X between them.
- Rewrite the plain-text as a sequence of two-letter pairs, using an X as a final filler for last letter-pair.
- If the two letters lie in the same row, replace each letter by the letter to its right (for example, CS becomes SB).
- If the two letters lie in the same column, replace each letter by the letter below it (TS becomes SK and PW becomes WA).
- If the two letters lie at corners of a rectangle embedded in the square, replace them by their counterpart in the same rectangle (TB becomes SH and CR becomes PB).

Using this cipher-key, the plain-text "The World Is Round" becomes first

TH EW OR LD IS RO UN DX

which when encoded, becomes

HE ZA PU BN CB UP ZU ZS.

The same cipher-key is used to decode this message, but the rules are interpreted in reverse. It is quite difficult to decode this cipher-text without access to both the keyword and the conventions to construct the common cipher-key, though very possible.

The problem with all substitution and transposition encryption systems is their dependence on shared secrecy between the coders and the intended decoders. To transmit plain-text via cipher-text and then decode it back to public-text successfully, both parties would have to know and use common systems, common keywords, and common visual arrangements. In turn, privacy is required, since these systems are of no value if the user learns the key-word or is able to use frequency techniques of word/letter patterns to break the code. A more complicated and secure encryption process was needed, but it was not invented until the 1970s.

The revolutionary idea in encryption was the idea of a public key system, where the encryption key is known by everyone (that is, the public). However, the twist was that this knowledge was not useful in figuring out the decryption key, which was not made public. The RSA public-key cipher, invented in 1977 by Ronald Rivest, Adi Shamir, and Leonard Adleman ("RSA" stands for the names of the inventors), all of whom have bachelor's degrees in mathematics and advanced degrees in computer science, is still used today thanks to powerful mathematics and powerful computer systems.

In a RSA system, the "receiver" of the intended message is the driver of the process. In lieu of the "sender," the receiver chooses both the encryption key and the matching decryption key. In fact, the "receiver" can make the encryption key public in a directory so any "sender" can use it to send secure messages, which only the "receiver" knows how to decrypt. Again, the latter decryption process is not even known by the "sender."

Because the problem is quite complex and uses both congruence relationships and modular arithmetic, only a sense of the process can be described as follows:

- As the "receiver," start with the product n equal to two very large prime numbers p and q.
- Choose a number e relatively prime to $(p-1)(q-1)$.
- The published encryption key is the pair (n,e).
- Change plain-text letters to equivalent number forms using a conversion such as $A=2, B=3, C=4, \ldots, Z=27$.
- Using the published encryption key, the "sender" encrypts each number z using the formula $m \equiv z^e \bmod(n)$, with the new number sequence being the cipher-text.
- To decode the text, the "receiver" not only knows both e and the factors of n but also the large primes p and q as prime factors of n.
- Then, the decryption key d is private but can be computed by the "receiver" using an inverse relationship $ed \equiv 1 \bmod (p-1)(q-1)$, which allows the decoding of the encrypted number into a set of numbers that can be converted back into the plain-text.

The RSA public key system works well, but the required primes p and q have to be very large and often involve more than 300 digits. If they are not large, powerful computers can determine the decryption key d from the given encryption key (n,e) by factoring the number n. This decryption is possible because of the fact that, while computers can easily multiply large numbers, it is much more difficult to factor large numbers on a computer.

Regardless of its type, a cryptographic system must meet multiple characteristics. First, it must reflect the user's abilities and physical context, avoiding extreme complexity and extraneous physical apparatus. Second, it must include some form of error checking, so that small errors in composition or transmission do not render the message into meaningless gibberish. Third, it must ensure that the decoder of the cipher-text will produce a single, meaningful plain-text. There are many mathematicians working for government agencies like the National Security Agency (NSA), as well as for private companies that are developing improved security for storage and transmission of digital information using these principles. In fact, the NSA is the largest employer of mathematicians in the United States.

Further Reading

Churchhouse, Robert. *Codes and Ciphers*. Cambridge, England: Cambridge University Press, 2002.

Kahn, David. *The Codebreakers: The Story of Secret Writing*. New York: Macmillan, 1967.

Lewand, Robert. *Cryptological Mathematics*. Washington, DC: Mathematical Association of America, 2000.

Smith, Laurence. *Cryptography: The Science of Secret Writing*. New York: Dover Publications, 1971.

JERRY JOHNSON

See Also: Bar Codes; Intelligence and Counterintelligence; Mathematics, Applied.

Cold War

Category: Government, Politics, and History.
Fields of Study: All.
Summary: The Cold War had a broad influence on mathematics, including education, coding theory, game theory, and many applied fields.

The Cold War was a 45-year-long period of bitter competition between two large groups of nations. It lasted from the end of World War II in 1945 to the collapse of the Soviet Union in 1991. The two groups never came to direct combat—hence the term "cold"—but it was a war in every other way, fought with deadly ferocity in the political, economic, ideological, and technological arenas. It was also a time of unprecedented investment in new mathematical ideas, driven in part by the desire of each side to dominate the other through nuclear intimidation, economic strength, espionage, and political control. The Cold War had a great impact on mathematics education, on the study of codes and code-breaking algorithms, and the development of new fields such as game theory. More generally, the term "cold war" can be applied to any fight-to-the-death competition between nations in which the two sides avoid direct military combat.

In the original Cold War (1945–1991), the two groups of nations divided along ideological lines. One side, the Soviet bloc, adhered to the communist political and economic philosophy of Karl Marx and Vladimir Lenin. The other side, the Western bloc, adhered to the older free-market capitalist philosophy originated by Adam Smith.

The two sides of the Cold War were essentially forced to avoid military conflict by the recent invention of the atomic bomb, because neither side wanted to risk combat that might give rise to an unstoppable military escalation. The inevitable result of such an escalation would have been worldwide nuclear war, with most large cities destroyed in an instant by nuclear warheads, followed by massive clouds of radioactive ash circling the globe and causing the death of hundreds of millions of innocent people.

Mutually Assured Destruction (MAD)

Prior to the development of the atomic bomb, there were no weapons capable of destroying the population of an entire city in a single blow. Wars were fought as purely military conflicts, without risking the life of civilization itself. This nature of conflict changed forever with the advent of nuclear weapons.

Prior to the Cold War, the dominant mathematical model of warfare was a simple predator–prey model invented by Frederick Lanchester during World War I. Perhaps not surprisingly, given the slow, grinding progress of World War I, the Lanchester model places primary emphasis on the rates of attrition of the military forces. The side that survives this deadly attrition process wins the battle. In a nuclear-armed battle, however, it is survival rather than attrition that is the vital concern, and the Lanchester equations are irrelevant.

In the earliest years of the Cold War, only the United States possessed the theory and technology to construct an atomic bomb. The presence of the bomb in the arsenal of one side but not the other made possible a strategy known as "nuclear blackmail." The owners of the atomic bomb could threaten to use the bomb if their adversaries did not comply with their demands. For example, newly declassified documents have revealed that in 1961, Great Britain threatened China with nuclear retaliation if China were to attempt a military invasion of the British Crown Colony of Hong Kong. The United States backed up this threat, and China refrained from invading Hong Kong.

Clearly, the ownership of the secret of the atomic bomb by just one nation in 1945 had destabilized the military balance of power among the victors of World War II. Great Britain and France allied themselves with the United States and were given access to atomic secrets. The Soviet Union chose to develop its own versions of the atomic bomb, or to steal the secrets through espionage. Thus arose the great division of the Cold War, between the respective allies of the Soviet Union—known as the Warsaw Pact—on one side, and the United States—through the North Atlantic Treaty Organization (NATO)—on the other.

The Soviet Union tested its first nuclear weapon in 1949, ending its four-year period of vulnerability to nuclear blackmail. The military doctrine that took its place was known as mutually assured destruction (MAD). As long as each side in the Cold War could assure the other that it would be utterly destroyed in a nuclear exchange, then—so it was hoped—military conflict could be prevented. MAD did indeed prevent the two nuclear powers from directly attacking each other, but it had two unfortunate consequences: the people of both sides lived in terror of nuclear annihilation, and both superpowers engaged in so-called proxy wars, using much smaller nations as their proxies in localized military conflicts.

Albert Wohlstetter was an influential and controversial strategist who was a major force behind efforts to deter nuclear war and avoid nuclear proliferation. He worked as a consultant to the RAND Corporation's mathematics division starting in 1951. Initially, he collaborated on problems related to modeling logistics, but then he was asked to turn his skills to a problem posed by the U.S. Air Force regarding the assignment and location of bases for Strategic Air Command (SAC). On the surface, it was a common logistics problem, but ultimately SAC's method of basing its medium-range manned bombers, which were one of the country's major deterrents against a Soviet invasion of western Europe, had far greater implications. This work drew him into global strategy. He and his wife. Roberta Wohlstetter, a historian and intelligence expert, received the Presidential Medal of Freedom in 1985. Wohlstetter was also reputedly one of the inspirations for the film *Dr. Strangelove*.

The Arms Race as a Nash Equilibrium

From its very outset, neither of the Cold War's two superpowers—the United States and the Soviet Union—believed that they could stop developing new and ever more powerful nuclear weapons. The MAD

In accordance with a post–Cold War agreement, the Titan II Missile silo doors can only open halfway.

doctrine applied only as long as the forces of each side could pose a credible nuclear threat to the other. Therefore, each side worked to create new weapons as fast as possible. Throughout the 1950s and well into the 1960s, both nations tested ever more powerful nuclear weapons. This became known as the "arms race."

In the mathematical theory of games, the military arms race brought about by the MAD doctrine is an example of a Nash equilibrium, named for mathematician John Nash. In the decades-long arms race between the two Cold War competitors, each side could be seen as playing a simple noncooperative game. Each player in this game has a choice: to construct new and more terrible nuclear weapons, or not. If either player chose not to develop further weapons, while the other did, then the first player would face the very real risk of eventually facing nuclear blackmail. Each player understood the other's dilemma all too well, and so both continued to develop new weapons as fast as possible.

The persistence of this behavior comes from the fact that neither player can benefit by changing strategy unilaterally. When this occurs in a game, then it is in a form of equilibrium whose existence was first proved in the general case by Nash in 1950.

Game theory itself was a child of the Cold War, having been created in 1944 by John von Neumann, a mathematician who also played a key role in the development of the first atomic bomb, and Oskar Morgenstern, an economist. Throughout the Cold War, the theory of games was studied and elaborated, both by the military and by economists, as a means for better understanding the fundamental nature of competition, cooperation, negotiation, and war.

The fundamental irrationality of the nuclear arms race, in which each side became able to kill every single person on the planet many times over, was apparent to almost everyone. This realization did little to stop the arms race, because of the power of the Nash equi-

Old Soviet anti-aircraft missile rockets, first deployed in 1957. Generations of children grew up with the ever-present threat of war during the Cold War (1945–1991) and were taught bomb threat procedures.

librium to trap the players of the game into modes of behavior that, individually, they deplored.

In some critical respects, an arms race resembles a famous game known as the Chicken: two cars race toward each other down a narrow road, with the driver who first swerves away to avoid a crash being the loser. The key to winning a game of chicken is to act in such a way that your opponent comes to believe that you are so irrational as to be willing to die before swerving. In other words, the "rational" solution to the game is to be utterly and convincingly irrational. The same principle holds in a nuclear arms race.

The game of Chicken and other apparent paradoxes of rationality within the theory of games led to the development in the 1970s of meta-game analysis. This and other mathematical forms of strategic analysis played an important role in the eventual winding down of the arms race with a series of strategic arms agreements between the major powers of the Cold War.

Political Competition in the Cold War

The bitter competition of the Cold War was at least as much political and economic as military, and new mathematical ideas contributed mightily to this competition. In the economic arena, the Cold War was fought between the proponents of multiparty, free-market economies on the Western side, and the proponents of single-party, command economies on the Soviet side.

Both sides claimed to be democratic in the Cold War, but they used different meanings for the word. In the West, the word "democracy" retained its historic meaning, a political system in which leaders are chosen in free elections. In the Soviet system, "democracy" meant a "dictatorship of the proletariat" in which all political power rested in a hierarchy of labor councils, known as *soviets*, and the supreme soviet could dictate any aspect of public affairs. Soon after the Russian Revolution, however, the Communist Party seized control of the soviets, and after that, no election in the Soviet system was free.

The intense political competition between these two systems of government led to great interest in the West in how to conduct elections in the fairest possible way. A large body of mathematical theory of elections emerged, much of it devoted to the study of election systems that come the closest to meeting a measure of fairness known as the Condorcet criterion. In an election, the Condorcet winner is the candidate who can beat any of the other candidates in a two-person run-off election. Many forms of preference balloting, in which voters rank the candidates, come quite close to the Condorcet criterion, but none is without problems. Arrow's Paradox, discovered and proved by Kenneth Arrow in 1950, states that when voters have three or more choices, then no voting system can convert the ranked preferences of the voters into a community-wide ranking that meets a particular beneficial set of criteria. This Cold War mathematical discovery is the starting point of the modern theory of social choice, the foundation of the mathematical theory of political science.

Economic Competition in the Cold War

There are many forms of socialism known in economic theory, but the form practiced by the Soviet bloc of nations was particularly severe. In its purest form, Soviet socialism entailed state ownership of all means of economic production: all industrial plants, all commercial businesses, all farms, and all financial institutions. Soviet socialism was a command economy, meaning that the state had to tell every plant, business, and institution how much to produce, and at what price they should sell their goods and services.

In order to come up with the enormous number of production and price commands that had to be sent out every month and year, the Soviet system of government employed a vast bureaucracy. The system used by these bureaucrats was developed in the 1920s, during the early years of the Soviet Union, without the benefit of mathematics. Known as the "method of balances," this system attempted to function so that the total output of each kind of goods would match the quantity that its users were supposed to receive.

In practice, the Soviet "method of balances" functioned very much like the U.S. War Production Board during World War II, and by its counterparts in the war economies of Great Britain and Germany. The first production decisions were made with respect to the highest priority items (ships, tanks, airplanes) and those were balanced with the available amounts of strategic resources (iron, coal, electricity), and so on down to the lowest-priority items. The command system was thought to be crude and error-prone, and its mistakes and imbalances were widely noticed.

In the West, the response of mathematicians to these failures of the wartime command economy was the development of the field of engineering known as operations research. The mathematical technique known as "linear programming"—originally a little-known Russian discovery—was successfully developed by George Dantzig and John von Neumann in 1947 to optimize production quantities under linear constraints on supplies. The Soviet economy was very slow to adopt these ideas, preferring for ideological reasons to stick with the inefficient and error-prone method of balances until very late in the Cold War.

After World War II, the nations of the West ended their wartime command economic systems and reverted back to using the free market to make price and production decisions. The Soviet Union and its allies, however, continued to rely on a large army of bureaucrats to make all economic decisions without the aid of good operational theory.

Wassily Leontief, a Russian economist working in the United States, solved one of the fundamental problems of a command economy in 1949 with his method of input–output analysis. This method required the creation of a very large matrix showing the contribution of each component or sector of the economy to every other component. When properly constructed, the required inputs of raw materials to the economy can be calculated from the desired outputs by matrix inversion. The Soviet Union failed to quickly see the significance of Leontief's achievement, and did not incorporate his ideas into its planning system for many decades.

It is one of the great ironies of the Cold War that the mathematical theories that were required to make a command economy function properly were perfected in the West, where they are now universally employed within industrial corporations—some now larger than the entire economy of the old Soviet Union—to run their operations in the most efficient way possible. In the end, the economy of the Soviet Union and its satellites was not able to keep pace with the West, and in 1991, it suffered catastrophic political and economic collapse.

Further Reading

Arrow, Kenneth. *Social Choice and Individual Values*. 2nd ed. New Haven, CT: Yale University Press, 1963.

Erickson, Paul. "The Politics of Game Theory: Mathematics and Cold War Culture." Ph.D. dissertation, University of Wisconsin-Madison, 2006.

Howard, Nigel. *Paradoxes of Rationality*. Cambridge, MA: MIT Press, 1971.

Johnson, Thomas. "American Cryptography During the Cold War." National Security Agency. http://www.gwu.edu/~nsarchiv/NSAEBB/NSAEBB260.

Karp, Alexander. "The Cold War in the Soviet School: A Case Study of Mathematics Education." *European Education* 38, no. 4 (2006).

Kort, Michael. *The Columbia Guide to the Cold War*. New York: Columbia University Press, 2001.

Leontief, Wassily. "The Decline and Rise of Soviet Economic Science," *Foreign Affairs* 38 (1960).

Loren Cobb

See Also: Atomic Bomb (Manhattan Project); Game Theory; Predator–Prey Models; World War II.

Combinations

See *Permutations and Combinations*

Comic Strips

Category: Arts, Music, and Entertainment.
Fields of Study: Communications; Connections.
Summary: Mathematics plays a role in comic strip formats and is sometimes even the subject of comics.

The comic strip is a combination of word and picture in a narrative structure, unique from forms of communication based solely on the one or the other. The standard comic strip format presents its creator with two unique mathematical puzzles: to tell a story that fits into the pattern unconsciously expected by the reader, and to organize the illustrations of a Sunday strip into a format of exacting geometric and narrative demands.

Origins

The term "comic strip" entered the English language in 1922, via a poem by Carl Sandburg, describing the single-strip, black-and-white cartoons published in daily newspapers. Scholars disagree on the origins of the comic strip. Daily strips first appeared circa 1903, as part of the racing tips section of the newspaper. This was some 20 years after the appearance of the first full- or half-page color comics in supplemental sections, a reaction to protests against publishing on Sundays.

Color cartoons were a continuation of the European tradition of sociopolitically inspired prints that date back to the widely circulated wood block broadsheets of fifteenth-century Germany. Some scholars go further, tracing ancestry as far back as La Tapisserie de La Reine Mathilde, narrative scrolls of China and Japan, Trajan's Column, Bronze Age logographs, or even ancient petroglyphs and cave paintings. Regardless of the exact origins, the current format presents the cartoonist with two challenges of special interest to the mathematician.

Story and Art

The original supplements carried full- or half-page features with detailed drawings and developed stories. Older strips would not be legible if published at current smaller sizes. As a result, the expansive serial strip has been replaced almost entirely by the gag strip.

Theorists suggest that humor is based on pattern recognition. If the audience recognizes a pattern, it begins to anticipate what will come next. A deviation from the pattern, if done correctly, is perceived as humorous. In gag strips, between two-thirds and three-quarters of the space is spent establishing the pattern. The deviation happens next, sometimes followed by a character reaction to the deviation, sympathetically reinforcing the audience reaction, or providing additional deviation.

Not only have comic strips become smaller, all color comics must fit into an extremely limited template. Syndicates require a minimum of six panels, but some newspapers elect not to publish one or both of the first panels. Character poses and scene layouts in the two-dimensional plane are designed to lead the reader's gaze from one point of interest to another, driving the story forward. Deciding on the orientation of visual images and their relationships can be difficult when the artist does not know where one panel will be in relation to the next.

Mathematics in Comic Strips

Mathematics is used not only to decide the layout and flow of comic strips but can be used within comic strips as an element of humor. Many comics reveal or satirize widely held societal attitudes and beliefs about mathematics. Bill Amend, creator of the widely circulated *Foxtrot* comic strip, has a degree in physics, and his strip frequently features mathematically based humor. The same is true of Randall Munroe's Web comic *xkcd*, which is subtitled "A webcomic of romance, sarcasm, math, and language." Comic strips may be used in classrooms as motivators for serious discussions about mathematics concepts and analysis of peoples' attitudes about mathematics. There are also entire comic books and graphic novels intended to teach mathematics. The work *Logicomix* dramatizes the life story of philosopher and mathematician Bertrand Russell, who spent his life trying to establish an indisputable logical foundation for mathematics. In the course of the novel, he encounters many mathematicians of note, including Gottlob Frege, David Hilbert, Kurt Gödel, and Ludwig Wittgenstein.

Further Reading

Doxiadis, Apostolos. *Logicomix: An Epic Search for Truth.* London: Bloomsbury Publishing PLC, 2009.

Emmer, Michele. *Mathematics and Culture IV.* Berlin: Springer, 2006.

Ksir, Amy, and Russell Goodman. "FoxTrot Brings Mathematics to the Comics Page." *Math Horizons* 13 (November 2005).

McCloud, Scott. *Understanding Comics.* Northampton, MA: Kitchen Sink Press, 1993.

John N. A. Brown

See Also: Communication in Society; Connections in Society; Representations in Society; Sequences and Series.

Communication in Society

Category: School and Society.
Fields of Study: Communication; Connections.
Summary: Communication helps mathematicians and others be informed of past and current research and to formulate and organize their own ideas.

Communication is fundamental to mathematics as a discipline, the mathematics community, mathematics education, and society as a whole, since communication is an essential part of everyday life and any social interaction. Effective communication is inherent in validating mathematics. Using a common language and a set of notions and drawing upon a shared body of knowledge, mathematicians communicate with each other—both orally and in writing—about their mathematical ideas, perceptions, or methods.

For example, mathematicians exchange ideas with their colleagues, write technical reports, publish original research papers and expository articles in professional journals, or give oral presentations. Some associate good mathematics communication with beautiful expository lectures or clear writing, while others focus on the quality of the interactions between people, such as those working in a group on mathematics. A peer review process is frequently part of mathematics communication and dissemination, ensuring some degree of consensus on what constitutes *appropriate* or *valid* mathematics. In this way, the standards of mathematics are socially developed. In addition to interacting with their colleagues, mathematicians need to communicate with the rest of the society using a language and terminology that are more familiar to the general public. For instance, mathematicians explain to the public how the discipline of mathematics contributes to society or demonstrate the various applications of mathematics in fields such as engineering, medicine, and communication technologies.

The role of communication in the education of mathematics is similar to the vital role communication plays in the discipline of mathematics. Drawing upon mathematical language and notation, teachers and students talk about mathematics; share, explain, and justify mathematical ideas; or analyze, discuss, and interpret mathematical concepts. Communication about mathematics and communication using mathematical language do not occur only in the mathematics community or in mathematics classrooms. Regardless of one's profession, wise decision making in personal lives and participation in civic and democratic life increasingly demand mathematical communication skills. For example, people need to communicate with mortgage companies when buying a house and interpret various mathematical concepts (such as percentage and rate) presented in the media. Thus, communication with mathematics and about mathematics is an essential part of daily life.

Communication Media

In the twenty-first century, there are a wide variety of electronic and print venues for communicating mathematics, and the evolution of electronic media and databases has vastly changed the way people access mathematics. Historically, mathematicians communicated by letters, during visits, or by reading each other's published articles or books once such means became available. Some mathematical concepts were developed in parallel by mathematicians working in different areas of the world, such as German Karl Friedrich Gauss and American Robert Adrain, who both made advances in the theory of the Normal distribution in the early nineteenth century. Some mathematicians were not aware of each other's progress because they did not have the venues of communication that are available in the twenty-first century. In an effort to increase the accessibility of mathematics research articles, reviews began appearing in print journals like *Zentralblatt für Mathematik*, which originated in 1931, and *Mathematical Reviews*, which originated in 1940. Since the 1980s, electronic versions of these reviews have allowed researchers to search for publications on a specific topic. In 2010, MathSciNet, the electronic version of *Mathematical Reviews*, listed more than 2 million items and more than 1 million links to original articles. In 2011, the database Zentralblatt MATH listed more than 3 million items from approximately 3500 journals and 1100 serials. Both contain work dating back to the early 1800s. There are also thousands of mathematics journals that are not listed in these collective databases, such as most mathematics education research.

Some mathematicians publish open access drafts of their papers on their personal Web pages before official publication in peer-reviewed and other journals, or in other online settings such as the ArXiv.org e-print archive. Co-authors from around the world can work together using e-mail or other Web-based collaborative tools. Mathematics students, teachers, and researchers often discuss mathematics ideas and share resources on blogs, through online chats, or using other forums. For instance, what began in 1992 as the Geometry Forum was extended in 1996 to become the Math Forum. There are many additional resources for sharing and teaching mathematics content, both in print and in electronic media. Some electronic examples include the National Council of Teachers of Mathematics Illuminations Web site; Wolfram MathWorld, which was developed by Eric Weisstein; and Math Fun Facts, developed by Francis Su. Social and historical context is also often addressed in sites such as The MacTutor History of Mathematics archive, developed by John O'Connor and Edmund Robertson, or Mathematicians of the African Diaspora, created by Scott Williams.

One important question related to online communication is how to represent and display mathematical notation, which is an important part of mathematical validity and understanding. Some Web pages contain fixed images for each equation or graph. Others use Java applets for dynamic display. The Mathematical Markup Language (MathML) is one way to encode mathematics. TeX was created by Donald Knuth in order to typeset scientific and mathematical research. TeX-based software such as LaTeX has become the standard in printing mathematics. Another issue is the validation of online resources, which may be created or published without peer review. On one level, this issue is an extension of the existing issue of peer review for print media, as mathematics journals already employ varying degrees of rigor when reviewing and publishing papers. At the same time, there is in increasing trend of creating printed works from electronic sources or using electronic sources as references, which creates an added difficulty in ensuring the collective accuracy of the body of mathematics communication.

With so many options available, the specific nature of mathematics communication depends in large part on the purpose and intended audience. There are some mathematics publications and communications aimed at a general audience, others aimed at students, and yet others intended for researchers. Mathematicians, educators, and other communications specialists work to match the form and venue of the mathematics communication to the need. Some careers that are regularly involved in communicating mathematics include technical writers or publication editors. The Society for Technical Communication and the Council of Science Editors are two professional associations that address this need. In 2007, Ivars Peterson became the director of Publications and Communications at the Mathematical Association of America, which, like other professional associations, publishes items for both the specialist and the nonspecialist. He previously wrote MathTrek for *Science News*. In 1991, he received a Joint Policy Board for Mathematics (JPBM) Communications Award for his "exceptional ability and sustained effort in communicating mathematics to a general audience." He also served as East Tennessee State University's Basler Chair of Excellence for the Integration of the Arts, Rhetoric, and Science in 2008 and taught a course there called Communicating Mathematics. In a talk on the topic of communication in mathematics, he noted:

> The importance of communicating mathematics clearly and effectively is evident in the many ways in which mathematicians must write, whether to produce technical reports, expository articles, book reviews, essays, referee's reports, grant proposals, research papers, evaluations, or slides for oral presentations.

National Council of Teachers of Mathematics

The National Council of Teachers of Mathematics (NCTM) emphasizes clear and coherent communication of mathematical ideas and thinking as a skill that students need to learn from pre-kindergarten through grade 12. Given the essential role of communication in teaching and learning of mathematics, NCTM has set forth process standards for communication for primary and secondary mathematics curricula. The Principles and Standards for School Mathematics (2000) states that instructional programs from pre-kindergarten through grade 12 should enable all students to

- Organize and consolidate mathematical thinking through communication
- Communicate their mathematical thinking coherently and clearly to peers, teachers, and others
- Analyze and evaluate the mathematical thinking and strategies of others
- Use the language of mathematics to express mathematical ideas precisely

Communication in Schools

Communication, both oral and written, is an essential part of mathematics education. The act of communication allows students to systematize and incorporate their mathematics thinking and understanding, both for learning mathematical theory and mathematical problem solving. For example, when students communicate their own mathematical thinking and understanding, they are required to rationalize and organize their reasoning and also formulate puzzling or complex questions well enough to present them as clearly as possible to a reader. As a result, the process guides students toward greater insights to their own thinking and learning. Focused reflection, which is conceptually intertwined with communication, helps students to increase the benefits of communicating their ideas with peers, teachers, and others. Written or oral reflections in which ideas are shared among peers, teachers, and others provide students multiple perspectives that sharpen ideas explored. The American Society of Mathematics (ASM), which is also known as the American Society for the Communication of Mathematics, sponsors problem-solving contests and the U.S. National Collegiate Mathematics Championship.

Proofs

One topic that illustrates the importance and the diverse nature of mathematics communication is the notion of proof. Researchers have proposed a wide variety of roles for proof in mathematics, such as establishing the truth of a statement, communicating mathematical knowledge, opening the way for further understandings and discoveries in mathematics, providing new techniques for doing mathematics, and organizing statements into systems of axioms and theorems. Throughout history, proofs and communication via proof have been incorporated in many different ways in mathematics education in the United States. The National Council of Teachers of Mathematics' (NCTM) 2000 Principles and Standards for School Mathematics emphasized the role of proof in mathematics learning for all students and helped to formalize its curricular importance and place in prekindergarten though high school education. Further, as proof became more systematized in K–12 education, some mathematics education researchers began to more deeply explore students' understanding of the definition or nature of proof, the role of proof as a mode of communication, and peer acceptance of the validity of a given proof, as well as how proof is taught in classrooms.

As the concept of proof came under investigation, an important issue was the conceptualization and the roles of proof in school mathematics. The NCTM defined proof in *Principles and Standards for School Mathematics* as "arguments consisting of logically rigorous deductions of conclusions from hypotheses." One element in the definition of proof is the acceptability of an argument as proof, which is referred to as "logically rigorous." An important question that NCTM's definition entails is who decides if a proof is logically rigorous enough to be accepted. To conceptualize the definition and identify the roles of proof in school mathematics, mathematics education researchers have referred to the qualifications and function of proof in the discipline of mathematics and investigated how it is implemented in mathematics classrooms.

Research has demonstrated the social nature of argumentation and justification in the classroom and beyond, and communication and validation by peers plays an important role in proof within and outside the classroom. This social dimension of proof is grounded in sociocultural theories of mathematical learning and is believed to reflect the process of becoming a mathematician. Yu Manin argued that within mathematics community, "A proof becomes a proof after the social act of 'accepting it as a proof.'" Erna Yackel and Paul Cobb concluded that acceptable justifications in mathematics education are interactively constituted by individual teachers and students in each classroom, where the teacher is the representative of the mathematical community. Mathematical justifications and argumentations are regulated by the general expectations and the regulations of the classroom community.

Thus, they are a part of the classroom norms and, more specifically, the sociomathematical norms, which are the extension of general classroom social norms to specifically focus on the normative aspects of mathematical discussions as students participate in mathematical activities. Yackel and Cobb argued that: "Normative understandings of what counts as mathematically different, mathematically sophisticated, mathematically efficient, and mathematically elegant in a classroom are sociomathematical norms.... The understanding that students are expected to explain

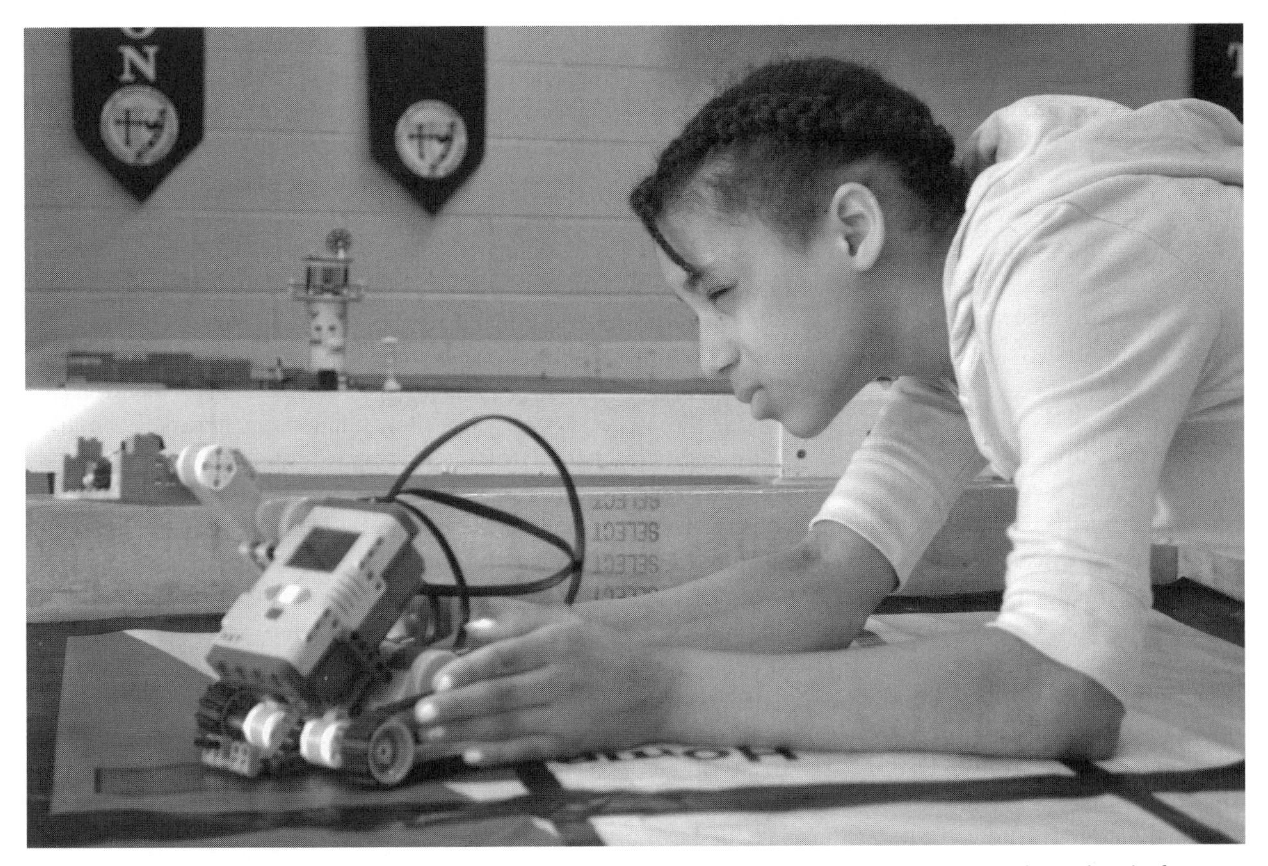

A student at a science, technology, engineering, and math (STEM) summer camp examines her robot before releasing it for a test. The camp encouraged K–12 students to pursue education and careers in the STEM fields.

their solutions and their ways of thinking is a social norm, whereas the understanding of what counts as an acceptable mathematical explanation is a sociomathematical norm." This idea plays a role in mathematics educator Andreas Stylianides's conceptualization of proof. He proposed four aspects that are required to consider an argument a proof: foundation, formulation, representation, and social dimension. He presented an example in which an elementary school student constructed a mathematical argument that was founded on definitions of mathematical constructs, formulated using deductive reasoning from these definitions, and then represented verbally. Regarding the social dimension of the proof, although the student's argumentation was logically rigorous and would have been accepted as a proof in the wider mathematical community, it generated counterarguments among her classroom peers and her argument was not accepted as a proof by the classroom community.

Indeed, the conceptualization of mathematics, in particular the social dimension that is appropriate for school mathematics, requires more research to develop. Mathematical discourse is an important factor in the development of shared understanding of mathematically valid justifications. However, students at various levels, particularly younger elementary school students, may have different levels of understand regarding the rules and norms of mathematical discourse, and understanding is not necessarily shared by all. Thus, as was the case in Stylianides' study, a valid mathematical argument was not accepted as valid by all students. In such cases, the teacher, acting as an authoritative representative of the mathematical community, could intervene and explain why the argument is indeed valid by broader standards. However, in some ways this action would negate the social dimension aspect that is used to evaluate mathematical acceptability, at least with respect to the classroom environment. Thus, the subtleties in what constitutes a valid argument within a mathematics classrooms and the relation to a teacher's role as the communicator of other mathematical norms as they acculturate students in the processes of proving need to be explored. It is important to note that teachers need to know when, how, and how much to intervene so as to not play an authoritarian role, thereby creating a learning environment in which students are forced into authoritarian schemes and communication is essentially unidirectional, from teacher to student.

Mathematical Applications in Communication Technologies

In the increasingly digital world of the twenty-first century, the safe communication of information has become a major issue for discussion and research in mathematics and science, in large part because of theft and fraud often perpetrated using new technologies. Mathematics plays an important role in making communication as safe as possible. Cryptology is a technique used to ensure that messages or data are transmitted safely to the receiver. Dating to the substitution ciphers used in ancient Rome and other civilizations, this field has always drawn heavily from mathematics. Research in mathematics and other disciplines, such as computer sciences and engineering, has resulted in an increasingly sophisticated array of coding techniques and technologies, as well as code-breaking methods. Some of the most common and known applications of cryptography include encryption of credit card numbers or passwords for electronic commerce and encryption of e-mail messages for secure communication. Confidentiality, authenticity, and integrity in electronic commerce or communication have become an apparent and sensitive issue for people who engage in online transactions such as buying or selling items online, online banking, and online communications, as well as for applications like medical records. If proper action is not taken for data transmission, information sent over an open network can be stolen by hackers. Such an action can reveal secret information or messages containing personal information, like a credit card number, a password, or online banking information, facilitating crimes like identity theft. A hacker can use digital data to clone a person's identity and use a victim's resources for the hacker's own good. Even worse, this information could be a national secret, and it may cause more serious problems. For that reason, the National Security Agency (NSA) uses its cryptologic heritage in the midst of challenging times to protect national security systems, and the NSA is one of the leading employers of mathematicians in the United States at the start of the twenty-first century.

Along with digital security, mathematics also plays a fundamental role in both the hardware and software that make the increasingly wireless, globally connected world possible. The *Advances in Mathematics of Communications* journal publishes research articles related to mathematics in communication technologies. Mathematicians and mathematical methods contribute to many aspects, including the Internet's computer server backbone and communications protocols; vast cell phone networks; and smartphones that act as mobile platforms for an array of communications methods, such as voice, text, photo, e-mail, and Internet. Music, movies, dance, art, theater, and many other methods people use to convey ideas to one another involve mathematics as part of the creative endeavor. Humans can communicate with neighbors next door, with people on the opposite side of the world, with satellites orbiting the planet, or even with probes that have been sent into the far reaches of the solar system thanks to mathematics. Some would in fact argue that mathematics is itself a universal language or method of communication.

Further Reading

Elliott, Portia, and Cynthia Garnett. *Getting Into the Mathematics Conversation: Valuing Communication in Mathematics Classrooms—Readings From NCTM's School-Based Journals*. Reston, VA: National Council of Teachers of Mathematics, 2008.

Manin, Yu. *A Course in Mathematical Logic*. New York: Springer-Verlag, 1977.

Mathematical Association of America. "JPBM Communications Award." http://www.maa.org/Awards/jpbm.html.

National Council of Teachers of Mathematics. *Principles and Standards for School Mathematics*. Reston, VA: National Council of Teachers of Mathematics, 2000.

Peterson, Ivars. "Writing Mathematics Well." http://sites.google.com/site/ivarspeterson/workshop1.

Stylianides, Andreas. "The Notion of Proof in the Context of Elementary School Mathematics." *Educational Studies in Mathematics* 65 (2007).

Yackel, E., and P. Cobb. "Sociomathematical Norms, Argumentation, and Autonomy in Mathematics."

Journal for Research in Mathematics Education 22 (1996).

Zeynep Ebrar Yetkiner Ozel
Serkan Ozel

See Also: Coding and Encryption; Professional Associations; Reasoning and Proof in Society; Universal Language; Wireless Communication.

Comparison Shopping

Category: Business, Economics, and Marketing.
Fields of Study: Geometry; Measurement; Number and Operations.
Summary: Both simple and complex algorithms are used to compare consumer prices and contextualize mathematics instruction.

The globalization of the marketplace has resulted in a plethora of choices for any given item at both the local store and via the Internet. People comparison shop for both very expensive items like a car or a plane ticket and fairly inexpensive purchases like a box of cereal. Comparison shopping is perhaps one of the most widely used applied mathematics lessons, both in K–12 and lower-level college courses.

Mathematics is at the forefront of comparison shopping through unit pricing, which makes use of division and fractions. Geometric methods can be used to compare volume or weight. Notions from pre-algebra and algebra model financial decisions such as purchasing a cell phone plan or taking out a car loan. Students explore parameters in order to make balanced and informed choices. Mathematics educators not only use these examples in classrooms, but they also study their effectiveness. Researchers and online shopping agents take advantage of mathematical methods to extract, compare, and mine huge amounts of data. Comparison techniques also include data envelopment analysis and multiple regression.

Unit Pricing

One method of comparing differently priced items in different sized containers is through unit pricing. Dividing the price by the quantity or amount of items, such as how many ounces, will yield a cost per unit term that can be used for comparison purposes. For example, an 11.5 oz box of cereal might cost \$4.49, while a 24 oz box of cereal costs \$4.99. The unit price of the first box is $\$4.49/11.5 \approx \$0.39$ per ounce, while the second box is $\$4.99/24 \approx \$0.21$ per ounce. Some items are already priced by their weight, like meats, fruits, vegetables, or coffee, and others are priced according to their volume, that is, by the container size. For those items that are not priced by weight or volume, unit pricing is listed on the shelf tag in many stores. However, the unit price is not the only important feature in comparison shopping. Personal preferences and other important factors must also be taken into consideration, like whether one will be able to use up a larger quantity before the expiration date. Unit pricing examples proliferate in lessons on fractions and in classes like pre-algebra and developmental mathematics. Students also compare scenarios in which sales occur or other discounts are applied.

Debt and Interest

Another common classroom scenario is found in comparing house and car purchases in financial mathematics segments. For instance, students can use the loan payment formula to calculate the monthly payment R in terms of the monthly interest rate r, the loan amount P, and the number of months, n

$$R = \frac{rP}{1-(1+r)^{-n}}.$$

Then they can calculate the total interest by multiplying the monthly payment and the number of months and subtracting the loan amount. One comparison scenario is determining how the monthly payment and total interest change as the price of the car or house changes or the interest rate fluctuates. Another is determining whether one should take out a smaller loan versus paying loan points to buy down the interest rate. Students also compare car prices to income level using the debt-to-income ratio. The debt-to-income ratio is the debt divided by the income, which is the percentage of debt. Banks use the debt-to-income ratio in making decisions about mortgage or car loans. From the Great Depression in the 1930s until the deregulation of banking restrictions in the 1970s, an upper limit of 25% was typical. However, that level rose after deregulation and with the increase in consumer credit card debt. In the twenty-first century, it is common for an upper limit to range between 33% and 36%. Given a monthly car payment, house payment, and other monthly debts, students can add up the total debt and solve for the necessary income level in order to stay below 36%. They can also compare the way that debt and the needed income change as the interest rates vary.

Contextualizing Instruction

Mathematics educators use purchasing scenarios in the classroom and study and debate their effectiveness. Some studies have found that the contextualization of mathematics using examples from shopping helps students. Terezinha Nunes, Analucia Schliemann, and David Carraher compared the mathematical abilities of children who were selling items in Brazilian street markets to questions in school. They found that the closer to the real-life situation, the more successful the student. Other studies have also found that there can be a disconnect between performance in the supermarket and performance in school. Some researchers assert that the contextualization may disguise the mathematics and be problematic in elucidating the underlying mathematical processes.

Mathematical Models for Comparison Shopping

Businesses and researchers employ a variety of mathematical techniques in order to compare large shopping data sets. Online shopping agents use mathematical methods in situations such as a Web search for airplane ticket prices or hotel rooms. Historically, dating back to at least the nineteenth century, travel agents sold vacations to consumers on behalf of suppliers. Travel agencies grew in popularity with the increase in commercial aviation after World War I. At the end of the twentieth century, the Internet vastly changed the way in which consumers compared and purchased vacation travel. Airlines, hotels, and other vacation companies offered online services directly to consumers, bypassing travel agents. In response, some travel agencies created travel Web sites that would compare options. Their computer programs extracted comparative price data from Web sites in order to build comparison shopping engines. Researchers continue to develop advanced comparison shopping techniques including methods in data mining, data envelopment analysis, and multiple regression. They create sophisticated algorithms to analyze data and find patterns. In data envelopment analysis, networks can be viewed as decision-making units, and efficient configurations are selected. In multiple regression, several variables are combined in an attempt to create a meaningful predictor or measure. Mathematical methods are also important in predicting shopping preferences and consumer behavior.

Marketplace globalization has resulted in increased choices at both local stores and via the Internet.

Further Reading

Berry, Michael, and Linoff, Gordon. *Data Mining Techniques for Marketing, Sales and Customer Support.* Hoboken, NJ: Wiley, 1997.

Boaler, Jo. "The Role of Contexts in the Mathematics Classroom: Do They Make Mathematics More 'Real'?" *For the Learning of Mathematics* 13, no. 2 (1993).

Devlin, Keith. *The Math Instinct: Why You're a Mathematical Genius (Along With Lobsters, Birds, Cats, and Dogs).* New York: Basic Books, 2005.

Herzog, David. *Math You Can Really Use—Every Day.* Hoboken, NJ: Wiley, 2007.

McKay, Lucia, and Maggie Guscott. *Practical Math in Context: Smart Shopping Math.* Costa Mesa, CA: Saddleback Educational Publishing, 2005.

Nunes, Terezinha, Analucia Schliemann, and David Carraher. *Street Mathematics and School Mathematics.* New York: Cambridge University Press, 1993.

Sarah J. Greenwald
Jill E. Thomley

See Also: Coupons and Rebates; Data Mining; Inventory Models; Market Research; Predicting Preferences.

Competitions and Contests

Category: Mathematics Culture and Identity.
Fields of Study: Communication; Connections; Problem Solving.
Summary: Mathematics competitions and contests help encourage students to practice and study mathematics and develop problem-solving abilities.

Well-designed mathematics contests provide excellent vehicles for students to hone their skills, expand their knowledge, develop their ability to focus, practice creative problem solving, and join a community of peers who love mathematical challenges. Mathematicians and educators organize competitions, help students prepare for them, participate on committees to grade the results, and assess contests' long-term impact. Some mathematics competitors are known as "mathletes."

MATHCOUNTS and USAMTS

There are a number of well-known mathematics competitions in the United States for middle school and high school students. MATHCOUNTS, a mathematics competition for sixth, seventh, and eighth graders, emphasizes problems from geometry, combinatorics, and algebra. The competition includes written and oral rounds with both individual and team competitions, and students advance from school, to chapter, to state, and to national levels. The USA Mathematical Talent Search (USAMTS) is an open mathematics competition for U.S. middle and high school students. USAMTS consists of two rounds of six problems per round and operates on the honor system, since participants are given a full month to work on the problems. The goal of USAMTS is to help students develop their proof writing ability, improve their technical writing abilities, and mature mathematically while having fun. The organizers strive to foster insight, ingenuity, creativity, and perseverance. The American Mathematics Competitions (AMCs) provide three levels of competitions. Students who perform well on the AMC 10 or AMC 12 exams, for students in grades 10 or 12 and below, respectively, are invited to participate in the American Invitational Mathematics Examination (AIME). Approximately the top 270 performers on the AIME and the AMC 12 advance to the United States of America Mathematical Olympiad (USAMO), which is the final round of the AMC series of contests. The top 230 AIME and AMC 10 only participants take part in the USA Junior Mathematical Olympiad (USAJMO). The top 30–40 performers on the USAMO, along with a dozen or so others from the USAJMO, attend the Mathematical Olympiad Summer Program, a training program from which the six members of the U.S. International Mathematical Olympiad (IMO) team are selected. Students who do well on the AIME typically receive scholarship offers from prestigious colleges and universities.

IMO

The International Mathematical Olympiad (IMO) is an annual two-day, six-problem, mathematical competition for pre-collegiate students that began in 1959. Approximately 100 countries send teams of up to six students. The problems are extremely difficult and involve ideas that are not usually encountered in high schools or colleges. Many IMO participants have become world-class research mathematicians, such as Noam Elkies, who eventually became the youngest full professor in Harvard University's history at the time of his promotion. Filmmaker George Csicsery documented the 2006 U.S. IMO team in Slovenia. The documentary also included segments on families and schooling, girls, and the Olympiad, as well as the problems and their solutions. Melanie Wood, who was the first female to represent the United States in the IMO, noted: "Math competitions are great. They introduce all these new ideas and in particular give students who are at school the first chance to see how you can be creative in solving a problem." She went on to obtain her Ph.D. in mathematics in 2009.

The Benet Academy Math Team poses with their trophy after a state math competition awards ceremony at the University of Illinois at Urbana-Champaign in 2007. Mathematics competitors are sometimes called "mathletes."

William Lowell Putnam Mathematical Competition

College students also participate in mathematical contests. The William Lowell Putnam Mathematical Competition is an annual mathematics competition for mathematically talented undergraduate college and university students in the United States and Canada administered by the Mathematical Association of America. The competition, in which both individuals and teams compete, consists of morning and afternoon three-hour exams, each with six problems. Although the problems are extraordinarily difficult and require highly creative thinking, they can typically be solved with only knowledge of college-level mathematics. The problems are so challenging that a median score for the 120-point exam is often 0 or 1. Of the more than 120,000 times the exams have been taken since the competition's inception in 1938, there have been only three perfect scores as of 2010.

In recent years, about 4000 students and 400 teams have participated. The top five teams and individual scorers receive thousands of dollars in prize money. Many top five scorers, named as Putnam Fellows, have become distinguished researchers in mathematics and other fields, including Fields Medalists (the highest award in mathematics for people younger than 40) John Milnor, David Mumford, and Daniel Quillen, and Nobel laureates Richard Feynman and Kenneth G. Wilson. Several Putnam Fellows have been elected to the National Academy of Science. In 2010, Putman Fellow David Mumford received the National Medal of Science, bestowed by President Barack Obama.

MCM

Unlike other competitions, which place a premium on speed and individual performance, the Mathematical Contest in Modeling (MCM) contest rewards teamwork, research skills, programming skills, organizing ability, writing ability, and stamina. The MCM is a 96-hour mathematics competition held annually since 1985 by the Consortium for Mathematics and Its Applications and sponsored by the Society of Industrial and Applied Mathematicians, the National Security Agency, and the Institute for Operations Research and the Management Sciences. Approximately 1000 international teams of three undergraduates each produce original mathematical papers in response to one of two open-ended modeling problems. The students may use any references and the Internet but are not permitted to discuss their problem with anyone not on their team. Approximately 1% to 2% of the teams are designated Outstanding Winners. The skills required in the modeling contest are those typically most valued by employers. Many students who do not excel in problem-solving contests excel in the modeling competition.

Value and Benefits

The value of mathematics competitions is that they pique interest in mathematics and encourage students to pursue intellectual activities. The benefits of participating in mathematics competitions are very much like the benefits derived from athletic contests or becoming accomplished in playing a musical instrument. The intention is that those engaged in such activities develop a sense of accomplishment and a positive self-image. On the other hand, some object to mathematics being presented as a competition. While some students may thrive in a competitive environment, others may be discouraged. For some, the competitive environment highlights mutual interests, which can help create lasting bonds and friendships.

Like sports, participants in mathematics contests may learn to set goals and work toward them, be highly motivated, be able to focus, have self-discipline, perform under pressure, cope with success and failure, and have a competitive spirit. As in music, participants in mathematics contests must learn self-discipline, develop the ability to concentrate, pay attention to detail, and practice many hours. Perhaps the important lesson learned from participating in mathematics contests is that success is the fruit of effort.

Further Reading

Csicsery, George. *Hard Problems: The Road to the World's Toughest Math Contest*. Washington, DC: Mathematical Association of America, 2008.

Flener, Frederick O. *Mathematics Contests: A Guide for Involving Students and Schools*. Restin, VA: National Council of Teachers of Mathematics, 1990.

Gallian, J. "The Putnam Competition from 1938–2009." http://www.d.umn.edu/~jgallian/putnam06.pdf.

Rusczyk, Richard. "Pros and Cons of Math Competitions." http://www.artofproblemsolving.com/Resources/articles.php?page=pc_competitions.

Joseph A. Gallian

See Also: Careers; Clubs and Honor Societies; Professional Associations; Succeeding in Mathematics.

Composing

Category: Arts, Music, and Entertainment.
Fields of Study: Algebra; Number and Operations; Representations.
Summary: Mathematics and music developed in tandem and composition is firmly grounded in mathematics.

Throughout the history of Western music, composers have utilized mathematical techniques in creating musical works. From Pythagoras, Plato, and Ptolemy in ancient Greece to the sixth-century music theorist Boethius, music was thought to be a corollary of arithmetic. With the widespread development of modern standardized musical notation thought to have begun in the Renaissance, compositional craft became more highly developed. Compositions intertwined with mathematical patterns were particularly highly regarded.

The eighteenth-century composer and theorist Jean-Philippe Rameau was unequivocal in his views on the connection between mathematics and music in his 1722 *Treatise on Harmony*, writing, "Music is a science which should have definite rules; these rules should be drawn from an evident principle; and this principle cannot really be known to us without the aid of mathematics." Fugal composition techniques in the high Baroque period were highly mathematical. The classical and romantic eras, characterized by a movement away from polyphonic music, produced less obvious mathematically oriented composition technique. In the twentieth century, however, mathematical formalisms were fundamental as replacements for the tonal structures of the romantic era. There are even subgenres of rock music (started in the 1980s) called "math rock" and "mathcore" (after metalcore, a fusion of heavy metal and hardcore punk), which uses complex and atypical rhythmic structures, angular melodies, unusual time signatures, and changing meters. Metalcore, in particular, also uses harmonic dissonance. In another example, Robert Schneider composed a mathematical score for a play in 2009. He said:

> I wrote a composition called 'Reverie in Prime Time Signatures,' that is obviously written in prime time signatures, that is, only prime numbers of beats per measure. Also the piece has kind of a sophisticated middle section that encodes some ancient Greek mathematics related to prime numbers in musical form, that I am proud of.

The Renaissance Canon

During the Renaissance, mathematical devices were developed to a considerable degree by Northern European composers. In the canons of Johannes Ockeghem, a single melodic voice provides the basis by which one or more additional voices are composed according to various mathematical transformations of the original: mirror reflection of musical intervals (inversion), time translation, mirror reflection in time (retrograde), or a non-unit time scaling (mensuration canon). Composers of this period understood the word "canon" to mean a rule by which secondary voices could be derived from a given melody, in contrast to our modern usage of the word, which means a simple duplication with later onset time, as in the nursery rhyme round "Row, Row, Row Your Boat."

Mathematical Transformations in Composition

In addition to standard musical notation, music can be represented mathematically as a sequence of points in an algebraic structure. A musical composition can be represented as a sequence of points from the module M over the cyclic groups of integers Z_p

$$M = Z_{p1} \times Z_{p2} \times Z_{p3} \times Z_{p4},$$

with the coordinates representing (respectively) onset time, pitch, duration, and loudness. For example, the 12

Bach: The Canon Master

Johann Sebastian Bach was a master of canonic composition. Bach's canons challenged performers to solve puzzles he set before them. Examples abound in *A Musical Offering* (BWV 1079), written in 1747. The first of two *Canon a 2* (canon for two voices) from *Musical Offering* appears to have two different clef symbols: one at the beginning of the first measure, and one at the end of the last. The first singer had to read from beginning to end, and the second had to start at the same time and read in the opposite direction. In this small piece, Bach provides an example of retrograde or cancrizan (crab) canon. The puzzle in the second *Canon a 2* is even more cleverly concealed: a single line with two clef signs in the first measure, one upside down. The cryptic instruction *Quaerendo invenietis* ("Seek and ye shall find") is inscribed at the top of the manuscript.

The second, inverted clef sign indicates that the second voice of the canon is to proceed in inversion, and the performer is left to "seek" the appropriate time translation at which the second voice should begin. Another example of Bach's masterful canonic treatment is BWV 1074: *Kanon zu vier Stimmen*, which with its numerous key signatures, clefs, and repeat signs can be played from any viewing angle.

notes of the chromatic scale would be represented in the second coordinate by Z_{12}. In this schematic, if a point (x_1, x_2, x_3, x_4) in a musical motif were repeated later at a different volume level, the repetition would differ in the first and last coordinate and would be represented as $(x+\alpha, x_1, x_2, x_3, x_4+\beta)$, where α is the time shift and β is the amount of the volume difference.

Inversion takes the form $(x_1, 2\alpha-x_2, x_3, x_4)$. Mensuration, as in the canons of Ockegham, is written $(x_1, x_2, \alpha \cdot x_3, x_4)$. Transformations of this form were used extensively in the Renaissance and Baroque eras and played a fundamental role in post-tonal era of the twentieth century.

Mathematical Structure in Atonal Music

At the turn of the twentieth century, music theorists and composers looked for new organizing principles on which atonal music could be structured. Groundbreaking composer Arnold Schoenberg turned to the idea of "serialism," in which a given permutation of the 12 chromatic pitches constitutes the basis for a composition. The new organizing principle called for the 12 pitches of this "tone row" to be used—singly, or as chords, at the discretion of the composer, always in the order specified by the row. When the notes of the row have been used, the process repeats from the beginning of the row.

Composers like Anton Webern, Pierre Boulez, and Karlheinz Stockhausen consciously used geometric transformations of onset time, pitch, duration, and loudness as mechanisms for applying the tone row in compositions. In the latter half of the twentieth century, set theoretic methods on "pitch class sets" dominated the theoretical discussion.

Predicated on the notions of octave equivalence and the equally tempered scale, Howard Hanson and Allen Forte developed mathematical analysis tools that brought a sense of theoretical cohesion to seemingly intractable modern compositions. Another mathematical approach to composition without tonality is known as aleatoric music, or chance music. This technique encompasses a wide range of spontaneous influences in both composition and performance. One notable exploration of aleatoric music can be seen in the stochastic compositions of Iannis Xenakis from the 1950s. Xenakis's stochastic composition technique, in which musical scores are produced by following various probability models, was realized in the orchestral works *Metastasis* and *Pithoprakta*, which were subsequently performed as ballet music in a work by George Ballanchine.

Further Reading

Beran, Jan. *Statistics in Musicology*. Boca Raton, FL: Chapman & Hall/CRC Press, 2003.

Forte, Allen. *The Structure of Atonal Music*. New Haven, CT: Yale University Press, 1973.

Grout, Donald Jay. *A History of Western Music*. New York: Norton, 1980.

Temperley, David. *Music and Probability*. Cambridge, MA: MIT Press, 2010.

Xenakis, Iannis. *Formalized Music: Thought and Mathematics in Composition*. Hillsdale, NY: Pendragon Press, 1992.

Eric Barth

See Also: Geometry of Music; Harmonics; Scales; Time Signatures.

Computer-Generated Imagery (CGI)

See *Animation and CGI*

Congressional Representation

Category: Government, Politics, and History.
Fields of Study: Algebra; Data Analysis and Probability; Number and Operations.
Summary: Though the Constitution dictates proportional representation by state, there are multiple methods for attempting to achieve fair apportionment.

Apportionment is the process of distributing a fixed resource on a proportional basis, particularly associated with government. The legislative branch of the

U.S. federal government—and most U.S. states—is bicameral, meaning that two separate bodies deliberate on laws. Reflecting a great political compromise of American government, these bodies are formulated on two distinct representative principles. The U.S. Senate has equal representation from each state to ensure that states have equal voices. For the House of Representatives, the U.S. Constitution requires that "Representatives...shall be apportioned among the several States which may be included within this Union, according to their respective Numbers."

This requirement ensures that larger states have a voice that fairly represents their greater constituencies. The primary mathematical challenge in most systems of representation is that typically not all representatives will represent the same number of citizens, and calculations rarely result in integers. Deciding a fair system of rounding for representative numbers for fractional constituencies has proven surprisingly challenging, and Congressional apportionment has generated substantial controversy throughout the history of the United States.

Numerous serious apportionment methods have been proposed. Most have names associated with the people who proposed them, such as third U.S. president Thomas Jefferson, and are generally classified as "divisor methods" or "quota methods." Many systems have been used in the United States, and mathematicians have long investigated fair apportionment. In 1948, at the request of the National Academy of Sciences, mathematicians Luther Eisenhart, Marston Morse, and John von Neumann recommended the Huntington–Hill method, proposed by mathematician Edward Huntington and statistician Joseph Hill. Apportionment is a prominent aspect of social choice theory, extensively studied by mathematicians such as Peyton Young and Michel Balinski. There have also been innovative links between apportionment and other areas of mathematics, like just-in-time sequencing and scheduling problems for manufacturing.

Apportionment Methods

A state's proportion of the total population of a country can be found by dividing the state's population by the total population. The state's fair share of the total seats in the nation's legislature, called its "standard quota," is the product of this proportion and the total seats. Alternatively, the standard quota can be found by using the standard divisor, which measures the average number of people per seat on a national basis, and is found by dividing the total population by the number of available seats.

For example, suppose that a small country consists of four states (A, B, C, D), with populations given as

State	Population
A	791
B	892
C	6987
D	530

The total population of this country is 9200, and State A has 791/9200 or approximately 8.6% of the population. If there are 25 seats in the country's legislature, then State A's standard quota is

$$\frac{791}{9200}(25) \approx 2.149.$$

seats. State A's population therefore warrants slightly more than two seats but less than three. The standard divisor in this case is 9200/25 = 368 people per seat, and State A's standard quota can also be represented as 791/368. Similarly, the standard quotas for States B, C, and D are calculated as 2.424, 18.986, and 1.440, respectively.

The requirement that each state be assigned an integer number of representatives forces a country to impose a systematic method for rounding standard quotas to whole numbers. It is reasonable to expect that any reasonable method will assign each state either its lower quota or its upper quota—its standard quota rounded down or up, respectively. This requirement is known as the Quota Rule. One method that arises naturally is to round up those standard quotas that are closest to the next number of seats. Specifically, one may choose to initially apportion each state its lower quota, which will always yield leftover seats. These surplus seats are distributed to the states whose standard quotas have the largest fractional part. This method is known as Hamilton's Method, Vinton's Method, or the Method of Largest Remainders, named for the first Secretary of the Treasury Alexander Hamilton and Congressman Samuel Vinton. In the above example, after assigning each state its lower quota, only 23 of the 25 seats have been apportioned. The first surplus seat is

Table 1: Distribution of Congressional Seats

State	Population	Standard Quota	Lower Quota	Apportionment
A	791	2.149	2	2
B	892	2.424	2	2
C	6987	18.986	18	19
D	530	1.440	1	2

assigned to State C, whose standard quota is very close to 19, while the other surplus seat is assigned to State D (See Table 1).

Some apportionment methods solve the problem of apportionment by using a specific rounding rule and modifying the standard divisor if necessary. In Jefferson's Method, for example, all quotas are rounded down to the integer part of the quota. As with Hamilton's Method, this yields unassigned seats in an initial apportionment. Rather than distributing those surplus seats as Hamilton's Method does, Jefferson's Method instead modifies the divisor by making it smaller. This method makes it easier for the states to obtain a seat and allows the states' quotas to grow larger. In a successful Jefferson apportionment, a modified divisor is found so that when the modified quotas are rounded down, the total number of seats apportioned is the desired number.

Adams' Method, named for sixth U.S. president John Quincy Adams, is similar to Jefferson's Method. Rather than rounding down, however, quotas are instead rounded up to the next largest integer. In this case, the initial attempt at apportionment results in too many seats being distributed and a modified divisor must be chosen that is larger than the standard divisor, reflecting a need to make it more difficult to obtain a seat.

Other divisor methods differ from the methods of Jefferson and Adams only in how the rounding is conducted. In Webster's Method, named for Senator and Secretary of State Daniel Webster, for example, all quotas are rounded conventionally—to the nearest whole number. If a state's quota has a fractional part that is 0.5 or greater, the quota is rounded up. Otherwise, it is rounded down. In other words, one can think of the tipping point for rounding in Webster's method as being the arithmetic mean of a state's lower and upper quota. In the Method of Huntington–Hill, a state's cutoff for rounding is the geometric mean of the state's lower and upper quotas. For this method, if a state's lower quota is L and its upper quota is U, then the cutoff for rounding is

$$\sqrt{LU}.$$

If the quota is less than the cutoff, then it is rounded down, otherwise it is rounded up. In Dean's Method, named for mathematician and astronomer James Dean, the cutoff for rounding is the harmonic mean of the lower and upper quota, expressed algebraically as

$$\frac{2}{\frac{1}{L}+\frac{1}{U}}.$$

Applying any of these methods requires searching for a modified divisor so that when the modified quotas are calculated and rounded according to the given rule, the number of seats distributed is the correct total.

Applying these divisor methods to the sample situation given above results in the apportionments seen in Table 2. A number of important aspects of apportionment can be seen in the table. First, the apportionment method makes a difference; different methods can yield different apportionments. Jefferson's Method has a substantial bias toward larger states. Adams's Method, on the other hand, is biased toward smaller states and can cause lower quota violations. Quota rule violations can occur with Webster's Method as well, though they are relatively rare. Webster's Method demonstrates little bias overall. The HuntingtonHill Method and, to a greater degree, Dean's Method, have biases toward smaller states.

History of U.S. Apportionment

The U.S. Constitution mandates a decennial census. Congressional representatives are reapportioned every 10 years based on the results. Though the Constitution provided for an initial distribution of U.S. Congressional representatives, it specified no particular apportioning method. Following Constitutional ratification in 1787 and the census of 1790, the first apportionment was carried out. In 1792, Congress passed a bill instituting Hamilton's Method, which George Washington vetoed. Congress then approved Jefferson's Method, which was used through 1832 when a Quota Rule violation was observed. New York State had a standard quota of 38.59 seats, so New York should have received either 38 or 39 seats. However, Jefferson's Method assigned New York 40 seats. John Quincy Adams and Daniel Webster immediately put forth separate bills calling for the adoption of the apportionment methods that carry their names. Though both bills failed, this was the last apportionment in which Jefferson's Method was used.

Webster's Method was used for the apportionment of 1842, but in 1852, Hamilton's Method was adopted as "permanent" by Congress. In 1872, Hamilton's Method was not applied correctly. In 1882, additional difficulties arose with the method itself. While considering different sizes for the House of Representatives, observers noted that with a House size of 299, Alabama would receive eight seats under Hamilton's Method, but with a House size of 300, Alabama would receive only seven seats. This flaw, whereby increasing the number of seats to apportion can, in and of itself, cause a state to lose a seat, became known as the "Alabama paradox." Congress sidestepped this issue in 1882 by increasing the size of the House to 325 seats, but the flaw led to their discarding Hamilton's Method in 1901. Webster's Method was used in the apportionments of 1901, 1911, and 1931, though no apportionment was completed after the 1920 census. In 1941, Congress adopted the Huntington–Hill Method as "permanent," with the House size of 435 seats, which is the method still in use at the start of the twenty-first century, though controversy continues.

Impossibility

Many mathematicians and others have asked whether there is an "ideal" apportionment method that solves the apportionment problem in a reasonable way and is free of flaws such as the Alabama paradox and quota rule violations. In the 1970s, Balinski and Young proved that no such method exists. Every apportionment method will either potentially violate the quota rule or cause either the Alabama paradox or another problematic paradox called the "Population paradox," whereby one state whose population is growing at a faster rate can lose a seat to a state with a slower growth rate. The search for perfection in apportionment is an inherently impossible task, but mathematicians continue to study the problems and paradoxes and seek new approaches to reduce bias.

Further Reading

American Mathematical Society. "Apportionment: Introduction." http://www.ams.org/samplings/feature-column/fcarc-apportion1.

American Mathematical Society. "Apportionment II: Apportionment Systems." http://www.ams.org/samplings/feature-column/fcarc-apportionii1.

Balinski, Michel, and Peyton Young. *Fair Representation: Meeting the Ideal of One Man, One Vote.* 2nd ed. Washington, DC: Brookings Institution Press, 2001.

Table 2. Dean's Method.

State	Standard Quota	Jefferson	Adams	Webster	Huntington–Hill	Dean
A	2.149	2	2	2	2	2
B	2.424	2	3	2	2	3
C	18.986	20	18	20	19	18
D	1.440	1	2	1	2	2
Valid Range of Divisors		333–349	396–411	357–358	365–374	396–411

Frederick, Brian. *Congressional Representation & Constituents: The Case for Increasing the U.S. House of Representatives.* New York: Routledge, 2009.

Stephen Szydlik
Matt Kretchmar

See Also: Census; Gerrymandering; Government and State Legislation; Voting Methods.

Conic Sections

Category: History and Development of Curricular Concepts.
Fields of Study: Algebra; Communication; Connections; Geometry.
Summary: Conic sections have many interesting mathematical properties and real-world applications.

Conic sections, or simply "conics," are the simplest plane curves other than straight lines. Students in the twenty-first century begin to study these curves in middle school. In coordinate geometry, they can be expressed as polynomials of degree 2 in two variables while straight lines are polynomials of degree 1 in two variables. Conic sections can further be divided into three types: ellipse, parabola, and hyperbola. Conics were named and systematically studied by Apollonius of Perga (262–190 B.C.E.). At that time, the study of conics was not merely to explore the intrinsic beauty of the curves but to develop useful tools necessary for applications to the solution of geometric problems. Today, the theory of conics has numerous applications in our daily lives including the designs of many machines, optical tools, telecommunication devices, and even the tracks of roller coasters.

Representations of Conics and Their Applications

One can generate a two-sheet circular cone by fixing a straight line as the axis of the cone in the space first. Choose a fixed point on it as the vertex of the cone. Rotating another straight line through the vertex that makes a fixed angle with the axis, we obtain the desired cone as the trace of the rotating line. Any straight line on the trace is called a "generating line" of the cone. Conic sections are obtained by intersecting the two-sheet cone with planes not passing through its vertex as shown in Figures 1A–C.

The three types of conic sections are generated according to the positions of the intersection.

Ellipse

When the intersecting plane cuts only one sheet of the cone and the intersection is a closed curve, an ellipse is created. A circle is obtained when the intersecting plane is perpendicular to the axis of the cone; an ellipse is obtained when the intersecting plane is not perpendicular to the axis of the cone. A circle, as such, can be considered as a particular case of an ellipse (Figure 1A). As illustrated in Figure 2, an ellipse is the collection of points in a plane that the sum of distances from two fixed points F_1 and F_2, the foci, to every point in the collection is constant.

On the coordinate plane, if the foci are located on the x-axis at the points $(-c, 0)$ and $(c, 0)$ and the constant distance between F_1 and F_2 is $2a$, the equation of an ellipse can be derived as

$$\frac{x^2}{a^2} + \frac{y^2}{b^2} = 1 \text{ with } b^2 = a^2 - c^2 < a^2.$$

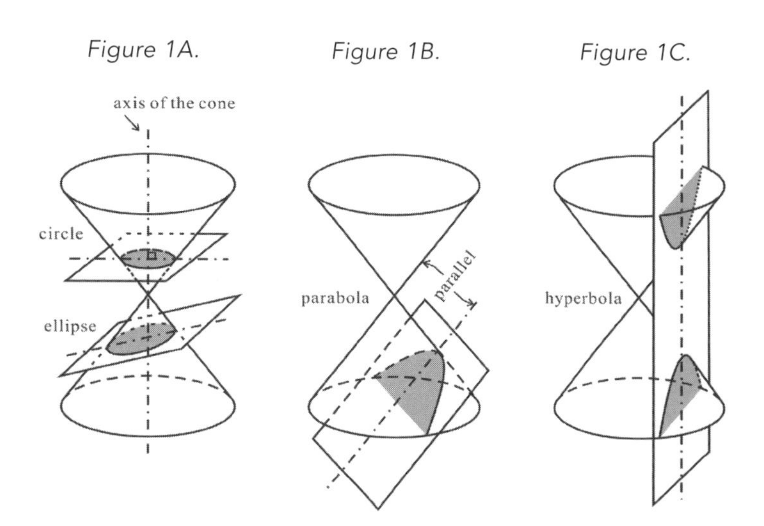

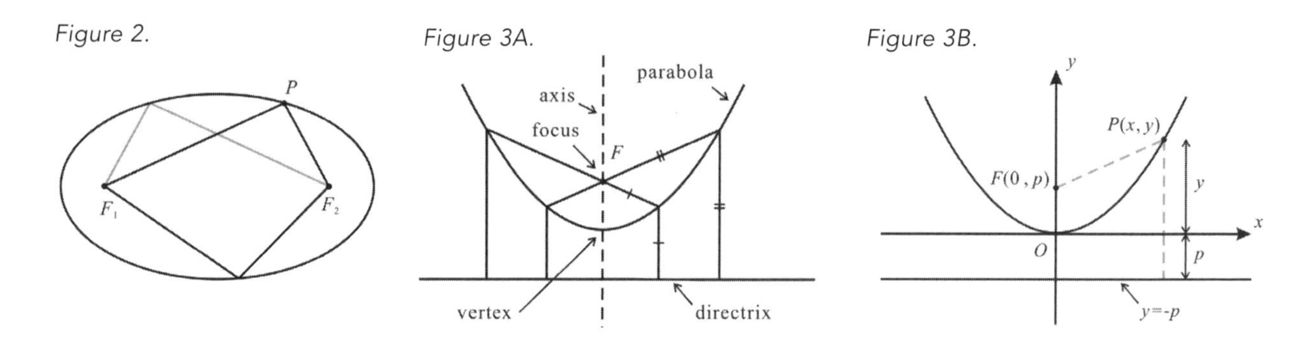

Parabola

When the intersecting plane cuts only one sheet of the cone and is parallel to exactly one generating line of the cone, the intersection is a non-closed curve—a parabola (Figure 1B). A parabola is the collection of points in a plane that are equidistant from a fixed point F (called "focus") and a fixed line (called "directrix"). The graph of the parabola is illustrated in Figure 3A. The graph is symmetric with respect to the line through the focus and perpendicular to the directrix. This line of symmetry is called the "axis" of the parabola. The intersection of the graph with the axis is called the "vertex" of the parabola. On the coordinate plane, if the vertex is located at the origin O, and the focus at the point $(0, p)$, then its directrix will be on the line $y = -p$ (Figure 3B), and the equation of the parabola can be derived as $x^2 = 4py$.

Hyperbola

When the intersecting plane meets both sheets of the cone, the intersection is a hyperbola, which consists of two identical non-closed parts, each located in one of the two sheets of the cone (Figure 1C). A hyperbola is the collection of all points in a plane that the difference of distances from two fixed points F_1 and F_2, the foci, to every point in the collection is constant. The graph of a hyperbola is drawn as shown in Figure 4A.

On the coordinate plane, if the foci $(-c, 0)$ $(c, 0)$ are located on the x-axis and the differences of distance is $\pm 2a$, then the equation of the hyperbola can be derived as

$$\frac{x^2}{a^2} - \frac{y^2}{b^2} = 1 \text{ with } c^2 = a^2 + b^2 \text{ (See Figure 4B)}$$

A Brief History of Conic Sections

Between 460 B.C.E. and 420 B.C.E., three famous geometry problems were posed by the ancient Greeks. These problems were (1) the trisection of an angle, (2) the squaring of the circle, and (3) the duplication of the cube. The last problem merely asks that given any cube of side length a, can one construct another cube with exactly twice the volume, $2a^3$. Hippocrates of Chios (circa 470–410 B.C.E.) had the idea of reducing that problem by finding two quantities x and y such that

$$\frac{a}{x} = \frac{x}{y} = \frac{y}{2a} \, .$$

Then, $x^2 = ay$, $y^2 = ax$, and $xy = 2a^2$.

As such, x is the required solution for the problem. This solution is equivalent to solving simultaneously any two of the three equations ($x^2 = ay$, $y^2 = 2ax$, and $xy = 2a^2$) that represent parabolas in the first two and a hyperbola in the third. However, no explicit construction of the conic sections was given. Menaechmus (380–320 B.C.E.) is believed to be the first mathematician to work with conic sections systematically, which is theorized to have arisen because of curves traced out by sundials. At his time, the conic sections were formed by cutting a right circular cone with a plane perpendicular to a side.

The sections were named according to whether the vertex angle was acute, right, or obtuse (Figure 5). Menaechmus constructed conic sections that satisfied the required algebraic properties suggested by Hippocrates and thus obtained the points of intersection of these conic sections that would lead to the solution of the problem of the duplication of the cube.

Figure 4A.

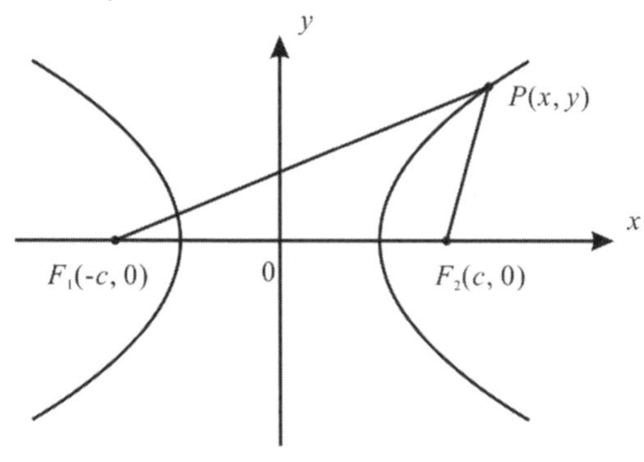

Figure 4B.

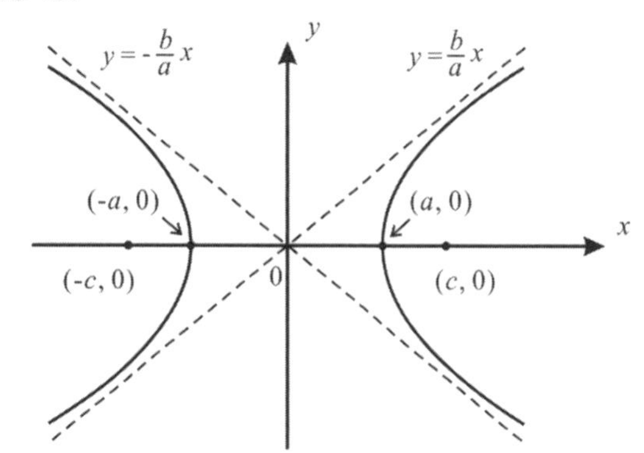

The breakthrough in the study of conics by the ancient Greeks was attributed to Apollonius of Perga. His eight-volume masterpiece *Conic Sections* greatly extended the existing knowledge at the time (one of the eight books has been lost to history). Apollonius' major contribution was to treat the conic sections as plane curves and use their intrinsic properties to characterize them. This method allowed conic sections to be analyzed in great detail by the ancient Greeks.

Abu Ali al-Hasan ibn al-Haytham studied optics using conic sections in the tenth and eleventh centuries. Omar Al-Khayyami (Omar Khayyam) authored *Treatise on Demonstration of Problems of Algebra* in the eleventh century. This work showed that all cubic equations could be classified using geometric solutions that involve conic sections. Later, in the seventeenth century, Gerard Desargues (1591–1662) and Blaise Pascal (1623–1662) connected the study of conic sections to developments from projective geometry. At the same time, René Descartes (1596–1650) and Pierre de Fermat (1601–1665) also connected it with the developments from coordinate geometry. Eventually, problems of conics in geometry could be reduced to problems in algebra.

Johan Kepler (1571–1630) revolutionized astronomy by introducing the notion of elliptical orbits. According to Isaac Newton's later law of universal gravitation, the orbits of two massive objects that interact are conic sections.

If they are bound together, they will both trace out ellipses; if they move apart, they will both follow parabolic or hyperbolic trajectories.

The Applications of Conic Sections

Besides applications in astronomy, conics have many other applications.

In an ellipse, any light or radiation that begins at one focus will be reflected to the other focus (Figure 6 on following page). This property can be used in theater designs. In an elliptical theater, the speech from one focus can be heard clearly across the theater at the other focus by the audience. It can also be applied in lithotripsy, a medical procedure for treating kidney stones. The patient is placed in an elliptical tank of water, with the kidney stone fixed at one focus. High-energy shock waves emitted at the other focus can be directed to pulverize the stone. Also, elliptical gears can be used for many machine tools.

Figure 5.

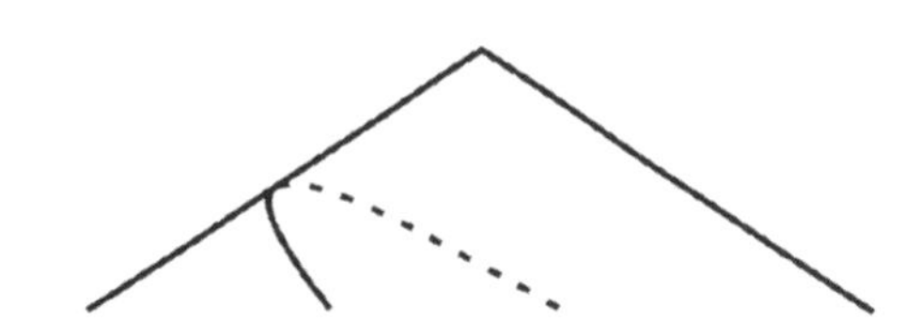

Figure 6.

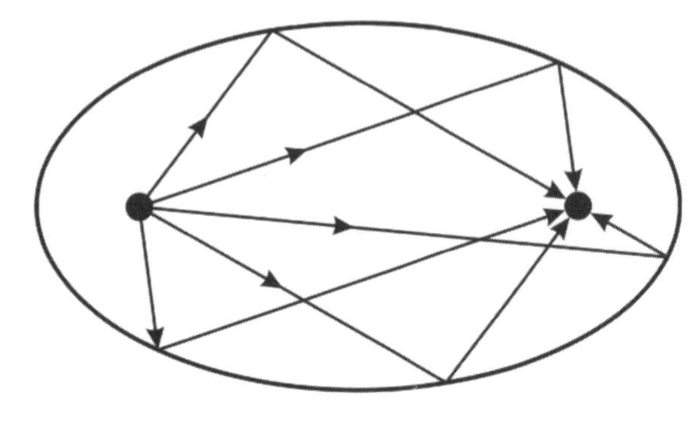

Figure 7.

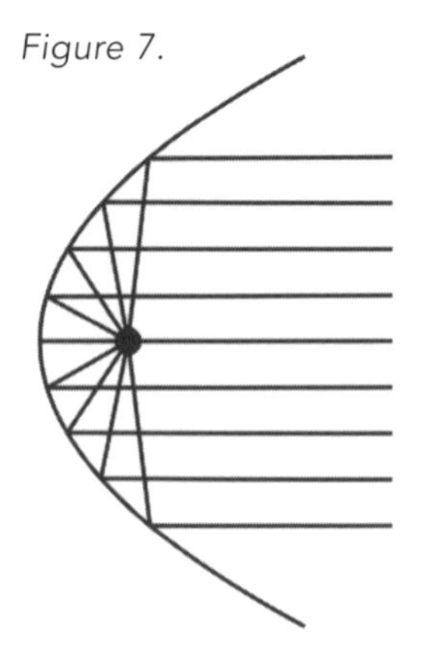

In a parabola, parallel light beams will converge to its focus (see Figure 7 on following page). Parabolic mirrors are used to converge light beams or heat radiations, and parabolic microphones are used to perform a similar function with sound waves.

In reverse, if a light source is placed at the focus of a parabolic mirror, the light will be reflected in rays parallel to said axis. This property is used in the design of car headlights and in spotlights because it aids in concentrating the parallel light beam. Hyperbolas are used in a navigation system known as Long Range Navigation (LORAN). Hyperbolic—as well as parabolic—mirrors and lenses are also used in systems of telescopes.

Further Reading

Akopyan, A. V., and A. A. Zaslavsky. *Geometry of Conics.* Providence, RI: American Mathematical Society, 2007.

Courant, R., and H. Robbins. *What Is Mathematics?* New York: Oxford University Press, 1996.

Downs, J. W. *Practical Conic Sections: The Geometric Properties of Ellipses, Parabolas and Hyperbolas.* Mineola, NY: Dover, 2003.

Kendig, K. *Conics (Dolciani Mathematical Expositions).* Washington, DC: The Mathematical Association of America, 2005.

Kline, M. *Mathematical Thought From Ancient to Modern Times.* New York: Oxford University Press, 1972.

Suzuki, Jeff. *A History of Mathematics.* Upper Saddle River, NJ: Prentice Hall, 2002.

Ka-Luen Cheung

See Also: Curves; Geometry of the Universe; Greek Mathematics.

Connections in Society

Category: School and Society.
Fields of Study: Connections.
Summary: An integrated approach to mathematics stresses the importance of making connections among various perspectives and applications.

While mathematics in educational settings is often separated out into differing subjects, it is important to understand that mathematics is an interconnected field of study. While most individuals are aware that they must be familiar with basic addition and subtraction to ensure the proper handling of money, very few individuals give any thought to the multitude of deeper mathematical connections they experience daily. In fact, both the National Science Foundation and the National Council of Teachers of Mathematics have recently begun to strongly advocate for the use of an interconnected curriculum in K–12 mathematics education. An integrated approach to mathematics education stresses the importance of making connections among mathematical perspectives, as in algebra and geometry, making connections to other fields, as in physics or religion, and connecting mathematical concepts to society as a whole, as in applications and usefulness in daily living.

The purpose of an interconnected curriculum is to help students better understand how the various branches of mathematics are connected and how mathematics is connected to the real world. By teaching mathematics as a unified whole, rather than multiple discrete subjects, students may better understand that mathematics is not a set of indiscriminate rules and isolated skills; rather, it involves a rich interplay between mathematical concepts, as well as complex interactions with other academic subjects. It is this integrated approach to mathematics that seeks to answer that question, "When are we ever going to use this in *real life?*" When this objective is met, students often show an increased appreciation and enthusiasm for mathematical principles.

People use many different interrelated approaches to process ideas, analyze objects, make decisions, or solve problems. For example, one might calculate the optimal viewing distance of a painting in order to see the depth that the artist intended, examine the surface of the painting to appreciate the finer details and glazes, or stand back to appreciate the overall effect and balance of colors. Real-life situations are not divided the way they are in textbooks by their applicability to a certain topic or technique, like exponential models. In fact, throughout the twentieth century, employers, such as engineering firms, complained about the lack of connections made in school between different subjects. Mathematician Eliakim Moore discussed this problem a century ago in his 1902 address as the president of the American Mathematical Society. In 1989, the National Council of Teachers of Mathematics published a set of national standards for mathematics that included connections as a significant component.

Whereas traditional mathematics curricula in the twentieth century separated subject areas like algebra and geometry, an integrated approach involves presenting mathematical subjects as one interrelated whole that also connects to other subjects and real-world experiences. In antiquity, the square of a number was defined as the area of a square with the same side length. People with interdisciplinary interests were perhaps more common at that time—Greek mathematicians were also astronomers, inventors, engineers, and philosophers. Throughout history, mathematicians such as Carl Friedrich Gauss contributed to so many areas of mathematics and to other fields, like geodesy; but in the twenty-first century, researchers who specialize in a subdiscipline are more common. However, connections among multiple mathematical perspectives are still important in the development of mathemat-

Mathematics and the FIFA World Cup Finals

Mathematics is also used in one of the most popular sporting events in the world: the FIFA World Cup Finals. Thirty-two teams qualify for the World Cup Finals, and they are assigned to eight groups of four teams. The top seven teams in the world and the host country's team are seeded based on the FIFA World Rankings and recent World Cup performances and put into the eight different groups. The rest of the teams are put into different pots based on their geographical location, and then teams from each pot are randomly assigned to the eight groups. In the group stage of the World Cup Finals, each team plays every other team in its group.

A team earns three points for a win, one point for a tie, and zero points for a loss. In each group, the two teams with the most points advance to the knockout stage. If teams are equal in points, the greatest goal difference, the greatest number of goals scored, and other various statistics can be used to determine the top two teams to advance to the next stage. Sixteen teams advance to the knockout stage, which is a single-elimination tournament. At the end of the tournament, FIFA crowns one World Cup champion, as well as several individual awards, such as the Golden Ball for the best player and the Golden Boot for the top goal scorer. The winner of the Golden Ball award is based on a vote of media members. The Golden Boot award is given to the player with the greatest number of goals scored, as well as with the greatest number of assists. Mathematics is used in important calculations of sports statistics, college and world rankings, tournament rankings, and awards for individual performances.

ics. Algebra and geometry remain linked and the field of algebraic geometry is active today. Many researchers use techniques from a variety of different mathematical fields. Geometers may heavily rely on concepts from analysis, linear algebra, number theory, or statistics, for example. Other researchers work in the intersection of fields like statistical analysis.

Mathematics can easily be connected to other scientific disciplines, like physics or biology. Mathematics is sometimes referred to as the "foundation" or "language" of science. However, there are many other types of links between mathematics and the sciences. Some researchers work on problems at the interface of mathematics and a scientific field, while others translate ideas from the sciences to solve problems in mathematics and vice versa. Scientific disciplines like physics are often referred to as partner disciplines for mathematics. Researchers have met for a conference named "Connections in Geometry and Physics" that explores the interdisciplinary facets. In geometry and physics there is a concept called a "connection," which is an operator that allows for comparison at different points in a space via parallel transport. Mathematics has been interwoven with physics since antiquity. There have also been historical linkages between mathematics and biology, but the interdisciplinary field of mathematical biology has grown rapidly in the early twenty-first century.

Students may have difficulty appreciating the importance of mathematics in nonscientific disciplines, but the connections between mathematics and subjects like business, art, music, or religion are multilayered and multifaceted. For example, mathematics has played a part in religious life since the earliest documented cultures. The ancient Mesopotamians, embracing a polytheistic faith, developed the time system we use today with bases of 60 (60 seconds make up a minute, and 60 minutes comprise an hour). Adherents of Christianity, Judaism, and Islam have all embraced elements of mathematics in the conceptualization of sacred time. Given the importance of religion today, this time is still of great value for humankind. Mathematics plays a key role in the calculation of religious celebrations around which many faiths flourish. The week and solar day provide a delineation of sacred days that are different from the others—Sunday for Christians, Saturday for Jews, and Friday for Muslims. In other ways, numeric or geometric symbolism plays a significant part of religious practice.

There are countless examples that highlight the importance of mathematics in daily tasks. In the twenty-first century, it is almost impossible to find a task that does not connect to mathematics, either directly or through the tools and technologies in which mathematics plays an important role. In turn, mathematicians formulate new theories and concepts in order to meet the needs of society.

Mathematics as a Universal Language

Many people consider mathematics as the only truly universal language, regardless of gender, culture, or religion. For example, while the precise number of digits that are used in applications may differ, the ratio of the circumference to the diameter of a circle is still π, irrespective of the cultural context. Calculating the cost of groceries involves the same mathematical processes whether one is paying for those groceries in dollars, pesos, or pounds. With the universal language of mathematics, regardless of the unit of exchange, humans are likely to arrive at similar mathematical results. In fact, there are many examples of researchers in different areas of the world who independently arrived at the same theorems. Thus, mathematics as a universal language provides a common ground, creating the capacity for human beings to connect to one another across continents and across time.

Nutrition Labeling

An important way that mathematics can be found in our everyday life is on nutrition facts panels, which are mandated by the Nutrition Labeling and Education Act of 1990 to be placed on nearly all multiple-ingredient foods. The nutrition facts label on foods must list the fat, saturated fat, trans fat, cholesterol, sodium, total carbohydrate, fiber, sugar, protein, Vitamin A, Vitamin C, calcium, and iron content of the food. Other nutrients may be listed voluntarily. These labels also include a column that lists the percent Daily Value (% DV) to help consumers decide whether the nutrient content of a serving of the food product is a lot or a little. Mathematics is used to calculate the calories per serving and the % DV of a serving listed on the nutrition facts label.

As shown in Figure 1, at the top of the nutrition facts label, the serving size, as well as the number of servings per container, is listed directly underneath "Nutrition Facts." In this case, a serving size is ½ cup and there are eight servings per container. This means that there are four cups (½ cup × eight servings = four cups) of food in this package. If a person consumed half the container, or two cups of food, he or she would have had four servings (the amount of food consumed divided by a serving size, or two cups divided by ½ cup per serving = four servings).

Next, the calories per serving and the calories from fat per serving are listed. In this food, there are 200 calories per serving and 130 calories from fat in one serving. If the person consumed four servings and there are 200 calories per serving, then he or she consumed 800 calories (four servings × 200 calories/serving = 800 calories). Similarly, this person consumed 520 calories from fat (four servings × 130 calories from fat/serving = 520 calories from fat).

Following the calorie content, the nutrition facts label also lists the number of grams of total fat, total carbohydrate, and protein, which are calorie-yielding nutrients. A gram of fat contains nine calories, which is listed at the very bottom of the label. In this food, a single serving contains 14 grams of fat, which yields 126 calories (14 grams of fat × 9 calories/gram of fat = 126 calories from fat). This calculation was done to create the number of calories from fat listed on the panel (they rounded up to 130). As previously mentioned, if a person ate four servings, he or she consumed about 520 calories from fat.

Figure 1. A Common Nutrition Facts Label.

Nutrition Facts
Serving Size 1/2 cup (about 82g)
Servings Per Container 8

Amount Per Serving

Calories 200 Calories from Fat 130

	% Daily Value*
Total Fat 14g	**22%**
Saturated Fat 9g	**45%**
Trans Fat 0g	
Cholesterol 55mg	**18%**
Sodium 40mg	**2%**
Total Carbohydrate 17g	**6%**
Dietary Fiber 1g	**4%**
Sugars 14g	
Protein 3g	

Vitamin A 10% • Vitamin C 0%
Calcium 10% • Iron 6%

*Percent Daily Values are based on a 2,000 calorie diet. Your daily values may be higher or lower depending on your calorie needs:

	Calories:	2,000	2,500
Total Fat	Less than	65g	80g
Saturated Fat	Less than	20g	25g
Cholesterol	Less than	300mg	300 mg
Sodium	Less than	2,400mg	2,400mg
Total Carbohydrate		300g	375g
Dietary Fiber		25g	30g

Calories per gram:
Fat 9 • Carbohydrate 4 • Protein 4

The number of calories from carbohydrates and proteins can also be calculated. Both carbohydrates and protein yield four calories per gram, which is also listed at the very bottom of the nutrition label. In this food, there are 17 grams of carbohydrates, which provides 68 calories (17 grams × 4 calories/gram = 68 calories). In four servings, a person would ingest about 272 calories from carbohydrates (68 calories/serving × 4 servings = 272 calories from carbohydrates). There are three grams of protein in one serving, which means there are 12 calories from protein in one serving (3 grams × 4 calories/gram = 12 calories) and 48 calories from protein in four servings (12 calories/serving × four servings = 48 calories).

On the right side of the nutrition facts panel, the % DV is also listed. These daily values are based on a 2000-calorie diet, which is stated on the label next to the asterisk. Near the bottom of the label, it lists the maximum number of grams or milligrams of total fat, saturated fat, cholesterol, or sodium that a person should consume per day if on a 2000-calorie diet. It also lists the number of grams of total carbohydrate and fiber a person should eat if on a 2000-calorie diet.

If there are 14 grams of fat in one serving of this food and a person on 2000-calorie diet should consume no more than 65 grams of fat per day, then one serving of this food yields 22% of a person's DV of fat (14 grams of fat/65 grams of fat = about 22%). If this person has consumed four servings, then he or she has eaten 88% of his or her DV of fat (22%/serving × four servings = 88%). The same calculations can be made for the saturated fat, cholesterol, sodium, total carbohydrate, and fiber. Similar calculations are also made for the vitamins listed on a nutrition facts panel.

As demonstrated, mathematics is used in the calculations surrounding calorie content and % DV on nutrition labels. The mathematics used can affect a person's choice of foods and, in turn, a person's health.

Sports

Mathematics is used in numerous other everyday activities, such as sports. It is common in popular sports to calculate statistics to measure performance. In baseball, a common statistic is a batting average. A batting average is a simple calculation: the number of "hits" divided by the number of "at bats." This statistic is used to estimate an individual's batting skills. In professional baseball, a batting average of .300 is considered an excellent batting average.

A similar statistic to the batting average is in volleyball, which is called a hitting percentage. However, it is slightly different because it tries to measure an individual's hitting or attacking skills and takes errors into accounts. It is calculated by taking the number of kills, subtracting the number of errors, and then dividing the difference by the number of attempts. A "kill" is when a hitter's attack results directly in a point (the ball falling into the opponent's area of the court, an opponent not being able to return the ball, or the opponent making a blocking error as a result of the attack). An "error" is when a player hits the ball and it goes into the net (does not cross to the opponent's side) or out of bounds. An "attempt" is anytime the player tries to attack the ball. For example, if a player had 10 kills, 3 errors, and 17 attempts, the player's hitting percentage would be about .412 ($(10-3)/17 = 0.412$), which would also be considered a good hitting percentage, similar to the guidelines to the batting average.

Mathematics is important in the calculation of college football Bowl Championship Series (BCS) rankings as well. A mathematical formula is used to calculate these rankings, which order the top 25 NCAA Division I-A football teams based on their performance during the prior week. At the end of the season, the top two teams play each other in the national championship bowl. Mathematical formulas are also used to calculate which teams will play in the other bowls, taking into consideration the conference the team comes from and how many fans and advertising dollars the team is likely to bring in as well.

More specifically, the main factors that go into these rankings are subjective polls, computer rankings, the difficulty of a team's schedule, and the number of losses. The subjective poll numbers come from the average of two rankings from the Associated Press (AP) and the *USA Today*/ESPN Coaches Poll Ratings. Sports writers and broadcasters vote in the AP poll and a select group of football coaches vote in the *USA Today*/ESPN Coaches poll on which football teams they think are the best, and then these two rankings are averaged. The computer rankings are based on eight different computer rankings that are calculated based on a team's statistics for that week (strength of the opponent, final score, win-loss record, and so forth). The strength of a team's schedule is based on a cumulative win-loss record of its opponents, as well as their opponent's opponents. The calculation of the number of losses is straightforward. Each loss that a team suffers corresponds to one point, which is added to its final score. Points from each category are assigned to the team, and then these values are added to create a team's final score. The team with the lowest point total is ranked "number one" in the rankings.

Speedometers

Mathematics is also used in cars. All cars have a speedometer, which is a device used to calculate an instantaneous speed of a vehicle. It is important for a driver to know the speed of the vehicle at all times to ensure the safety of passengers and pedestrians and to abide by local traffic laws. In the United States, speedometers are read in terms of miles per hour. The calculation of the speed of the vehicle requires significant mathematics.

In many vehicles, an eddy current or mechanical speedometer is used, which is the speedometer with a needle that points to the speed that the vehicle is travelling. In these cars, there is a drive cable that runs from the speedometer to the transmission, which has a gear that tracks the rotational speed of the wheels. In other words, the gear tracks the number of revolutions the wheel makes within a certain time frame. Digital speedometers calculate miles per hour slightly differently, using a vehicle speed sensor. The vehicle speed sensor is in the transmission and also tracks the rotations of the wheels. From this information, the vehicle's speed is calculated and displayed on either a digital screen or a traditional needle-and-dial display.

The calculation of a vehicle's speed is dependent on the size of the tire as well. For example, if the tire rotates x times per minute, then the vehicle's speed can be calculated in miles per hour. Knowing the diameter of the tire, the circumference of the tire can be calculated (diameter $\times$ π). Therefore, the vehicle travels the distance of the number of revolutions times the circumference of the tire, within a certain time frame. This ratio can then be converted to miles per hour by converting the units. Because all of these calculations are based on an assumed tire diameter and circumference, it is very important for drivers to ensure that the correct size tires are on their vehicle. If a car's wheels are too large or too small, the speedometer will read slower or faster than the vehicle's actual speed, which may lead to accidents, speeding tickets, or just slower driving.

Conclusion

Mathematics can be found in everyday situations that have a real and important effect on our lives. All areas of one's life are in some way connected to mathematical principles. Only a small number of examples have been presented here—the list can be expanded infinitely. In fact, one would be hard pressed, in today's technologically advanced world, to present even a handful of activities that do not involve some mathematical concepts, if even at the unconscious level. By bridging the disconnect between "school mathematics" and "real-life mathematics," individuals gain a greater appreciation for—and curiosity of—mathematical applications.

By viewing mathematics as an integrated whole and understanding its connectedness to society, individuals become active participants, rather than passive recipients, of information. When one becomes aware of mathematical connectedness, rather than viewing math as a series of isolated and disconnected concepts to be learned though rote memorization, an individual develops the understanding of mathematics as a crucial and meaningful tool that can aid in the understanding, predicting, and quantifying of the world around us.

Further Reading

Brookhart, Clint. *Go Figure: Using Math to Answer Everyday Imponderables.* Chicago: Contemporary Books, 1998.

Cuoco, Al. *Mathematical Connections.* Washington, DC: The Mathematical Association of America, 2005.

Garland, Trudi H., and Charity V. Kahn. *Math and Music: Harmonious Connections.* Palo Alto, CA: Dale Seymour Publications, 1995.

House, Peggy, and Arthur Coxford. *Connecting Mathematics Across the Curriculum.* Restin, VA: National Council of Teachers of Mathematics, 1995.

Martin, Hope. *Making Math Connections: Using Real World Applications With Middle School Students.* Thousand Oaks, CA: Sage, 2007.

Lee Anne Flagg
Matthew West
Kristi L. Stringer
Casey Borch

See Also: Mathematics, Applied; Mathematics, Utility of; Mathematics and Religion; Mathematics Research, Interdisciplinary; Painting.

Continuity

See *Limits and Continuity*

Contra and Square Dancing

Category: Arts, Music, and Entertainment.
Fields of Study: Communication; Geometry; Representations.
Summary: Square and contra dancing employ many mathematical principles, including symmetries and permutations.

Square dance is geometry and combinatorial mathematics in motion. A caller directs the dancers through a set of choreographed dance movements unique to each type of square dancing. The dancers are sorted and shuffled in a myriad of ways by the caller and then returned to their original positions. Not only do the participants create mathematical forms as they move, mathematics is used to analyze different aspects of square dancing and its related form, contra dancing. For example, graph theory, matrix theory, and group theory can be used to represent the various structures and symmetries. Mathematics has also been used to analyze optimal calling patterns depending on the specific combinations of movements in the dance. Square dancing is a popular pastime for many people with an interest in mathematics. Several colleges have square dancing clubs, such as the Square Roots at North Central College in Illinois. That college has also offered a course called "The Mathematics of Square Dancing," which combined advanced dance patterns with discussions of mathematics theory, including parallelogram or hexagon dancing.

The Basic Square

The basic square consists of four couples. A square is symmetric under rotations of 90, 180, 270, and 360 degrees. Some or all of the dancers in the basic square can rotate in a circular movement according to these symmetries. Including the mirror reflections about each of the two lines of symmetry passing through the

Decomposing Squares Into Columns and Lines

Besides arranging dancers in squares and circles, the caller can also arrange them into columns and lines. A column arrangement occurs when all the couples are aligned one behind the other. A caller can shuffle this arrangement into any of 24 possibilities. A column of dancers can then be bisected longitudinally into two lines or crosswise into two smaller squares. There are two kinds of lines: one in which all dancers face the same direction, and another (a wave) in which they alternate the direction they are facing.

center of the square and parallel to an edge, there are six different targets of movement for the dancers. Further, in respect to each male (m)-female (f) pair, there are 10 possible movements. Thus, f_1 could be directed to replace either f_2 or m_2, m_1 could replace either f_2 or m_2, or both f_1 and m_1 could replace f_2 and m_2. Since there are four pairs, there are 240 possible movements among the dancers ($6 \times 10 \times 4 = 240$). Dance is about movement and not positions; thus, dance movements are not transitive. A movement of f_1 to f_2 is not the same as a movement of f_2 to f_1, although the outcome is the same arrangement. The two cases differ in respect to who initiates the action and who must react to the other's actions.

Secondary Squares

Besides the basic square, several other squares are part of square dancing. First, each m-f pair is a square. Several calls direct the movements of these dancers relative to one another. Thus, in a Do-Si-Do, the two members dance a square around one another and return to their initial positions. Alternatively, the basic square can be divided into a square within which a circle is inscribed. Four of the dancers constitute the square, while the remaining dancers move inward so that they are contained by the larger square. These can then be instructed to move according to the four symmetries. This arrangement can be inverted. The pairs can move toward a center point and form the radii of a circle, while the square that contains the circle is implicit. Again, the four symmetries constrain these movements. Instead of being expanded, the square can be constricted. The larger square can be divided into two smaller squares, each with four dancers. The dancers can be instructed to form smaller squares with the pair on the right, the pair opposite, or the pair on the left.

Further Reading

Mathematical Association of America. "Square Dancing Takes a Mathematical Spin." http://mathdl.maa.org/mathDL/pa=mathNews&sa=view&newsId=230.

Mui, Wing. "Connections Between Contra Dancing and Mathematics." *Journal of Mathematics and the Arts* 4, no. 1 (2010).

MICHAEL K. GREEN

See Also: Ballet; Permutations and Combinations; Polygons; Similarity; Symmetry.

Cooking

Category: Arts, Music, and Entertainment.
Fields of Study: Number and Operations; Measurement.
Summary: A good cook must be able to compute conversions, costs, and measurements.

In his Renaissance play, *The Staple of News,* Ben Jonson likens a master cook to—among other things—a mathematician. Although many people would think this comparison is an exaggeration, the mathematical requirements placed on the modern cook are significant.

In the past, cooking skills were passed on orally and through apprenticeship from generation to generation; today, inexperienced cooks are expected to learn to cook from recipes, which consist of a list of measured ingredients followed by instructions that refer to temperatures, times, and possibly more esoteric measurements. In addition to being able to scale recipes, the cook in our global world encounters many interesting recipes from diverse cultural traditions, which use a variety of systems of measurement.

Cooks must also be able to plan healthy and cost-effective menus.

Measurement of Ingredients

In recipes written in the United States, quantities for both liquid and dry ingredients are often specified by volume, and are measured in terms of teaspoons, tablespoons, or cups, in which there are 3 teaspoons to 1 tablespoon, 16 tablespoons to 1 cup, and 2 cups to 1 pint. Special measuring cups are made that permit the leveling of dry ingredients to ensure precise measurement. For measuring liquid ingredients, different cups are used that have graduation marks down the side and a convenient pouring spout. Measuring spoons are used for smaller quantities of both liquid and dry ingredients. For an experienced cook, the quantities given in recipes serve as general indications; however, in baking, when certain chemical reactions are expected to be balanced, precision is needed.

For more consistent outcomes, quantities are specified by weight. Ingredient densities vary. For example, a cup of water weighs 8 ounces, whereas a cup of flour—depending on how it was scooped—weighs about 5 ounces. Tables to assist in conversion between weight and volume can be found on the Internet. There can be confusion with the word "ounces," which can refer to either weight or volume. Ounces used for dry ingredients refer to one-sixteenth of a pound. Ounces measuring liquid ingredients refer to either one-sixteenth of a pint or to one-twentieth of a pint, depending on what is being measured.

Modern recipes written outside the United States provide measurements in the metric system. Liquid ingredients are specified in liters (volume) while dry ingredients are specified in grams (mass). Since kitchen scales actually measure weight, most cooks view grams as measuring weight. One liter of water weighs approximately 1000 grams. A liter is 1000 cubic centimeters, or about 1.057 quarts. A kilogram, 1000 grams, is approximately 2.205 pounds. A deciliter is one-tenth of a liter and is often used for recipes designed for home use. The metric system—based on multiples of 10—is designed to simplify calculations and scaling of measurements and is becoming the preferred system for cooks.

Scaling a Recipe

Recipes often specify the number of portions that they produce. To alter the number of portions generated, the recipe is scaled. This involves multiplying the quantity of each ingredient by a scale factor. To double a recipe, the scale factor is 2, while to halve a recipe, the scale factor is 1/2. At times, a more complex scaling is required. For example, imagine a baker is following a recipe that calls for 125 grams of pre-fermented dough. The recipe to make pre-fermented dough calls for 1000 grams of flour, 10 grams of yeast, and 0.6 liters of water and results in 1610 grams of dough. Since only 125 grams of pre-fermented dough are needed, the required scale factor is $125/1610 = 0.078$.

A naïve scaling results in 78 grams of flour, 47 grams of water (.047 liters), and the absurdly small amount (0.78 grams) of yeast. An experienced cook would add more yeast. Most recipes written for home use can only be scaled up or down by less than a factor of 4. Additionally, some ingredients, like spices, gelatin, and leavening agents, should not be scaled proportionately. Most good general cookbooks will give advice on scaling recipes. A good collection of professional recipes for large numbers of portions is available from the Armed Forces Recipe Service.

Measuring Temperature

Controlling temperatures on most modern stovetops is easier than doing so on wood-burning stoves. However, techniques vary significantly among gas, electric, and induction cookers and are best described by the manufacturer. In some instances, such as deep fat frying or candy making, temperature on the stovetop is measured by a thermometer. In making candy sugar syrup, temperature can also be measured by "feel" or by the way a drop of the syrup interacts with cold water. Books on making candy describe the relationships among these methods. The temperature of an oven is accurately monitored by a thermostat, which can be set. Often, an oven thermometer is also used to check the oven thermostat. Most recipes give the required temperature in either Fahrenheit or Celsius (previously called centigrade). The formula for converting from Fahrenheit to Celsius is given by

$$C = \frac{5}{9}(F - 32)$$

and from Celsius to Fahrenheit by

$$F = \frac{9}{5}(C + 32).$$

Thus, an oven temperature of 350 degrees Fahrenheit is about 177 degrees Celsius. Temperatures in some older British recipes are given in gas mark settings, in some older French recipes in numbered settings, in some older German recipes as Stufe settings, and in some much older recipes as verbal descriptions such as Very Slow or *Doux*. Tables showing conversions among these various approaches to measuring temperature can be found in general cookbooks and on the Internet.

Other Important Measurements

Other important quantities that need to be measured when cooking include time, acidity, and density. Time measured in seconds, minutes, and hours—a system based on 60—is now probably universal. Because estimating the passage of time is fraught with error, early recipes specified important times "as measured by the clock." Acidity is measured on the pH scale. Water, which is neutral, has a pH of 7. An acidic solution, like orange juice, might have a pH of 3, while a basic solution of baking soda in water might have a pH of 9. In home cheese making, the conversion of lactose to lactic acid is tracked by monitoring pH levels of the milk; however, traditional cheese makers will use the Dornic scale.

Measuring the density of a solution is important in wine and beer making, and in candying fruits. For example, the density of fresh grape juice indicates the ripeness of the grapes and the alcohol content of the finished wine. Candying fruit in sugar water can take many days. The daily gradual increase of sugar in the syrup where the fruit is steeping maximizes the amount of sugar absorbed by the fruit. The density of the syrup is carefully checked to ensure the correct increase of sweetness. Density of syrups is measured with a hydrometer, and a variety of scales, including Brix, Baumé, and specific gravity, have been used in

Molecular gastronomy is a new trend in cooking with a scientific slant. A chef plates a dish called strawberry ravioli created using reverse spherification, and places so-called caviar spheres of sauce with chop-sticks.

recipes. Although older French recipes will refer to the Baumé scale, since the 1960s, most recipes have used specific gravity. For syrups that are denser than water, a simple approximate conversion from Baumé to specific gravity (sg) is given by:

$$sg = \frac{145}{145 - {}^\circ B}.$$

Menu Planning and Budgeting

Cost and nutrition are also important factors for cooks. Many modern recipes, in addition to giving calories per serving, will give grams of carbohydrates, protein, fat, cholesterol, sodium, and calcium. This information, along with labels on prepared food, helps guide the cook in making nutritional choices. A cook might also be interested in knowing the cost of a portion size. For example, consider a portion of boneless chicken breast. The cost as purchased is what the chicken breast with bone costs per pound. Once the breast has been boned, what remains weighs less and results in a higher cost per pound of the edible portion. During cooking, the breast will shrink, resulting in an even higher cost per pound of the breast as served. Being aware of these costs, along with labor costs and inventory costs, helps the cook determine the cost of each item served. Although the home cook probably does not go through all these computations, a good home cook will have an idea of monthly food expenditures and how these costs are distributed among the various kinds of food served.

Further Reading

Bilheux, R., and Alain Escoffier. *Creams, Confections, and Finished Desserts.* Hoboken, NJ: Wiley, 1998.

Haines, R. G. *Math Principles for Food Service Occupations.* 3rd ed. Albany, NY: Delmar Publishers, 1996.

Jones, T. *Culinary Calculations: Simplified Math for Culinary Professionals.* Hoboken, NJ: Wiley, 2004.

Labensky, S. R. *Applied Math for Food Service.* Upper Saddle River, NJ: Prentice-Hall, 1998.

Reinhart, P. *The Bread Baker's Apprentice: Mastering the Art of Extraordinary Bread.* Berkeley, CA: Ten Speed Press, 2001.

Carl R. Seaquist
Catherine C. Galley

See Also: Connections in Society; Measurement, Systems of; Number and Operations in Society.

Coordinate Geometry

Category: History and Development of Curricular Concepts.
Fields of Study: Algebra; Communication; Connections; Geometry.
Summary: The development of coordinate geometry revolutionized mathematics, has a wide variety of applications, and is now widely used in many areas of mathematics.

The discovery that plane geometric configurations could be entirely described by real number pairs and two-variable equations revolutionized geometry and many other important fields of mathematics that emerged later, including real analysis, vectors, calculus, linear algebra, and matrix theory. Also referred to as "analytic geometry" or "Cartesian geometry," named for the great philosopher and mathematician René Descartes, the subject of coordinate geometry is the study of geometry using the Cartesian coordinate system with algebraic operations. In twenty-first century classrooms, children in primary school begin to examine coordinate systems and create plots on graph paper.

The level of sophistication of knowledge builds through high school and college through the use of various coordinate systems including Cartesian, polar, and spherical systems and by representations in two- and three-dimensional geometry. Some calculus courses are titled "Calculus and Analytic Geometry." Various coordinate system standards are in use in physics or mathematics, for surveyors, or at the state or company level. High school and college students learn to convert between some of these representations. Coordinate geometry has many applications and is used in every conceivable area of mathematics, science, and engineering to calculate precise locations and boundaries, distances and bearings from reference points, and to define graphs and curves using a point location, radius, and arc-lengths.

The fundamental building block of coordinate geometry is the Cartesian coordinate system, which includes an infinite collection of points on a plane determined by an ordered pair of numerical coordinates (x, y). The x-coordinate (called "abscissa") represents the horizontal position, and the y-coordinate (called "ordinate") represents the vertical position. These positions can be expressed as signed distances from the origin $(0, 0)$, a point that is at the intersection of two perpendicular reference lines called the "coordinate axis" (see Figure 1).

Once points are determined by ordered pairs (x, y) on the coordinate plane, one can then obtain analytic formulas for various geometric quantities on the plane. For example, an application of the Pythagorean theorem then yields the distance between any two points (x_1, y_1) and (x_2, y_2) given by

$$d = \sqrt{(x_2 - x_1)^2 + (y_2 - y_1)^2}.$$

Figure 1.

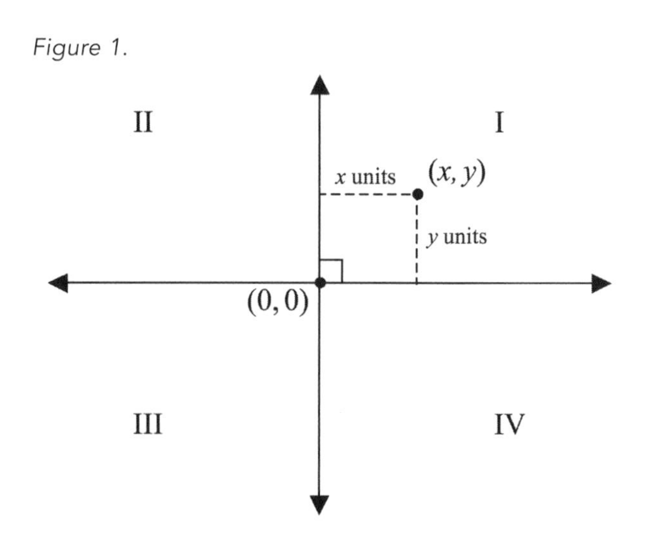

Early Variations

Coordinate-like types of systems arose in cartography well before Descartes. Maps with grids date back to ancient times, including those by Dicaearchus of Messana and Eratosthenes of Cyrene. Claudius Ptolemy attempted to create coordinates of well-known places in the world, essentially their latitude and longitude, from spherical projections, although the astronomical and mathematical methods to accurately calculate these would not be completely developed until much later. Islamic Mathematicians in the medieval Islamic world, such as Abu Arrayhan Muhammad ibn Ahmad al-Biruni, who compared the work of Ptolemy and Abu Ja'far Muhammad ibn Musa Al-Khwarizmi, provided coordinates for more than 600 geographical locations. Al-Biruni also used rectangular coordinates to represent three-dimensional space as well as ideas that some consider as a precursor to polar coordinates. In the twenty-first century, the global positioning system calculates the coordinates of a user from a system of satellites.

Other aspects of coordinate geometry can also be found in various early contexts. Some have noted that the mathematical work of ancient Greek mathematician Menaechmus could be interpreted as one that used coordinates. However, there was no algebra in ancient Greece, and others have highlighted the challenge that mathematics historians face in judging historical works. Coordinate geometry is a natural leap for the historians but probably not for Menaechmus, critics assert. Graphing techniques were developed in the fourteenth century in publications of Nicole d'Oresme and a work titled *De latitudinibus formarum* (The Latitudes of Forms), which some attribute to d'Oresme. Others assert that this attribution is an error and that the author is unknown. These works may have influenced coordinate geometers.

Transformations of coordinate-like systems developed along with perspective drawing techniques of curves and shapes, like in the works of Leone Battista Alberti and Piero della Francesca. Polar coordinates were motivated through the work of mathematicians such as Bonaventura Cavalieri on spiral curves like the Archimedean spiral, named for Archimedes of Syracuse.

Development

Descartes and Pierre de Fermat are both credited with independently introducing coordinate geometry. They each introduced a type of single-axis system or ordinate geometry. Distances could be measured at a fixed angle to the reference line. In Fermat's work, curves are generated as loci rather than by plotting points. Historian of science Michael Sean Mahoney noted: "There is connected with the system an intuitive sense of motion or flow wholly in keeping with the intuition which underlies the notion of an algebraic variable." Descartes' published work on coordinate geometry dates to 1637 in the appendix (*La Géométrie*) of a short book entitled *Discourse on the Method*. Descartes defined the five algebraic operations of addition, subtraction, multiplication, division, and extraction of square roots as geometric constructions on line segments and showed how these operations could be performed in the Euclidean plane by straightedge-and-compass constructions. He also developed geometric techniques for solving polynomial equations by intersecting curves, such as conic sections, with each other or with lines to obtain solutions algebraically. Coordinate geometry helps to classify conic sections, which are curves corresponding to the general quadratic equation

$$ax^2 + bxy + cy^2 + dx + ey + f = 0$$

where a, b, c, d, e, and f are constants and a, b, and c are not all zero. Coordinate geometry became useful in a wide variety of mathematical and physical situations. Sir Isaac Newton and others investigated various coordinate systems as well as how to convert between them. In the nineteenth century, Christof Gudermann investigated the sphere, and Julius Plücker published numerous volumes on analytic geometry.

Variations

In situations where there is no obvious origin or reference axes, mathematicians developed local coordinates or coordinate-free approaches. For instance, the Frenet–Serret frame is named for Jean Frédéric Frenet and Joseph Serret. It is a type of coordinate axis system for a curve in three-dimensional space and represents the twists and turns of a curve as three vectors that move along the curve. Jean-Gaston Darboux explored the analog for a surface.

Another example is "isothermal coordinates" on surfaces in the work of mathematicians, like Carl Friedrich Gauss. Engineer, mathematician, and physicist Gabriel Lamé is noted as the first to use the term in his 1833 work on heat transfer. August Möbius introduced barycentric coordinates, which utilizes notions related to the center of mass and the centroid of a triangle, and these coordinates can be found in computer graphics. Möbius' work used both the position and magnitude.

Other mathematicians developed similar systems, including vectors, which allowed for compact notation. Hermann Grassmann and William Hamilton created the algebra of vectors. The development of vectors was especially useful when extending the geometry or physics to higher dimensions. A point (x, y) in the plane can also be represented by a vector as $r = x\hat{\imath} + y\hat{\jmath}$ where $\hat{\imath}$ and $\hat{\jmath}$ are unit vectors. Mathematicians including Jean-Victor Poncelet and Michel Chasles developed synthetic projective geometry, which focused on axioms instead of coordinates. Gregorio Ricci-Curbastro and Tullio Levi-Civita explored a coordinate-independent calculus, which led to the development of tensor analysis that later became important in general relativity. Bernhard Riemann's work on geodesics and Riemannian geometry led to geodesic coordinates, which also became important in relativity.

Education

Coordinate geometry took on an increased prominence in schools in the nineteenth and twentieth centuries. One reason was the development and curricular use of graph paper. A patent for printed graph paper dates back to Dr. Buston in the late eighteenth century. Graph paper makes it easier to plot points and create curves, and it was found to be useful in surveying and civil engineering projects.

Mathematicians in the nineteenth century, like E. H. Moore, advocated the use of paper with "squared lines" in algebra classes. Coordinate geometry topics were also included in algebra textbooks and in textbooks devoted to the subject.

One notable textbook was published by Scottish mathematician Robert J. T. Bell in 1910. His treatise on coordinate geometry in three dimensions became a very successful textbook on the subject and was translated into numerous languages.

Further Reading

Boyer, Carl B. *History of Analytic Geometry*. New York: Dover Books on Mathematics, 2004.

Mahoney, Michael Sean. *The Mathematical Career of Pierre de Fermat, 1601–1665*. 2nd ed. Princeton, NJ: Princeton University Press, 1994.

Sasaki, C. *Descartes's Mathematical Thought*. New York: Springer, 2010.

Padmanabhan Seshaiyer

See Also: Algebra and Algebra Education; Equations, Polar; Geometry and Geometry Education; Geometry in Society; GPS; Graphs; Maps; Relativity.

Coral Reefs

Category: Weather, Nature, and Environment.
Fields of Study: Connections; Data Analysis and Probability; Geometry.
Summary: Mathematics helps describe and explain the formation of coral reefs.

Coral reefs are complex stony structures made of exoskeletons of coral polyps. Colonies of polyps form corals, with their stony parts consisting of calcium carbonate. All polyps in a single coral are genetically identical. Polyps get their energy from photosynthesis of their internal symbionts, one-cell algae living in the polyps. Some corals also have stinging tentacles for catching plankton, and can be painful for people to touch. The development and growth of coral reefs and atolls was fiercely debated in the nineteenth and early twentieth centuries. Charles Darwin argued in his 1842 publication *Structure and Distribution of Coral Reefs*, based on his personal observations, that the geometry of coral reefs resulted from the natural geological subsidence of oceanic islands.

In other words, coral reefs formed around islands, growing as the islands sank away. Darwin's chief opponent in this debate was Alexander Agassiz, who advocated the theory that coral reefs were not wholly dependent on subsistence for their formation but rather arose from a variety of geological and biological factors. Agassiz collected data from nearly every coral reef on Earth before his death in 1910, but none of his research had been published at that time. Contemporaries of both Darwin and Agassiz were inhibited by the inability to collect data other than observations and relatively shallow rock samples. In the 1950s, geologist Harry Ladd conducted tests in conjunction with the U.S. War Department, including boring thousands of holes in the coral of Eniwetok Atoll. Ladd's drill went to a depth of nearly 5,000 feet before finally passing completely through the coral into the soil below, confirming in many scientists' minds that the atoll had been built up as the land had sunk away. Ladd purportedly erected a sign on Eniwetok that read, "Darwin was right!"

Measurements and Variables

The shape of a coral reef is determined by the sea floor and the historical changes in sea levels. Reef scientists recognize three main shape types: fringing reefs, barrier reefs, and atolls. Fringing reefs stay close to shores, and their shape is determined by the shore they circle. Barrier reefs start as fringing reefs, but as the water levels rise relative to the shore, there are deep, large lagoons separating the shore and the reef. When volcanic islands completely subside underwater, their fringing or barrier reefs can stay near the surface, forming a circular lagoon. Such reefs are called "atolls."

Reefs need clear waters for photosynthesis and can be modeled as interesting hyperbolic structures.

In most places, sea levels rise over the land. The speed of reef growth depends on multiple variables, including temperature, water salinity, water clarity necessary for photosynthesis, and wave action. Reefs can grow up to 25 centimeters (about 10 inches) per year in height. Reefs cannot grow faster than sea levels rise, because the polyps can survive out of water only for a short time—for example, during the low tide. When the speed of reef growth matches the rise of the sea level, they are called "keep-up reefs." When the speed of reef growth is slower than the rise of the sea level temporarily, reefs may become either "catch-up reefs" when the speeds eventually match, or "drowned out" reefs that die as they are submerged too deeply. Global warming threatens to increase the rate at which sea levels rise beyond the speed of reef growth.

Because reefs need clear waters for photosynthesis, they grow in the parts of the ocean that are relatively nutrient-poor. However, reefs themselves support rich and diverse ecosystems—the contradiction called "Darwin's paradox." Reefs underlie less than 1% of the world's ocean beds but host about 25% of the marine species. They are called "underwater rainforests" because of their active biomass production, measured in weight per area per day.

Coral reefs have high fractal dimensions; in other words, their surface is rough, wrinkled, and uneven. This characteristic explains why corals thrive in moving waters. The fractal-like coral surfaces break the still water barrier surrounding them, with any agitation of water creating and amplifying turbulence. This turbulence means more water moves through the polyps, delivering nutrients to them and removing sediments that could prevent photosynthesis.

Mathematical Models

Coral reefs are vulnerable to storms, tsunamis, and other strong natural events. By modeling reef damage, it is possible to intervene, and to preserve some reefs that would otherwise be destroyed. Existing models include equations that measure the forces applied to reefs, and the forces reefs can withstand.

The ratio between the area of attachment of a reef and its total surface area plays a role in the models. The higher the surface area of the reef, the higher the pressure storms apply to it. On the other hand, the higher the area of attachment, the more force it takes to detach the reef. By modifying these variables, as well as the force of the storm, oceanologists can predict what happens to particular reefs. Moreover, with more computation power comes the opportunity to model detailed shapes of reefs, individual currents, and other local variables, making predictions more precise.

Dynamic systems of differential equations are the area of mathematics applicable to complex ecosystems such as coral reefs. More deterministic models such as algebraic or simple differential equations do not capture the reality as well.

Hyperbolic Crochet Coral Reef Project

The crocheted coral reef is a collaborative project with hundreds of contributors and several exhibits worldwide, and is coordinated by the Institute for Figuring. It demonstrates hyperbolic geometry, which is a non-Euclidian geometry discovered about 200 years ago and found in nature—including corals. "Hyperbolic crocheting," the process for modeling corals, was first described in the late 1990s. It involves a simple repeating algorithm with introduced "mutations" that produce varied forms.

The models explore mathematical entities that can be found in coral reefs, such as the hyperbolic radius of curvature, pseudospheres, hyperbolic planes, and geodesics.

Further Reading

Dobbs, David. *Reef Madness: Charles Darwin, Alexander Agassiz, and the Meaning of Coral.* New York: Pantheon Books, 2005.

Institute for Figuring. "Hyperbolic Crochet Coral Reef." http://crochetcoralreef.org.

Sale, Peter. *Coral Reef Fishes: Dynamics and Diversity in a Complex Ecosystem.* San Diego, CA: Academic Press, 2002.

Maria Droujkova

See Also: Animals; Crochet and Knitting; Geometry of the Universe; Surfaces; Tides and Waves; Transformations.

Counterintelligence

See *Intelligence and Counterintelligence*

Coupons and Rebates

Category: Business, Economics, and Marketing.
Fields of Study: Algebra; Measurement.
Summary: Mathematical differences between coupons and rebates provide different rewards to consumers.

Offering price reductions through coupons and rebates is a popular means of increasing the number of sales of a product, attracting customers to retail stores (both physical and online), and promoting public awareness of a brand name or product. One of the first known instances of a coupon was in 1894, when the Coca-Cola Company gave out handwritten tickets for samples of its new soft drink. The next year, Charles Post, of Post Cereal, started issuing coupons to help sell groceries. By the 1930s, these coupons were increasingly popular for saving money during the Great Depression. Some researchers claim that by the mid-1960s, half of American households used coupons.

In the twenty-first century, coupons are available on the Internet, or as permanent discount cards, in addition to their traditional paper form. In the age of online shopping, coupons for free shipping are cited by some as one of the most important factors in determining where to shop. For the customer, coupons and rebates bring savings on regularly purchased items and provide an incentive to try new products or services. In general, coupons are small discounts (a few dollars or cents) redeemed at the time of payment. The term "rebate" generally refers to larger reimbursements or discounts where the price reduction is either applied at the time of sale (an instant rebate) or reimbursed after required documents are mailed in by the customer. The specific type of coupon or rebate affects the calculation of both the discount and any applicable sales tax.

Coupons

Coupons may be issued by a product manufacturer or a store, and the redemption is somewhat different for the two types. When a customer presents a manufacturer coupon to a retailer, the customer pays any applicable tax on the full price of the item before the coupon is applied. For example, if a retailer charges $50 for a product with an 8% sales tax and a customer presents a manufacturer coupon for $10, then the cost to the customer at the time of purchase would be $44; that is, the original $50, plus sales tax of $4 ($50 × 0.08), minus the $10 coupon. Typically, the manufacturer reimburses the retailer for the amount of the coupon plus handling.

Sometimes a retailer like a grocery store, pizza restaurant, or automobile detailer will offer its own store coupons or rebates on its products or services. When a customer uses a store coupon, tax is computed on the balance after the coupon is deducted. If the $10 manufacturer's coupon is replaced by a $10 store coupon, the cost to the customer would be less: $43.20 versus $44.00.

Many retailers issue plastic cards that customers present to take advantage of weekly card specials or to receive a certain percentage discount on purchases made with the card. These cards not only allow shoppers to save money, but also the data collected when these cards and the associated purchases are scanned allow stores to better track their sales and inventory, and sometimes offer additional discounts tailored to a specific buyer's purchasing patterns. Sometimes these cards are free, but other times they require an initial or annual fee.

A retailer may offer a card at a cost of $10 that can be used for a 10% discount on all purchases at that store for one year. If a first-time customer checks out with a balance of $110 before tax, the customer can determine whether to purchase the card and take the 10% reduction. Although the card costs $10, the customer would save $11 on the initial balance (10% of the $110 total), resulting in a final cost of $109. The tax would be marginally less as well, since the total was reduced. Thus, the card would pay for itself at the first purchase, even before any other savings occur.

Another form of coupon is a card that is stamped each time the customer purchases a specified type of product, until a certain number of stamps are accrued. The customer then receives the next purchase of the specified product type free of charge, except for—possibly—sales tax. This form of coupon may be offered by certain restaurants, food markets, coffee shops, or bookstores.

Rebates

For a manufacturer rebate, an electronics retailer may sell a computer for $1,500, together with a free $100 printer after a mail-in rebate. The customer pays the tax on both the computer and the printer, and the manufacturer reimburses the customer $100 after the rebate is processed. With 8% sales tax, the cost to the customer after the rebate would be $1,628 (where the sales tax was 8% of $1,600, or $128).

Historically, economists have viewed consumer spending as a function of income. Politicians often cite this principle when pushing for tax rebates, believing they will increase consumption. However, there is little empirical evidence to support this notion, and in some cases there is contrary evidence. In 2001, the U.S. Congress enacted a tax rebate, giving $300 to anyone who had paid income taxes the previous year ($600 for couples). Economic indicators showed no associated increase in spending but rather a spike in saving. A survey of a sample of households that received a rebate reported that roughly one in five of those asked said they would spend the money. The *Wall Street Journal* ran the headline "Rebates Boost Incomes, But Not Spending." A study of the 2008 rebate found similar results.

Coupon Collector's Problem

There is a classic probability problem known as the Coupon Collector's Problem, which has been explored by a number of mathematicians, including the prolific Paul Erdos. The problem supposes that there is some number of different coupons (n) a person needs to collect to win a prize and asks how many coupons will he or she have to acquire, one at a time, to get a complete set. Usually, the coupons are equally likely to be drawn, and getting one of the n coupons does not prevent another of the same type from being drawn. Solutions to the problem can be found in a number of ways, including harmonic numbers, probability generating functions, and simulation. Extensions of the Coupon Collector's Problem are very useful in manufacturing quality control, for situations in which a number of product types must be sampled.

Further Reading

Better Business Bureau. "Mail-In Rebates: Now Available in Paper or Plastic." http://www.bbb.org/us/article/mail-in-rebates-now-available-in-paper-or-plastic-13249.

Spencer, K., and S. Rose. *How to Shop for Free: Shopping Secrets for Smart Women Who Love to Get Something for Nothing*. Philadelphia: Da Capo Press, 2010.

Barbara A. Shipman

See Also: Budgeting; Comparison Shopping; Market Research.

Credit Cards

Category: Business, Economics, and Marketing.
Fields of Study: Algebra; Number and Operations; Data Analysis and Probability.
Summary: Credit card issuers use mathematical models to determine credit lines and interest rates, as well as to detect fraud and analyze offers.

Credit card issuers use statistical analysis in a wide variety of ways. Statistical models of risk help the banks decide whom to approve for card membership and what interest rate to charge. Models also help issuers manage the risks of their existing customers and detect fraudulent transactions. Credit card issuers use designed experiments to help decide which offers have the largest potential to be profitable. Typically, the bank tries out the new offer on a sample of people (while leaving others in a control group) before deciding whether the new offer will be successful if given to the entire customer base. Data mining techniques help banks look at customers' past transactions in order to model future uses of the card and to help decide which customers are most likely to want which other products and services that the bank offers.

History

The first credit card was born when businessman Frank McNamara realized that he had forgotten his wallet at a New York City restaurant. After his wife rescued him by bringing cash to the restaurant, he vowed he would never face that embarrassment again. The Diners Club card was born a few months later in 1950 and became the first widespread alternative to cash. The first businesses honoring Diners Club purchases were charged 7% of each transaction (typical costs are now 2% to 5%), and subscribers were charged $3 per year.

Bank of America pioneered its BankAmericard program in Fresno, California, in 1958, and American Express issued the first plastic card in 1959. Carte Blanche was another early card. The idea of a credit "card" really gained momentum when a group of banks formed a joint venture to create a centralized system of payment. National BankAmericard, Inc. (NBI) took ownership of the credit card network in 1970 and for simplicity and marketability changed its name to Visa in 1976. (One reason for the name "Visa" is that it is pronounced nearly the same way in every language.)

Fraud Detection

Credit card banks use statistical algorithms to detect fraudulent use of credit cards. During the few seconds that it takes to approve or deny a credit card transaction at a merchant's site, information about the card is sent to a processing center. Typically at this point, only cards that are known to be stolen, fraudulent, delinquent, or other states that can be looked up will cause a denial. After the transaction has been approved, algorithms examine transactions to see if the pattern is suspicious. The cardholder may be contacted, usually by telephone, to verify that the transaction was made by the cardholder. The algorithms that identify a suspicious transaction can be quite sophisticated and are based on the past behavior of the cardholder.

That year, Visa processed 679,000 transactions—a volume that is processed on average every four minutes today. The Visa system is currently able to handle a load of about 6800 transactions per second, a capacity nearly exceeded on December 23, 2005, during the height of the Christmas shopping season. Visa is the largest merchant network, although MasterCard, American Express, and others process many transactions as well.

The Fair Isaacs Company (FICO) has grown in parallel with the credit card industry. It was founded in 1956 by mathematician Earl Isaac and engineer Bill Fair with the idea that data, used intelligently, can be used to make better business decisions. The next year, Conrad Hilton hired FICO to design and implement a complete billing system for his Carte Blanche card. FICO next developed the methodology to "score" the credit rating of customers but was unable to sell the idea to credit card banks until the 1970s. By the early 1990s, nearly every credit card bank was using some form of credit card scoring to help decide which customers to approve for credit and at what price. In 1995, both Fannie Mae and Freddie Mac, the two largest mortgage brokers in the United States, recommended using FICO scores for use in evaluating U.S. home mortgages. Today, U.S. citizens can access their various credit scores through online credit bureaus and, in fact, the U.S. government developed a policy allowing consumers to find out their scores once a year for free.

Credit Scoring

Credit bureaus use statistical analysis on past transactions, as well as income and other demographic information, to generate a credit score, usually referred to as a FICO score. This number is on an arbitrary scale that generally runs from 350 to 850 (with slight variations). The three main credit bureaus are Experian, TransUnion, and Equifax. Credit scores on the same individual may differ among the credit bureaus because of slight variations in the statistical model used to generate the number and slightly different data reported to the various bureaus. In all cases, the credit score is a prediction of how likely a borrower is to pay back the loan. For credit card companies, the score is used to decide both whether to issue the card, and what price (annual percentage rate) to charge on a balance that's carried over from month to month.

Data Mining

Credit card transactions, while vital to the running of the credit card bank, also contain information on the cardholder's spending patterns. These databases are very large, containing the records of tens of millions of customers, and dozens to hundreds of transactions per record. Using statistical models (often logistic regression models), banks can use these vast data repositories to identify the customers who are predicted to have the highest probability of enrolling for a new product or service. These offers may be made via a number of different channels. The offer may be given while the cardholder is calling a call center (800 number) with an issue concerning his or her card (in which case, the statistical algorithm will notify the operator that this customer should get the specific offer), by e-mail, by an outbound telemarketing call, by a targeted ad that pops up while the customer is visiting the issuer's Web site, or as direct marketing (so-called junk mail).

Experimental Design

To evaluate whether a new type of offer (the so-called "challenger") will be more effective (as measured by higher enrollment, revenue, profit, or other criteria) than the current offer (the "champion"), banks often use statistically designed experiments. The simplest such experiment is randomized at two levels, also known as a champion/challenger design. In this design, a sample is selected at random from the entire customer database. A proportion of those are chosen as the control group. They receive the current offer (the champion), and the rest are chosen to receive the challenger. The data are then collected, and the differences in response between the two groups are evaluated. The design can be complicated by blocking (stratification) on card type, region, income, or other demographic variables. Designs can be complicated by adding more factors, more levels, and by asymmetries introduced by infeasible treatment combinations. In the credit card industry, analysis is also complicated by the fact that one cardholder may be getting more than one experimental treatment (offer) simultaneously from different groups within the same organization and from different organizations. Capital One Bank claims to run upward of 40,000 such experiments a year on its cardholders.

Further Reading

Box, G. E. P., J. S. Hunter, and W. Hunter. *Statistics for Experimenters*. 2nd ed. Hoboken, NJ: Wiley Interscience, 2005.

McNamee, Mike. "Credit Card Revolutionary." *Stanford Business* 69, no. 3 (2001).

Paterson, Ken. "Credit Card Issuer Fraud Management." Mercator Advisory Group, 2008. http://www.sas.com/new/analyts/mercator_fraud_1208.pdf

Richard De Veaux

See Also: Accounting; Budgeting; Data Mining; FICO Score; Money.

Crime Scene Investigation

Category: Government, Politics, and History.
Fields of Study: Data Analysis and Probability; Geometry; Measurement; Number and Operations; Problem Solving.
Summary: Crime scene investigation uses sophisticated mathematical models to determine what events took place at a crime scene, based on the available physical evidence.

Crime scene investigation (CSI) is the rigorous preservation and documentation of physical evidence at a specific location related to a criminal event. Investigators meticulously collect and measure crime-related evidence for scientific and mathematical analysis, reconstruction, and courtroom presentation. Overall, crime scene investigation and reconstruction involve the application of basic mathematical formulas and equations, as well as physics, geometry, and analytical thinking. Applied mathematical procedures based on well-collected data produce accurate results that generate reliable evidence for presentation in a criminal trial. Analysis of bullets, blood patterns, fingerprints, vehicle skid marks, chemical traces, and other data yield quantitative results that are invaluable in finding, arresting, and convicting suspects.

According to mathematician Chris Budd, "Many of the mathematical techniques used by forensic scientists are similar to those used in medical imaging for brain tumors, oil prospecting and remote sensing by satellites.…It is remarkable how often ideas which might be thought of as pure mathematics often find very real and important applications."

Mathematical Modeling of Projectiles

An automated ballistic identification system (ABIS) is a computer system designed to capture, store, and compare digital images of bullets and cartridge casings. A scanner captures images of bullets and cartridges so that a mathematical algorithm can extract their unique shapes, marks, and striation patterns (signatures), which are compared to a vast database of stored images. Both wavelets and statistical correlation techniques play a role in these analyses. Forensic ballistics involves the study of a projectile in motion, from the time of shooting to the time of impact with the target. Mathematics is used to analyze and describe a projectile's path though both the air and any obstructions, such as a body, as well as the mechanical characteristics of the weapon that fired the projectile.

Unobstructed projectile motion through air is typically parabolic, but a bullet may trace a complex path if deflected or stopped by an object, which requires more advanced mathematics, such as fractional differential equations, to describe. Blood droplets are another sort of projectile found at crime scenes, and the blood spray patterns are analyzed with geometric and trigonometric methods to determine the point of origin and other crucial characteristics. Along with ballistics and blood spatter, precise wound descriptions, which are closely related to fields like surveying and topography, can be mathematically modeled to suggest the type of weapon or bullet most likely to have made the wound.

Locations and Relationships

The locations and relative relationships among the various pieces of evidence are also important in making sense of a crime scene. Precise measurements allow investigators to place every item of evidence in its original location with some degree of certainty. These may be represented in a two-dimensional diagram, or in a computer reconstruction that uses two- or three-dimensional representation. Newer laser technology can record distance very quickly and precisely, as well as compute height using trigonometry. Mathematical computer algorithms can then combine data from multiple measures of a single object, taken from many angles, to produce three-dimensional models with minimal error. Another example of imaging used to solve a famous ancient "murder mystery" is the case of King Tutankhamen. X-rays from the 1960s, which could only provide two-dimensional images, were inconclusive. However, using CAT scans, which can mathematically construct three-dimensional images, scientists concluded that the king probably died from an infection in a broken leg.

Probability

Though the phrase "innocent until proven guilty" is often heard in connection with criminal investigations, in many cases the available evidence allows only a statement of what probably happened versus absolute certainty. Homicide investigators must logically infer or deduce what transpired at the crime scene by using evidence to reconstruct events and by matching a crime scene's characteristics to other examples. They may hypothesize a timeline or scenario and then apply scientific analysis to verify or refute the sequence of events to a high degree of probability.

This process requires critical scientific thinking and logical analysis. Investigators may use controlled experimentation, such as firing several bullets from the same weapon to look for variations in the pattern. Increasingly, they can also use computerized reconstructions of crime scenes and data to manipulate critical variables and conduct multiple "what if" simulations to eliminate unlikely scenarios and narrow the set of possible suspects and causes. Probability also comes into play in DNA analysis, where results are given the form of probability matches, and scientific tests such as gunshot residue, which are not 100% accurate and may occasionally result in a false outcome.

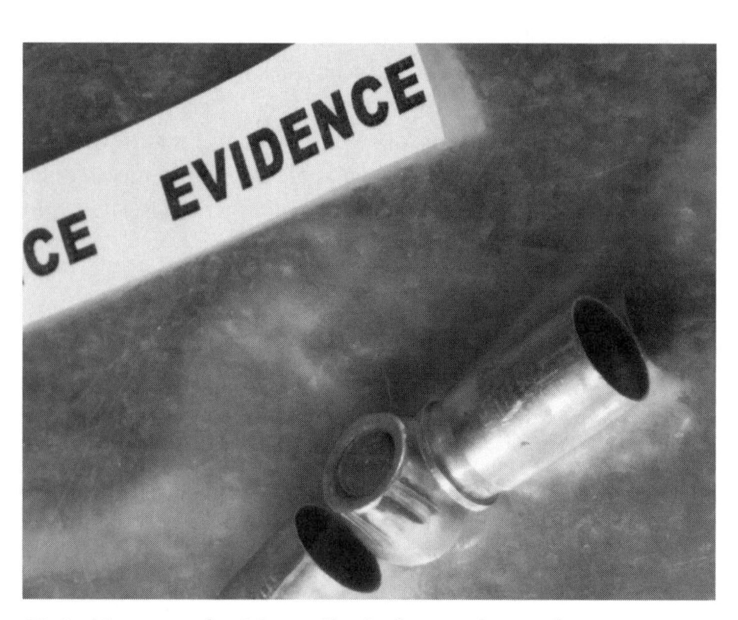

Digital images of evidence like bullets and cartridge casings are processed in an automated ballistic identification system (ABIS).

Conclusion

In summary, crime scene investigation requires investigators to apply many scientific and mathematical analyses to determine an accurate sequence of events and reconstruct what actually happened at a crime scene. Physical evidence helps investigators focus on a suspect and the manner in which the crime was committed. Successful crime scene investigations, reconstructions, and interpretations are the result of sound hypothesis formulation, experimentation, laboratory examination, and logical analysis. Applied mathematics provides the logic and rational simulations for scientific reasoning and assumptions.

Further Reading

Adam, C. *Essential Mathematics and Statistics for Forensic Science.* Hoboken, NJ: Wiley, 2010.

Budd, C. "Crime Fighting Maths." *+Plus Magazine* 37 (2005).

Devlin, Keith, and Gary Lorden. *The Numbers Behind NUMB3RS.* New York: Penguin, 2007.

Thomas E. Baker

See Also: Fingerprints; Intelligence and Counterintelligence; Mathematics, Applied; Measuring Time; Medical Imaging; Probability.

Crochet and Knitting

Category: Arts, Music, and Entertainment.
Fields of Study: Geometry; Measurement; Representations.
Summary: Crochet and knitting can be used to create models of mathematical surfaces.

Crochet and knitting are techniques for turning one-dimensional yarn or thread into two-dimensional fabric by knotting it in a regular pattern. Both produce flexible, elastic fabric, although crochet is firmer than knitting. Historically, crochet and knitting were used to produce both functional and ornamental textiles by hand, but both are now hobby pursuits.

Since both techniques produce regular arrays of stitches, they can be used to display a wide variety of symmetric patterns. Furthermore, both can be used to make intrinsically curved fabrics. This allows mathematicians and others to approximate or replicate the geometry of hard-to-visualize objects, including models of two-dimensional mathematical curved surfaces, such as spheres, tori, or sections of the hyperbolic plane. Crocheting and knitting circles have been held at professional mathematics conferences for both recreation and serious discussion of mathematical concepts. Mathematician Carolyn Yackel has noted, "Knitting and crocheting are helping us think about math we already know in a different light."

A hexagonal medallion made by crocheting in rounds. Crochet can be also worked in rows.

Crochet

In crochet, stitches are made by pulling loops of yarn through each other with a hook. One stitch is worked at a time. Every crochet stitch is attached at its base to an earlier stitch. Varying the type of stitch and the way new stitches are worked into earlier stitches can produce many different patterns. Crochet can be worked back and forth in rows or in circular rounds. Working two stitches into one base stitch increases the number of stitches and makes the fabric wider; decreasing the number of stitches reduces the width of the fabric. Placing increases or decreases at the edges of the work makes flat fabric with curved edges. Placing increases or decreases in the middle of the fabric makes it intrinsically curved.

The origins of crochet are not well understood. Few—if any—samples are known from before the nineteenth century. At that time, it was generally worked in fine cotton or linen thread and used for lace edgings,

Knit and purl stitches combined to create a basketweave pattern.

doilies, and other household textiles. From the middle of the twentieth century on, crochet has generally been worked in thicker yarn. It is often used to make blankets known as "afghans." The hobby of crocheting stuffed animals, known as "amigurumi," has spread around the world in recent years; because of the curved shape that these toys are crocheted in, they have few seams.

Several mathematicians have designed crocheted models of mathematical curved surfaces. As mathematician Daina Taimina has pointed out, it is especially simple to crochet negatively curved surfaces, such as a hyperbolic plane; the crocheter simply works an increase (an extra stitch) once every two or three (or n) stitches in every row. These increases cause the fabric to fold back on itself rather than lie flat. The closer together the increases are, the more ruffled the fabric.

The Hyperbolic Crochet Coral Reef, a project by the Institute for Figuring in Los Angeles, is intended to increase awareness of global warming issues by bringing together mathematicians, marine biologists, and community crafters in a highly visible way. The project asks volunteers to crochet models of coral reef life forms using Taimina's patterns. This effort and other mathematical crochet or knitting projects have been used successfully by mathematics educators in their classrooms.

Knitting

In knitting, as in crochet, stitches are made by pulling loops through each other. Knitting can also be worked in either rows or rounds. Two (or more) needles are used and many stitches are held on the needles simultaneously. The most basic stitches are "knit" and "purl" and there are techniques for increasing, decreasing, and making textural elements such as holes, cables, or bobbles. Knitting produces a flatter, stretchier fabric than crochet. (Indeed, most elastic fabric produced today is machine knitted.) As with crochet, increases and decreases allow the knitter to change the shape and curvature of the fabric. The shaping and elasticity make knitting ideal for garments such as socks, hats, gloves, and sweaters where both fit and comfort are important.

Hand knitting was once an important industry in Europe. Medieval guilds produced stunning garments for the wealthy in the Middle Ages, and a large cottage industry knitted stockings in the eighteenth and nineteenth centuries. Written patterns become available in the nineteenth century, and ornate knitting in fine thread became a popular pastime for ladies.

Hand knitting resurged in popularity in the first decade of the twenty-first century. Many current designers of garments and home textiles take their inspiration from mathematics, using symmetry and geometry to create attractive garments and household items.

Like crochet, knitting can be used to produce curved mathematical surfaces. Wide, soft, knitted Mobius bands are often knitted for use as scarves.

Further Reading

Belcastro, Sarah-Marie, and Carolyn Yackel, eds. *Making Mathematics With Needlework.* Wellesley, MA: A K Peters, 2008.

Bordhi, Cat. *A Treasury of Magical Knitting.* Friday Harbor, WA: Passing Paws, 2004.

Gaughan, Norah. *Knitting Nature: 39 Designs Inspired by Patterns in Nature.* New York: STC Craft/Melanie Falick Books, 2006.

Obaachan, Annie. *Amigurumi Animals: 15 Patterns and Dozens of Techniques for Creating Cute Crochet Creatures.* New York: St. Martin's Press, 2008.

Osinga, Hinke, and Bernd Krauskopf. "Crocheting the Lorenz Manifold." *Mathematical Intelligencer* 26, no. 4 (2004).

Taimina, Daina. *Crocheting Adventures With Hyperbolic Planes.* Wellesley, MA: A K Peters. 2009.

Elizabeth L. Wilmer

See Also: Crystallography; Escher, M.C.; Origami; Sculpture; Textiles.

Crosswords

See *Acrostics, Word Squares, and Crosswords*

Crystallography

Category: Weather, Nature, and Environment.
Fields of Study: Geometry; Measurement; Number and Operations; Representations.
Summary: Various mathematical principles are inherent in the structure of crystals and are used to study and classify them.

Crystallography is the study of the periodic structural arrangements of particles in solids. The first discoveries of the crystallographic structure of materials were made in the early twentieth century with the X-ray diffraction technique pioneered by Max van Laue. Solids that have crystal structures have a sharp melting point, which distinguishes them from amorphous substances, such as glass, which has neither a sharp melting point nor a crystal structure.

All matter tends to crystallize, since a crystal form is the lowest energy state. In reality, most physical crystals will have flaws rather than a perfect geometric structure. The chemical composition of a substance does not determine its crystal form. Calcareous spar, for example, has at least three distinct crystal types. Although crystals exist in three dimensions, some substances, such as graphite, form strong bonds between molecules in a plane, and only weak bonds between parallel planes. Mathematics is inherently connected to crystallography, as mathematicians describe and classify crystal structures and also use crystallographic methods to solve mathematical questions, such a packing problems. Despite almost a century of the existence of the modern science of crystallography, scientists do not have a good understanding of how local ordering principles produce large-scale order.

Lattices

The first consideration in crystal structure is the lattice, also known as the Bravais lattice, after August Bravais. There are 14 types of lattices. In a crystal structure, a translation is a motion in space in a certain direction through some distance. The arrangement of atoms, ions, and molecules must be periodic, and there must be three nonunique axes of translation. An axis of translation specifies a direction in which the structure repeats. If the whole structure is moved the proper distance in the direction of an axis, it will exactly cover itself. The lattice can be considered to be all the points to which any given particle can be translated by a translation, which also moves the entire crystal structure onto itself. Thus, the lattice consists of all the points that a given point or particle is moved to by a translation. From every point in the lattice, the view of the rest of the crystal is exactly the same. The portion of the crystal obtained by starting with a particle and moving it the smallest possible distance in each of the three translation directions is known as the unit cell.

Symmetries in Crystals

The geometry of a crystal structure is characterized by its symmetries. Besides translations, other symmetries include reflections in a plane, rotations through an angle about an axis, glide reflections (translation combined with a reflection), and screw translations (translation with a rotation). A crystal structure can only have rotations that are one-half, one-third, one-fourth, or one-sixth of a complete revolution. Mathematically speaking, two crystallographic structures are the same if their symmetries are the same. A collection of symmetries for an object is called a "symmetry group." Yevgraf Federov and Arthur Schoenflies, in the late 1800s, independently discovered that there are 230 distinct crystallographic symmetry groups in three-dimensional space.

Other Crystals

Wilson Bentley provided a wealth of insight into the structure of snow crystals using a photographic microscope, taking thousands of photographs of individual snowflakes over the course of 50 years. His photographs show that although snowflakes always have a basic hexagonal symmetry, they exhibit an endless variety of detail and seem to have a limitless number of forms. The simpler snowflakes grow slowly at high altitudes in low temperatures, and the more complex ones form at higher temperatures at greater humidity. Besides direct examination, information about the structure of snowflakes has been deduced by the forms of halos that they cause around the sun and moon.

In recent years, substances such as various aluminum alloys have been discovered to have regularity of structure but no translational symmetry. These substances are called "quasicrystals," and unlike true crystals, they can have 5-fold, 8-fold, 10-fold, or 12-fold rotational symmetry.

Further Reading

Bentley, W. A., and W. J. Humphries. *Snow Crystals*. New York: Dover Publications, 1962.

Burke, John G. *Origins of the Science of Crystals*. Berkeley: University of California Press, 1966.

Engels, Peter. *Geometric Crystallography*. Dordrecht, Holland: D. Reidel, 1986.

Kock, Elke, and Werner Fischer. "Mathematical Crystallography." http://www.staff.uni-marburg.de/~fischerw/mathcryst.htm.

Lord, Eric A., Alan L. Mackay, and S. Ranganathan. *New Geometries for New Materials*. Cambridge, England: Cambridge University Press, 2006.

Steven R. Edwards

See Also: Molecular Structure; Nanotechnology; Polyhedra; Symmetry.

Cubes and Cube Roots

Category: History and Development of Curricular Concepts.
Fields of Study: Algebra; Communication; Connections; Geometry.
Summary: Cubes and cube roots have been the subject of classical problems in mathematics, some of which were not solved for centuries.

Cubes and cube roots of numbers have played an important part in the development of mathematics. Middle school students are taught cubes and cube roots in order to solve equations and to calculate volumes of solids. In calculus, the cube root function is a common example of a function that is continuous everywhere but has an infinite derivative at one of its points. In addition, the cube function is an example of a function that is strictly increasing everywhere but has a point where the derivative is zero. The cube rule relates the percentage of popular vote in an election with the expected percentage of seats won in a two-party election. The power needed to overcome wind resistance is directly proportional to the cube of the wind speed. One model shows the heart rate in mammals is inversely proportional to the cube root of the weight of the animal. People in many different cultures have studied cubes and cube roots, and numerous interesting stories are found in its history. These simple objects have also generated many new ideas and new fields of mathematics.

Definition

To cube a particular number x, multiply it by itself 3 times—this is denoted x^3. If x is a number such that $x^3 = y$ for some other number y, then x is a cube root of y, written as $x = \sqrt[3]{y}$. Since $(-5)^3 = -5 \times -5 \times -5 = -125$, -5 is the cube root of -125, and the notation is

$$\sqrt[3]{-125} = -5.$$

The cube of any real number is unique; however, every real number has exactly one cube root that is a real number and two cube roots that are complex numbers.

Early History

As with squares, the earliest uses of cubes of numbers involved common geometric objects, specifically the cube, which is a three-dimensional object with six sides, all of which are congruent squares. The volume of a cube is the cube of the length of one of its sides. The volume of a sphere is directly proportional to the cube of its radius. One of the classical problems in Greek mathematics was the problem called "Duplication of the Cube." The problem was to find the length of an edge of a cube that has double the volume of a given cube using the tools of the time, the ruler and compass. It is now known that if x is the length of the side of the given cube, then $\sqrt[3]{2} \times x$ is the length of the cube with twice the volume. One possible origin of this problem is that, in 430 B.C.E., it was proclaimed through the oracle at Delos that the cubical altar to Apollo was to be doubled in volume in order to alleviate a plague that had befallen the people. Another possibility is that the Pythagoreans successfully doubled the square and doubling the cube was a natural extension. In any event, many great mathematicians throughout history worked on this problem, and, in the nineteenth century, it was proven that a solution was impossible.

Cube roots can be exact numbers if the cube root is an integer or a fraction. However, the cube root of most numbers is irrational (it has a infinite non-repeating decimal expansion) and its value can only be approximated. The easiest method to approximate the real cube root of a real number is to raise the number to the 1/3 power on a calculator. Obviously, the calculator is a recent invention, and other methods have been developed for approximating a cube root. Some of the earliest known methods are found in the Chinese text *Nine Chapters on the Mathematical Art* (c. first century C.E.) and in the book *Aryabhatiya* by the Indian mathematician Aryabhata (b. 476 C.E.). Both methods use the formula $(a+b)^3 = a^3 + 3a^2b + 3ab^2 + b^3$ repeatedly to generate the successive digits of the cube root. Approximations to cube roots can also be computed with the Chinese abacus, the *suànpán*, which dates to 200 B.C.E. In many cases, scribes would create tables of cube roots, which people would use to look up values for use. Barlow's Tables, named for mathematician Peter Barlow, who originally published in 1814, give the value of cubes and cube roots to nine decimal places and are still in print in the twenty-first century. Recreational mathematicians have found it fun to devise ways to compute cubes and approximations to cube roots in their head without outside assistance.

Cubic Equations

Cubic equations are equations that involve positive integer powers of *x* where the highest power is 3. Mathematicians have been trying to solve these equations from the earliest times. The Babylonian text *BM 85200* (c. 2000 B.C.E.) contains many problems that compute the volume of an excavated rectangular cellar by setting up and solving a cubic equation. Another Babylonian tablet contains, among other things, a table of integers and the sum of each integer's square and cube, and it was presumably used to solve cubic equations.

Archimedes of Syracuse (c. third century B.C.E.) considered the problem of passing a plane through a sphere such that the volumes of the two pieces had a certain ratio. This problem gives rise to what would be a cubic equation. A manuscript, thought to have been written by Archimedes, was found centuries later that gave a detailed solution to the problem that involved finding the intersection of a parabola with a hyperbola. Omar Khayyam (c. eleventh century C.E.) was the first to find a positive root of every cubic equation having one. Before this time, numbers were thought of as specific quantities of objects, so very little was done with negative numbers—and certainly not complex numbers. As with Archimedes, Khayyam's solutions involved intersecting conic sections.

By the fifteenth and sixteenth centuries, negative numbers and zero were accepted, and many of the Greek mathematical texts were translated to Latin. The field of algebra had been developed, and people could study equations as expressions with variables that can be manipulated (as is done in the twenty-first century). As the solution of the general quadratic equation had been discovered, Italian mathematicians focused their attention to the solution of the general cubic equation $ax^3 + bx^2 + cx + d = 0$. During this period, academic reputations and employment were based on public problem-solving challenges, and discoveries were kept secret so they could be used to win one of these challenges.

The solution of certain cubic equations provided the backdrop to one of the more entertaining chapters in the history of mathematics. On his deathbed in 1526, Italian mathematician Scipione del Ferro told one of his students, Antonio Maria Fior, how to solve a specific type of cubic equation. Nine years later, Fior submitted 30 cubic equations of this type to mathematician Niccolo Tartaglia in a public challenge. During the contest, Tartaglia himself discovered the solution and won the contest. After hearing of the contest, Girolamo Cardano contacted Tartaglia to inquire about his method. Tartaglia told him his solution, only after Cardano agreed to keep it secret as Tartaglia indicated he was going to publish it (thinking he was the first to discover it). Years later, Cardano found out that del Ferro actually discovered the formula and published it as del Ferro's method in addition to solutions to the cubic equation in all cases that he and his assistant, Lodovico Ferrari, discovered. Tartaglia was extremely angry and felt Cardano had broken his promise. In the twenty-first century, the formula for the solution of the cubic is known as the Cardan(o)–Tartaglia formula.

Uses and Applications

The cubic equation also played an essential role in the formulation of complex numbers. In his 1572 text, *Algebra*, Rafael Bombelli considered the equation $x^3 = 15x + 4$.

Applying the formula of Tartaglia and Cardano, one obtains a solution

$$x = \sqrt[3]{2+\sqrt{-121}} + \sqrt[3]{2-\sqrt{-121}}.$$

However, Bombelli knew the solution was actually 4 and that, somehow, the square root of −121 could be manipulated in a way to reduce this expression to 4. He developed an algebra for working with these roots of negative numbers (thought to be of no use to earlier mathematicians), and complex numbers and the field of complex analysis was born. With complex numbers, one can show that any cubic equation has exactly three solutions, two of which must be complex and one real.

In 1670, it was discovered that French mathematician Pierre De Fermat claimed that for all natural numbers $n > 2$ there are no nontrivial solutions of positive integers *a*, *b*, *c* such that $a^n + b^n = c^n$. Andrew Wiles proved this theorem in 1994 using objects called "elliptic curves." Elliptic curves are defined by a cubic equation of the form $y^2 = x^3 + ax + b$, whose graph has no cusps or self-intersections. These curves are studied in the twenty-first century and used in both number theory and cryptography (the study of coding information). Even though Fermat's equation has no positive integer solutions for $n = 3$, other problems involving sums of cubes have been studied.

In 1770, Edward Waring proposed the following question: for every positive natural number *k*, does there exist a natural number *s* such that every natural number *N* can be written as the sum of at most *s* numbers which are *k*th powers? If $k = 3$, the question becomes: can every positive number be written as a sum of at most *s* cubes? Some examples are $5 = 1^3 + 1^3 + 1^3 + 1^3 + 1^3$ and $23 = 2^3 + 2^3 + 1^3 + 1^3 + 1^3 + 1^3 + 1^3 + 1^3 + 1^3$.

As 23 shows, one requires at least 9 cubes. In 1909, David Hilbert proved that 9 is the maximum number of cubes that are required for any positive natural number. The Waring-Goldbach problem asks a similar question, except it requires at most *s* cubes of prime numbers. Some progress has been made, but this question remains unsolved as of 2010.

One of the more interesting recent mathematicians is Srinivasa Ramanujan from India (1887–1920). He was mostly self-educated and was able to prove theorems in number theory that shocked one of the eminent mathematicians of the time, G. H. Hardy. Once when Hardy visited Ramanujan, he mentioned he arrived in a cab numbered 1729, which did not seem very interesting. Ramanujan responded that 1729 is a very interesting number in that it is the smallest positive integer that can be represented by a sum of two cubes in two different ways, $1729 = 1^3 + 12^3 = 9^3 + 10^3$, which is correct. The taxicab numbers are generalizations of this idea. The *n*th taxicab number, denoted $Ta(n)$, is the smallest positive integer that can be written as two different cubes in *n* different ways. By Ramanujan's comment, $Ta(2) = 1729$. It is also true that $Ta(1) = 2$, since $2 = 1^3 + 1^3$ and $Ta(3) = 87{,}539{,}319$, since

$$\begin{aligned} 87{,}539{,}319 &= 167^3 + 436^3 = 228^3 + 423^3 \\ &= 255^3 + 414^3. \end{aligned}$$

The first 6 taxicab numbers are known, but $Ta(7)$ and beyond are all unknown as of 2010.

Further Reading

Burton D. *A History of Mathematics: An Introduction.* 7th ed. New York: McGraw-Hill, 2011.

Dunham, William. *Journey Through Genius*, New York: Penguin Books, 1991.

Katz, Victor. "The Roots of Complex Numbers." *Math Horizons* 3 (1995).

Washington, Lawrence. *Elliptic Curves: Number Theory and Cryptography*. 2nd ed. Boca Raton, FL: CRC Press, 2008.

Gregory Rhoads

See Also: Algebra and Algebra Education; Measurements, Volume; Numbers, Complex; Squares and Square Roots; Units of Volume.

Currency Exchange

Category: Business, Economics, and Marketing.
Fields of Study: Algebra; Measurement; Representations.
Summary: Mathematical models seek to price financial products in the foreign exchange market.

The term "currency exchange" refers to the business transaction that trades one currency for another. Such a transaction happens in the foreign exchange (FX) market and is measured by foreign exchange rates, which are often called exchange rates. Exchange rates fluctuate all the time. There are many factors that influence the movements of exchange rates. After all, foreign exchange rates are largely determined by the supply and demand in the FX market. Numerous mathematical models have been proposed by financial mathematicians and financial engineers to price different financial products in the FX market. Some of them have been used successfully by practitioners.

Exchange Rate Definition

There are many different currencies in the world. A measurement of the value of one currency in terms of another is called a (foreign) "exchange rate" or a "currency rate." In simple terms, an exchange rate of *K* currency *X* to currency *Y* means the value of *K* units of currency *X* is equivalent to the value of 1 unit of currency *Y*. It is often quoted as the price of currency X divided by currency Y is K. For example, the price of "euros/U.S. dollars is 1.3578" denotes an exchange rate of 1.3578 U.S. dollars to euros. In other words, it means the value of 1 euro is the same as that of 1.3578 U.S. dollars.

Types of Exchange Rates

A fixed exchange rate (also known as "pegged rate") means one currency is pegged to a major currency such as the U.S. dollar. Usually, the government or the central bank of a country will intervene in the market to peg its currency to a major currency to maintain a fixed exchange rate.

In contrast, a floating exchange rate is determined by the market forces of demand and supply.

Exchange Rate Fluctuation

Fluctuation of exchange rates, like fluctuation of stock prices, interest rates, and many other economic indices, is a ubiquitous phenomenon. Many factors drive the exchange rates up and down. These factors include but are not limited to capital flows, international trades, speculation, political factors, government or central bank intervention, and interest rates. However, the fundamental driving force is the invisible hand—the demand and supply—of the market.

Besides those quantifiable drivers of the FX market, there are other nonquantifiable ones such as the expectation of the investors. Attempts have been made by economists to account for those driving forces as well. Some economists have put the theory of exchange rate into a behavioral finance framework. Others used information theory and game theory.

FX Markets and FX Financial Products

The FX market is where the currency exchange happens, and is one of the largest financial markets in

Having exchange rates for national currencies allows us to consistently express the value of an item across borders of countries and cultures.

the world. Its major participants include commercial banks, investment banks, companies, investors, hedgers, speculators, traders, governments, and central banks. A variety of financial instruments are traded in the FX market, including currencies, currency forward contracts (also known as "FX forward contracts"), currency futures contracts (also known as "FX futures contracts"), currency options (also known as "FX options") and currency swaps (also known as "FX swaps"). Thus, the FX market has several important submarkets: the FX spot market, the FX forward market, the FX futures market, the FX options market, and the FX swaps market.

Although hundreds of financial products exist in the FX market, the basic ones are currencies, currency forward contracts, currency futures contracts, currency options, and currency swaps. Currencies are priced by the exchange rates. Both currency forward contracts and currency futures contracts are agreements made between two parties to exchange a specified amount of currency for a specified price at a specific future date. The main difference is that a currency forward contract is traded over the counter, whereas a currency futures contract is traded on an exchange. They both are financial derivatives. Their prices can be determined using simple algebra and are expressed in terms of exponential functions. Currency options and currency swaps are also financial derivatives. A currency call/put option gives one party the right—but not the obligation—to buy or sell a specific amount of the currency at a price (called "strike price") at a specific time in the future.

A European option can be exercised only at maturity, whereas an American option can be exercised at any time up to maturity. The cash flows of currency options are more complicated than those of the currency forward and currency futures contracts. The pricing requires sophisticated mathematical tools from stochastic calculus. Fisher Black, Myron Scholes, and Robert C. Merton made fundamental contributions in option pricing by giving the basic pricing formulas of European options. Scholes and Merton were awarded the Nobel Prize in Economics for this accomplishment in 1997 (Black was not awarded the prize because he had passed away).

A currency swap is an agreement between two parties to exchange the principal and interests of one currency at an interest rate for the principal and interests of another currency at another interest rate for a certain period of time. For example, suppose party A enters into a currency swap contract with party B today. For the next five years, party A will pay party B the interest of a principal of $1 million at an annual interest rate of 5%. In return, party B will pay party A the interest of a principal of 95 million Japanese yen at an annual interest rate of 4.5%. The two parties will also exchange the principals at the end of the fifth year. Like currency forward and currency futures contracts, the currency swap can also be priced using simple algebra.

Further Reading

De Grauwe, Paul. *The Exchange Rate in a Behavioral Finance Framework*. Princeton, NJ: Princeton University Press, 2006.

Driver, Rebecca, Peter Sinclair, and Christoph Thoenissen. *Exchange Rates, Capital Flows and Policy*. New York: Routledge, 2005.

Hull, John C. *Options, Futures and other Derivatives*. 7th ed. Upper Saddle River, NJ: Pearson Education, 2008.

McDonald, Robert L. *Derivatives Markets*. 2nd ed. Upper Saddle River, NJ: Pearson Education, 2006.

Rosenberg, Michael R. *Exchange Rate Determination: Models and Strategies for Exchange-Rate Forecasting*. New York: McGraw-Hill, 2003.

Weithers, Tim. *Foreign Exchange*. Hoboken, NJ: Wiley, 2006.

Liang Hong

See Also: Connections in Society; Money; Risk Management.

Curricula, International

Category: Mathematics Around the World.
Fields of Study: All.
Summary: Comparisons of mathematics curricula worldwide help facilitate growth and development.

A long history exists of comparisons between undergraduate mathematics curricula in other countries and the United States, and in recent decades, similar comparisons are being made at the primary and secondary levels. A recent movement in mathematics education has shifted the focus of how mathematics at all grade levels is taught. This movement was in large part spurred by the results of international testing. Since 1995, the Trends in International Mathematics and Science Study (TIMSS) has collected data on student achievement for fourth-, eighth-, and twelfth-grade students around the world.

The TIMSS was designed to allow for international comparisons, and has motivated educators to examine more closely those countries that consistently show success in educating students. Another international assessment, the Programme for International Student Assessment (PISA), focuses on measuring the mathematical literacy of 15-year-olds. The results of the PISA reflected those of the TIMSS, prompting educators in less successful nations to explore how some countries, such as Singapore, Japan, and Korea, educate students in mathematics. One area that has been explored as a result of the TIMSS and PISA is that of curricula. Liping Ma's 1999 book *Knowing and Teaching Elementary Mathematics*, which compared teaching methods in the United States and China, has also spurred numerous discussions about curricula and teaching methods, including teacher education and preparedness of teachers for presenting mathematical concepts at all levels.

It is important, first of all, that a distinction be made between curriculum and instructional programs. "Curriculum" is generally defined as a set of standards or objectives that guides what is taught at a particular age or grade level. "Instructional programs," on the other hand, are resources that are available to teach the curriculum, such as textbooks. On the international stage, a variety of instructional programs exist and are in use, but mathematics curricula across nations remain surprisingly similar.

An analysis of 16 countries' curricula conducted by Graham Ruddock demonstrated that different nations used the same basic mathematical principles as a foun-

Since 1995, the Trends in International Mathematics and Science Study (TIMSS) has collected international data on student achievement in mathematics for fourth-, eighth-, and twelfth-grade students.

dation for building mathematics curricula: number, algebra, geometry, measures, probability, and statistics. While some of the principles may be combined together into a single topic (for example, probability and statistics), these basic principles existed in the curricula of all nations that were studied. However, Ruddock pointed out that it is important to realize that just because nations use the same label, it does not mean that the content included in the principles is consistent across nations, nor does it mean that each nation explores each of the principles with equal rigor.

Nations also generally agree which principles of mathematics should be taught in the lower grades. Number is the primary focus for younger students, with a shift in focus toward algebra as students move into the middle grades. Nations vary widely in their mathematics curricula for upper grades, because of the nature of the different educational systems. For example, Japan uses an integrated approach to mathematics through the upper grades, where all principles are taught in varying degrees at all grade levels, while the United States utilizes a traditional division of mathematics topics (for example, algebra, geometry, calculus as separate courses).

Recent Pedagogical Changes

Interestingly, most nations at the beginning of the twenty-first century incorporate what is known as a "spiral curriculum," which is designed so that students revisit topics that were previously learned. This form of curriculum represents a shift in thinking in mathematics education that occurred during the 1990s. The purpose of the spiral curriculum is to assist students in making connections between mathematical ideas as well as ensure that students retain the knowledge that has been previously taught. A well-designed spiral curriculum is designed to encourage students to view mathematics as an integrated whole, rather than as discrete, unrelated topics.

An additional pedagogical shift has come as mathematics educators consider the value of conceptual understanding versus procedural understanding. Curricula in various nations have been adapted to include a stronger focus on the conceptual understanding of mathematics, rather than rote memorization and mastery of basic math skills. For example, curricula in Japan, Korea, and Singapore, all of which have consistently performed well on the TIMSS and PISA, have shifted from the learning of basic skills through rote memorization to an emphasis on problem solving and critical thinking. Curricula in other nations have followed this example.

National Mathematics Curricula

Some nations, such as England, France, Italy, and Japan, have required national mathematics curricula. Other nations, such as the United States, Australia, Canada, and Germany, view education as a local responsibility; therefore, a national mathematics curriculum does not exist. However, organizations such as the National Council of Teachers of Mathematics have developed national standards as suggested guidelines for what mathematics should be taught at different grade levels.

The greatest difference between nations regarding curricula is that of implementation. Curricula implementation varies widely among different nations, with some nations, like Hungary and Spain, placing a focus on local implementation while Japan has national guidelines for how teachers are to implement the curricula into their classrooms. From this variety of approaches comes the question of intended versus enacted curriculum. In other words, are teachers implementing the mathematics curriculum as it was designed? While the intended curricula across nations appear to have some strong similarities, especially at the lower grades, the enacted curricula may be quite different, thus resulting in substantial differences in student learning.

Current Trends in Curriculum Approaches

In recent years, the Singapore mathematics curriculum has garnered a great deal of attention because of the impressive performance of Singapore students on the TIMSS. The Singapore curriculum focuses on developing concept mastery through an in-depth exploration of a few mathematical topics each year. Also emphasized are the use of visual strategies in problem solving and establishing connections between mathematical topics. The Singapore mathematics curriculum has undergone a variety of changes since it was first developed in 1981, with the latest version including the introduction of calculators at a younger age and a reduction in emphasis on mental mathematics. Several countries, including the United States and Canada, have begun to implement curricula that mirror the Singapore mathematics curricula in the hopes of acquiring similar levels of student achievement on national and international assessments.

The International Baccalaureate (IB) Programme has also gained in popularity in recent years. The IB is designed to be a broad-based international curriculum, and is offered at three different levels: the Primary Years Programme (PYP), the Middle Years Programme (MYP), and the Diploma Programme (DP). While the IB does not focus specifically on mathematics, all three levels include mathematics as an integral part of the IB experience, as "mathematics is a universal language with diverse applications." Mathematics in the IB is viewed as a key connection to students' understanding of culture and history, and as a primary method of developing students' logic and critical thinking skills.

Since World War II, a growing number of foreign-educated students in mathematics and other related fields have chosen to attend graduate school or seek postdoctoral positions at American universities, with the largest growth occurring in the 1990s. For example, studies show that in 2002, nearly one-third of all graduate students enrolled at U.S. universities came from abroad. Many reasons are cited for this effect, including the quality of research universities, the availability of funding, and the existence of desirable job opportunities. A phenomenon colloquially known as "brain drain" reflects the significant migration of students with mathematical and technical skills away from their native countries, diminishing these countries' ability to compete in the global marketplace. In response, countries are beginning to expand their efforts to retain these students. For example, China has reorganized some current universities and built new ones, as well as engaged in significant curriculum reform. This reform includes new partnerships, such as a new Danish-Chinese University Centre for collaborative technology research, which was formalized in 2010.

Further Reading

Committee on Policy Implications of International Graduate Students and Postdoctoral Scholars in the United States. "Policy Implications of International Graduate Students and Postdoctoral Scholars in the United States." Washington, DC: The National Academies Press, 2005.

International Baccalaureate. "Academic Programmes." http://www.ibo.org/general/what.cfm.

Ma, Liping. *Knowing and Teaching Elementary Mathematics.* Mahwah, NJ: Lawrence Erlbaum Associates, 1999.

Mansfield, C. S., N. A. Pateman, and N. Bednarz. *Mathematics for Tomorrow's Young Children: International Perspectives on Curriculum.* Berlin: Springer, 2010.

National Center for Education Statistics. "Program for International Student Assessment (PISA)." http://nces.ed.gov/surveys/pisa.

National Center for Education Statistics. "Trends in International Mathematics and Science Study (TIMSS)." http://nces.ed.gov/timss.

Ruddock, G. "Mathematics in the School Curriculum: an International Perspective." http://inca.org.uk/pdf/maths_no_intro_98.pdf.

"The Singapore Math Story." http://www.singaporemath.com/Singapore_Math_Story_s/10.htm.

Calli A. Holaway

See Also: Connections in Society; Curriculum, K–12; Succeeding in Mathematics.

Curriculum, College

Category: School and Society.
Fields of Study: All.
Summary: Collegiate mathematics education is determined by the student's choices within the constraints of graduation and department requirements.

For thousands of years, mathematics has been considered an important part of a liberal arts education. Examples of this idea abound, including schools and scholars in ancient Greece, China, and the medieval Islamic world, as well as in the rise of North American colleges in the seventeenth century.

Debate has existed for decades about which topics should be a part of the college curriculum and how best to teach them. Common curricula, such as geometry, or educational tools, like the abacus, have been replaced by other focuses as societies' needs have changed and technology has advanced.

New discoveries in mathematics and emerging disciplines also result in curriculum changes. In the twenty-first century, the mathematics curriculum at the university level varies depending on the educational goals of the student. In the United States, the types and the number of mathematics courses required in the curriculum are typically based on a student's major subject of concentration. In this regard, there tend to be three broad categories into which a typical college student may be classified: a student who needs to fulfill a general education requirement in the mathematical sciences; a student majoring in a partner discipline, such as the physical sciences, the life sciences, computer science, engineering, economics, business, education, and the social sciences; and a student whose major is in the mathematical sciences, including pure (theoretical) or applied mathematics, statistics, actuarial sciences, and mathematics education. At most colleges and universities, curriculum is approved by both internal governing bodies, such as curriculum committees, and external accrediting agencies. Local, national, and specialized accrediting agencies may approve programs at the department or college level.

History

There is a rich history of mathematics in higher education contexts. From the schools of ancient Greece to the universities of the Renaissance, mathematics was an important component of the seven liberal arts, and mathematics was seen as a way to understand reality. Three of the liberal arts, the Trivium, consisted of grammar, rhetoric, and logic. In the Quadrivium, the other four liberal arts, arithmetic was the study of numbers, geometry was the study of numbers in space, music was the study of numbers in time, and astronomy or cosmology was the study of numbers in space and time. The first college in the United States was Harvard University, founded in 1636. Harvard and other institutions of higher learning included mathematics in their curriculum. Around the time of the Revolutionary War, advanced mathematics included topics in surveying, algebra, geometry, trigonometry, and calculus. In 1776, Congress advised that disabled veterans, "[w]hen off duty, shall be obliged to attend a mathematical school, appointed for the purpose, to learn geometry, arithmetic, vulgar and decimal fractions, and the extractions of roots." This obligation led to the official founding of the United States Military Academy in 1802. After World War II and the beginnings of the Cold War, the growing emphasis on computer technology greatly impacted the mathematics curriculum in the United States.

Teachers have long explored different methods to help students succeed in mathematics. The philosopher Socrates is known for the Socratic Method, and in the early part of the twentieth century, topologist Robert Lee Moore developed a Socratic style of teaching that became widely known as the Moore Method. Versions of the Moore Method, or a modified Moore Method, continue to be used in twenty-first-century undergraduate and graduate mathematics classrooms. In some implementations, students work on problems and present proofs or solutions they develop on their own, with the class being responsible for corrections and the teacher acting as a guide. In the 1980s, a calculus reform movement that is often referred to as the "calculus wars" spurred debates among mathematicians regarding various aspects of teaching, including the use and balance of lectures, technology, and rigor in calculus classrooms.

Calculus education had already undergone many changes in the twentieth century, such as a shift to calculus being taken earlier in the college curriculum. Following the ethos of "calculus should be a pump, not a filter," educators explored many different approaches, often based on empirical studies. Some campuses embraced new approaches, while others soundly rejected them. In the early twenty-first century, mathematicians continue to discuss and refine the calculus course as well as other mathematics courses. There are also discussions at both the college and federal level of the possibility of standardized college mathematics assessments.

General Education Mathematics Requirement

For the college student majoring in a subject area that does not require specific mathematics courses, the extent of the mathematics curriculum may consist of mathematics courses that satisfy general education core requirements. At most colleges and universities, these courses enroll almost twice as many students as all other mathematics courses combined. These students represent a broad variety of majors, including students from the humanities, fine arts, elementary education, and several branches of the social sciences.

Courses that fall into this category may be termed or described as one of the following: quantitative literacy; liberal arts mathematics; finite mathematics; college algebra with modeling; or introductory statistics. These courses are designed to have students learn to think effectively, quantitatively, and logically, and may actually also be requirements for a student's major. Such courses often serve as students' final experience of college mathematics. While these courses may be terminal, such courses could also entice students to study mathematics further, and therefore, such course offerings may act as a springboard or gateway through which a student chooses to continue the study of the mathematical sciences.

There is a wide variety of topic options in these courses. Some professors incorporate topics directly from daily life, like financial mathematics, while others focus on algebraic or statistical techniques that might be important in future coursework. General education courses are also seen as the final place to impact students' perceptions about mathematics and its role in society. In the same way that a survey course on important literature might include works by William Shakespeare, some mathematicians select course topics from the masterpieces of mathematics, which might include great theorems, like Euler's theorem, named for Leonard Euler; interesting applications, like Chvátal's art gallery theorem, named for Václav Chvátal; interdisciplinary topics, like fractals, perspective drawing, or the philosophy of mathematics; or beautiful mathematical topics, like the golden mean. Some classes focus on the breadth of mathematics, while others try to cover a few topics in depth. There is also a wide variety of teaching methodologies and pedagogy. In some classrooms, the focus is on lectures, while in others it is on discussion or presentations. Technology may be a fundamental part of the class, or the class might focus on pencil-and-paper methods.

Mathematics and Partner Disciplines

During the second half of the twentieth century and into the twenty-first, there has been an enormous growth and development of scientific and technological disciplines, and, consequently, the role of mathematics is increasing in an expanding array of subject areas and professional programs. Students may be required to take specific mathematics courses that complement their major field of study. These partner or client disciplines include physics, chemistry, biology, computer science, engineering, business, finance, economics, nursing, psychology, and education. Partner and client discipline courses may impact mathematics as well as the respective discipline.

Some of these courses are taught in mathematics departments; others are taught as a quantitative course in the major, as in some psychology departments. This system provides numerous opportunities for faculty and students in mathematics departments to collaborate with their counterparts in other academic departments on campus. It is not uncommon for students who major in these partner disciplines to also study advanced mathematics, often resulting in dual majors or a minor in mathematics. One emerging area in the twenty-first century has been calculus for the life sciences. Cutting-edge pedagogies may come from mathematics or a client discipline. Faculty in either mathematics or a client discipline may lead efforts in interdisciplinary curricular development, or departments may resist changes because of staffing or philosophical considerations, sometimes leading to friction between departments.

Such courses need not be limited to calculus-based courses. For example, students in the sciences often benefit from the skills and techniques used in introductory statistics and discrete mathematics courses, which may not have a calculus prerequisite. The ability to visualize in three dimensions is also valued by partner disciplines, and courses that emphasize geometric and graphical reasoning, linear systems, and vector analysis may also be required.

Precise, logical thinking is an essential part of mathematics. While it remains a component of the mathematics courses taken by students who study the aforementioned partner disciplines, additional needs specific to such fields of study are also imbedded in the courses taken by these students. Logical and deductive reasoning skills may need to be developed in a specific context, and certain disciplines may or may not have a need for the use of formal proof found in the mathematics courses. Also, the level and type of logical reasoning may vary depending on discipline. For example, business majors may require more quantitative or statistical analysis, while engineering students need to engage in more formal analysis in a course like multivariable calculus. Students studying the natural sciences benefit from heuristic arguments and data analysis, while computer scientists and software engineering students need the ability to use logic to write simple proofs. The courses that bridge various other subject areas with mathematics attempt to balance the rigorous proof and deductive reasoning inherent to mathematics with the skills these partner disciplines require of their students.

Students who are preparing to teach elementary or middle school mathematics also fall into this category. The curriculum designed for future primary mathematics teachers varies state by state, with many requirements set by schools or teacher program accrediting agencies. They often aim to provide these students with a firm foundation in various mathematical topics, such as number and operation, algebra and functions, geometry and measurement, and data analysis, probability, and statistics. These topics are studied at a level above and beyond that which they will eventually teach. Courses are designed to provide students with an understanding of these broad areas as well as an ability to make connections among various mathematics topics and with other subjects taught in the elementary and middle school curriculum. The intent is that the future teachers will be able to guide their students in ways that instill mathematical breadth and depth and "plant the seeds" of ideas that will come later. From 2003 to 2009, the Mathematical Association of America ran a program that was funded by the National Science Foundation called Preparing Mathematicians to Educate Teachers (PMET). PMET strove to improve the mathematics education of teachers by targeting the development of faculty awareness and teaching as well as instructional materials.

Concentration in the Mathematical Sciences

For students who choose to major in the mathematical sciences, their college curriculum is centered on this goal of study. Major programs in the mathematical sciences include courses that focus on pure (theoretical) mathematics, applied mathematics, statistics, actuarial sciences, or secondary mathematics education. Depending on the college or university, the programs and faculty for statistics or applied mathematics, as well as actuarial sciences and mathematics education, may be housed in a department distinct from the traditional mathematics department.

The actual course of study for mathematics majors will differ depending on the specific college or university. In general, students in their first years of study will take a sequence of courses in calculus consisting of single- and multivariable calculus, which include the topics of differentiation and integration, sequences and series, vector analysis, and differential equations. Beyond calculus, mathematics students often take a transition course that includes an introduction to proof-writing techniques demonstrated by a study of various foundational topics in mathematics, such as logic, set theory, functions and relations, and cardinality.

Other commonly required courses for the mathematics major include linear algebra, abstract algebra, and real analysis (or advanced calculus). Several other advanced courses in mathematics that make up the major include ordinary differential equations, partial differential equations, discrete mathematics, probability and statistics, modern geometry (Euclidean and non-Euclidean geometry), complex analysis, topology, combinatorics, and number theory. Students who are interested in learning more applied mathematics may take courses in dynamical systems, numerical analysis, cryptanalysis, and operations research. A course in the history of mathematics may also be offered, especially for those students preparing to teach mathematics.

Because there are numerous topics that connect mathematics with other disciplines, various interdisciplinary courses may also be offered by mathematics departments in conjunction with other academic departments on campus. Some schools use common syllabi or exams for certain courses, and other schools allow more flexibility in what is taught and how it is taught. Regardless, there are at least some common expectations because mathematical definitions, ideas, and proofs build upon one another across courses, and so earlier courses in the major impact later ones. For example, a single-variable calculus class impacts multivariable calculus, and an analysis course impacts courses in complex analysis and topology.

The curriculum for students majoring in mathematics is designed so that there is a progression from the study and practice of computational methods and procedures toward an extensive understanding of the subject, which may include logical reasoning, generalization, abstraction, sophisticated applications, and formal proof. Students majoring in mathematics are also encouraged to demonstrate their mathematical knowledge in both written and oral formats. Students should also gain experience in the analysis of data, gaining the ability to move between context and abstraction—an especially important ability for students whose course of study focuses on applied areas of mathematics as well as for those becoming mathematics teachers. While mathematics students may prefer one area of mathematics over another, they are encouraged to gain a broad view of the subject, recognizing the complementary nature of the following concepts: theory versus application; discrete versus continuous; algebraic versus geometric; and deterministic versus probabilistic.

In addition to specific mathematics courses, students majoring in mathematics may take courses in computer science. On some college and university campuses, mathematics and computer science are classified in the same department or division. The natural affinity between the skills used by mathematicians and computer scientists makes this partnering possible, since the application of logical reasoning to the task of programming enhances the learning of both disciplines. For mathematics majors who are preparing to enter the nonacademic workforce, experience with teamwork, creativity, and problem synthesis skills is enhanced by computer programming coursework.

Undergraduate Research in Mathematics

Many mathematical science departments require their mathematics majors to engage in some form of research at the undergraduate level. This research can take many forms, such as a capstone course, a thesis, or some other form of a project during the senior year of college. The

Mathematics Departments

Mathematics departments need to serve all students well—not only those who major in the mathematical or physical sciences. The following steps will help departments reach this goal.

- Design undergraduate programs to address the broad array of problems in the diverse disciplines that are making increasing use of mathematics.
- Guide students to learn mathematics in a way that helps them to better understand its place in society: its meaning, its history, and its uses. Such understanding is often lacking even among students who major in mathematics.
- Employ a broad range of instructional techniques, and require students to confront, explore, and communicate important ideas of modern mathematics and the uses of mathematics in society. Students need more classroom experiences in which they learn to think, to do, to analyze—not just to memorize and reproduce theories or algorithms.
- Understand and respond to the impact of computer technology on course content and instructional techniques.
- Encourage and support faculty in this work—a task both for departments and for administrations.

The CUPM Guide 2004 presents six general recommendations to assist mathematics departments in the design and teaching of all of their courses and programs:

1. Understand the student population and evaluate courses and programs.
2. Develop mathematical thinking and communication skills.
3. Communicate the breadth and interconnections of the mathematical sciences.
4. Promote interdisciplinary cooperation.
5. Use computer technology to support problem solving and to promote understanding.
6. Provide faculty support for curricular and instructional improvement.

From the Introduction of "Undergraduate Programs and Courses in the Mathematical Sciences: CUPM Curriculum Guide 2004" by Committee on the Undergraduate Program in Mathematics (CUPM) of The Mathematical Association of America (MAA).

The Hutlee'/Umyuarchdelee program is funded by the National Science Foundation to help Natives and rural Alaskans succeed in college and pursue careers in science, technology, engineering, and mathematics.

area of study for such research may connect knowledge of previous courses in an advanced manner. Such research often culminates in both a written paper and an oral presentation. This presentation provides the opportunity for mathematics students to not only study the mathematics, but write and speak about their results in the fashion conventional to the discipline.

Separate from major program requirements, research in mathematics at the undergraduate level can also be performed at National Science Foundation (NSF) programs, such as Research Experiences for Undergraduates (REUs) held at various schools across the country, often during the summer months. These opportunities allow students to become actively involved in current mathematical research projects under the guidance of faculty, and thus demonstrate how mathematical research is done and how it differs from research done in other fields. Programs such as REUs demonstrate how the activities of a professional mathematician are performed, including the various stages: formulating and solving a problem, writing a mathematics paper, communicating the results in a talk or poster (perhaps at a local or national mathematics conference), and possibly publishing a research article. The topics of study in REUs go beyond the standard undergraduate curriculum and also draw upon previous coursework and experience. By conducting research before they graduate from college, students get a taste of what happens in graduate school programs in mathematics, specifically the research component of the dissertation requirement.

Two-Year Colleges

A significant percentage of students who receive a bachelor's degree in the mathematical sciences have taken some of their mathematics courses at two-year colleges. While many college students may fulfill their general education requirement in mathematics by taking such courses at a two-year college prior to attending a four-year college or university, many potential mathematics majors complete a variety of mathematics courses that satisfy requirements in the major program. Such courses include developmental mathematics, precalculus, introductory calculus, multivariable calculus, linear algebra, differential equations, discrete mathematics, and statistics. While an associate's degree in mathematics may not be obtainable from a two-year college, it is becoming more common that future mathematics majors are beginning their mathematics career at these schools, including students who are also preparing to become mathematics teachers at the various school levels.

Technology

With the advances made in science and technology during the latter half of the twentieth century, many new instructional techniques are being designed and utilized in the mathematics classroom at all levels, including the collegiate. With an emphasis on critical thinking and deductive reasoning and a movement away from rote memorization of mathematical theories and algorithms, there has been an increase in the use of technology for teaching and learning advanced mathematics. Accurate visualization of graphs and geometric objects and easy manipulation of algebraic constructs are some of the benefits of current technology available for mathematics education.

Computational technology changed rapidly during the latter part of the twentieth century. At the beginning of the twenty-first century, computer algebra systems (CAS), such as Mathematica, MATLAB, and Maple, are often helpful tools for both in-class demonstrations and independent student assignments. These software packages are commonly implemented in a variety of courses, such as calculus, linear algebra, differential equations, statistics, real analysis, and complex analysis.

Other software packages, such as Geometer's Sketchpad and Exploring Small Groups (ESG), are more course-specific to geometry and group theory, respectively. In addition to desktop or laptop computer technology, the development of handheld graphing calculators, such as the various models produced by Texas Instruments (TI-83+, TI-84, TI-86, and TI-89), has also influenced the use of this technological tool in the classroom. Computer programs and graphing calculators are also being used at the secondary school level, and the transition to using such technology in the mathematics classroom at the collegiate level is often a smooth experience for the mathematics student.

Further Reading

American Mathematical Association of Two-Year Colleges (AMATYC). *Crossroads in Mathematics: Standards for Introductory College Mathematics Before Calculus.* Memphis, TN: AMATYC, 1995. http://www.amatyc.org/Crossroads/CROSSROADS/V1/index.htm.

American Mathematical Society (AMS) and The Mathematicians and Education Reform (MER) Forum. "Excellence in Undergraduate Mathematics: Confronting Diverse Student Interests." http://www.math.uic.edu/~mer/pages/Excellencepage/index.html.

Cajori, Florian. *The Teaching and History of Mathematics in the United States.* Washington, DC: Government Printing Office, 1890. http://www.archive.org/details/teachingandhist03cajogoog.

Committee on the Undergraduate Program in Mathematics (CUPM) of The Mathematical Association of America (MAA). *Undergraduate Programs and Courses in the Mathematical Sciences: CUPM Curriculum Guide 2004.* Washington, DC: MAA, 2004. http://www.maa.org/cupm/cupm2004.pdf.

Friedler, Louis. "Calculus in the U.S.: 1940–2004." *Studies in College Mathematics* 8, no. 3 (2005). http://gargoyle.arcadia.edu/mathcs/friedler/Calculus1940-2004.pdf.

Steen, Lynn Arthur. *Achieving Quantitative Literacy: An Urgent Challenge for Higher Education.* Washington, DC: Mathematical Association of America, 2004.

Zitarelli, D. "The Origin and Early Impact of the Moore Method." *American Mathematical Monthly* 111 (2004).

Daniel P. Wisniewski

See Also: Calculus and Calculus Education; Mathematics, Applied; Mathematics, Theoretical; Pythagorean School.

Curriculum, K–12

Category: School and Society.
Fields of Study: All.
Summary: Curricular standards of mathematics have undergone a series of changes in the twentieth and twenty-first centuries in response to various national concerns.

The term "curriculum" has been variously defined as a coherent program of study in a specific subject area consisting of a set of courses and learning experiences provided by an educational institution. Since the beginning of the twentieth century, the school curriculum (grades K–12) of the United States has undergone numerous changes, particularly in mathematics. Indeed, the U.S. school curriculum has been more profoundly influenced by the demands of society than any other country in the world.

Twenty-first-century debates over the school curriculum are essentially debates on how best to prepare students to live in a just, democratic society; be competitive in a global economy; and thrive in a technologically literate workforce. Employers in both explicitly technical and nontechnical fields seek candidates who can use mathematics content, logical reasoning, and other problem-solving skills that are ideally acquired by the end of high school. Both government and private employers have publicized the various requirements for different career paths. Differences of opinion often stem from opposing views about the nature of learning, the needs of society, and the purposes of schooling. The evolution of the school curriculum in the United States and the changing nature of its mathematics component are most easily understood from an historical context.

Change in the Mathematics Curriculum

Since the inception of the United States, the predominant view has been that education is necessary for the common good of society and the survival of democracy. Although mathematics has always been recognized as an essential component of the K–12 curriculum, the role of mathematics in public schools and the nature of its content have fluctuated over the years. Benjamin Franklin was one of the nation's first leaders to understand the need for mathematics instruction beyond basic arithmetic and measurement. For example, in 1751, he helped institute an academy in which geometry and algebra were among curricula designed to meet the practical needs of merchants, seamen, builders, and artisans.

Common School Movement

Throughout the 1800s, the curricular and educational trends of Europe influenced the mathematics curriculum of the United States. The focus of the mathematics curriculum was on basic arithmetic skills in the early grades and algebra and geometry in the upper grades. In 1837, the *Common School* movement was instituted by Horace Mann (1796–1859) from Massachusetts. He worked to develop a statewide common-school (public school) system. The philosophy was that education is a major "human equalizer" that balances the social structure of a country. For this reason, Horace Mann is often considered the "father of American public education." The curriculum of the common schools of the 1800s served to reflect the values and needs of a democratic society and instituted free education for all U.S. citizens, making mathematics education much more broadly available

The Progressive Movement

For the majority of the twentieth century, U.S. educators consistently promoted a "Progressive Education" agenda, spearheaded by John Dewey (1859–1952). Progressive educators believed that the school curriculum should be determined primarily by the needs and interests of children. Dewey advocated a school curriculum that encouraged students to be thinkers and problem solvers. He encouraged instructional methods that were experiential and child-centered, covering content arising naturally within the child's environment. This method is in contrast to traditional instruction which is usually classroom-based and teacher-centered, covering predetermined content. During the Progressive movement, mathematics instruction emerged primarily when needed within the real-life experiences of the child and was thus widely varying

By the 1940s, an alternate version of Progressive Education called "Life Adjustment" had gained popularity among some U.S. schools. The curriculum of Life Adjustment schools was designed to prepare many of the students for the working world and everyday living, though some opponents claim it was motivated by anti-intellectual philosophies. These students focused on practical concerns, such as home budgeting, consumerism, taxation, health, and citizenship, and mathematics courses, such as algebra, geometry, and trigonometry, were deemphasized.

By the end of World War II, rapid societal and technological changes abruptly came to the forefront. Public knowledge of the impact of atomic energy, radar, cryptography, and other scientific and technological advances underscored the need for a strong national curriculum in mathematics and science to maintain national security, to retain the nation's lead in technology, and to prepare students for jobs in the sciences. As a consequence, Progressive Education came under severe attack following World War II.

New Math

A momentous event occurred in 1957 that impacted the nation's mathematics and science curricula at all levels. U.S. society was stunned by the launching of the first space satellite, Sputnik, by the Soviet Union. Sputnik was considered a national embarrassment and a potential security threat. Its mere existence suggested that the Soviet Union was technologically superior and had a military capacity of launching offensive missiles at the United States. It also underscored an overall weakness in the U.S. educational system, particularly in mathematics, science, and technology.

The U.S. Congress responded to the nation's panic and the implications of a security threat by passing the 1958 National Defense Education Act, intended to increase the quantity and quality of mathematics and science professionals. That same year, the American Mathematical Society (AMS) established the School Mathematics Study Group (SMSG), headed by Edward G. Begle of Yale University, to develop a new mathematics curriculum for the nation's high schools. The aim was to produce the most highly capable mathematics students in the world, with a view toward regaining the nation's technological superiority and bolstering its defense system against the Soviet Union. This marked the beginning of the New Math movement.

Funded by the National Science Foundation, the SMSG created a new, more-rigorous high school mathematics curriculum for college-bound students and wrote textbooks supporting the new curriculum. The SMSG curriculum was developed by mathematics professionals consisting of working mathematicians, university professors, high school teachers, and school supervisors. The SMSG soon expanded its curriculum to include mathematics for grades K–12. Similar mathematics curricula emerged in the early 1960s, modeled after the original work of SMSG. These curricula were products of other federally funded projects such as the Ball State Project, Greater Cleveland Mathematics Program, the University of Maryland Mathematics

U.S. Mathematics Curriculum Evolution

The evolution of the mathematics curriculum in the United States has been unstable. Over the years, it has responded to the demands of society, professional educators, and national organizations, often at the expense of the needs of the country. While unpopular in some segments of society, the New Math movement responded favorably to the national panic following the Soviet Union's launch of Sputnik. The New Math movement was responsible in large measure for regaining the nation's international lead in technology and winning the race to the moon. Since the 1980s, the National Council of Teachers of Mathematics has contributed significantly to the nation's mathematics curriculum by developing a series of well-articulated standards that have informed the curriculum and assessment strategies for every state.

Student assessment has also gained a more prominent role in federal funding and curriculum development. Specifically, the success of states in meeting national education standards will be measured by students' performance on high-stakes tests. A positive contribution of the Common Core State Standards, initiated by the nation's state governors, will be the establishment of a national school curriculum. According to U.S. Secretary of Education Arne Duncan, "For the first time, a child in Mississippi and a child in Massachusetts will be judged by the same yardstick."

Project, and the Minnesota School Science and Mathematics Center. Each curriculum mirrored the rigorous mathematics content and educational philosophy of the New Math movement.

The New Math curriculum included advanced content that had never before been covered in public schools, such as set theory, Boolean algebra, base arithmetic, field axioms, algebraic structures, and formal math language and symbolism. The curriculum was designed to provide the theoretical foundations for studying calculus and abstract algebra in college, with the intent of producing as many mathematics, science, and engineering majors as possible.

Even though there were numerous successes in the New Math movement, after a decade of implementation, it was slowly removed from the nation's public schools. Some believed its downfall was a result of excessive rigor and mathematical formalism at the expense of basic skills and problem solving. Many parents and school administrators were confused by the unfamiliar mathematics content and advanced symbolism. More importantly, a large number of the nation's older teachers were unable to implement the New Math curriculum because they, themselves, were not academically prepared to deal with the content. Notwithstanding, 12 years following the launching of Sputnik, the United States succeeded in placing the first man on the moon with its 1969 Apollo 11 mission.

Back to Basics

By the early 1970s, the New Math movement was over. The National Science Foundation discontinued its funding of New Math programs, and the U.S. public called for a return of a "Back to Basics" curriculum, under a Progressive agenda. Mathematical rigor and advanced symbolism were discouraged; teachers experimented with child-centered instructional approaches, such as Individualized Instruction, Continuous Progress, and Open School environments. The failures of such practices were soon exposed as standardized mathematics test scores of U.S. students steadily declined throughout the decade.

Fueled by the failure of the Back to Basics movement, the lowering of college entrance requirements, and reduced enrollments in higher-level mathematics courses, by the end of the 1970s, the U.S. school curriculum was once again under severe public attack. Two publications by U.S. federal agencies had a significant impact on public perceptions of the U.S. educational system, *An Agenda for Action* and *A Nation at Risk*.

An Agenda for Action

Based on multiple national assessments, the National Council of Teachers of Mathematics (NCTM) produced the 1980 *An Agenda for Action*, which provided recommendations for reform in mathematics education. Its primary recommendation was that problem solving should be the primary focus of the mathematics curriculum, supported by the following instructional practices:

- Calculators and computers should be used in K–12 classrooms.
- Estimation and approximation should be an integral part of instruction.
- Team efforts in problem solving should be encouraged in the elementary classroom.
- Manipulatives should be used to develop new mathematical concepts and skills.
- Instructional strategies should provide for situations requiring student discovery and inquiry.
- Mathematics programs and student performance should be evaluated on a broader range of measures than conventional testing.

A Nation at Risk

Although *An Agenda for Action* provided innovative and lofty recommendations for reform in the school mathematics, it was overshadowed by the 1983 publication of *A Nation at Risk,* a report by the National Commission on Excellence in Education. In graphic terms, it warned Americans, "The educational foundations of our society are presently being eroded by a rising tide of mediocrity that threatens our very future as a nation and a people," and, "If an unfriendly foreign power had attempted to impose on America the mediocre educational performance that exists today, we might well have viewed it as an act of war."

A variety of educational issues and specific weaknesses in the mathematics curriculum were addressed. Specifically, the commission found that the textbooks used for instruction were void of rigorous content, the curriculum lacked continuity and depth, and high school teachers were typically underprepared in mathematics and academically weak. Despite out-dated information, *A Nation at Risk* is still often quoted in the twenty-first century and remains an influential publication.

National Standards-Based Curriculum

The publication of *A Nation at Risk* and similar reports of the dismal performance of U.S. students on international assessments have all served to provoke U.S. society and government to demand higher academic standards in public schools. International assessments provided strong evidence that mathematics teaching and its school curriculum must change if U.S. students are to be competitive in the global economy and able to deal with the complex decisions they will confront as responsible citizens and members of a technologically literate workforce.

In 1989, NCTM took a giant step in recommending a national agenda for curriculum reform, resulting in NCTM's 1989 publication of *Curriculum and Evaluation Standards for School Mathematics*. This document initiated a national standards-based curriculum movement, influenced by its earlier work reported in An Agenda for Action.

Within five years, NCTM also produced two supporting documents: the 1991 *Professional Standards for Teaching Mathematics* and the *1995 Assessment Standards for School Mathematics*. These documents recommended teaching standards, instructional methodologies, and an array of assessment strategies for accommodating the new standards-based curriculum.

An updated version of NCTM's original 1989 *Standards* was published in 2000, having a new title: *Principles and Standards for School Mathematics*. These three standards documents continue to be profoundly influential in the twenty-first century in matters of curriculum and assessment decisions for U.S. school mathematics. Specifically, Principles and Standards for School Mathematics provides six principles for school mathematics, five process standards, and five content standards.

The six principles for school mathematics are as follows:

- Equity—high expectations and strong support for all
- Curriculum—courses and learning experiences focused on important mathematics, well articulated across the grades
- Teaching—instruction that is challenging, supportive, and focused on what students know and need to learn
- Learning—develops understanding by building new knowledge on students' experiences and prior knowledge
- Assessment—provides useful information to teachers and students and supports the learning of important mathematics
- Technology—essential for teaching and learning mathematics, influences the mathematics that is taught, and enhances student learning

The five process standards considered essential for teaching all mathematics, are problem solving, reasoning and proof, communication, connections, and representation. These processes were expected to be integrated into the teaching of all of mathematics, regardless of the topic or the grade level.

The five content standards include each of the following: number and operations, algebra, geometry, measurement, and data analysis and probability. Each content area is expected to be covered to some degree of integrity at every grade level.

Overall, given the breadth of mathematics applications found in modern society, both in work and everyday life, schools were encouraged to widen their mathematics offerings. For example, one recommendation for the high school curriculum was that calculus should not always be the primary goal for the mathematics curriculum. Instead, discrete mathematics, probability, and statistics should also be considered valuable goals. The curriculum must prepare students for a variety of career paths that use mathematics; for example, actuarial science (probability), engineering and electronics (technical mathematics), economics and behavioral science (statistics and decision theory), theoretical or nuclear physicist (calculus), and numerous others.

In 2006, NCTM released another supporting document, *Curriculum Focal Points for Prekindergarten through Grade 8 Mathematics: A Quest for Coherence*, which articulated the specific topics that should be the focus for each grade level pre-K–8. The curriculum focal points acknowledged that NCTM's five content standards are not equally weighted and should have greater emphasis at different grade levels. These topics are identified in this document.

In 2009, NCTM released *Focus in High School Mathematics: Reasoning and Sense Making.* This publication was designed to provide teachers with curriculum guidance and content focal points for high school mathematics, modeled after NCTM's 2006 pre-K–8 document, *Curriculum Focal Points.* In this 2009 document, NCTM stresses that reasoning and sense making should be the focus of all high school mathematics, spanning all content areas, and evident in the teachers' instructional strategies and assessment practices. The goal is for mathematics to be viewed as a logical, problem-solving tool, rather than a set of meaningless procedures, disconnected from everyday life and decision making. It is stressed that students should have experiences with reasoning and sense making within a broad curriculum that may deviate from the textbook. Such experiences should be designed to meet students' future needs and prepare them for citizenship, the workplace, and future careers.

Twenty-First-Century Mathematics Curriculum

History has shown that as national needs and societal perceptions change, so does the school curriculum. Pervasive and radical changes have occurred throughout the world since the 1990s, such as genetic engineering, nanotechnologies, global economies, environmental disasters, global warming, depleting energy sources, and countless others. It is clear that U.S. citizens must be prepared to deal creatively and competently with a multitude of rapid changes and to tackle complex problem situations. The school curriculum must respond accordingly to provide students with the content knowledge, problem-solving skills, and learning experiences that are necessary for students to meet these immense challenges.

Unfortunately, the U.S. mathematics curriculum still has a long way to go in preparing students to meet these challenges. In fact, international assessments report serious deficiencies in the mathematical performance of U.S. students. In 2009, the Programme for International Student Assessment (PISA) reported that 15-year-olds from the United States ranked 18 among 33 developed nations in mathematical literacy and problem solving. In sum, compared to other developed nations, students from U.S. schools score in the lower 50th percentile in mathematics. Furthermore, the National Center for Education Statistics reported in 2003 significant racial achievement gaps in the United States. Societal concerns for economic stability, national security, and equity in instruction all demand immediate and substantial reforms in the U.S. mathematics curriculum and educational system as a whole.

No Child Left Behind

Research has shown that the school curriculum is closely tied to assessment. One governmental attempt to address the school curriculum and the lagging academic achievement of U.S. students is the No Child Left Behind (NCLB) Act of 2002. NCLB includes a number of mandates designed to promote significant gains in student achievement and to hold states and schools accountable for meeting curricular goals. NCLB supports a standards-based curriculum and was founded on the belief that setting high standards and measurable objectives would result in improved teaching and learning in the nation's schools.

As a provision for federal funding, the NCLB Act requires that states develop assessments in basic skills for students at specific grade levels and that each state set its own curriculum, content standards, and achievement benchmarks. The Act further mandates that 100% of the students in each school be proficient in reading and mathematics by the year 2014. As a consequence, any school not showing significant progress toward meeting these goals will be subject to sanctions, culminating in the closing of the school and termination of the faculty and staff.

The underlying theory is that schools will show significant improvement if children in grades 3–8 are held accountable for their academic achievement, as measured by their test scores every year. As of 2011, more than 10,000 schools have been labeled as "failures"; thousands of teachers have been fired; and numerous schools, heretofore considered "very good," are being forced to close. To meet NCLB goals, many schools have eliminated studies in art, history, science, foreign languages, physical education, and geography from their offerings. The majority of school time is now devoted to preparing students for high-stakes tests in the basic skills, the results of which will determine if the school remains open for the following year.

The consequences of students' test performance are so punitive that some districts have experienced record amounts of cheating. Some states have even lowered the passing score on their annual mathematics exams to increase the pass rates for their schools. Reactions such as these to the mandates of the NCLB Act underscore the fact that testing alone will not increase student achievement nor improve instruction. Regardless of how well the curriculum is constructed, meaningful instruction will be abandoned for the sake of test preparation.

Common Core State Standards

Several of the nations to which the United States is often compared academically do have national curricula, such as Great Britain, Germany, France, and Japan. Even though NCTM has provided national guidelines for mathematics education, until 2010, nearly every state had its own unique set of mathematics standards and curriculum for each grade level. In some cases, decisions about curricula were made by county and local school districts and boards. Consequently, state mathematics standards have varied considerably from state to state, and valid comparisons are difficult to make with respect to student performance. Because of the absence of a common set of standards among states, 48 of the nation's state governors and their chief school officers set forth to create the Common Core State Standards (CCSS), released in 2009.

The CCSS were developed in collaboration with content experts, college professors, public school teachers, school administrators, and parents. They are designed for a curriculum that includes rigorous content and applications; requires high-order thinking skills; and prepares students to succeed in a global economy. They are also aligned with the mathematics curricula of top-performing countries in the world. As of 2011, 41 of the 50 states have adopted CCSS.

Race to the Top

The rapid adoption of the CCSS by nearly every state in the nation was surely spurred by the Race to the Top (RTTT) program funded the Educational Recovery Act of 2009. RTTT is a $4.35 billion U.S. Department of Education program offering competitive grants designed to promote educational reforms in state education. The underlying federal agenda is to establish national standards, tests, and curricula. Even though the principle of states' rights ensure that individual states have total control over their educational systems, the promise of RTTT's discretionary funding of hundreds of millions of dollars is a huge incentive for states to adopt the CCSS, which is prerequisite to RTTT funding. When states receive RTTT discretionary federal funding, they must agree to implement the CCSS as well as comply with other stipulations.

Consistent with NCLB, state assessments for RTTT's funding are highly reliant on students' test scores as the sole measure of student achievement. Additionally, many states intend to use students' test scores to evaluate their teachers' performance and determine salaries and bonuses. As of 2011, there is also a rising movement among state governors to push for an end to teachers' unions, tenure, and rights to due process, many of which have existed since at least the early twentieth century. It appears that if these movements continue in the twenty-first century, teachers will soon have no organized voice for addressing teaching conditions, budgetary concerns, or program and curricular issues.

Further Reading

Alter, Jonathan. "Obama's Lesson Planner." *Newsweek* (April 11, 2011).

The Conference Board, Corporate Voices for Working Families, The Partnership for 21st Century Skills, and the Society for Human Resource Management. "Partnership for 21st Century Skills. Are They Really Ready to Work? " http://www.p21.org/documents/FINAL_REPORT_PDF09-29-06.pdf.

Council on Competitiveness. Innovate America: National Innovation Summit and Report. "Thriving in a World of Challenge and Change." Washington, DC: 2004.

Klein, David. "A Brief History of American K–12 Mathematics Education in the 20th Century." 2003. http://www.csun.edu/~vcmth00m/AHistory.html.

National Council of Teachers of Mathematics (NCTM). *An Agenda for Action: Recommendations for School Mathematics of the 1980s*. Reston, VA: NCTM, 1980.

National Council of Teachers of Mathematics. "Standards and Focal Points." http://www.nctm.org/standards/default.aspx?id=58.

Reys, Barbara. "The Intended Mathematics Curriculum as Represented in State-Level Curriculum." 2006. http://www.mathcurriculumcenter.org/ASSM_report.pdf.

Smith, Karl. *Mathematics: Its Power and Utility*. 9th ed. Bellmont, CA: Brooks-Cole, 2009.

United States Bureau of Labor Statistics. "Occupational Outlook Handbook." http://www.bls.gov/oco.

Sharon Whitton

See Also: Careers; Competitions and Contests; Curricula, International; Curriculum, College; Educational Manipulatives; Educational Testing; Government and State Legislation; Learning Models and Trajectories; Math Gene; Mathematicians, Amateur; Mathematics Literacy and Civil Rights; Professional Associations; Schools.

Curves

Category: History and Development of Curricular Concepts.
Field of Study: Algebra; Calculus; Communication; Connections; Geometry.
Summary: Curves have many different definitions and applications in various fields of mathematics.

Intuitively, a curve might be thought of as a path, like that of a curveball. A curve is viewed and defined in several ways depending on the branch of mathematics. A curve can be defined as the one-dimensional continuous trajectory of an object in space moving in time, the intersection of two surfaces in space, the image of the unit interval under a continuous function, or the graph of a solution of a polynomial equation. Each of these approaches captures the intuitive idea of a curve in their respective domains; the first is more physical, the second is geometric, the third is topological, and the last is an algebraic view of a curve. Curves can be used to create figures, model paths of motion, or express relationships between variables.

There are many types of curves that are the focus of classroom investigations, including yield curves, which are important to investors, and the normal distribution or bell-shaped curve. Felix Klein is noted to have said, "Everyone knows what a curve is, until he has studied enough mathematics to become confused through the countless number of possible exceptions." In education, a "learning curve" is a phrase that is meant to informally capture the notion of the change in knowledge over time. Algebraic and geometric curves are also important in school. Children study lines and circles in primary and middle school. They investigate their lengths and areas. By high school and college, they learn about parametric equations of curves and the area under a curve. In order to enrich classroom learning, mathematicians and mathematics historians created the National Curve Bank Web site.

Early History of the Study of Curves

The Greeks initiated the study of curves and discovered numerous interesting curves. Apollonius of Perga studied conic sections as the intersections of a plane and a cone by changing the angle of intersection. Diocles of Carystus invented the cissoid curve and used it in his attempts to solve the problem of doubling the cube. Nicomedes invented the conchoid curve and used it in his attempts to solve the problems of doubling the cube and trisecting an angle. Some have noted that aspects in the design of the columns of the Parthenon may resemble a conchoid of Nicomedes, although others present different curves as the model. Canon of Samos invented the spiral that was eventually called the "Archimedean spiral."

This curve was utilized by Archimedes of Syracuse as a method to attempt to trisect an angle and square the circle. The Greek view of curves was geometric, since Greek mathematics was, essentially, geometry-centered. Hence, their study of curves usually was through some elaborate and often ingenious methods of construction. Besides the lack of analytical tools, their insistence of having concrete or mechanical methods of construction, and—more importantly—their attempts to solve some important problems of antiquity that later were shown to be unsolvable by ruler and compass constructions are some of the factors contributing to the Greek concept of curves.

Mathematicians, philosophers, and others introduced and investigated the geometry of many interesting curves long after the ancient Greeks. For example, Nicholas of Cusa lived in the fifteenth century. He is noted as the first of many to explore the cycloid, which was eventually known as the path of a point on a wheel as the wheel rolls along a straight line.

Developments Since the Seventeenth Century

With the introduction of analytic geometry in the seventeenth century, the theory of curves received a new impetus—expressing curves by equations would make their study much easier compared to doing it via elaborate geometrical constructions. Analytic geometry enabled mathematicians to focus on the intrinsic features of curves; discover and investigate new curves; study curves in a more systematic way, leading to their classification into algebraic versus transcendental categories; and apply the results to various physical problems, such as the long-standing problem of determining the orbits of planets or solving the problem of a hanging chain, which was posed by Jacob Bernoulli. Gottfried Leibniz, Christiaan Huygens, and Jacob Bernoulli's brother Johann Bernoulli responded to the elder Bernoulli's challenge with the equation of the catenary. In the eighteenth century, Guido Grandi investigated rhodonea curves that resemble roses and what was later to be known as the Witch of Agnesi, named because of a mistranslation of the example in Maria Agnesi's famous calculus textbook.

Beginning with the seventeenth century, smooth curves have been an intense subject of investigation leading to determination of various features. Smooth curves, like lines, circles, parabolas, spirals, and helices, possess properties that make them amenable to numerous applications besides lacking any jagged behavior. For example, younger students learn that a straight line is the shortest path between two points in the plane, and mathematicians in the seventeenth century wondered about an analog for surfaces. A geodesic curve is locally a minimizing path; as a result, it is important in advanced mathematics and physics classes. Leonhard Euler published differential equations for geodesics in 1732. Mathematicians also investigated the classification of smooth curves. One invariant is the length of a curve. In general, length does not distinguish two different curves. It turns out that two other invariants, called the "curvature" and "torsion," work much better for this purpose. Broadly speaking, at any point on the curve, the curvature measures the deviance of the curve from being a straight line, and the torsion function measures the deviance of the curve from being a plane curve. Furthermore, the fundamental theorem of curves states that these invariants determine the curve, a result that is proved in twenty-first-century college differential geometry classrooms using the Frenet–Serret Formulas. These are named for Jean Frédéric Frenet and Joseph Serret, who independently discovered them in the nineteenth century.

With further investigations by prominent mathematicians, like Carl Friedrich Gauss, Gaspard Monge, Jean-Victor Poncelet, and their students, the theory of curves, particularly smooth curves, matured into an active field of research. The findings in the theory of curves not only enriched the realm of curve studies, they also contributed to the development of new ideas that ended up revolutionizing mathematics in the nineteenth century. Broadly speaking, the general definition of a curve is topological; namely, a curve is defined as a continuous map from an interval to a space. Curves can be algebraic (those defined via algebraic equations). For instance, a plane curve can also be expressed by an equation

$$F(x, y) = 0$$

and a space curve can be expressed by two equations

$$F(x, y, z) = 0 \text{ and } G(x, y, z) = 0.$$

A curve is algebraic when its defining equations are algebraic—a polynomial in x and y (and z). The cardioid, a heart-shaped curve whose Cartesian equation is

$$\left(x^2 + y^2 - 2ax\right)^2 = 4a\left(x^2 + y^2\right)$$

and the asteroid, whose Cartesian equation is

$$x^{\frac{2}{3}} + y^{\frac{2}{3}} = a^{\frac{2}{3}}$$

where a is a constant, are algebraic curves.

Before analytic geometry, each of these curves had been expressed using geometric investigations; for example, a circle turning around a circle that sweeps out the cardioid, or wheels turning within wheels that form the asteroid. Transcendental curves cannot be defined algebraically and include the brachistochrone curve, also known as the "curve of fastest descent"; very complicated looking fractal curves, such as the Koch snowflake, named for Helge von Koch, who explored the geometry in a 1904 paper; and paradoxical sounding space-filling curves, discovered by Giuseppe Peano in 1890. The last two types of curves can be extremely jagged curves with no smooth components.

An algebraic curve of the form $y^2 = x^3 + ax + b$, where a and b are real numbers, satisfying the relation $4a^3 + 27b^2 \neq 0$, is called an "elliptic curve." Geometrically, this condition ensures that the curve does not have any cusps, self-intersections, or isolated points. On the points of elliptic curves (including the point at infinity), one can define an operation by three

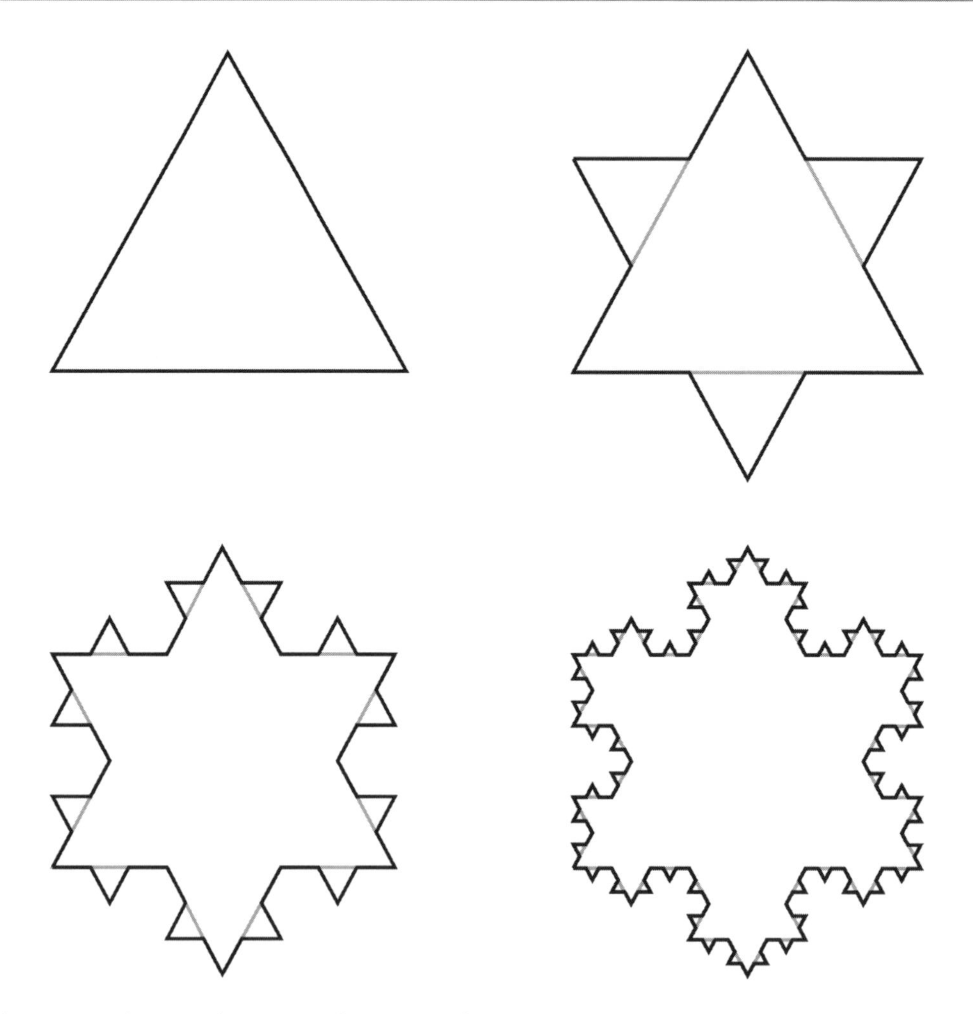

Curves can be extremely jagged curves with no smooth components, such as the complicated Koch snowflake fractal curve named for Helge von Koch who explored the geometry in a 1904 paper.

points sum to zero, if and only if they are collinear. This interesting feature of elliptic curves, besides being an important algebraic structure to be studied on its own, also has found some astonishing applications, such as in cryptography for developing elliptic curve-based public-key cryptosystems. Elliptic curves are also important in number theory; they are effective tools in integer factorization problems. They also turned up as an instrumental tool in the proof of Fermat's Last Theorem, named for Pierre de Fermat.

Further Reading

Boyer, C. B. "Historical Stages in the Definition of Curves." *National Mathematics Magazine* 19, no. 6 (1945).

Lockwood, E. H. *A Book of Curves*. New York; Cambridge University Press, 1963.

National Curve Bank. http://curvebank.calstatela.edu/home/home.htm.

O'Connor, John, and Edmund Robertson. "MacTutor: Famous Curves Index." http://www-history.mcs.st-andrews.ac.uk/Curves/Curves.html.

Dogan Comez
Sarah J. Greenwald
Jill E. Thomley

See Also: Conic Sections; Limits and Continuity; Normal Distribution; Planetary Orbits; Polynomials.

D

Dams

Category: Architecture and Engineering.
Fields of Study: Algebra; Geometry; Measurement.
Summary: Mathematics is vital to the design, monitoring, maintenance, and safety of dams.

Dams are embankments across a waterway for control of water or for water storage; they have served many functions in societies throughout history. The earliest dams were primarily used for irrigation and as a water source for livestock. Today, smaller dams provide water for livestock, fish and wildlife habitat, and recreation. Larger dams can provide flood control in places below sea level, like New Orleans and the Netherlands; municipal and industrial water supply; irrigation for crops; hydroelectric power; commercial navigation; and recreation. They are typically earthen dams, concrete structures, or some combination. Older dams were sometimes made of timber, masonry, or steel. Mathematicians and engineers investigate many aspects of the construction and maintenance of dams using geometry, trigonometry, and stochastic and limit-state analyses. For instance, Boris Galerkin, who had degrees in applied mathematics and mechanics, studied stress in dams, and Pelageia Polubarinova, who had a degree in mathematics, contributed to the theory of seepage flow of groundwater through porous materials that included earth dams. Some well-known dams are the Itaipú Dam in Brazil and Paraguay, the Hoover Dam in the United States, the Aswan Dam in Egypt, and the Dneproges Dam in the Ukraine.

Considerations for building a dam must take into account both positive and negative impacts. There are a variety of benefits of a dam that are closely related to its uses—providing water supply, flood control, hydroelectric power, and navigation. Hydroelectric power provides an important source of electrical power around the world. Commercial navigation through river systems provides efficient and economical transportation of agricultural products and commercial goods. Many dams that control flood plains provide farmers with an increased crop yield because land that would once have been flooded is now controlled upstream by the dam. Negatively, some dams may hinder fish movement; for example, along some streams, salmon are not able to get back to their native spawning areas because of the dam. Additionally, dams affect the natural order of a stream—its sediment load and flooding characteristics.

Purposes and Design

Dams are constructed with a definite purpose in mind based on the function(s) they are to serve. Dams are built to control watershed areas (all the area upstream of the dam, which provides runoff to the structure).

Engineers use a variety of mathematics skills as they plan, design, construct, and operate a dam. During the planning stage, engineers work with sponsors to scope out the needs and develop a basic design for the structure including design issues such as location, height, and base flow of the structure. Base flow is calculated with the formula $Q = v \times A$ where Q is the base flow rate, v is the velocity of water, and A is the area. Another important part of the planning stage is determining the economic feasibility of building the dam by calculating a benefit-to-cost ratio. Using a mathematical model, both the benefits of the dam over its life and the total cost of building and maintaining the dam are calculated. Ideally, for the construction of a dam to be feasible, the benefit-to-cost ratio needs to be greater than 1.

As a part of the design process, engineers must create detailed blueprints for the structure and an accompanying cost sheet that includes items such as quantities or volumes of a variety of materials (for example, cubic yards of concrete) and the cost of the removal and placement of earthen materials, which can be millions of cubic yards in the case of large dams. During the construction of the dam, the blueprints must be followed with precision and detail to ensure the integrity of the dam. Once the dam is constructed, regular monitoring is important to ensure the most efficient use of the available storage. Engineers monitor the amount of water leaving the dam through its spillway, as well as the amount of water entering the watershed. These inflows and outflows must be balanced in order to maintain storage needs and prevent flooding or low flows in the river downstream.

Safety

A major consideration in the planning, design, construction, and maintenance of any dam is safety. Engineers determine a hazard rating for each dam, with the highest hazard rating dealing with potential loss of human life. A breach in a dam can be catastrophic. A breach in a dam can be caused by a flaw in the design of the structure, extreme rainfall, lack of or poor maintenance of the structure, or a geological occurrence. Regular inspection and maintenance are important to ensure the safety of those downstream from the dam.

Further Reading

Hiltzik, Michael. *Colossus: Hoover Dam and the Making of the American Century*. New York: Free Press, 2010.

Macy, Christine. *Dams*. New York: W. W. Norton & Company, 2009.

Prabhu, N. U. *Stochastic Storage Processes: Queues, Insurance Risk, Dams, and Data Communication*. 2nd ed. New York: Springer, 1998.

JULIANA UTLEY

See Also: Engineering Design; Floods; Water Distribution.

Data Analysis and Probability in Society

Category: School and Society.
Fields of Study: Connections; Data Analysis and Probability.
Summary: Today, most industries depend on data analysis for some aspect of their work.

Data analysis can be thought of as the process of collecting, transforming, summarizing, and modeling data, usually with the goal of producing useful information that facilitates drawing logical conclusions or making decisions. Virtually any field that conducts experiments or makes observations is involved in data analysis.

There are many mathematical data analysis methods, including statistics, data mining, data presentation architecture, fuzzy logic, genetic algorithms, and Fourier analysis, named for mathematician Joseph Fourier. Probabilistic statistical methods are among the most widely applied tools, and they are what many people think of when they hear the term "data analysis." The use of probability, statistical analysis, and other mathematical data analysis methods is widespread, especially given technological advances and computer software that facilitate rapid, automated data collection and efficient, effective processing of massive data sets. According to forecasts included in the U.S. Bureau of Labor Statistics' 2010–2011 *Occupational Outlook Handbook*, the demand for statisticians and individuals with mathematical data analysis skills is expected to grow. Jobs that involve data collection, probabilistic modeling, statistical data analysis, data interpretation, and data dissemination are found in both the public and private sector, as well as in a diverse array of disciplines, including agriculture, biology, computer science, digital imaging, economics, engineering, education, forestry, geography, insurance, law, manufacturing, marketing, medicine, operations research, psychology, and pharmacology.

Many specialized data analysts are known by job titles or classifications other than statistician, such as actuary, biostatistician, demographer, econometrician, epidemiologist, or psychometrician. In the twenty-first century, both probability and data analysis are components of U.S. primary and secondary mathematics education, usually starting in the earliest grades and continuing through high school. This curriculum has been advocated by the National Council of Teachers of Mathematics in its *Principles and Standards for School Mathematics*, published in 2000, as well as by professional organizations such as the American Statistical Association and the Mathematical Association of America.

Professional Education

The first college statistics department was founded in 1911 at University College London. Other departments in universities around the world followed. In the twenty-first century, more than 200 colleges and universities in the United States offer undergraduate statistics degrees, and many more schools offer minors and courses in probability and statistics, data mining, and other mathematical data analysis methods. These courses may be taught either in mathematics and statistics departments or, often, in one of many partner disciplines, such as psychology, biology, or business. Graduate degrees in statistics do not necessarily require an undergraduate degree in statistics or mathematics, but most graduate degree programs prefer strong mathematical or statistical backgrounds with courses in areas like differential and integral calculus, mathematical modeling, probability theory, statistical methods, vector analysis, linear algebra, and mathematical statistics. Historically, computational methods were a primary focus of statistics education. With the evolution of technology and the growing role of statistics in everyday life, statistics education has shifted to focus on conceptual understanding, analysis of real data in context, survey sampling and experimental design methods, technology for analysis and presentation, communication of methodology and results to both technical and nontechnical audiences, and statistical thinking or literacy. People with bachelor's degrees in mathematics, statistics, or related mathematical fields, like operations research or decision sciences, can often find entry-level data analysis positions in government and industry, but research-related jobs and teaching at the community college level typically require master's degrees. Teaching or research-related jobs at four-year colleges and universities usually require doctoral degrees. Work experience or qualifying exams, such as those administered by the Society of Actuaries, are often necessary for employment in some industries. Training and certification programs like Six Sigma Black Belt also signify a certain level of data analysis skill and knowledge.

Government

Virtually all federal organizations have data analysis specialists or entire statistical subdivisions that use mathematical and statistical models. Since ancient times, governments have collected data and used mathematical methods to perform necessary functions. Archaeological evidence suggests that many ancient civilizations conducted censuses to enumerate their populations, often for taxation or military recruitment. Livestock, trade goods, and other property were sometimes counted in addition to people. Mathematics facilitated decisions regarding the distribution of resources like land, water, and food. The German word for this process of "state arithmetic" is cited as the origin of the English word "statistics," which first appeared in *Statistical Accounts of Scotland*, an eighteenth-century work by politician John Sinclair that included data about people, geography, and economics. In the United States, counting of the population is required by the U.S. Constitution, and congressional representation for the U.S. House of Representatives is determined by the decennial census population values.

Over the decades, many mathematicians and statisticians worked on planning and implementing the census, like Lemuel Shattuck, who also co-founded the American Statistical Association in 1839. Since its creation in 1902, the duties and activities of the U.S. Census Bureau have grown beyond the mandated 10-year census to include collecting and analyzing data on many social and economic issues, and the U.S. Census Bureau is one of the largest employers of mathematicians and statisticians in the country. At the start of

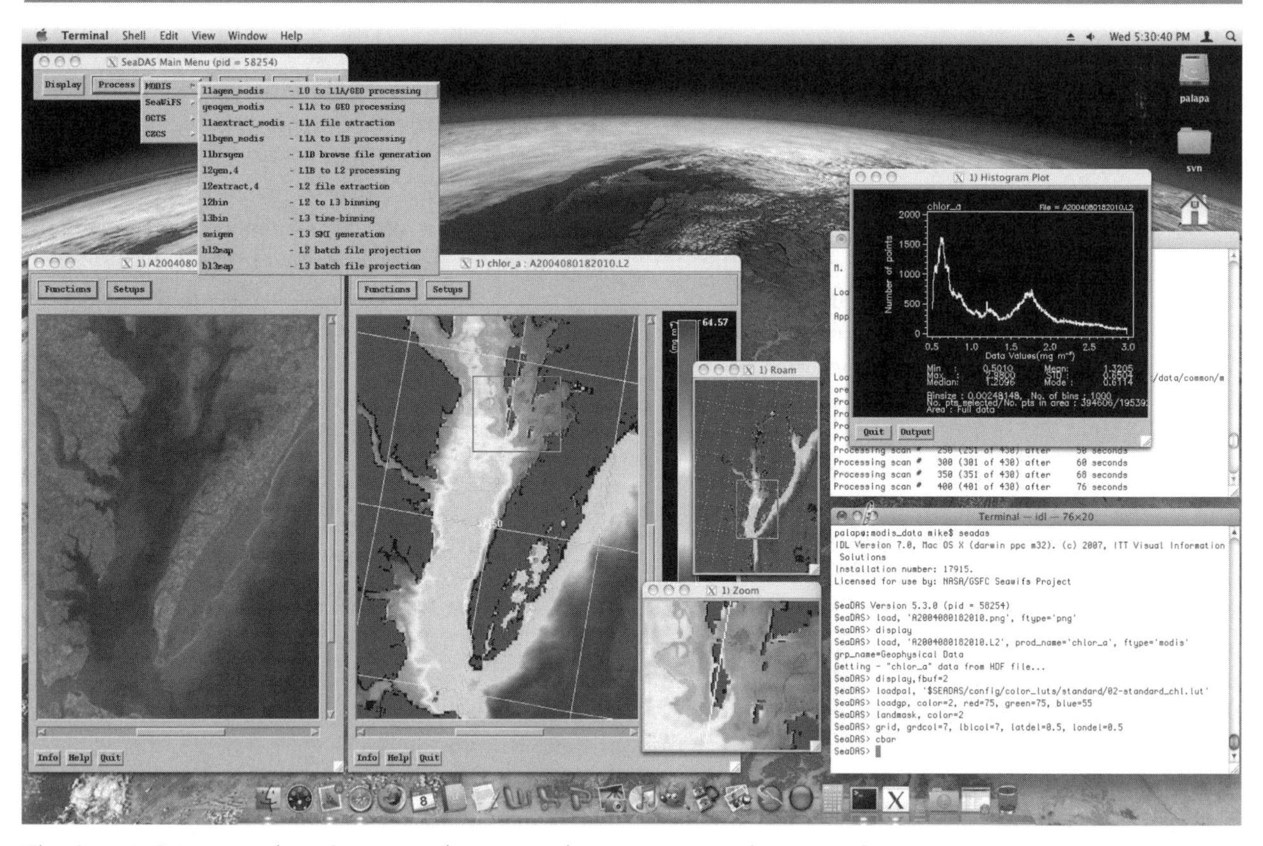

The SeaWiFS Data Analysis System at the National Aeronautics and Space Administration (NASA) is an image analysis package for the processing, display, analysis, and quality control of ocean color data.

the twenty-first century, various agencies of the U.S. government employed approximately 20% of the statisticians in the country. An additional 10% were employed by state and local governments, including state universities.

Statisticians and other mathematical data analysts working within many federal agencies are also responsible for developing new and innovative methods for gathering, validating, and analyzing data, especially the massive, messy, or incomplete data sets that are increasingly common in technological and industrialized societies. They also work to reduce bias and more accurately model issues that affect individuals and organizations. Many countries and governing entities around the world have agencies that perform similar functions. One major area of interest for most governments is the economic health of the country and the well-being of its workers. In the United States, the Bureau of Labor Statistics measures and forecasts factors such as labor market activity, productivity, price changes, spending, and working conditions. They began collecting data at the federal level in 1884. The *Current Population Survey*, implemented by the U.S. Census Bureau, is a monthly survey of about 50,000 households that has been conducted for more than 50 years, and the *Current Employment Statistics Survey* gathers data from about 410,000 worksites to summarize variables such as hours worked and earnings.

While the Bureau of Labor Statistics focuses mostly on manufacturing and services, the U.S. Department of Agriculture's Economic Research Service, established in 1961, is responsible for data about farming, natural resources, and rural development, addressing issues like food safety, climate, farm employment, and rural economies. Its online *Food Environment Atlas* includes indicators that describe the U.S. "food environment" and model concepts like people's geographic proximity to grocery stores or restaurants and food prices. The National Agricultural Statistics Service, also established in 1961, conducts the *Census of Agriculture*. It can be traced in part to a 1957 Congressional decision to approve probability survey methods for agriculture research. The U.S. Internal Revenue Service's Statistics Income Division, created in 1916, was among the first federal agencies to use stratified random sampling and machine summarization of data, both in the 1920s. In the twenty-first century, it assesses the tax impact of federal legislation.

Beyond their workforces, governments are also typically interested in the overall health, safety, and education of members of the broader society. The U.S. National Center for Health Statistics, established in 1960, compiles public health statistics, tracks federal health initiatives, and helps assess trends related to health care and health behaviors. For example, it has monitored efforts to reduce obesity and teen pregnancy. Other data include health care delivery and changes, such as the use of prescription medications and emergency rooms. The Bureau of Justice Statistics, founded in 1980, is primarily responsible for crime and criminal justice data collection, analysis, and dissemination in the United States. One of its principal reports is the annual *National Crime Victimization Survey*. The Federal Bureau of Investigation, founded in 1908, creates the annual *Uniform Crime Report*. The National Center for Education Statistics is mandated by the 2002 Educational Sciences Reform Act to collect and analyze "statistics and facts as shall show the condition and progress of education in the several states and territories" of the United States. The U.S. Congress uses data from this agency to plan education programs and to apportion federal funds among states.

In the twentieth century, issues like the energy crisis of the 1970s, climate change, and concerns over the future availability of oil focused more attention on U.S. energy resources and infrastructure. The Energy Information Administration (EIA) was established in 1977 to independently and impartially collect and analyze data to disseminate information about energy resources, uses, infrastructure, and flow, as well as their impacts on and responses to economic and environmental variables. The goals are to assist in creating policies and making energy decisions as well as educating the public about all aspects of energy. The EIA's "Energy Kids" Web site contains educational materials for primary and middle school students, and its "Energy Explained" Web site is aimed at the older students and the general public.

While government is one of the largest producers and users of statistics, not everyone agrees on their validity or utility. Many have criticized politicians for selectively using or deliberately misusing data and statistics, while others have suggested that the issue is insufficient training or understanding of mathematical data analysis—though statistical methods are increasingly part of political science degree programs. Former North Carolina Representative Lunsford Richardson Preyer once said: "Statistics do not always lie, but they seldom voluntarily tell the truth. We can argue any position on this bill on a set of statistics and some study or another." At the same time, some propose that effective democracy depends on citizens being able to access and understand current statistics. The burden and responsibility to produce credible information then rests with both the public, which has an obligation to provide valid data and seek to understand the outcomes, and the government, which must collect, analyze, and publicize information in a reliable, timely, and nonpartisan manner.

Industry and Manufacturing

The notion of interchangeable parts—pioneered by individuals like eighteenth-century army officer and engineer Jean-Baptiste Vaquette de Gribeauval and inventor Eli Whitney—followed by the mass production of goods during the Industrial Revolution, ushered in a new era of data collection and analysis to ensure the quality of manufactured products. In the early twentieth century, physicist and statistician Walter Shewhart pioneered data analysis methods in manufacturing that led some to call him the "father of statistical quality control." Among other accomplishments, he developed specialized charts using data and probability to sample and track the variability in processes to identify both natural, random process deviations and nonrandom deviations in order to eliminate the latter and thus improve consistency in the product.

W. Edwards Deming expanded on these notions to help develop the industrial management practice known as "continuous quality control" or "continuous quality improvement." Deming is credited with significant contributions to Japan's post–World War II reputation for high-quality products, and his data-based control methods have been widely adopted in the United States. For example, Motorola's Six Sigma program, founded in the 1980s, focused on training managers and employees at various levels in statistical methods and practices designed to identify and remove causes of product defects with the overall goal of minimizing process variability. The program name derives from statistical notation: sigma (σ) is commonly used to represent standard deviation, a measure of variability. Six standard deviations on either side of the mean in a bell-shaped or normal curve encompasses virtually all of the data values. If there are six standard deviations between the process mean and the nearest product specification limit, only three or four items per million produced will fail to meet those specifications. General Electric and other companies adapted and evolved the original Six Sigma ideas by merging them with other management strategies. For example, in the 1990s, concepts from a manufacturing optimization method known as "lean manufacturing" resulted in a hybrid program called "Lean Six Sigma."

Data analysis and probability are also used in advertising and market research. Many of the common market research practices used in the twenty-first century are traced to the work of engineer and pioneer television analyst Arthur Nielson. These practices include data analysis to quantify market share and determining sales patterns by combining consumer surveys with sales audits.

Medicine and Pharmacy

In the nineteenth century, some in the medical community began to investigate the idea of using data analysis for medical applications. Physician William Farr applied data analytic methods to model epidemic diseases. He is often credited as the founder of epidemiology. Physician John Snow gathered data to trace the source of an 1854 cholera outbreak in London. Along with his census work, Shattuck helped implement many public health measures based on data analyses. Florence Nightingale invented her own graphical data presentations in order to summarize data on the health impacts of poor hygiene in British military hospitals. In the twenty-first century, agencies like the U.S. Centers for Disease Control and Prevention and the World Health Organization collect, analyze, and model data in order to, among other goals, track the spread of infectious disease; assess the impact of preventive measures, like vaccinations; and test the virulence of infectious agents.

Clinical trials or experiments are also performed to determine the effectiveness and safety of new medical procedures and drugs. In the eighteenth century, physician James Lind tested remedies for scurvy aboard a British navy ship, which can be cited as one of the first recorded cases of a controlled medical trial. Statistician and epidemiologist Austin Bradford Hill helped pioneer randomized, controlled clinical trials in the twentieth century and also worked to develop the Bradford-Hill criteria, a set of logical and mathematical conditions that must be met to determine causal relationships. Approval and patenting of pharmaceuticals and medical devices by federal agencies like the Food and Drug Administration, part of the U.S. Department of Health and Human Services, require extensive experimentation and data analysis. For example, when a television commercial for a drug states that it is "clinically proven," this usually means that it has gone through experimental testing and that appropriate analyses of data have determined that it is very probably effective and safe, according to measures like the Bradford-Hill criteria.

Finance and Insurance

Probability is essential for quantifying risk, a concept that underlies most financial ventures and drives interest, credit, loan, and insurance rates. Data analysis can be used to derive probabilities and create financial models or indices like Fair Isaac Corporation (FICO) scores, the Dow Jones Industrial Average, and nations' gross domestic products. Engineer and economist William Playfair is considered to be one of the creators of graphical data analysis. Beginning in the eighteenth century, he researched trade deficits and other types of economic and financial data. Mathematician Louis Bachelier is known as the "father of financial mathematics" for his use of Brownian motion to model stock options at the turn of the twentieth century. Brownian motion, named for botanist Robert Brown, is a stochastic (probabilistic or random) process. The international Bachelier Financial Society is named for Louis Bachelier. Its goal is "the advancement of the discipline of finance under the application of the theory of stochastic processes, statistical and mathematical theory," and it is open to individuals in any discipline. Actuarial scientists or actuaries are also widely employed to develop models of the financial impact of risk. For example, they may use a combination of theoretical probability and data analysis to determine appropriate premiums for life or health insurance using variables such as life expectancy, which is adjusted for characteristics or behaviors that modify risk, like gender or smoking. Astronomer and mathematician Edmund Halley, for whom Halley's

Comet is named, is also often cited as the founder of actuarial science. He calculated mortality tables using data from the city of Breslau, Germany (now Wrocław, Poland). Published in 1693, these tables are the earliest known works to mathematically quantify the relationship between age and mortality.

Entertainment and Gambling

Archaeological evidence suggests that games of chance have existed since antiquity. Probability appears in different forms in written works throughout the centuries, like the body of Talmudic scholarship and the 1494 treatise of mathematician and friar Luca Pacioli known as *Summa de arithmetica, geometria, proportioni et proportionalita*. The mathematical study of probability as it is known in the twenty-first century is traditionally traced to seventeenth-century mathematicians Blaise Pascal and Pierre de Fermat, who were inspired to formulate their mathematical "doctrine of chances" by problems in gambling. In the twenty-first century, gambling is a multibillion dollar industry. In Las Vegas and other places, oddsmakers use probability to determine risks, point spreads, and payoff values for games of chance, sporting events, and lotteries. Players often use betting systems that are based on data analysis or probability to attempt to beat the odds and increase their chances of winning.

One example was a group of students from the Massachusetts Institute of Technology and other schools who used card counting techniques and mathematical optimization strategies in blackjack, which was the basis of the 2008 movie *21* and a television documentary *Breaking Vegas*. The television game show *Deal or No Deal*, which has aired versions in approximately 80 countries around the world, has been studied by mathematicians, statisticians, and economists as a case of decision making involving probability and data analysis concepts, like expected value. Probability-based random number generation is incorporated into many popular video games to increase realism and create multiple scenarios, while moviemakers are exploring probability-based artificial intelligence systems to generate realistic behavior in large, computer-generated battle scenes. The pioneering *Lord of the Rings* movies used a program developed by computer graphics software engineer Stephen Regelous and named Multiple Agent Simulation System in Virtual Environment (MASSIVE), which uses probabilistic methods like

World Statistics Day

The growing importance of statistical data analysis in global twenty-first century society was highlighted by the first World Statistics Day, which was held on October 20, 2010 ("20.10.2010" in common international date notation). In his letter to world leaders, United Nations Secretary General Ban Ki-moon emphasized the importance of data and statistical analysis to the current and future welfare of global society: "Let us make this historic World Statistics Day a success by acknowledging and celebrating the role of statistics in the social and economic development of our societies and by dedicating further efforts and resources to strengthening national statistical capacity." More than 130 countries and areas, as well as professional statistical organizations, universities, and other groups, held celebrations. Several international organizations also hosted a World Statistics Day conference in Geneva, Switzerland. That gathering brought together data analysis professionals from academia, government, and business, along with various end-users of statistics, to discuss the essential role of statistical data analysis in everyday life and in solving humanity's most pressing social, economic, and environmental issues. In the United States, President Barack Obama cited the importance of such methods: "Statistical data drives countless decisions which impact our nation. It guides representation in the United States Congress; informs our economic, social service, and national security outlook; and helps determine where infrastructure like schools, hospitals, and roads should be built.

fuzzy logic, derived from the fuzzy set theory of computer scientist and mathematician Lotfi Zadeh. Most sports collect a wide variety of data about their players, but in the latter twentieth century, advanced mathematical modeling, such as sabermetrics, developed by statistician George William "Bill" James, gained popularity for analyzing player and team performance and making predictions.

Further Reading

Best, Joel. *Damned Lies and Statistics: Untangling Numbers From the Media, Politicians, and Activists*. Berkeley: University of California Press, 2001.

Davenport, Thomas. *Competing on Analytics: The New Science of Winning*. Cambridge, MA: Harvard Business School Press, 2007.

Mlowdinow, Leonard. *The Drunkard's Walk: How Randomness Rules Our Lives*. New York: Vintage Books, 2009.

Murphy, Megan, ed. "World Statistics Day." *Amstat News* 400. (October 2010). http://magazine.amstat.org/blog/2010/10/01/world-statistics-day.

Rosenthal, Jeffrey. *Struck by Lightning: The Curious World of Probabilities*. Washington, DC: Joseph Henry Press, 2008.

Salsburg, David. *The Lady Tasting Tea: How Statistics Revolutionized Science in the Twentieth Century*. New York: Holt Paperbacks, 2002.

Taleb, Nassim. *Fooled by Randomness: The Hidden Role of Chance in Life and in the Markets*. New York: Random House, 2008.

United States Bureau of Labor Statistics. "Occupational Outlook Handbook." http://www.bls.gov/oco/.

Wainer, Howard. *Picturing the Uncertain World: How to Understand, Communicate, and Control Uncertainty through Graphical Display*. Princeton, NJ: Princeton University Press, 2009.

Sarah J. Greenwald
Jill E. Thomley

See Also: Baseball; Census; Congressional Representation; Deming, W. Edwards; Diseases, Tracking Infectious; Energy; Expected Values; Forecasting; Gerrymandering; Industrial Revolution; Inventory Models; Life Expectancy; Market Research; Normal Distribution; Probability; Quality Control; Sample Surveys; Statistics Education; Stock Market Indices.

Data Mining

Category: Business, Economics, and Marketing.
Fields of Study: Data Analysis and Probability; Measurement; Number and Operations.
Summary: Data mining is the relatively recent practice of using algorithms to distill patterns, summaries, and other specific forms of information from databases.

Advances in technology in the latter half of the twentieth century led to the accumulation of massive data sets in government, business, industry, and various sciences. Extracting useful information from these large-scale data sets required new mathematical and statistical methods to model data, account for error, and handle issues like missing data values and different variable scales or measures. Data mining uses tools from statistics, machine learning, computer science, and mathematics to extract information from data, especially from large databases. The concepts involved in data mining are drawn from many mathematical fields such as fuzzy sets, developed by mathematician and computer scientist Lotfi Zadeh, and genetic algorithms, based on the work of mathematicians such as Nils Barricelli. Because of the massive amounts of data processed, data mining relies heavily on computers, and mathematicians contribute to the development of new algorithms and hardware systems. For example, the Gfarm Grid File System was developed in the early twenty-first century to facilitate high-performance petascale-level computing and data mining.

History

Data mining has roots in three areas: classical statistics, artificial intelligence, and machine learning. In the late 1980s and early 1990s, companies that owned large databases of customer information, in particular credit card banks, wanted to explore the potential for learning more about their customers through their transactions. The term "data mining" had been used by statisticians since the 1960s as a pejorative term to describe the undisciplined exploration of data. It was also called "data dredging" and "fishing." However, in the 1990s, researchers and practitioners from the field of machine learning began successfully applying their algorithms to these large databases in order to discover patterns that enable businesses to make better decisions and to develop hypotheses for future investigations.

Partly to avoid the negative connotations of the term "data mining," researchers coined the term "knowledge discovery in databases" (KDD) to describe the entire process of finding useful patterns in databases, from the collection and preparation of the data, to the end product of communicating the results of the analyses to others. This term gained popularity in the machine learning and AI fields, but the term "data mining" is still used by statisticians. Those who use the term "KDD" refer to data mining as only the specific part of the KDD process where algorithms are applied to the data. The broader interpretation will be used in this discussion.

Software programs to implement data mining emerged in the 1990s and continue to evolve today. There are open-source programs (such as WEKA, http://www.cs.waikato.ac.nz/ml/weka and packages in R, http://www.r-project.org) and many commercial programs that offer easy-to-use graphical user interfaces (GUIs), which can facilitate the spread of data mining practice throughout an organization.

Types of Problems

The specific types of tasks that data mining addresses are typically broken into four types:

1. Predictive Modeling (classification, regression)
2. Segmentation (data clustering)
3. Summarization
4. Visualization

Predictive modeling is the building of models for a response variable for the main purpose of predicting the value of that response under new—or future—values of the predictor variables. Predictive modeling problems, in turn, are further broken into classification problems or regression problems, depending on the nature of the response variable being predicted. If the response variable is categorical (for example, whether a customer will switch telephone providers at the end of a subscription period or will stay with his or her current company), the problem is called a "classification." If the response is quantitative (for example, the amount a customer will spend with the company in the next year), the problem is a "regression problem." The term "regression" is used for these problems even when techniques other than regression are used to produce the predictions. Because there is a clear response variable, predictive modeling problems are also called "supervised problems" in machine learning. Sometimes there is no response variable to predict, but an analyst may want to divide customers into segments based on a variety of variables. These segments may be meaningful to the analyst, but there is no response variable to predict in order to evaluate the accuracy of the segmentation. Such problems with no specified response variable are known as "unsupervised learning problems."

Summarization describes any numerical summaries of variables that are not necessarily used to model a response. For example, an analyst may want to examine the average age, income, and credit scores of a large batch of potential new customers without wanting to predict other behaviors. Any use of graphical displays for this purpose, especially those involving many variables at the same time, is called "visualization."

Algorithms

Data mining uses a variety of algorithms (computer code) based on mathematical equations to build models that describe the relationship between the response variable and a set of predictor variables. The algorithms are taken from statistics and machine learning literature, including such classical statistical techniques as linear regression and logistic regression and time series analysis, as well as more recently developed techniques like classification and regression trees (ID3 or C4.5 in machine learning), neural networks, naïve Bayes, K-nearest neighbor techniques, and support vector machines.

One of the challenges of data mining is to choose which algorithm to use in a particular application. Unlike the practice in classical statistics, the data miner often builds multiple models on the same data set, using a new set of data (called the "test set") to evaluate which model performs best.

Recent advances in data mining combine models into ensembles in an effort to collect the benefits of the constituent models. The two main ensemble methods are known as "bootstrap aggregation" (bagging) and "boosting." Both methods build many (possibly hundreds or even thousands of) models on resampled versions of the same data set and take a (usually weighted) average (in the case of regression) or a majority vote (in the case of classification) to combine the models. The claim is that ensemble methods produce models with both less variance and less bias than individual models in a wide variety of applications. This is a current area of research in data mining.

Applications

Data mining techniques are being applied everywhere there are large data sets. A number of important application areas include the following:

1. *Customer relationship management (CRM)*. Credit card banks formed one of the first groups of companies to use large transactional databases in an attempt to predict and understand patterns of customer behavior. Models help banks understand acquisition, retention, and cross-selling opportunities.
2. *Risk and collection analytics*. Predicting both who is most likely to default on loans and which type of collection strategy is likely to be successful is crucial to banks.
3. *Direct marketing*. Knowing which customers are most likely to respond to direct marketing could save companies billions of dollars a year in junk mail and other related costs.
4. *Fraud detection*. Models to identify fraudulent transactions are used by banks and a variety of government agencies including state comptroller's offices and the Internal Revenue Service (IRS).
5. *Terrorist detection*. Data mining has been used by various government agencies in an attempt to help identify terrorist activity—although concerns of confidentiality have accompanied these uses.
6. *Genomics and proteomics*. Researchers use data mining techniques in an attempt to associate specific genes and proteins with diseases and other biological activity. This field is also known as "bioinformatics."
7. *Healthcare*. Data mining is increasingly used to study efficiencies in physician decisions, pharmaceutical prescriptions, diagnostic results, and other healthcare outcomes.

Concerns and Controversies

Privacy issues are some of the main concerns of the public with respect to data mining. In fact, some kinds of data mining and discovery are illegal. There are federal and state privacy laws that protect the information of individuals. Nearly every Web site, credit card company, and other information collecting organization has a publicly available privacy policy. Social networking sites, such as Facebook, have been criticized for sharing and selling information about subscribers for data mining purposes. In healthcare, the Health Insurance Portability and Accountability Act of 1996 (HIPAA) was enacted to help protect individuals' health information from being shared without their knowledge.

Further Reading

Berry, M. A. J., and G. Linoff. *Data Mining Techniques For Marketing, Sales and Customer Support*. Hoboken, NJ: Wiley, 1997.

De Veaux, R. D. "Data Mining: A View From Down in the Pit." *Stats* 34 (2002).

———, and H. Edelstein. "Reducing Junk Mail Using Data Mining Techniques." In *Statistics: A Guide to the Unknown*. 4th ed. Belmont, CA: Thomson, Brooks-Cole, 2006.

Piatetsky-Shapiro, Gregory. "Knowledge Discovery in Real Databases: A Workshop Report." *AI Magazine* 11, no. 5 (January 1991).

Richard De Veaux

See Also: Data Analysis and Probability in Society; Forecasting; Neural Networks; Predicting Preferences; Statistics Education.

Daubechies, Ingrid

Category: Mathematics Culture and Identity.
Fields of Study: Algebra; Number and Operations; Representations.

Summary: The first female president of the International Mathematical Union, Belgian Ingrid Daubechies revolutionized work on wavelets.

Ingrid Daubechies is a physicist and mathematician widely known for her work with time frequency analysis, including wavelets, and their applications in engineering, science, and art. Some people even refer to her as the "mother of wavelets." In 1994, Daubechies became the first tenured woman professor in the Mathematics Department of Princeton University, and in 2004 she was named the William R. Kenan, Jr. Professor of Mathematics at Princeton. Daubechies has achieved many honors internationally and was the first woman to receive a National Academy of Sciences Award in Mathematics. In 2010, she became the first woman president of the International Mathematical Union.

Physicist and mathematician Ingrid Daubechies is the first woman president of the International Mathematical Union.

Daubechies was born in Houthalen, Belgium, in 1954. As a child she enjoyed sewing clothes for dolls, saying about her experiences, "It was fascinating to me that by putting together flat pieces of fabric one could make something that was not flat at all but followed curved surfaces." She also computed powers of two in her head before sleeping, a childhood activity that coincidentally her future husband also engaged in. She had the support of her parents, which she appreciated. Her father, a coal mine engineer, answered her mathematical questions, and she tried to do the same with her own children. She attended a single-sex school and was not exposed to the idea that there might be gender differences in mathematics, saying, "So it didn't occur to me.…Later on, I did meet people who felt or even articulated very clearly that women were less 'suited' for mathematics or science, but by then I was confident enough to take this as a sign of their narrow-mindedness rather than let it influence me." She earned her bachelor's degree in 1975 in physics, and her Ph.D. in physics in 1980 from the Free University (*Vrije Universiteit*) in Brussels, Belgium. She held a research position at the Free University until 1987, when she accepted a position as a member of the technical staff of the Mathematics Research Center at AT&T Bell Laboratories in the United States. She remained at the Bell Labs until 1994, although she took two leaves of absence for research: one for six months at the University of Michigan, and another for two years at Rutgers University

Wavelet Analysis

Daubechies is best known for her work in wavelet analysis, a cross-disciplinary field that allowed her to combine her interest in mathematics with her training in physics. She has stated that she now considers herself a mathematician rather than a physicist because her work in physics was always highly theoretical and mathematical, and because she is interested in applications outside physics, particularly in engineering. A wavelet is an oscillation that has an amplitude that moves from zero to some point and then decreases back to zero (similar to an oscillation on a heart monitor). A wavelet transform is a mathematical function, similar to a Fourier transform, which allows data to be divided into frequency components and may be used to analyze signals that contain discontinuities and spikes. Jean Morlet and Alex Grossman developed the continuous wavelet function in the 1980s, and Daubechies, working with Yves Meyer and Alex Grossman, developed a discrete approach that allowed the reconstruction of wavelets from discrete values.

Applications of Wavelet Analysis

Wavelet analysis has many practical applications, particularly in creating and storing digital images. For instance, the U.S. Federal Bureau of Investigation (FBI) has used wavelet analysis since 1993 to encode digitized fingerprint records. This application is due in large part to the fact that a wavelet transform of an image reduces the amount of computer memory required to store it by as much as 93% compared to conventional image storage methods. Another application of wavelet analysis is in medical imaging systems, such as magnetic resonance imaging and computerized tomography. These technologies use scanners to collect digital information that is then assembled by a computer into a two- or three-dimensional picture of some internal aspect of the patient's body. Data processing methods involving wavelet transforms "clean up" and smooth digital information to yield a sharper image. Using wavelet transforms in medical scanning also reduces the time used to take the scan (thus reducing the patient's exposure to radiation) and makes the process of acquiring usable images faster and cheaper.

Further Reading

Case, Bettye Anne, and Anne Leggett. *Complexities: Women in Mathematics*. Princeton, NJ: Princeton University Press, 2005.

Haunsperger, Deanna, and Stephen Kennedy. "Coal Miner's Daughter." *Math Horizons* 7 (April 2000).

National Academy of Sciences. "InterViews: Ingrid Daubechies." http://www.nasonline.org/site/PageServer?pagename=INTERVIEWS_Ingrid_Daubechies.

SARAH BOSLAUGH

See Also: Animation and CGI; Digital Images; Digital Storage; Women.

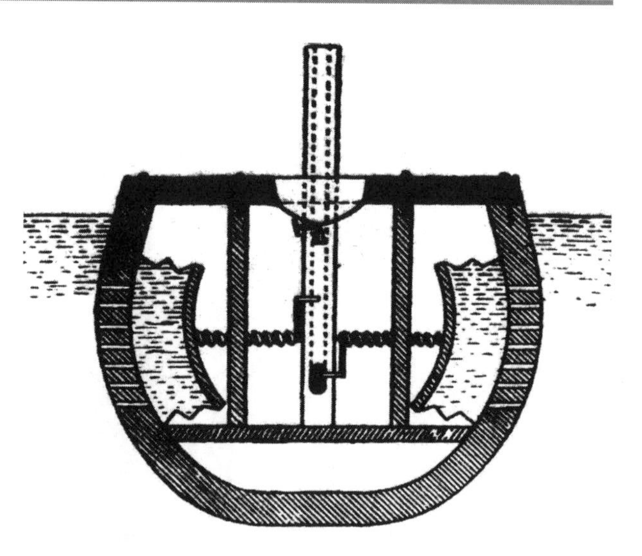

William Bourne's 1578 design was one of the first recorded plans for an underwater navigation vehicle.

Deep Submergence Vehicles

Category: Travel and Transportation.
Field of Study: Algebra; Measurement; Number and Operations.
Summary: Submergence vehicles must be carefully designed to take into account undersea conditions.

Deep submergence vehicles are primarily designed to aid researchers in exploring the depths of Earth's oceans. Much is unknown about the suboceanic environment, and exploration of these depths requires transport vehicles that can withstand tremendous pressures. Modern submergence vehicles can not only dive to great depths but can also stay submerged for hours at length, and are equipped with external lights and tele-operated robotic manipulators to gather deep sea samples for further research. Besides researching marine life, deep submergence vehicles also play vital roles in the oil exploration and the telecommunications industries where robotic submarine vehicles known as "autonomous underwater vehicles" detect faulty cables and help in oil field exploration. English mathematician William Bourne may have been the first to record a design for an underwater navigation vehicle in 1578. In addition to mathematics and mathematicians impacting deep submergence vehicles, submarines have also impacted the development of mathematics. Mathematicians examined the optimal way for airplanes to search for submarines, and the field of operations research was born.

Physical Characteristics of the Abyss

Pressure. At any given depth under the sea level, the pressure on a body can be calculated as

$$P = \rho \times g \times h$$

where P is pressure, ρ is the density of the seawater, g is the acceleration because of gravity, and h is the depth at which the measurement is being taken.

The atmospheric pressure at sea-level is about 100 kPa (~ 14.6 psi), the same amount of water pressure at about 10 meters (33 feet) below the surface, making the combined pressure experienced by a body at a 10 meter depth almost double of that at the surface.

Light. Most of the visible light entering the ocean is absorbed within 10 meters (33 feet) of the water's surface. Almost no light penetrates below 150 meters (490 feet). Solid particles, waves, and debris in the water affect light penetration. The longer wavelengths of light, red, yellow, and orange, penetrate to 15, 30, and 50 meters respectively, while the shorter wavelengths—violet, blue, and green—can penetrate further. The depth of water where sunlight penetrates sufficiently for photosynthesis to take place is called the Euphotic Zone and is normally around 200 meters (655 feet) in the ocean. The zone where filtered sunlight only suffuses in the water is known as the Disphotic Zone and extends from the end of the Euphotic Zone to about a depth of 1000 meters. Below that, no sunlight ever penetrates, and this is known as the Aphotic Zone.

Temperature. There is a significant difference in the temperatures between the Euphotic and Aphotic zones. However, in the Aphotic Zone, the temperature remains almost constant, hovering around 2 to 4 degrees Celsius. The only exception occurs when deep-sea volcanoes or hydrothermal vents exist, which cause significant warming of the waters.

History

The earliest deep-sea submersibles were known as "bathyspheres" (from *bathys*, Greek for "deep"). They were raised in and out of the water by a cable. They were fitted with oxygen cylinders inside to provide air to the divers, and had chemicals to absorb the expelled carbon dioxide. The early bathyspheres were not maneuverable—the only degree of freedom they had enabled them to go up and down.

The notable Swiss physicist Auguste Piccard (1884–1962) was influential in making the next design iteration to the bathysphere, called the "bathyscaph." The vessel was not suspended from a ship but instead attached to a free-floating tank filled with petroleum liquid. This tank made it buoyant (lighter than water). The bathyscaph had metal ballasts that, when released, allowed the vessel to surface. Auguste and his son Jacques designed the next generation bathyscaph, the Trieste. The Trieste set a new world record when it reached the lowest point on Earth, the Marianas Trench (35,800 feet).

Improvements in electronics and materials engineering have led to the design of Alvin, a deep-sea vessel capable of accommodating up to three people and diving for up to nine hours. Alvin sports two robotic arms that can be customized depending on the mission it is undertaking. Alvin's most notable contribution was its role in exploring the RMS Titanic.

Further Reading

Arroyo, Sheri, and Rhea Stewart. *How Deep Sea Divers Use Math*. New York: Chelsea House, 2009.

Morse, Philip and George Kimball. *Methods of Operations Research*. Kormendi Press, 2008.

Mosher, D. C., Craig Shipp, Lorena Moscardelli, Jason Chaytor, Chris Baxtor, Homa Lee, and Roger Urgeles. *Submarine Mass Movements and Their Consequences*. New York: Springer, 2009.

Ashwin Mudigonda

See Also: Coral Reefs; Marine Navigation; Robots; Tides and Waves.

Deforestation

Category: Weather, Nature, and Environment.
Fields of Study: Algebra; Measurement; Problem Solving.
Summary: Mathematicians study and model many aspects of deforestation.

Deforestation is the removal of forests by logging or burning. While some deforestation can occur accidentally as a result of wildfires, most is deliberate. Trees may be sold for lumber or charcoal, and land may be cleared for housing, farming, or pasturing livestock. Trees may also be removed for beneficial purposes, such as directing water flow or controlling future forest fires. Many people believe that deforestation is a significant factor in climate change and biodiversity loss, and research has shown that deforested regions are much more vulnerable to soil erosion and desertification.

While logging is linked to deforestation in the popular imagination, the United Nations Framework Convention on Climate Change actually found that in the early twenty-first century, logging actually accounted for less than 20% of deliberate deforestation. In contrast, commercial agriculture claimed about one-third of deforested lands and subsistence farming nearly one-half. This statistic indicates one reason why deforestation is increasing primarily in relatively poorer countries. However, within an industrialized country, like the United States, logging and clearing land for housing or other real estate development account for far more deforestation than subsistence farming, which few Americans have practiced since the dawn

Many people believe that deforestation is a significant factor in climate change and biodiversity loss, and research has shown that deforested regions are much more vulnerable to soil erosion and desertification.

of the twentieth century. Mathematicians study and model many aspects of deforestation, including possible causes and the biological, geological, social, and economic effects; uses of deforested land; patterns of regrowth and biodiversity in areas where the forest has been allowed to return; and spatial mapping and visualizations of geographical regions before, during, and after deforestation. Data collection, statistical analyses, and spatial dependency analyses, as well as stochastic spatial modeling, linear programming, geometry, and digital image analysis, are all mathematical methods that have played a role in such analyses.

Environmental Effects

Deforestation is implicated in numerous environmental problems. The relationship between the forest and atmospheric carbon dioxide, for instance, is complicated. While they are alive and actively growing, trees remove carbon dioxide from the atmosphere, store it as carbon, and release oxygen back into the atmosphere through respiration. This process reduces the amount of greenhouse gases in the atmosphere, and this basic dichotomy—plants breathing in carbon dioxide and releasing oxygen, while humans and animals do the opposite—has long been taught to schoolchildren as the critically interdependent relationship between flora and fauna on Earth. In the early twenty-first century, the world's forests store roughly three-quarters or greater of aboveground and soil carbon. When trees are cut down and burned, they release their stored carbon back into the atmosphere. When trees die and decay, they do the same, as fungi and bacteria break down the carbon products into carbon dioxide and methane. Their effect on the world's oxygen supply is actually very minor—the amount of oxygen they release is not as significant as the amount of carbon dioxide involved in a tree's lifespan.

But cutting trees down and turning them into long-lived products (using them to build houses, for instance) stores the carbon just as efficiently. For forests to continue to take carbon dioxide in from the atmosphere, the trees must be harvested regularly—with new trees planted—so that there are always actively growing trees. Left to their own devices, mature forests cycle through periods as carbon dioxide sources (when the carbon dioxide released by decaying or wildfire-burned trees exceeds that taken in by growing trees) and sinks (when the net carbon dioxide release is negative).

The greatest amount of carbon dioxide is taken in by deciduous trees when spring leaves are growing, which results in an observable dip in the Keeling Curve (a graph that tracks variation in the concentration of atmospheric carbon dioxide from 1958 onward). The dip is mirrored by a rise corresponding to the release of carbon dioxide back into the air every fall when these leaves fall and decay. The curve is named for Charles Keeling, a University of California, San Diego, oceanographer whose observations helped bring global attention to anthropogenic climate change. Measurements continue to be taken at Mauna Loa, in Hawaii, and those data have shown a roughly 20% to 25% increase in the amount of atmospheric carbon dioxide between 1958 and 2010. There have been no declining trends in that time, countering the pre-Keeley claim that an apparent rise in carbon dioxide atmospheric concentration was the result of random fluctuations. Periodic local decreases and increases of about 1% to 2% are associated with seasonal cycles.

Anti-Deforestation Efforts

Recent efforts to reduce greenhouse gas emissions, and international agreements binding countries to do so, have brought more focus to the task of accurately measuring those emissions. It came to light in 2010 that Australia's efforts to reduce emissions in order to comply with the Kyoto Protocol goals were hampered by their inaccurate measurement of deforestation emissions. Since 1990, Australia has had the highest rate of deforestation in the developed world, and thus is the only developed country targeting deforestation emissions as its primary way of reducing overall emissions. But its inability to generate an accurate figure of what those emissions currently are, to establish a baseline, or reliably measure them in the future, has thrown a wrench in its efforts.

Data Collection and Mathematical Modeling

The highly complex nature of forest ecosystems and even individual trees makes it virtually impossible to collect complete data on the system dynamics of natural forests. As a result, investigations of long-term dynamics rely heavily on scientific inference. One way of making any estimate, heavily relied on when considering the environmental costs of possible actions, is through ecosystem modeling, which constructs mathematical representations of ecosystems. The entire ecosystem need not be represented (though this leaves open the possibility of unforeseen consequences in parts of the ecosystem not modeled). Typically, models are constructed to examine the inventory of a specific chemical in the environment, like carbon, nitrogen, phosphorous, or a toxin. The ecosystem is reduced to a set of state variables that describe the state of a dynamic system, like the population of a specific species or the concentration level of a particular substance.

Mathematical functions define the relationships between those variables, such as the relationship between new leaf growth and carbon dioxide intake. A usable model typically requires many variables and much fine-tuning to affirm that the relationships have been defined accurately, and, in some cases, a model may be constructed simply to test a hypothesis about those relationships, by comparing the behavior of the model ecosystem to the real one. For example, mathematician and ecologist Nandi Leslie developed mathematical models using techniques such as spatial statistics, mean field and pair approximation, and the theory of interacting particle systems to investigate questions about forest fragmentation and degradation, ecology and biodiversity in lands reclaimed by forests, and landscape-level impact of land-use activities in Bolivia and Brazil. Leslie is included on a Web site called Mathematicians of the African Diaspora and is the daughter of mathematician Joshua Leslie, who has published widely in the fields of algebraic and differential geometry. The applications of modeling in deforestation are as broad as the types of models. Some mathematicians have used calculus to measure tree density, including the number of trees per acre and the quantity of foliage. Logistic functions have been used to estimate insect density or infestations. Many linear and nonlinear modeling techniques, like regression analysis, are widely employed to help reveal and explain associations between multiple variables, such as social choices and government policies; economic measures; environmental measures; geographic features, like altitude and slope; and human constructions, like roads. These models are then frequently used to forecast important quantities of interest, like deforestation rates and the overall proportion of deforested land. However, inappropriate extrapolations and generalizations can lead people to make inaccurate predictions or conclusions. For example, extrapolations from exponential models tend to lead to overestimation of future values. This has an impact on contentious and world-reaching scientific debates, such as global warming.

Further Reading

Babin, Didier. *Beyond Tropical Deforestation: From Tropical Deforestation to Forest Cover Dynamics and Forest Development*. Paris: UNESCO, 2005.

Fowler, Andrew. *Mathematical Models in the Applied Sciences*. New York: Cambridge University Press, 1997.

Harte, John. *Consider a Spherical Cow: A Course in Environmental Problem Solving*. Sausalito, CA: University Science Books, 1988.

Shugart, Hurman. *A Theory of Forest Dynamics: The Ecological Implications of Forest Succession Models*. Caldwell, NJ: Blackburn Press, 2003.

Bill Kte'pi

See Also: Floods; Forest Fires; North America; South America.

Deming, W. Edwards

Category: Business, Economics, and Marketing.
Fields of Study: Connections.
Summary: W. Edwards Deming (1900–1993) was an American applied statistician who revolutionized Japanese management as part of the rebuilding effort after World War II.

W. Edwards Deming was an internationally renowned consultant whose work led Japanese industry into new principles of management and revolutionized quality and productivity in Japanese companies. He believed: "Innovation comes from freedom. It comes from those who are obligated to no one. It comes from people who are responsible only to themselves." He was born October 14, 1900, and died on December 20, 1993. His undergraduate degree in engineering was from the University of Wyoming in 1921. He earned a M.S. in physics and mathematics from the University of Colorado in 1925 and a Ph.D. in physics from Yale University in 1928. He had a wide and varied career, which included his first scientific paper on the nuclear packing of helium, mathematical and statistical work for the Department of Agriculture in Washington, D.C., and work on sampling issues for the U.S. Census Bureau. He noted that he also worked on many different studies, including: "application of statistical theory to problems that arise in industrial production, in tests of physical materials . . . motor freight, rail freight, accounting . . . average life of returnable bottles, comparison of medical treatments, comparison of methods of diagnosis, social, and demographic problems created by physical or mental handicaps. My part in any study is the design thereof, followed by evaluation of the statistical reliability of the results." He received many honors and awards and was also an active member in professional societies, such as the American Statistical Association and as president of the Institute of Mathematical Statistics in 1945.

In 1999, the *Los Angeles Times* recognized him as one of the 50 people who most influenced business in the twentieth century because of his work in rebuilding Japan after World War II. He urged Japanese companies to concentrate on constant improvement, improved efficiency, and doing things right the first time. The essence of his ideas was based on the concepts of statistical process control, a process originally developed by Walter A. Shewhart in the 1920s. It has since been expanded to include the total quality management approach.

The essence of Deming's process was to record the number of product defects, statistically analyze why those defects occurred, institute changes to correct the defects, record how much the quality then improved, and to continue to refine the production process until it was done correctly. He said: "If you don't have a method, you were goofing off. A system must be managed and must have an aim."

Deming first successfully applied his ideas in the United States during World War II in improving the manufacture of munitions and other strategically important products. As mentioned above, he brought those same ideas to Japan in the 1950s and early 1960s. During that time period, "Made in Japan" went from being a joke and a synonym for poor quality to a symbol of some of the highest quality products. The focus on quality that he emphasized was defined as the ratio of results of work efforts with total costs. If a company or manager focuses on quality, Deming's work demonstrated that, over time, quality will increase and costs will fall. On the other hand, if the focus is primarily on costs, then costs will rise and quality will decline.

Two major publications have outlined his theories and the processes he developed. In his 1982 book *Out of the Crisis*, Deming discusses his 14 key principles for management for transforming business executives. Deming felt that if his 14 points were applied in a meaningful way, they would lead to a process of continual improvement. *The New Economics*, published in 1993, emphasized that the solution to problems comes from cooperation, not competition. This concept is accomplished through a new type of management, which Deming identified as profound knowledge and which includes four parts: appreciation for a system, knowledge about variation, theory of knowledge, and psychology.

Deming also had an interest in music. He composed several pieces, mostly liturgical. He also composed a new rendition of the *Star Spangled Banner* with the same words set to a different tune. He had always felt that the "pub" music of the original version was not appropriate for a national anthem.

Further Reading

Deming, W. Edwards. *Out of the Crisis.* Cambridge, MA: Massachusetts Institute of Technology Center for Advanced Engineering Study, 1982.

———. *The New Economics.* Cambridge, MA: MIT Center for Advanced Engineering Study, 1993.

———. "The W. Edwards Deming Institute: Dr. Deming's 1974 CV—from his Study." http://deming.org/index.cfm?content=621.

Paton, Scott. "Four Days With W. Edwards Deming." http://deming.org/index.cfm?content=653.

Jim Austin

See Also: Data Analysis and Probability in Society; Quality Control; Scheduling; Statistics Education.

Diagnostic Testing

Category: Medicine and Health.
Fields of Study: Data Analysis and Probability; Measurement; Number and Operations.
Summary: Diagnostic tests rely on statistics from clinical research to predict the presence or severity of a disease in a specific patient.

The ability of humans to detect and treat diseases has advanced considerably in the past two centuries, with the discovery of underlying causes, such as microorganisms, and treatments, like antibiotics, as well as methods for diagnosing injury and disease. In medicine, a diagnostic test is in an instrument used to detect or predict the presence or absence of disease or the severity of disease.

The instrument used may take a variety of forms, including a patient inventory or a mechanical device. In clinical research, it is common practice to assess the quality of such instruments relative to established gold standards.

Here, the intention is often to replace a traditional method by a newer one that offers greater benefits to health providers or patients, including cost reduction and less physical or psychological discomfort.

It may be of interest to use the diagnostic tool to predict outcomes based on existing symptoms. In this case, the gold standard is used to confirm patient outcomes for comparison with test predictions based on surrogate measures.

Common measures of instrument quality include reliability, validity, sensitivity, specificity, positive predictive value (PPV), and negative predictive value (NPV). Strictly speaking, these measures apply specifically to the scores forthcoming from the instruments rather than the instruments themselves, as they are based on studies applied to a specific sample of patients. Mathematicians and statisticians are essential partners in creating many diagnostic tools, such as magnetic resonance imaging, as well as for developing and refining the measures that allow clinicians and researchers to determine the efficacy of diagnostic instruments. They also help design experiments in which new instruments are tested and compared.

Nursing and other healthcare education programs frequently require courses in mathematics or statistics, and the field of biostatistics is one of the fastest-growing occupations in the late twentieth and early twenty-first century.

Reliability represents the reproducibility of the test outcomes. A simple case involves estimation of the extent of chance-corrected agreement in the interpretation of categorical findings from medical images derived from patients. Here, agreement might be measured across different clinicians based on a single imaging procedure or alternatively, across different imaging procedures. In such cases, an appropriate choice of Kappa statistic or intra-class correlation coefficient may prove helpful. For continuous data, the Bland–Altman method has also proved particularly popular in measuring agreement across different methods. This is especially so within medicine, where for example, there may be a need to compare residual tumor sizes obtained using magnetic resonance imaging, and pathologic findings (the gold standard) in breast cancer patients who have undergone neoadjuvant (preoperative) chemotherapy.

The remaining measures above represent the accuracy of the test outcomes. Validity, which is a function of reliability, represents the extent to which the diagnostic test measures what is intended and is particularly relevant in psychological testing. Sensitivity (specificity) measures the proportion of genuine instances of disease (absence of disease, respectively), which are detected as such by the diagnostic test. By contrast, the PPV (NPV) measures the proportion of cases diagnosed by the test as instances of disease (absence of disease, respectively) which are, or will turn out to be, genuine. In assessing test accuracy, it can prove misleading to focus exclusively on sensitivity and specificity.

The PPV and NPV for a disease are influenced strongly by disease prevalence (the pre-test probability that a randomly chosen person from the study cohort has the disease). The PPV increases with increasing prevalence and where prevalence is particularly low (less than 5%), the PPV can be markedly improved by moderate increases in test specificity. In interpreting a published PPV, it is essential not only to consider the CI but also to verify whether disease prevalence for the published study is representative of that for the types of patient currently under consideration. This requirement is also particularly true of the NPV.

Further, it is typically the case that an initial stage has occurred whereby diagnostic test measurements in continuous form have been classified into categories. This categorization requires the derivation of a threshold value for differentiating between diseased and non-diseased patients. The clinician may be interested in finding the threshold value that offers an optimal combination of values for sensitivity and (1-specificity). Examples of scores that have been used in this way include

- The GRACE (Global Registry of Acute Coronary Events) score in predicting death and myocardial infarction for patients with Acute Coronary Syndrome
- The APACHE (Acute Physiology and Chronic Health Evaluation) II score and GS (Glasgow Severity) score in the prediction of each of onset of severe pancreatitis, MODS (multiorgan dysfunction syndrome), and death in patients presenting with acute pancreatitis
- The MELD (Model of End-Stage Liver Disease) and UKELD (United Kingdom MELD) scores in the assessment of risk of acute liver failure and hence the prediction of waiting list mortality in patients awaiting liver transplants

The underlying procedure for deriving the threshold value involves the segregation of the test instrument scores into two groups, as determined by the gold standard, namely those who do and those who do not have the condition of interest. The accuracy of the diagnostic test is in turn assessed on the basis of these two groups. This assessment involves generating a series of threshold values and corresponding values for sensitivity and 1-specificity. The ROC curve (Receiver Operating Characteristic) involves a plot of sensitivity versus 1-specificity. If the intention is to compare the performance of competing diagnostic tests, ROC curves for the different tests can be plotted on the same graph. For any one plot, the numerically optimal combination of sensitivity and specificity values is represented by the point on the curve that is closest to the top left-hand corner. However, the trade-off between sensitivity and specificity must also be carefully weighed.

For example, if the test is confirmatory, as might be the case in human immunodeficiency virus (HIV) testing, it may be preferable to choose a slightly different point, which further reduces the proportion of false positives (1-specificity) with a small cost to sensitivity. In comparing the accuracy of two tests by means of ROCs, it is common to use the area under the curve (AUC).

Where the diagnostic test identifies cases falling into the upper (lower) range of a test score, the AUC may be interpreted as a measure of the likelihood for a randomly chosen diseased patient and disease-free patient that the diseased patient will have a higher value (lower value, respectively) than the disease-free patient.

Where ROCs do not overlap, therefore, the greater the area under the curve, the more effective the diagnostic tool. Where they do overlap, the curve with the lower overall AUC may have a peak at an optimal combination of sensitivity and specificity values not attained by the other curve. It may therefore make sense to compare the partial areas under the curves within one or more ranges of specificity values.

Further Reading

Fox, Keith A., et al. "Prediction of Risk of Death and Myocardial Infarction in the Six Months Following Presentation With ACS: A Prospective, Multinational, Observational Study (GRACE)." *British Medical Journal* 333 (2006).

Lasko, Thomas A., et al. "The Use of Receiver Operating Characteristic Curves in Biomedical Informatics." *Journal of Biomedical Informatics* 38, no. 5 (2005).

Modifi, Reza, et al. "Identification of Severe Acute Pancreatitis Using an Artificial Neural Network." *Surgery* 141, no. 1 (2007).

Neuberger, James, et al. "Selection of Patients for Liver Transplantation and Allocation of Donated Livers in the UK." *GUT* 57 (2008).

Obuchowski, Nancy A. "Receiver Operating Curves and Their Use in Radiology." *Radiology* 229 (2003).

Pan, Jian-Xin, and Kai-Tai Fang. *Growth Curve Models and Statistical Diagnostics*. New York: Springer, 2002.

Partidge, Savannah C., et al. "Accuracy of MR Imaging for Revealing Residual Breast Cancer in Patients Who Have Undergone Neoadjuvant Chemotherapy." *American Journal of Roentgenology* 179 (2002).

Ward, Michael E. "Diagnostic Tests." www.chlamydiae.com/restricted/docs/labtests/diag_examples.asp.

Wilson, Edwin B. "Probable Inference, The Law of Succession, and Statistical Inference." *Journal of the American Statistical Association* 22, no. 158 (1927).

Zhou, Xiao-Hua, Donna McClish, and Nancy Obuchowski. *Statistical Methods in Diagnostic Medicine*. Hoboken, NJ: Wiley-Interscience, 2002.

Margaret MacDougall

See Also: Chemotherapy; Data Analysis and Probability in Society; Diseases, Tracking Infectious; HIV/AIDS; Medical Imaging; Probability; Psychological Testing; Surgery; Transplantation.

Dice Games

Category: Games, Sport, and Recreation.
Fields of Study: Algebra; Data Analysis and Probability; Number and Operations.
Summary: Probability is the key factor for winning any dice game.

Dice games use one or more dice as central components of the activity, which excludes board games using dice solely as random devices to determine moves. The definition can be murky, as in the case of Backgammon, dice outcomes determine a player's moves and are integral parts of game strategies. Historically, dice games involving gambling led to the creation of probability.

History

Archaeological evidence from as early as 6000 B.C.E. shows that dice games were part of early cultures, where dice were cast to invoke personal divinations. The notion of "luck" was not involved, with the dice rolls controlled by the gods. Gamblers still refer to Fortuna, the Roman goddess and Jupiter's daughter, as their "Lady Luck."

The ancient die differed from the six-sided cube bearing pips, as the number of sides varied with the materials used, including fruit pits, nut shells, pebbles, and animal knucklebones. The latter, with four sides involving different probabilities, led to the phrase "rolling the bones."

Compulsive gambling and dice games have always been connected, being traced to Egyptian pharaohs, Chinese leaders, Roman emperors, Greek elite, European academics, and English kings. On the request of professional gamblers in the fifteenth and sixteenth centuries, mathematicians such as Fra Luca Bartolomeo de Pacioli and Girolamo Cardano began to study the probabilities of winning dice games. In the seventeenth century, correspondence between Blaise Pascal and Pierre Fermat ultimately solved the "problem of points" and established basic principles of probability.

The problem of points involves a dice game between two players; multiple rounds are played with each player having an equal chance of winning on each roll. If the game was interrupted before either player had won the necessary number of rounds, gamblers could not determine the "fair" division of stakes based on current scores. Fermat and Pascal's solution analyzed the probability of dice rolls and each player winning the pot.

Types of Dice Games

The simplest dice game involves a single die, where the winner is the person rolling the highest number. This can be extended to rolls of multiple dice, with the player's score being the sum or product of the numbers shown. Since these dice games involve only luck, gamblers prefer variations with elements of strategy.

The dice game craps involves strategy, as the "shooter" controls the number of dice rolls and betting options. Though craps is complex, key elements can be explained. Mathematically, each roll of two dice has 36 possible outcomes with shown totals ranging from "2" to "12". However, the probabilities of the totals vary, as the probability of a "2" (known as "snake eyes") or "12" (known as "boxcars") is 1/36, while the probability of a "7" is 6/36. Prior to the first "come out roll," players bet on the "Pass Line" or "Don't Pass Line." If the "shooter" then rolls a "7" or "11," the "Pass Line" bet wins double their amount and the "Don't Pass Line" bet is lost. However, if the initial roll is a "2," "3," or "12," the "Pass Line" bet is lost, while the "Don't Pass Line" bet is doubled if a "2" or "3" shows and is returned if a "12" ("push") shows. A sum of "4," "5," "6," "8," "9," or "10" becomes the "point" number, which the shooter tries to duplicate on the second roll. If the point number is made, the point bet is won and additional rolls can be made. But, if a "7" is rolled before the point number, the shooter "craps out" and a different shooter starts a new round. Craps games involve many other options, such as "Come/Don't Come Bets" and "Horn Bets."

Other dice games are used for gambling, each with their own multiple versions and strategies. For example, in the dice game Ship, Captain, and Crew, a players gets three rolls of five dice to gain a ship ("6"), a captain ("5"), and a crew ("4") in that order (or simultaneously). When those special numbers are rolled, that die is removed from play, with a successful player's score being the sum of a roll of the two remaining dice.

In Buck Dice, a player throws one die to determine the "point number." Another player then rolls three dice, continuing the rolls as long as one of the dice equals the point number. When this doesn't occur, the player's score for that round is the number of rolls.

The dice game craps is thought to have developed from a simplification of the Old English game Hazard.

A "big buck" occurs when all three dice equal the point number, and the player withdraws from the game. A "little buck" occurs if all three die do not equal the point number, which adds 5 points to the player's score. Any player with exactly 15 points withdraws from the game; any score forced higher than 15 nullifies a roll, and the player must reroll. The loser is the last person without reaching 15.

In Aces, a player starts with at least five dice, which he or she loses according to the numbers thrown. All rolled "1"s are placed in the table's center and eliminated. All rolled "2"s are passed to the player on the left, while all "5"s are passed to the player on the right. Turns continue with rolls of the remaining dice until players either do not throw a "1," "2," or "5," or have lost all of their dice. Play continues around the table until the last die rolled is a "1," and the player who threw it is the winner.

Farkle begins with a player rolling six dice. Each "1" adds 100 points, each "5" adds 50 points, and if three dice show the same number, the player adds 100 times that shown number. A player can stop after any roll and keep the current total. Alternately, a player can roll again to possibly increase his or her score. But, if the next dice do not produce a positive score, the player lose all accumulated points for that round. The winner is the first to reach 10,000 points. Some variations of Farkle give 1000 points for shown runs of "1–5" or "2–6."

In line with their history, multiple versions of dice games exist and will continue to be used by gamblers. Thus, the players who understand the probabilities involved will always have the advantage.

Further Reading

Barboianu, Catalin. *Probability Guide to Gambling: The Mathematics of Dice, Slots, Roulette, Baccarat, Blackjack, Poker, Lottery, and Sport Bets*. Craiova, Romania: INFAROM Publishing, 2008.

Bell, R. C. *Board and Table Games From Many Civilizations*. New York: Dover Publications, 1979.

Devlin, Keith. *The Unfinished Game*. New York: Basic Books, 2008.

Mohr, Merilyn. *The Game Treasury*. Shelburne, VT: Chapters Publishing, 1993.

Jerry Johnson

See Also: Betting and Fairness; Board Games; Game Theory.

Digital Book Readers

Category: Communication and Computers.
Fields of Study: Geometry; Measurement; Number and Operations; Representations.
Summary: The twenty-first-century surge in e-books began with the advent of "electronic ink" and future innovations include sketchpad-like functionality.

People have been reading digital content on computer screens since the 1970s, but the technology used for most computer screens at the end of the twentieth century made them somewhat less useful for replacing paper books, magazines, and newspapers. In 1971, volunteers started digitizing and archiving books for Project Gutenberg, whose goal was to encourage the development of electronic books. Research on electronic paper began in the 1970s. Many open and proprietary digital document formats were devised for potential use in e-books, like Adobe's Portable Document Format (PDF), created by mathematician and engineer John Warnock. However, most early attempts at digital books were unsuccessful or aimed at niche technical audiences.

In the early twenty-first century, the E Ink company introduced electronic ink technology, which revolutionized digital books. The company was co-founded by several individuals, including physicist Joseph Jacobson and Russell Wilcox, who has a degree in applied mathematics. The resulting "electronic paper" has a high contrast ratio similar to standard paper, and for most users it closely matches the experience of reading on standard paper. One early application was flexible, changeable store signs. The 2004 Sony Librié, released in Japan, was the first e-reader to make the technology widely available, while the Amazon Kindle is credited with popularizing it in the United States. As of 2010, there were many variations on e-readers with the ability to display multiple e-book formats. Some of the most popular included the Sony Reader, the Amazon Kindle, and the Barnes & Noble Nook. Motorola's FONE F3 was the first portable phone to include this technology.

Electronic Ink

Electronic ink technology is based on microcapsules, which were already in use for applications like scratch-and-sniff stickers and time-release medications. Rotating microcapsule spheres for electronic ink are filled with a clear liquid containing a mix of small, electrically

A third-generation Amazon Kindle, showing text from Herman Melville's novel Moby-Dick.

charged black and white particles. Some implementations contain on the order of 100,000 spheres per square inch. Electronic paper is a sheet of plastic coated with millions of microcapsules and equipped with an electronic device to draw the black and white particles into desired patterns of black and white dots.

When viewed from a distance, the patterns create words and pictures. The dots can also be mixtures of black and white, resulting in a range of grayscale tones. To change the image, computer programs in the reader send an electronic pulse to rearrange the pattern. Microcapsules are bistatic, which means they stay in place once they are arranged without drawing continuous electrical power. This factor contributes to long battery life. Electronic paper also has no backlighting like personal computer screens; it uses light reflection for viewing, just like ordinary paper. Scientists are investigating red, green, and blue filters to produce full-color electronic ink images. A version of the Barnes & Noble Nook released in 2010 uses a liquid crystal display (LCD) screen for color and touch-screen functionality. Some praise this, while others consider it to be a step backward in e-reader technology.

Early twenty-first-century digital book readers embody several other features that make them well-suited alternatives for leisure reading and textbooks in schools. One important aspect is their portability with high-capacity storage. Typical readers have the capability to store hundreds of books, so all required textbooks could be stored in single digital reader. Connectivity via wireless networking allows the downloading of a variety of books or teacher-created documents, including RSS feeds for blogs and Web content. RSS was developed by programmers like David Winer, who has degrees in mathematics and computer science. The reading experience is customizable; some e-readers have touch-screen navigation, adjustable font levels, the ability to take notes directly on screen or highlight text sections, built-in dictionaries, or search functions.

Readers that debuted in 2010 featured applications to allow users to write or draw, like a tablet PC, which would be important for many mathematical subjects, like geometry. Some mathematics educators have explored the use of electronic ink to support mathematics distance education. For example, electronic ink tools in a chat program allowed students and instructors to post and edit mathematical formulas, diagrams, and graphs while communicating in real time.

Further Reading

Howard, Nicole. *The Book: The Life Story of a Technology*. Westport, CT: Greenwood Press, 2005.

Kipphan, Helmut. *Handbook of Print Media: Technologies and Production Methods*. New York: Springer, 2001.

Serkan Ozel
Zeynep Ebrar Yetkiner

See Also: Cell Phone Networks; File Downloading and Sharing; Personal Computers.

Digital Cameras

Category: Arts, Music, and Entertainment.
Fields of Study: Geometry; Measurement; Number and Operations; Representations.

Summary: Rapid advances in digital camera technology have led to their widespread use.

From the invention of modern photography in the 1800s to the rise of digital photography in the twenty-first century, the function of the camera has been the same: to record patterns of light. The word "photography," coined by Sir John Herschel in 1839, is from the Greek *phos* (light) and *gráphein* (to write). Simple pinhole cameras were described as early as the fourth and fifth centuries B.C.E. by Chinese philosopher Mo Ti and Greek mathematicians Aristotle and Euclid of Alexandria. Mathematician and physicist James Maxwell created the first color photograph in 1861. Not long after, American inventor and Kodak founder George Eastman developed inexpensive equipment and film that made photography practical for common use. Until recently, cameras recorded images on media coated in photosensitive compounds. Incoming light was registered as a chemical change that could be seen upon development in specialized photochemistry. Digital cameras use an electronic chip that is sensitive to light. The chip, either a charge coupled device (CCD) or complementary metal-oxide-semiconductor (CMOS), converts the light into an electrical signal, and a small computer in the camera then transforms that signal into the "ons" and "offs" (or "1"s and "0"s) of binary code for storage on a digital storage device. The digital information that represents an image can easily be copied onto a computer, manipulated, published electronically, and printed. Researchers also investigate mathematical questions like how many images one should shoot in order to be reasonably confident that no person in the photograph blinks. For groups under 20 people, the number of images is approximately equal to the number of people divided by one-half.

The Lens and the Shutter

Like most cameras, a digital camera begins the process of taking a photograph by letting light in through a lens (a curved piece of glass or plastic) that bends light through the principle of refraction and focuses the image. The light then passes through an opening called an "aperture" whose size can be adjusted to let more or less light pass. Apertures are described by an f-stop number, which is proportional to the focal length of the lens over the diameter of the entrance pupil. Since it is a ratio, larger f-stop numbers refer to smaller apertures. Each doubling or halving of the f-stop number translates to a change in amount of light let in by a factor of four. Thus, an f-stop of 11 lets in four times more light than an f-stop of 22. Finally, the light comes to a shutter that opens for a period of time when the shutter release is triggered, allowing light into the camera body. Usually, a shutter speed is a fraction of a second, though long shutter speeds can be used in low light, or for a variety of effects. With especially long shutter speeds, heat can build up in the CCD or CMOS, causing electrical interference that interferes with accurate, binary recording of the image, resulting in error or "noise," though camera manufacturers are developing a number of processes that have made this less of a problem over time.

The CCD or CMOS

In order to capture an image, the light that comes into the camera falls on the CCD or CMOS chip, which changes the image into electric current. A CCD is made up of tiny regions called "picture elements" or "pixels" that will correspond to the points in the photograph. A CMOS works similarly, though the specific underlying technology is a bit different. Some cameras have one CCD for all three primary colors of visible light, red, green, and blue; each pixel records only color of light from the scene. More advanced cameras use three different CCDs, one for each primary color, resulting in a more accurate image. Ultimately, the electric current from the CCD is encoded by a small computer in the camera into a stream of binary information in the form of "ons" and "offs" that ultimately will be stored on a flash memory card.

Sensitivity and ISO

In film cameras, different formulations of film are used for different light conditions, with more sensitive films employed in low light. In a digital camera, the signal from the CCD can be boosted to handle low light levels; however, doing so introduces noise in the signal. The setting for camera sensitivity is described as its "ISO number," an international standard for measuring the speed of color film. It uses both an arithmetic and logarithmic scale to combine two previous film standards. In the arithmetic scale, which is commonly the only one given, each doubling of the ISO representing a doubling of the sensitivity. Thus, a camera set to ISO 100 will be half as sensitive to light—and will require twice as long an exposure for a given scene to achieve the same result, given the same f-stop setting—as one with an ISO of 200.

Pixel Dimensions

One of the factors that determines the picture quality of an image produced by a digital camera is the number of pixels it records. This is especially relevant when images are blown up to large dimensions, as the individual pixels begin to become visible. Pixels are the individual binary units into which the image is "broken up" and stored during the electronic conversion process by the camera's chip. For example, the Droid Incredible phone, released in 2010, contains an 8 megapixel camera, which means its photographs are composed of about 8 million individual pixels, with each picture having a possible resolution of roughly 3264 pixels wide by 2468 pixels high. However, in practice, there are many factors that affect picture quality. The size of the electronic chip plays a large role. When the photosensitive regions of a camera's chip are packed too tightly together, they create electronic interference in their neighbors, potentially affecting the binary storage, and ultimately affecting the accuracy and quality of the stored image.

Further Reading

Stone, J., and B. Stone. *A Short Course in Digital Photography.* Upper Saddle River, NJ: Prentice Hall, 2009.

Svenson, Nic. "Velocity Science in Motion: Blink-free Photos, Guaranteed." http://velocity.ansto.gov.au/velocity/ans0011/article_06.asp.

Jeff Goodman

See Also: Digital Images; Digital Storage; Movies, Making of.

Cameras in Mathematics Classrooms

Cameras have grown in popularity since Eastman first made them readily available. Digital cameras are relatively inexpensive, and, in fact, are now standard features on many cell phones. Educators have seized on the digital camera as a very useful tool in the classroom for introducing concepts, making connections, and enriching educational experiences in a very hands-on way. For example, students in middle grades and above have been asked to use digital cameras to record their own examples of geometric concepts found in the world. They can then use the photographs, along with various mathematics concepts such as scaling and trigonometry, to answer questions like "Is the Houston Astrodome really round?" or "What is the slope of a roof?" In other cases, students use photos to record and measure themselves and their classmates—either once or repeatedly over time—to provide data for many interesting mathematics activities and discussions, such as variability and the importance of repeated sampling.

Digital Images

Category: Arts, Music, and Entertainment.
Fields of Study: Algebra; Geometry; Number and Operations; Representations.
Summary: Digital images are recorded as a binary account of pixels, which algorithms may compress.

Digital images are not images at all but rather are visual information encoded as binary data. Viewing a digital image requires a computer to decode binary information and display it on a screen in the form of an array of discrete lights called "picture elements" or "pixels." The first computer-generated digital images were produced in the early 1960s. The needs of the Cold War, medicine, and the space race drove many developments in digital imagery, some of which were achieved in the context of projects on satellite imagery, medical imaging, optical character recognition, and photo enhancement. The advent of microprocessors in the 1970s and advances in digital storage and display technologies made possible sophisticated imaging tools, like computerized axial tomography (CAT scanning).

The degree of mathematical sophistication that CAT scans introduced into medical imaging, such as integral geometry, optimal sampling, and transport equations, was unheard of at that time. It is reflected in further advances such as magnetic resonance imaging as well as developments in other fields that use similar imaging techniques, like seismic and electron microscopy. At the same time, scanners to digitize analog images began to be used in a diverse array of fields, such as archaeology and law enforcement. The first fully digital camera was released in 1995, and by the end of the twentieth century, charge-coupled devices (CCDs) largely displaced analog film and tape for photography and videography. Willard Boyle and George Smith shared the 2009 Nobel Prize in Physics for their invention of the CCD, an idea they first brainstormed at Bell Labs in 1969. Improved computing power also allowed for production of near-photorealistic images. All areas of digital imagery (creation, compression, restoration, recognition, and display) involve mathematics. In the twenty-first century, digital images are regularly used in both mathematics research and teaching.

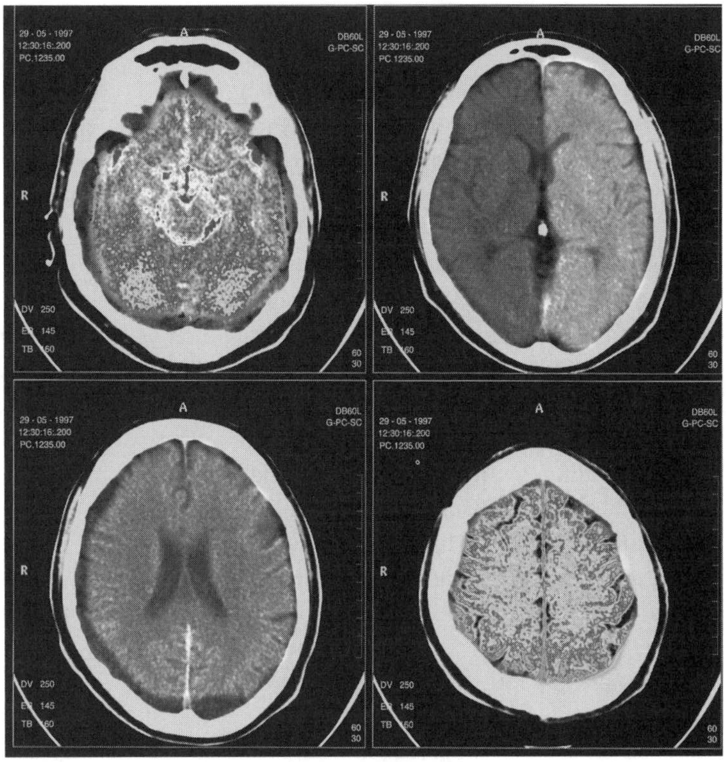

CAT scanning of the head is typically used to detect tumors, calcifications, hemorrhage, and bone trauma.

Bitmap Graphics

In most digital images, each pixel has been defined numerically and this number has been converted into a string of "1"s or "0"s. This system is the approach of "bitmap graphics" (also known as "raster graphics"), and it is how digital cameras work. Depending on the number of bits used to represent each pixel, more or less color information is given. For example, a one-bit system would allow only a black or white pixel, as the only choices would be a "0" or a "1." A two-bit system would gives four choices per pixel, "00" (black), "01" (dark grey), "10" (light grey), and "11" (white). Typically, in photo editing programs of the early twenty-first century, each pixel is described by 24 bits of information, yielding more than 16 million possible colors.

Resolution

Bitmap images contain information for a given number of pixels. The larger the pixel number, the more information is in the image and the higher the resolution; typically, this also results in a bigger file. Screens are all made of pixels, whether they are on computers or cell phones; if an image is viewed at full size, each pixel in the image will show up as one pixel on the screen. However, if a viewer zooms in beyond this point, the pixels in the file are actually represented by big blocks of pixels on the screen, and the image is said to become "pixelated."

Thus, if an image is to be viewed on a screen, it will ideally have the same number of pixels as the size one wants it on the screen; any more than that is wasted file space, and any fewer will result in an image that appears pixelated. If images are going to be printed, however, more pixels will translate into sharper pictures, limited only by the resolution of the printer. Again, the larger the print, the more pixels you will need for a sharp print.

File Types and Compression

Bitmap graphics can be stored in a variety of file formats depending on how they will be used. Raw files, which store all the raw data for the light that hits each CCD pixel, are commonly used by photographers who wish to have maximum flexibility and are not worried about file size. In order to make files smaller, computers use mathematical algorithms to compress the files. For example, instead of recording values for each pixel, the values for some could be calculated by the difference between a pixel and its surroundings, thus yielding substantial file size savings where blocks of pixels are the same as their neighbors. Some kinds of compression are considered "lossless," because all the information from the original can be re-created when the file is decompressed. However, there are a number of compression schemes such as the popular jpeg format in which the mathematical approximations do not quite match the original. In these cases, accuracy is sacrificed in order to save file size, and these approaches are said to be "lossy." However, the algorithms used to compress and decompress files are generally so good as to be unnoticeable in many cases. The JPEG 2000 image compression standard for both lossless and lossy compression uses biorthogonal wavelets, which extends from the work of mathematician Ingrid Daubechies, known as the "mother of wavelets."

Vector Graphics

Certain kinds of images, especially those created in computer graphics programs, use a different method for describing the content of the image. Instead of denoting each pixel with a number, these vector graphics are described mathematically as a set of equations representing the lines and curves that make them up. When a viewer zooms in on a vector graphic, the image does not become pixelated, because the computer recalculates the curve or line based on the new image size. While vector graphics are not appropriate for photographs, photo editing programs may use them when overlaying text or graphics on a digital image.

Image Reconstruction

The basic problem of image reconstruction is to build a "best-guess" object out of averaged data and then estimate how close the reconstruction is to the actual object. For example, in a single-angle X-ray of a person, the amount of radiation going in and coming out the other side can be measured and visualized on X-ray film. The difference between the values is how much was absorbed, but there is limited information about the inner structures that blocked the radiation. This limitation can make diagnoses difficult. However, if the same person is X-rayed from several directions and angles, the resulting information can be compiled, averaged, or mathematically modeled to estimate what the internal structure looks like.

Further Reading

Alsina, Claudi. *Math Made Visual: Creating Images for Understanding Mathematics*. Washington, DC: The Mathematical Association of America, 2006.

Hoggar, S. G. *Mathematics of Digital Images: Creation, Compression, Restoration, Recognition*. Cambridge, UK: Cambridge University Press, 2006.

JEFF GOODMAN

See Also: Daubechies, Ingrid; Digital Camera; Digital Storage; Medical Imaging; Televisions.

Digital Storage

Category: Communication and Computers.
Fields of Study: Algebra; Measurement; Number and Operations.
Summary: Information can be stored digitally—a process that requires information to be translated into binary code.

Digital information is information in binary code. In order to create, manipulate, and store this digital code, it must be created in physical form. This creation is done by using media that can exist in one of two distinct states and assigning one state to each of the two digits ("0" and "1") in binary code. Within a computer, the "1"s and "0"s are represented as "ons" and "offs"; on a magnetic hard disk, they are tiny magnets pointing one way or another; and on a CD, the two states are shiny and dull spots. Engineers used metal tape on reel-to-reel machinery to record audio signals in the early twentieth century. In 1952, IBM introduced a tape drive with iron oxide–coated plastic tape. Reel-to-reel tape drives were the standard for data storage by the mid-1970s. IBM also created magnetic hard disks in the late 1950s, but it took decades to overcome size and access speed issues to make hard disk drives (HDDs) feasible for applications like personal computers. Solid-state drive (SSD) technology, such as flash memory, was the necessary next step to overcoming the lagging mechanical speeds of HDDs. Mathematicians in many fields have been essential in all stages of development and continue to address emerging issues. Ingrid Daubechies, "the mother of wavelets," is perhaps best known for her work with wavelet-based algorithms for compressing digital images. Irving Reed and Gustave Solomon developed algebraic error-detecting and error-correcting codes. These Reed–Solomon codes are widely used in digital storage and communication, from satellites to CDs.

Bits and Bytes

The smallest unit of stored digital information, corresponding to a single "1" or "0," is called a "bit." The term "bit," a contraction of "binary digit," is commonly attributed to statistician John Tukey, working in conjunction with mathematician John von Neumann. Bits are collected into 8-unit chunks called "bytes," and these collections of 8 bits can represent various types of information. The lowercase letter "a," for instance, can be represented as 01100001, and "b" as 01100010. The music on a compact disc is encoded as a set of 44,100 reading (or samples) per second, with each reading represented by 2 bytes containing 16 bits.

Storage Size

Sizes of files, and the capacity of storage devices, are often referred to as multiples of the byte. A kilobyte (KB) is approximately 1000 bytes, enough information to store about 150 words, or about half a page of text from a paperback book. As larger units are used, the naming system employs other metric prefixes, with each step up representing a multiple of either 1000 or 1024, depending on the device. Thus, a megabyte (MB) is approximately 1000 KB, and a gigabyte (GB) is approximately 1000 MB. Units beyond the gigabyte include the terabyte (TB), petabyte (PB), and exabyte (EB).

Magnetic Storage

Since grains in a magnetic medium can be magnetized with the north pole pointed in either of two directions, magnetism is an ideal medium for representing binary information. In addition, since information stored in this way is relatively stable, it is useful for long-term storage. Finally, since this magnetism can be reset easily using an electromagnet, magnetic media are easy to erase and rewrite.

A magnetic hard disk employs one or more spinning platters coated in a magnetic medium. An arm with tiny electromagnetic heads floats over the surface of the disk and is used to magnetize regions of the disk corresponding to the "1"s and "0"s of binary code. To retrieve information, the disk spins past the heads, generating current that corresponds to the code stored on the disk. While the principle is straightforward, it has been a remarkable feat of engineering to create disks that spin up to 7200 revolutions per minute with arms that can travel across the surface of a platter 50 or more times per second as they seek and write information. Even so, writing and retrieval speeds have not increased over time at the same exponential rate as the amount of information that can be stored on such disks, resulting in undesirable lags.

Even in the early twenty-first century, long-term backup of computer information is often done on low-cost magnetic tape, with bits of information laid down as magnetic regions on moving tape. However, since the information is laid down on a long piece of tape, there can be no random access of information, limiting its usefulness in everyday applications. Until recently, digital camcorders used magnetic tape to record video; however, the desire to have random access of footage and recent advances in hard drive and other storage techniques have brought on a new generation of tapeless camcorders.

CDs, DVDs, and Flash Memory

Both CD and DVD players are optical devices that use lasers to read the shiny and dull spots encoded on a plastic disk. Information is recorded by burning nonreflective pits into the surface of the disk to represent "0"s and leaving the reflective surface to represent "1"s. When the disk is played, it spins past a laser. When the light encounters a pit, it is not reflected, and the player registers an "off" signal ("0"), and when the light bounces back off a shiny region, the player registers an "on" signal ("1"). This information is interpreted by a small computer in the player.

Many devices, including digital cameras, camcorders, video game consoles, and cell phones, use flash memory, which can store large amounts of information on small cards that have no moving parts. This technology employs an array of microscopic transistors

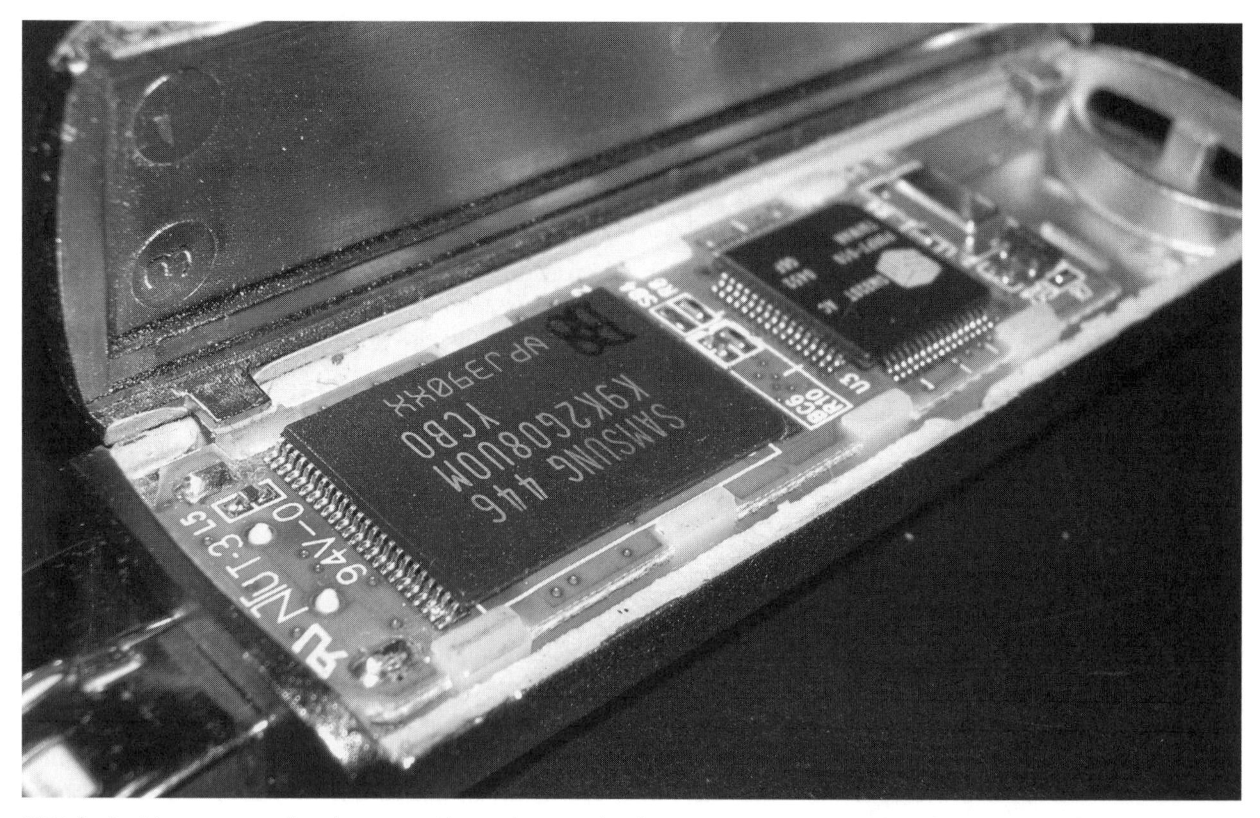

USB flash drives are smaller, faster, and have thousands of times more capacity than floppy disks or CD-ROMs. Flash memory stores and retrieves information accurately most of the time, but the devices are not problem-free.

through which current may pass. Whether this current passes through or not is controlled by what is called a "floating gate," and the path through the transistor can be electrically opened or closed. This method allows the transistor to have the two states needed for binary code. Sections of flash memory can easily be reset (erased) by flushing out the electrons trapped in the floating gate. One of the primary benefits of this technology is that information can be stored on a card with no moving parts, improving both access speed and portability.

Data Rot and Error Correction

Tape, hard disks, CDs, and flash memory store and retrieve information accurately most of the time, but they are not problem-free. Errors and noise can happen in an electromechanical recording system—"1"s that should have been "0"s, and vice versa—which diminish information accuracy. Mathematical methods are used to check for and correct errors. For example, cyclic redundancy check (CRC) coding algorithms calculate a fixed-length binary sequence (code) for each block of data using polynomial division in a finite field. The codes and data blocks are stored together, and they can be checked after transmission or retrieval. CRC was invented by mathematician W. Wesley Peterson, who also devised many error-correcting codes.

Even if the recording is perfect, the media that hold binary code can degrade in a variety of ways over time. For instance, magnetic media can lose their magnetic orientation, especially if they are subjected to a strong magnetic field. In addition, the substrates on which the magnetism is stored—the platters on hard drives and plastic backing on magnetic tape—will invariably degrade over time. Even the plastic on CDs and DVDs will begin to break down, and flash memory floating gates will ultimately leak the electrons that maintain data in their flash memory transistor states. Even if the storage media and binary information survive over time, there is a real chance that in the future there may not be hardware available to read information encoded in an outdated media.

Further Reading

Somasundaram, G., and Alok Shrivastava. *Information Storage and Management.* Hoboken, NJ: Wiley, 2009.
Wicker, Stephen. *Error Control Systems for Digital Communication and Storage.* Englewood Cliffs, NJ: Prentice Hall, 1994.

JEFF GOODMAN

See Also: Digital Images; MP3 Players; Personal Computers.

Disease Survival Rates

Category: Medicine and Health.
Fields of Study: Data Analysis and Probability; Number and Operations.
Summary: Sophisticated mathematics is used to calculate disease survival rates and to help doctors and patients make treatment decisions.

Disease survival rates indicate the seriousness of a certain disease, and the prognosis of a person with the disease based on the experience of others in the same situation (in terms of the stage of the disease, gender, and age). "Overall survival rate" is defined as the percentage of people who are alive after a specific period of time after diagnosis with the disease, which is computed using the following formula: Overall Survival Rate = 100 (Number alive at the end of a time period ÷ Number alive at the start of a time period).

Standard time periods such as one, five, and 10 years are often used. For instance, the five-year overall survival rate for stage-I breast cancer is said to be "95%" if 95% of all people who are diagnosed with stage-I breast cancer live for at least five years after diagnosis. Conversely, 5% of these people die within five years.

Survival rates depend on many factors, including both the type and stage of disease, as well as age, gender, health status, lifestyle, and treatment. Doctors and researchers use survival rates to evaluate the efficacy of a treatment, compare different treatments, and develop treatment plans. For example, the treatment having the highest survival rates over time is usually chosen. If treatments have similar survival rates but different numbers of side effects, the treatment with the fewest number of side effects is often selected.

Other Types of Survival Rates

Overall survival rates have some limitations. First, they do not distinguish causes of mortality within a given time period. For instance, a death may be caused by a car accident rather than by the disease. Second, they fail to indicate whether the disease is in remission or not at the end of the time period. Moreover, they do not directly provide the prognosis for a specific patient. For instance, the 95% five-year survival rate for stage-I breast cancer does not guarantee that every patient will survive more than five years. When considering only deaths caused by the disease, relative survival rate or cause-specific survival rate is often used. Relative survival rate is the ratio of the overall survival rate for people with the disease to that for a similar group of people in terms of age and gender without the disease.

One advantage is that relative survival rates do not depend on the accuracy of the reported causes of death. On the other hand, cause-specific survival rate is computed by treating deaths from causes other than the disease as withdrawals so that they do not deflate the survival rate due to the disease. When using this rate, there is no need to involve a similar group of people without the disease. Sometimes more detailed survival rates in terms of the status of a disease after a given period of time, such as disease-free survival rate and progression-free survival rate, are of interest. The computation for disease-free survival rate is similar to that of the overall survival rate except that the numerator is the number of patients who are cured at the end of the time period. Similar computation applies to the progression-free survival rate except that the numerator is the number of people who are alive and still have the disease, but the disease is not progressing at the end of the time period. As before, disease-free and progression-free survival rates can be adjusted by filtering out the effect of deaths from causes unrelated to the disease.

Survival Function

Related to survival rates, the survival function is a mathematical function that uniquely determines the probability distribution of a random variable. In survival analysis, the random variable of interest is survival time or time to a certain event, denoted by T. For instance, survival time could be time until recovery from a disease, or time to death. The survival function for T is a function of time point t defined as

$$S(t) = P(T > t)$$

which is the true probability that the survival time of a subject is beyond time t. The survival rates with an adjustment for deaths because of unrelated causes are estimates of the survival function at some t based on existing data. For a study with n patients, the survival function can be estimated by the empirical survival function:

$$S_n(t) = \text{Number of patients not experiencing the event up to } t/n.$$

In follow-up studies, however, a patient with a certain disease may withdraw, die from other causes, or still be alive at the end of the study. In such cases, the survival time T of the patient is not exactly observed but only known to be greater than a certain time (withdrawal time, death time, or time at the end of the study) called "censoring time." Then T is said to be right-censored, and the resulting set of data is called right-censored data. Based on right-censored data, the survival function can be estimated by

$$S_{KM}(t),$$

the K-M estimator developed by statisticians Edward Kaplan and Paul Meier in 1958. As a special case,

$S_{KM}(t)$ coincides with $S_n(t)$ when there is no censoring.

When estimating survival probability,

$$P(T > t) \text{ at a given time } t,$$

$S_{KM}(t)$ or a cause-of-death-adjusted survival rate introduced earlier can be used. Taking a more sophisticated approach, $P(T > t)$ can be estimated using a confidence interval. For example, one may conclude that, with 95% confidence, $P(T > t)$ is between two numbers, say 0.80 and 0.90. Such a confidence interval can be constructed using $S_{KM}(t)$ and its variance estimate from statistician Major Greenwood's formula based a normal distribution.

Further Reading

Gordis, Leon. *Epidemiology*. 4th ed. Philadelphia: Saunders Elsevier, 2009.

Kalbfleisch, John D., and Ross L. Prentice. *The Statistical Analysis of Failure Time Data*. 2nd ed. Hoboken, NJ: Wiley, 2002.

Klein, John P., and Melvin Moeschberger. *Survival Analysis: Techniques for Censored and Truncated Data*. New York: Springer-Verlag, 1997.

Marks, Harry. "A Conversation With Paul Meier." *Clinical Trials* 1 (2004).

Qiang Zhao

See Also: Data Analysis and Probability in Society; Functions; Statistics Education.

Diseases, Tracking Infectious

Category: Medicine and Health.
Fields of Study: Communication; Data Analysis and Probability.
Summary: Physicians and mathematicians have long worked together to develop and use models that track the spread of infectious diseases in order to develop appropriate countermeasures and responses to halt the disease spread.

The health of societies relies on quickly and correctly tracking and predicting the growth and spread of disease in populations. Epidemiology is a mathematically rich area. Exposure and infection are both probabilistic processes, and tracking infectious diseases is a dynamic application of mathematics. The World Health Organization (WHO) and other organizations concerned with public health use mathematical models in their decision-making, such as when WHO analyzed the risks and benefits of travel restrictions during the early twenty-first-century H1N1 (swine flu) epidemic. Epidemiology has a long history with important societal connections. Some trace one early use of mathematical modeling for disease to eighteenth-century mathematician Daniel Bernoulli. He presented an analysis of smallpox morbidity and mortality to demonstrate the efficacy of vaccination.

Nineteenth-century physician William Farr is often called the "father of epidemiology" and was responsible for the collection of official medical statistics in England and Wales. His most important contribution was to set up a system for routinely recording causes of death. Physician John Snow is frequently cited as using graphical methods to propose a mechanism of transmission and the source of a cholera epidemic in nineteenth century London. Epidemiologists using mathematical and statistical models have been influential in research, treatment, and some methods of prevention for potentially devastating diseases, like tuberculosis, smallpox, typhus, and malaria.

Infectious diseases are a leading cause of death for humans. In order to understand the dynamics of tracking infectious disease at the population level, it is important to understand the responsible mechanisms at the individual level. Infectious disease is caused by a pathogenic agent (for example, a virus, bacterium, or parasite) transmitted through one of many methods, such as air or body fluids. One method scientists have developed for investigating why outbreaks of disease take place and how to contain or end them is to design a system of surveillance and data collection from individual cases, which can then be used to model the infection's trajectory through a population. Other

This microfluidic labchip was used in a CDC bioanalyzer to evaluate Mycobacterium cosmeticum *strains. There are some 115 species of* Mycobacterium*, causing infectious diseases like tuberculosis and leprosy.*

times, they may use data from past similar situations to extrapolate possible solutions.

Surveillance of Infectious Disease

Central public health institutions have created computer systems to monitor emerging outbreaks of illnesses. Traditional notification has relied on disease reporting by laboratories and hospitals. However, the first indications of an outbreak usually occur before a formal diagnosis. People respond to illness with a variety of behaviors to illness that can often be tracked; for example, the number of visits to emergency rooms, or purchases of over-the-counter drugs. Other people's behaviors are more difficult to track, such as those people who continue their daily routines even though they feel sick. Systems of surveillance may compile data from many sources to look for unusual patterns or significant increases in activities like emergency room visits. Another approach, based on Internet search queries, collects disease-related searches. The searches are linked to geographic mapping tools and are used to identify clusters of symptoms. Further analysis and modeling using mathematical and statistical methods are needed to estimate the potential impact of a disease outbreak.

Modeling Infectious Disease

Quantitative analysis describes probable disease trajectories for predicting impact over time. The parameters may include the variables of time, geographic location, population density, contact rate, and saturation, as well as the personal characteristics of those who contract the disease. For example, eighteenth-century mathematician Daniel Bernoulli created mathematical models for smallpox to support the use of inoculations. At the turn of the twentieth century, British physician Ronald Ross began to develop mathematical models to help him understand malaria's trajectory, rate of progression, and probability of infection. He received the Nobel Prize in Physiology or Medicine in 1902, indicating the importance of his mathematical contributions to epidemic theory. Another early twentieth-century model is the Reed–Frost epidemic model, which was developed by scientists Lowell Reed and Wade Hampton Frost. It models disease transmission via person-to-person contact in a group and includes concepts like a fixed probability of any person coming into contact with any other individual in the group.

Quantitative research continued throughout the twentieth century and continues to be active in the twenty-first century. There are many large agencies that use epidemiological models, such as WHO and the U.S. Centers for Disease Control and Prevention (CDC). As medicine and technology advance, new variables become important in models; for example, global air travel, which brings previously isolated populations into greater contact with one another, along with new vaccinations and vaccination policies. Differential use of longtime practices, like quarantining sick and potentially exposed individuals, may also be a factor.

Other models incorporate seasonal information, such as varying contact rates, which can be affected by societal structures, such as school schedules. In the latter twentieth century, computer networking and the subsequent spread of computer viruses have led mathematicians and others to extend epidemiological models to research and model the spread of computers worms and viruses using mathematical techniques, such as directed graphs and simulation. In such an active field of research, new technologies and methods for quickening the pace of identifying patterns of disease are expected to be developed.

Further Reading

Diekmann, O., and J. A. P. Heesterbeek. *Mathematical Epidemiology of Infectious Diseases: Model Building, Analysis and Interpretation*. Hoboken, NJ: Wiley, 2000.

Hethcote, H. W. "The Mathematics of Infectious Diseases." *Society for Industrial and Applied Mathematics* 42 (2000).

Keeling, Matt, and Pejman Rohani. *Modeling Infectious Diseases in Humans and Animals*. Princeton, NJ: Princeton University Press, 2007.

Douglas Rugh

See Also: Disease Survival Rates; HIV/AIDS; Mathematics, Applied; Viruses.

Division

See *Multiplication and Division*

Domes

Category: Architecture and Engineering.
Fields of Study: Geometry; Measurement.
Summary: Domes have been used throughout history to cover open spaces.

In architecture, a dome is a hemispherical structure with a circular, polygonal, or elliptical base that is usually used to cover a large open space. It developed as a generalization of the full revolution of an arch around a vertical axis.

Early domes appeared in small buildings and tombs of the ancient Middle East, India, and the Mediterranean. Because these domes consist of horizontal layers of materials progressively cantilevered inward until they reach the top of the roof, they are considered "false domes" and called "corbel domes." True domes present the characteristic of having a continuously changing slope ranging from being vertical at the base to horizontal at the top, which requires adaptable roofing materials.

Large-scale masonry domes were first introduced by the Romans in public buildings, such as baths, temples, mausoleums, and basilicas. With an interior diameter of 142 feet, the Pantheon, built during the second century in Rome, remained the largest dome until 1881 and is still the world's largest unreinforced concrete dome. Built on a rotunda, it exerts tremendous thrusts on the perimeter walls. It is not only an engineering triumph but also a tremendous achievement in sacred geometry and cosmography. Its hemispherical ceiling has regularly been compared to the vault of heaven.

Carried on four pendentives, the 102-foot central dome that covers Hagia Sophia in Istanbul is a masterpiece of Byzantine architecture. Built in the sixth century, it exemplifies the full development of the pendentive (a triangular segment of a sphere) as a constructive solution allowing the construction of a dome over a square nave. Volumetric transitions and intersections are critical components of the geometry of architecture.

Built from 1420 to 1436 under the direction of Brunelleschi, the dome of Santa Maria del Fiore in Florence succeeded Hagia Sophia as the largest masonry dome in the world, a record it stills holds. Brunelleschi designed an eight-sided double dome shell without exterior buttresses that did not require any supporting framework during construction. Standing at about 165 feet above ground level, the interior of the dome is approximately 100 feet tall and spans 139 feet. The dome weighs more than 40,000 tons and required the use of more than 4 million bricks.

Domes became increasingly popular during the Renaissance, the Baroque era, and the nineteenth century. Influenced by the Pantheon and Santa Maria del Fiore Bramante, Michelangelo designed St. Peter's Basilica in Rome.

It has the world's tallest dome, and it inspired one of the most famous landmarks of Baroque architecture: the dome of the Invalides designed by Mansart. Completed in 1711, Wren's three-layer dome for St. Paul cathedral in London influenced the construction of the dome of

Hagia Sophia was built in the sixth century and is a masterpiece of Byzantine architecture.

the U.S. Capitol in Washington, D.C., which ultimately inspired the design of most U.S. state capitols.

Since the late nineteenth century, different materials, such as steel, wood, membrane, and reinforced concrete, have allowed the building of domes covering much larger spaces. The two monumental arches supporting the retractable roof of the Cowboys Stadium in Arlington, Texas, reach a height of 292 feet and span of 1225 feet—making it the largest domed stadium in the world.

Geodesic domes represent another modern type of dome, rejecting the classical arch principle. This type of dome is usually a partially spherical structure constituted of a network of triangular or polygonal facets that are in tension and compression. Because the thrust is equal in all directions, the dome can be anchored directly on the ground. Because of their cost-effectiveness and structural strength, hundreds of thousands of geodesic domes have been built all over the world—most often as a solution to provide shelter for poor families in developing countries, or to house people in extreme weather conditions. In 1960, Buckminster Fuller, who developed the mathematics of this type of dome, designed a geodesic dome two miles in diameter and one mile high at its top that would have covered Midtown Manhattan, and provided the whole district with permanent climate control.

Further Reading

Blackwell, William. *Geometry in Architecture.* Boston: John Wiley and Sons, Inc., 1984.

Hammond, Victoria. *Visions of Heaven: The Dome in European Architecture.* New York: Princeton Architectural Press, 2005.

Kenner, Hugh. *Geodesic Math and How to Use It.* 2nd ed. Berkeley: University of California Press, 2003.

King, Ross. *Brunelleschi's Dome: How a Renaissance Genius Reinvented Architecture.* New York: Penguin Books, 2000.

McDonald, William L. *The Pantheon: Design, Meaning, and Progeny.* 2nd ed. Cambridge, MA: Harvard University Press, 2002.

Catherine C. Galley
Carl R. Seaquist

See Also: City Planning; Engineering Design; Symmetry.

Doppler Radar

Category: Weather, Nature, and Environment.
Fields of Study: Algebra; Geometry; Representations.

Summary: Doppler radar uses the mathematical characteristics of waves to track and predict weather patterns.

Radio detection and ranging, commonly known by the acronym "radar," was initially developed to detect and determine the distance of enemy aircraft when visual methods were insufficient, such as in poor weather or at night. It is commonly traced to the nineteenth century work of physicist Heinrich Hertz, who investigated the reflection of radio waves from metallic objects. Doppler radar is a type of radar that uses the Doppler effect to judge the speed and direction of distant objects. The Doppler effect (also known as "Doppler shift") is a physical property that applies to all types of waves, including sound and light. Mathematician and physicist Christian Doppler presented a paper on this effect in 1842, describing how frequencies of waves change in correspondence to the relative movement between source and observer. In 1948, Hippolyte Fizeau independently discussed the shift in the wavelength of light coming from a star in similar terms. Doppler radar has applications in many fields including aviation, meteorology, sports, and traffic control. For example, Doppler radar is widely used for detecting severe weather, and it is a critical component in wind-shear detection and warning systems for airports.

Mathematics of Waves

The Doppler effect relies on the mathematical properties of waves. Transverse waves, which disturb a medium perpendicular to the direction the wave is traveling, are described in terms of their wavelength and amplitude. Wavelength is the distance between two wave crests or troughs, while amplitude is the height of the wave. An example of this is light. Longitudinal waves produce a series of compressions and rarefactions in a medium and are described by their amplitude and frequency. An example is sound, where amplitude corresponds to intensity (or "loudness") and frequency corresponds to pitch.

A car with a siren emits a series of sound waves of constant frequency. If the car moves toward a stationary observer, the waves will seem to be "bunched up" (to have greater frequency), thus a higher pitch. The same siren moving away will have waves that appear "stretched out," with lower frequency and pitch. Similarly, an oncoming light source will appear more blue, while one moving away will appear more red, corresponding to higher and lower frequencies on the electromagnetic spectrum. The amount of change in frequency is relative to both speed and direction of the moving object. The speed of a moving object can be measured by shooting waves of a known frequency at the object, and then observing the frequency of the waves that bounce from the object to the source. The difference between the outgoing and incoming frequencies is used to calculate speed. Common examples are the handheld radar guns used to measure the speed of automobiles or a thrown baseball. Edwin Hubble, for whom the Hubble Space Telescope was named, used the Doppler effect to help measure the distances to other galaxies. Light from other galaxies looks more red, indicating they are moving away. This "redshift" is commonly used as evidence in favor of the Big Bang theory of the origin of the universe.

Weather Detection

Many consider Doppler radar to be the best tool available for detecting tornadoes, hurricanes, and other extreme weather in the twenty-first century. Weather stations commonly emit radio waves that strike objects like clouds or heavy rain, and reflect back. Meteorologists use this data to determine the speed and direction of a weather system, as well as for probabilistic models to predict the path and potential severity of a storm in a given geographic area. Mathematical algorithms produce color-coded weather maps, weather animations, and other visualizations for new programs or Web sites, indicating how a storm system is predicted to move through a geographic area. Some researchers have used input data from a single radar station and knowledge of the mathematical structure of hurricanes to construct three-dimensional maps.

In the twenty-first century, a system of 21 Atlantic and Gulf coast radar stations, starting in Maine and ending in Texas, gathers real-time data to mathematically estimate the characteristics of hurricanes within 120 miles of the coast. Previously, forecasters had to fly aircraft into oncoming hurricanes and throw instruments overboard to collect data, giving them a lead time of about half a day before hurricane landfall. Other mathematicians have explored numerical weather prediction using Doppler radar and a technique known as "four-dimensional variational data assimilation," which estimates model parameters by optimizing the fit between the solution of a given model and a set of observations the model is intended to predict.

Further Reading

Harris, William. "How the Doppler Effect Works." http://www.howstuffworks.com/science-vs-myth/everyday-myths/doppler-effect.htm.

O'Connor, J. J., and E. F. Robertson. "Mactutor History of Mathematics Archive: Doppler Biography." http://www-groups.dcs.st-and.ac.uk/~history/Biographies/Doppler.html.

Schetzen, Martin. *Airborne Doppler Radar: Applications, Theory and Philosophy*. Reston, VA: American Institute of Aeronautics and Astronautics, 2006.

Sarah Boslaugh

See Also: GPS; Traffic; Weather Forecasting.

Drug Dosing

Category: Medicine and Health.
Fields of Study: Algebra; Measurement; Number and Operations.
Summary: Mathematicians and scientists calculate optimal drug dosages to help ensure patient health.

Drug dosing is the administration of a particular amount of medication according to a specific schedule. There are two kinds of drugs: prescription drugs and nonprescription drugs (over-the-counter medicine). For prescription drugs, medical doctors normally prescribe the amount and time to take the medication. For over-the-counter drugs, information of drug dosing is usually recommended on the label of the medicine. Drug dosing is common in everyday life, but an error in drug dosing may claim lives or create serious medical burdens. According to a conservative estimate in 2006, drug errors injure more than 2 million Americans per year.

Dosage Measurements

Some drug dosing errors stem from inaccurate measurements and administering improper amounts of chemical compounds to the patient. The first mathematics-related issue is the measurement systems in treatment dosing. Drug dosing normally utilizes the metric system, the apothecary system, or the household system. These are the three main forms of measurement systems in the pharmaceutical industry.

The apothecary system is historically the oldest system in medicine measurement. It consists of grains, drams, ounces, and minims.

60 grains (gr) = 1 dram	8 drams = 1 ounce (oz)	1 fluid dram = 60 minims.

Although the apothecary system was widely used during earlier times, it is rarely used in the twenty-first century. The most widespread dosing measurement in liquid drugs in the twenty-first century is the household system, which is rooted on the apothecary system but uses relatively common items as measurement units. The household system primarily consists of teaspoons (tsp), tablespoons (tbsp), ounces (oz), pints (pt), juice glasses, coffee cups, glasses, measuring cups, drops, quarts (qt), and gallons (gal).

1 tablespoon (tbsp) = 3 teaspoons (tsp)	1 teaspoon (tsp) = 60 drops
1 ounce (oz) = 2 tablespoons	1 juice glass = 4 ounces
1 coffee glass = 6 ounces	1 glass = 8 ounces
1 measuring cup (c) = 8 ounces	1 pint (pt) = 2 measuring cups
1 quart (qt) = 2 pints	1 gallon (gal) = 4 quarts

The household system is convenient and commonly understandable, but it is just an equivalent measure without specific precision; for instance, the size of a coffee cup may vary. A more scientific and precise way is to measure with the metric system. The metric system is accurate, simple, and popular in most scientific experiments, including drug measurements, even though it is not as handy as the household system. It essentially consists of length, volume, and weight measures.

The basic metric length measure is meter (m). Along with the meter are the following:

1 kilometer (km) = 1000 meters	1 decimeter (dm) = 0.1 meter
1 centimeter (cm) = 0.01 meter	1 millimeter (mm) = 0.001 meter

The basic metric volume measure is liter (L). Along with the liter are the following:

1 kiloliter (kL) = 1000 liters	1 milliliter (mL) = 0.001 liter

The basic metric weight measure is gram. Along with the gram are the following:

1 kilogram (kg) = 1000 grams	1 milligram (mg) = 0.001 gram
1 microgram (mcg) = 0.001 milligram	

Each measuring system has its advantages and disadvantages. Administering a drug with a wrong measurement system could result in a fatal error. It is critical to distinguish the different systems and use them appropriately. The following are some basic conversions among the three drug measuring systems.

480 grains = 1 ounce (oz)	1 minim = 1 drop
1 milliliter (mL) = 15–16 drops	1 tablespoon = 15 milliliters

Dose Response, Drug Dosing, and Statistics

Besides dosage measurement, another important aspect in drug dosing is to understand that because of the immune system and drug resistance, efficacy does not necessarily increase as dosage increases. Factors such as body weight and age affect the shape of the dose response curve for each individual. To take account of population diversity, the expected effect within a population is principally considered as the guideline for the recommended drug dosage. For example, over-the-counter medication normally uses age or body weight of the patient as the guide to recommend efficient dosages.

Similar to the efficacy of a drug, for some medicines, side effects or toxicity of a drug need to be simultaneously considered in drug dosing. If the side effect or toxicity is too strong, administering the medicine may kill (rather than cure) the patient. In this regard, it is necessary to identify the maximum tolerated dose of a drug. The maximum tolerated dose is the largest dosage at which the toxicity/side effect has not reached the level to cause the specific harm to the patient, while the minimum effective dose is the smallest dosage to reach the expected treatment effect of the drug. If the minimum effective dose exceeds the maximum tolerated dose, the drug is normally not permitted. If the minimum effective dose is smaller than the maximum tolerated dose, the dosage range in which the drug is both safe and effective is called the therapeutic window of the drug. For example, if the minimum effective dose of a drug is 5 mg daily and the maximum tolerated dose is 12 mg daily, then the therapeutic window of the drug is 5–12 mg daily.

To make an inference on the efficacy and toxicity of a drug at the same time, statistical methods are used. After clinical trials (such as the double-blind experiment), simultaneous inference methods are used to estimate the minimum effective dose and the maximum tolerated dose. One of the well-known methods in identifying dose effects is Dunnett's method for multiple comparisons with a control, developed by statistician Charles Dunnett in the mid-twentieth century. Other effective techniques for identifying the therapeutic window of a drug have been explored by mathematicians and statisticians since that time.

Shelf Life

Mathematics and statistics also intertwine with drug dosing on the shelf life of a drug. For medications that emit chemical compounds over time, the drug effect may be affected by chemical half-lives well before the expiration date. In the United States, the Food and Drug Administration (FDA) requires companies to conduct stability analyses to establish the shelf life of new products. The same is true in many other countries around the world. The conclusions are generally based on statistical sampling and mathematical modeling of data, using estimation methods such as simultaneous confidence segments over time.

Further Reading

Boyer, Mary. *Math for Nurses: A Pocket Guide to Dosage Calculation and Drug Preparation*. 7th ed. Hagerstown, MD: Lippincott Williams & Wilkins, 2008.

Chow, Shein-Chung, and Jun Shao. *Statistics in Drug Research: Methodologies and Recent Developments.* Boca Raton, FL: CRC Press, 2002.

Lacy C., et al. *Drug Information Handbook*. 19th ed. Hudson, OH: Lexi-Comp. Inc., 2010.

John T. Chen

See Also: Data Analysis and Probability in Society; Measurement in Society; Probability.

DVR Devices

Category: Communication and Computers.
Fields of Study: Measurement; Representations.
Summary: Mathematics is essential to the functioning of DVRs, including image processing, compression, and error correction.

Digital video recording devices (DVRs) have become an increasingly prominent factor in the television industry in the twenty-first century. The basic function of a DVR is to record television to a digital format on a disk drive, allowing it to be played back later. Combined with the timer and basic replay functions, this feature allows standard DVRs to perform many functions: store and play back television shows; automatically record specific television programs; and buffer live television to allow pausing and skipping.

Many DVRs can play and record the same program at the same time, a function earlier video recording devices lacked. A 1991 patent by father–daughter team

Eric and Romi Goldwasser is one of the earliest known for digital video recorders. One well-known brand of DVR is TiVo, introduced by engineer Michael Ramsay and computer scientist James Barton in 1999. Both previously worked at Silicon Graphics, Inc., which was a pioneer of computer workstations. Because of these roots in computers and the technology they utilize, some consider DVRs to be computers. Mathematics is essential to the functioning of DVRs, including image processing, compression, and error correction. It also plays a role in digital watermarking, which is widely used to enforce copyright laws.

Statistical analyses of television viewing habits by companies, such as ACNielsen, suggest that DVR use combined with online viewing are significantly changing the pattern of television delivery and assessments of popularity and marketability in the early twenty-first century. TiVo's Ramsay noted, ". . . it's forcing the industry to embrace the Internet . . . and once they embrace it, they will find that their business models change and new opportunities will arise."

Process and Functions

In DVRs, images are captured and stored in binary form. This process differs from older electromechanical systems, like videocassette recorders (VCRs). Raw video files tend to be very large and require sizable storage space, so DVRs use mathematical compression algorithms. Files must then be decompressed before viewing. Decompression is accomplished by hardware or software codec technology, which implements specific formats or standards. Motion Pictures Experts Group (MPEG) created the MPEG-1 format for digital storage in 1993 and MPEG-2 in 1994, which made high-definition television (HDTV) and digital versatile discs (DVDs) possible. The MPEG-4, released in 1999, facilitated digital video for Internet streaming and replaced some proprietary codecs in DVRs to facilitate file transfer.

MPEG compression is typically asymmetric; algorithmic encoders are more complex than their paired systematic decoders. Optimized compression to preserve image quality is achieved by mathematically controlling bit rates subject to constraints on variables like file size or transmission bandwidth. Quality applies not only to individual frames but also to the smoothness of transitions between frames, which affects the user's visual experience of motion. This approach can be formulated as a Lagrange minimization problem, named for mathematician Joseph Lagrange. Two- or three-pass encoding schemes are often used. A first pass collects complexity data for the entire video. Subsequent passes perform the actual encoding based on the information. Algebraic structures known as Galois fields, after mathematician Evariste Galois, are helpful in coding and error correction, and are sometimes paired with Fourier transforms, named for mathematician Joseph Fourier. This pairing is especially true in recorders that incorporate nonbinary, cyclic error correction, such as Reed–Solomon codes, named for mathematicians Irving Reed and Gustave Solomon, as well as for pseudo-random digital dither and randomized channel codes. Recording and compression are also affected by digital watermarking, where extra visible or invisible information is embedding in a digital signal. It can be used to identify ownership, track the file, and prevent recording. Watermarks may be classified by the embedding method, like quantization-type watermarks, which rely on quantization matrices.

Perhaps the best-known brand of DVR is TiVo, introduced in 1999. One of TiVo's features is its ability to employ statistical techniques, such as data mining, to generate recommendations. Viewers can rate shows they watch, and TiVo tracks the ratings, which are then examined for patterns. As of 2004, TiVo had accumulated more than 100 million user ratings on 30,000 different programs. The TiVo algorithm uses a collaborative filtering architecture, which relies on comparing viewer profiles and a viewer's past patterns using several thousand key details, like favorite actors and genres.

However, some users have complained about unusual or extreme matches resulting from this methodology and have intentionally subverted the algorithms by giving false or contradictory ratings. The server architecture is scalable and throttleable, which means as more server resources and user data become available, the system is faster for everyone and perhaps more efficient in finding recommendations for harder-to-match viewers.

Further Reading

Ali, Kamal, and Wijnand van Stam. "TiVo: Making Show Recommendations Using a Distributed Collaborative Filtering Architecture." Knowledge Discovery and Data Mining (KDD) conference paper, 2004.

Davis, Philip J., and Reuben Hersh. *The Mathematical Experience*. Boston: Houghton-Mifflin, 1981.
Littlewood, J. E. *Littlewood's Miscellany*. New York: University of Cambridge Press, 1986.
Schaeffler, Jimmy. *Digital Video Recorders: DVRs Changing TV and Advertising Forever*. Oxford, England: Focal Press, 2009.
Watkinson, John. *The Art of Digital Video*. 4th ed. Woburn, MA: Focal Press, 2008.

Bill Kte'pi

See Also: Digital Storage; MP3 Players; Nielsen Ratings; Predicting Attacks; Televisions.

E

Earthquakes

Category: Weather, Nature, and Environment.
Fields of study: Data Analysis and Probability; Measurement.
Summary: Earthquakes are measured in several ways, the most famous of which is the logarithmic Richter scale.

Earthquakes are the movements of Earth's crust resulting from tectonic plates colliding against each other. This sudden release in energy causes seismic waves that cause destruction. Depending on their severity, earthquakes range from being barely noticeable to causing permanent damage to infrastructure along with a significant loss of life. Most earthquakes are caused by the action of geological faults but they can also be caused by mine blasts, volcanic activity, and subterrestrial activity, such as injecting high-pressure water for geothermal heat capture. The focal point of the earthquake is called the "hypocenter." The point on the ground directly above the hypocenter is known as the "epicenter" of the earthquake. Philosophers, mathematicians, and scientists have long attempted to understand earthquakes. Thales of Miletus thought that earthquakes occurred because Earth rested on water. Mathematician, astronomer, and geographer Zhang Heng invented the first seismograph for measuring earthquakes in the second century. Mathematician Harold Jeffrey theorized that Earth's core is liquid after analyzing earthquake waves. Geologists use statistical methods to try to predict earthquakes.

Seismic Waves

A tremendous amount of energy is released from the epicenter radially outward. As the energy spreads, it is manifested in three forms: compression waves (P waves), shear waves (S waves), and surface waves.

P waves are felt first and do minimal damage. S waves follow the P waves and do minimal damage. It is the slower surface waves (also known as "Love waves") that cause the majority of the damage.

Measurement

The goal of earthquake measurement has been to quantify the energy released. Seismographs are highly sensitive instruments employed to record earthquakes. Conventionally, earthquake magnitudes are reported in the Richter scale. The Modified Mercalli Intensity Scale is commonly used to ordinally quantify (or rank) the effects of an earthquake on humans and infrastructure. Body wave or surface wave magnitudes are also used to measure earthquakes.

Richter Scale

The Richter scale quantifies the amount of seismic energy released during a quake. It is a base-10 logarithmic scale, which means that the difference between an earthquake of rating 2.0 on the Richter scale and 3.0 correlates to a tenfold increase in measured amplitude. Specifically, the Richter scale is defined as

$$M_L = \log_{10} A + B$$

where A is the peak value of the displacement of the Wood–Anderson seismograph (mm) and B is the correction factor. The wave intensity measurements are also logarithmic functions, using variables such as the ground displacement in microns, the wave's period in seconds, and distance from the earthquake's epicenter.

Modified Mercalli Intensity Scale

The Modified Mercalli Intensity Scale has 12 gradations: instrumental, feeble, slight, moderate, rather strong, strong, very strong, destructive, ruinous, disastrous, very disastrous, and catastrophic.

Further Reading

Brune, James. *Tectonic Stress and the Spectra of Seismic Shear Waves From Earthquakes.* La Jolla, CA: Institute of Geophysics and Planetary Physics, UCSD, 1969.

Gutenberg, B., and C. F. Richter. *Earthquake Magnitude, Intensity, Energy and Acceleration.* Pasadena, CA: Bulleting of the Seismological Society of America, 1956.

Hough, Susan. *Predicting the Unpredictable: The Tumultuous Science of Earthquake Prediction.* Princeton, NJ: Princeton University Press, 2009.

Ashwin Mudigonda

See Also: Exponentials and Logarithms; Geometry and Geometry Education; Weather Forecasting; Weather Scales.

Educational Manipulatives

Category: School and Society.
Fields of Study: Connections; Problem Solving; Representations.

Summary: Some educators use objects to engage students' attention and encourage them to learn sensorially and experientially.

Educational manipulatives are physical, technological, or virtual objects that are intended to help students learn concepts by taking advantage of tactile and visual explorations.

Mathematical tools and technologies are common in mathematics education. Entire companies and sales catalogues are devoted to such mathematical products, and national and state curriculum standards emphasize their importance in schools. There is a rich history of tools and manipulatives in mathematics classrooms, and these have changed over time along with curricular, industrial, and technological needs and innovations. For instance, in the seventeenth century, the slide rule replaced logarithmic tables in scientific calculation and mathematics classrooms but these in turn became obsolete in the twentieth century because of calculators and computers. Educators, including classroom teachers and university researchers, along with professional designers, continue to create and refine manipulatives and research their effectiveness. Some also work for companies to develop or market these products and materials.

History

Two early developers of collections of learning manipulatives included Friedrich Fröbel (1782–1852) and Maria Montessori (1870–1952). Fröbel was a German educational researcher who is also referred to as the "inventor of kindergarten." He developed a set of manipulative tools called the Fröbel Gifts, which were intended for kindergarten play and learning in the nineteenth century. The fuller development of the manipulatives occurred after Fröbel's death. The Fröbel Gifts set contained objects such as balls, cubes, tiles, sticks, and framed figures that were built out of toothpicks and peas. They were designed to help young students explore mathematical concepts in two and three dimensions. Some of the surfaces were hung from string in order to highlight their cross-sections and symmetries. One focus of Frobel's kindergarten philosophy was free play, which was also carried out in different settings with other objects. For example, the Milton Bradley Company, an American game company established in 1860, sold a curvilinear set of pieces that could form a cylinder. In addition to free play, students ultimately learned to draw what they observed. Ideally, children would revisit concepts they learned using the manipulatives in increasingly sophisticated ways as they progressed through school. For example, in 1869, Edward Wiebe, who was an early proponent of kindergarten education in the United States, suggested that children could explore concepts like the Pythagorean theorem, named for Pythagoras of Samos, long before they understood the square of a number. Frank Lloyd Wright acknowledged the influence of Fröbel's Gifts on his career as an architect. Aspects of Fröbel's legacy continue to be found in manipulative design and in schools, although they have been greatly modified and adapted.

In the twentieth century, Italian physician and educator Maria Montessori, who is well known for the Montessori method of education, also focused on the importance of manipulatives in classrooms. She developed an integrated set of sensorial learning materials that included cylinders, cubes, rods, circles, triangles, polygons, boxes, and binomial and trinomial cubes. Montessori designed activities with educational outcomes in mind. Her ideas became popular in the United States and are still used in the twenty-first century. Montessori schoolteachers challenge students to arrange objects in specific ways so that the students will uncover concepts.

Examples

There have been a wide number and variety of other educational manipulatives created in the twentieth and twenty-first centuries, including polyhedral dice with varying numbers of sides for studying probability; multiplication blocks; algebra tiles that represented polynomials and polynomial operations; multicolored and interlocking Unifix cubes intended to teach number and operations concepts; pattern blocks for studying tessellations and fractions; tangrams for exploring geometry; and geoboards, which are pegged boards on which rubber bands could be placed and stretched to investigate concepts like perimeter and area. The abacus or counting frame that had been in use since antiquity found its way into U.S. schools in the nineteenth century. While it has mostly disappeared from twenty-first century classrooms, it remains important in a few educational contexts, like in classrooms for visually impaired children. Virtual manipulatives have replaced physical objects in some cases. There are even applets that mimic some of the physical manipulatives like pattern blocks, which teach similar concepts while providing different sorts of tactile and visual stimulation.

Effectiveness

There are diverse opinions regarding the effectiveness of manipulatives in mathematics education. In 2005, mathematician David Klein warned, "Too much use of them runs the risk that students will focus on the manipulatives more than the math and even come to depend on them.... Yet many state standards recommend and even require the use of a dizzying array of manipulatives in counterproductive ways." In the final report of the National Mathematics Advisory Panel in 2008, the panel cautioned that

> Despite the widespread use of mathematical manipulatives such as geoboards and dynamic software, evidence regarding their usefulness in helping children learn geometry is tenuous at best. Students must eventually transition from concrete (hands-on) or visual representations to internalized abstract representations. The crucial steps in making such transitions are not clearly understood at present and need to be a focus of learning and curriculum research.

Developmental psychologists and educators David Uttal, Kathyrn Scudder, and Judy DeLoache noted that

> . . . the sharp distinction between concrete and abstract forms of mathematical expression may not be justified. We believe instead that manipulatives are also symbols; teachers intend for them to stand for or represent a concept or written symbol.

Other researchers and teachers counter the claims that there is insufficient evidence; they cite a vast amount of educational literature and anecdotes regarding the benefits of hands-on activities, software, and manipulatives. Many students report that they enjoy the tactile manipulation. Students may also feel satisfied when they discover or confirm mathematical relationships, and this may help them connect to mathematics. Mathematics educators continue to study the effects of various manipulatives and the potential differences between physical and virtual manipulatives on student learning.

Further Reading

Burns, Marilyn. "How to Make the Most of Math Manipulatives: A Fresh Look at Getting Students' Heads—and Hands!—Around Math Concepts." *Instructor* 105, no. 7 (1990). http://teacher.scholastic.com/lessonrepro/lessonplans/instructor/burns.htm.

Kidwell, Peggy, Amy Ackerberg-Hastings, and David Roberts. *Tools of American Mathematics Teaching, 1800–2000*. Baltimore, MD: John Hopkins University Press, 2008.

Klein, David. *The State of State MATH Standards*. Washington, DC: Thomas B. Fordham Foundation, 2005. http://www.math.jhu.edu/~wsw/ED/mathstandards05FINAL.pdf.

Moyer-Packenham, Patricia. *Teaching Mathematics With Virtual Manipulatives, Grades K–8*. Rowley, MA: Didax, 2010.

National Mathematics Advisory Panel. "Foundations for Success: Final Report of the National Mathematics Advisory Panel." March 13, 2008. http://www2.ed.gov/about/bdscomm/list/mathpanel/index.html.

Uttal, David, Kathyrn Scudder, and Judy DeLoache. "Manipulatives as Symbols: A New Perspective on the Use of Concrete Objects to Teach Mathematics." *Journal of Applied Developmental Psychology* 18, no. 1 (1997).

Sarah J. Greenwald

Jill E. Thomley

See Also: Calculators in Society; Measuring Tools; Puzzles; SMART Board; Software, Mathematics; Visualization.

Educational Testing

Category: School and Society.
Fields of Study: All.
Summary: Mathematicians and researchers are constantly exploring the validity and reliability of educational testing.

Purpose of Testing

Educational testing is pervasive in modern education at the local, state, and federal levels, and mathematics is one of the most frequently tested areas. The purpose of educational testing is broad and multifaceted: to assess student progress and school accountability; to identify students' strengths and weaknesses, as well as their eligibility or need for special services; to make educational decisions about individuals and groups of students; to choose curriculum and instructional techniques; to reward teachers or schools for performance; and to formulate educational legislation and policies. Students are often placed in courses and special programs as a result of educational testing and may be required to pass tests to graduate from high school or be admitted to schools at all levels, especially colleges and universities.

While some educators, parents, and politicians cite standardized tests for their presumed objectivity in measuring achievement and other skills or attributes, these tests are frequently a source of anxiety and competitive pressure for students. There is an entire industry dedicated to helping students prepare for and pass or score well on these tests. At the same time, researchers are constantly exploring the validity and reliability of tests with regard to fairness for subgroups of students, as well as their actual predictive ability. For example, there is a broad body of research on whether measures like high school grade point average, SAT math scores, or mathematics placement tests are predictors of success in college mathematics courses.

Types of Testing

The decisions that can be made based on testing information depend on the type of test that is administered. There are two different types of tests that provide different types of information: norm-referenced tests (NRT) and criterion-referenced tests (CRT).

NRTs are created for the purpose of comparing students to a norming group, which is composed of students who are similar to the student being tested. The scores of the norming group create the very familiar normal (bell-shaped) curve. NRT scores are typically reported as percentiles, which indicate that a student scored above a certain percentage of the norming group. For example, a student at the 84th percentile scored the same or higher than 84% of the students in the norming group. It is a common misconception with NRTs that students are compared to all other students who have taken the test; however, most NRTs are normed every several years using a new norming group

Students may be placed in courses and special programs as a result of educational testing.

with which test-takers are compared. NRTs are typically very general in nature, covering a broad range of objectives. Items that have a variety of difficulty levels are chosen for NRTs, as these types of items encourage a wide variability in the scores, thus allowing evaluators to more accurately determine how a student compares to others. The SAT and many intelligence tests such as the Wechsler Intelligence Scale for Children are norm-references tests.

Unlike NRTs, which are used to compare students to each other, CRTs are used to determine if a student has mastered a given set of standards. CRTs are typically narrow in focus, testing only a few objectives, and are generally focused on those objectives that are deemed most important. Scores for CRTs are typically reported as percentage correct or as scaled scores. Proficiency is determined by comparing a student's score to an established cut point. Many schools regularly administer end-of-grade or end-of-course tests through which student achievement in mathematics subjects is measured.

Issues in Educational Testing

Two primary concerns with educational testing are the validity and reliability of the assessment. "Validity" in this context refers to whether a test is appropriate for the population being tested, as well as whether it appropriately addresses the content it is intended to measure. Educators from around the United States have expressed concern as to whether the tests that are currently being used to measure student achievement are valid and reliable. In an effort to address this concern, many states have undergone revisions of their tests in the past several years.

An additional concern with educational testing is in how student progress is measured over time. Statisticians have developed a variety of growth models to determine if individual students are improving as they move through school. These models may focus on improvement from grade level to grade level, or they may focus on student progress within a single school year (referred to as "value-added" or "teacher impact"). An ongoing issue with measuring student progress over time lies with the relationship between the assessments and the statistical measures that are used to analyze assessment data. Growth models are all based on certain assumptions about the assessments, which may or may not be met. In order to determine the impact of schools on student learning, one must ensure that the assessments and the statistical models used to analyze the data are compatible.

Test Analysis

Standardized educational tests undergo a variety of analytical procedures to evaluate their effectiveness at measuring a construct. Item analysis is frequently conducted to determine if items are functioning the way test developers intended. This analysis of student responses to items provides the difficulty index and the discrimination index. The difficulty index is simply the ratio of the number of students who answered the item correctly to the number of students who attempted the item; a higher difficulty index indicates an easier item. The discrimination index provides information on how well an item differentiates between students who performed well on the test and students who did not. A positive discrimination index indicates that those students who performed well overall on the test were more likely to answer the item correctly, while a negative discrimination index indicates that those students who performed poorly overall were more likely to answer the item correctly. For NRTs, item discrimination is particularly important, and test developers attempt to develop items that will have a high discrimination index.

Modern test analysis also uses a process called "item response theory" (IRT) to determine the effectiveness of a test or test item. IRT evaluates items based on the parameters of item difficulty, discrimination, and guessing and provides test developers with the probability that a student with a certain ability level will answer an item correctly. In addition, IRT allows for a more sophisticated measure of a test's reliability.

Trends in Educational Testing

Recent trends in educational testing have been focused around making international comparisons of student achievement. The most well known of these comparisons are the Third International Mathematics and Science Study (TIMSS), conducted in 2007, and the Program for International Student Assessment (PISA), conducted in 2006. The TIMSS included fourth-grade students from 36 countries and eighth-grade students from 48 countries. Participating countries submitted items for the test and the test was developed by a committee of educational experts from various nations. The TIMSS also collected information on students' background, including attitudes toward mathematics and science, academic self-concept, home life, and out-of-classroom activities. The PISA focused on problem solving in mathematics and science and on reading skills. The 2006 PISA included 15-year-olds from 57 countries. The goal of PISA is to determine students' abilities to analyze and reason and to effectively communicate what they know. Additional international studies involving educational testing include the International Adult Literacy Survey, the Progress in International Reading Literacy Study, and the Civics Education Study.

In the United States, the National Assessment of Educational Progress (NAEP) is used to compare student achievement across states. NAEP includes students from grades 4, 8, and 12 and is designed to provide an overall picture of educational progress. Schools are randomly chosen to participate and students within those schools are also randomly chosen. The NAEP tests students in mathematics, reading, science, writing, civics, economics, and history.

The public focus on educational testing in the United States sharpened with the implementation of the No Child Left Behind (NCLB) Act in 2002. For the first time in American history, schools were publicly designated as "meeting" or "failing to meet" state standards, and issues of educational testing were brought to the forefront. Organizations like Achieve began closely examining how schools were preparing students for college and the work force and began working with state officials and business executives to improve student achievement. Educational testing is a valuable tool for these types of organizations, providing information on the effectiveness of American schools.

Controversies in Educational Testing

Not everyone believes educational testing is useful or meaningful, and there are many arguments against the use of such tests. For example, studies have suggested that the SAT is both culturally and statistically biased against African Americans, Hispanic Americans and Asian Americans. Others have found that socioeconomic status is correlated with performance on the SAT, which is believed to be related to the fact that students from wealthier families can afford expensive test preparation courses or multiple retakes of the test, both of which have been demonstrated to improve test scores in some cases. Others have documented a gender gap in SAT mathematics scores that is not easily explained by issues like the difference in the number of male and female test takers.

On many tests, stereotype threat or vulnerability has also been shown to affect test scores when race, gender, or culture are cued before a test. In response, some have advocated that self-identification should occur after a test. Researchers have also shown that the structure or methodology of the test can have an effect on performance. For example, female test scores on tests of spatial ability can improve when "I don't know" is removed as an answer, or when ratio scoring or un-timed tests are used. Finally, there are many who believe that there are concepts that cannot be adequately measured by standardized assessments, even when the answers are not exclusively multiple choice and that using standardized tests as a primary method of assessment leads to "teaching to the test" rather than a broader educational experience for students.

Further Reading

Allerton, Chad. *Mathematics and Science Education: Assessment, Performance and Estimates*. Hauppauge, NY: Nova Science Publishers, 2009.

Crocker, Linda, and James Algina. *Introduction to Classical and Modern Test Theory*. Chicago, IL: Harcourt College Publishers, 1986.

Kubiszyn, Tom, and Gary Borich. *Educational Testing & Measurement: Classroom Application and Practice*. 9th ed. Hoboken, NJ: Wiley, 2010.

Mertler, Craig A. *Interpreting Standardized Test Scores: Strategies for Data-Driven Instructional Decision Making*. Thousand Oaks, CA: Sage, 2007.

Wright, Robert J. *Educational Assessment: Tests and Measurements in the Age of Accountability*. Thousand Oaks, CA: Sage, 2008.

Calli A. Holaway

See Also: Curriculum, K–12; Diagnostic Testing; Learning Exceptionalities; Learning Models and Trajectories.

EEG/EKG

Category: Medicine and Health.
Fields of Study: Algebra; Data Analysis and Probability.
Summary: EEGs and EKGs visually convey important information about a patient's heart and brain.

Electrocardiography (ECG or EKG) and electroencephalography (EEG) are graphic representations of bioelectric activities of the heart and brain, respectively. EKG quantifies the rhythm of heart contraction—measurements that can be used to identify damage to various myocardial muscles. EEG is used in the diagnosis of epilepsy, seizure, and encephalopathy. The production of EKG and EEG signals is grounded in mathematical analysis. Diverse mathematical and statistical techniques, including applications of calculus and chaos theory, are also used to analyze and interpret signals related to conditions such as sleep disruptions, seizures, and mental illness.

EKG

EKG is a graphic representation of the myocardial contraction (systole) and relaxation (diastole) caused by depolarization of the heart. In the myocardial muscles, depolarization is an increase of membrane potential, and repolarization is a decrease of membrane potential. A typical EKG consists of P, Q, R, S, and T waves. Atrial depolarization normally begins at the SA node and is represented as the P wave. The depolarization proceeds to ventricles, which causes the ventricular depolarization (QRS complex) and then ventricular repolarization (T wave).

EKG was first systemically studied in humans by Augustus Walker in 1887. In 1903, Willem Einthoven created a reliable EKG device based on the galvanometer. Einthoven was awarded a Nobel Prize in 1924 for his invention. EKG provides information on heart contraction and the abnormality of EKG has been used to diagnose the area of myocardial damage. Heart rate variability is a quantification of fluctuations of EKG complex; a healthier heart has higher variability.

The production of EKG signals can be explained by an idealized model in which both intracellular and extracellular currents are confined to the direction parallel to the propagation of the plane wavefront. When there are no external currents, the relationship between the potential inside the membrane V_i and the potential outside the membrane V_o can be represented as

$$V_i = \frac{r_i}{r_i + r_o} U_m \quad \text{and} \quad V_o = -\frac{r_o}{r_i + r_o} U_m$$

where r_i and r_o are the intracellular axial resistance and extracellular axial resistance, respectively; and U_m is the membrane potential. During the depolarization, the transmembrane current i_m is

$$i_m = \frac{1}{r_i + r_o} \frac{\partial^2 V_m}{\partial x^2}$$

where the direction of positive current is defined as the direction of the positive *x*-axis. For the depolarization of cardiac tissue, a double layer appears at the wavefront with the dipole orientation in the direction of propagation.

A pair of electrodes can be used to produce one EKG signal; the output from the pair is called a "lead." Usually more than two electrodes are used and combined into pairs. Clinically, a 3-lead or 12-lead EKG is used to diagnose heart diseases. For a traditional 3-lead EKG, leads I, II, and III are defined as

$$\mathrm{I} = V_{LA} - V_{RA}$$

$$\mathrm{II} = V_{LL} - V_{RA}$$

$$\mathrm{III} = V_{LL} - V_{LA}$$

where *LA, RA,* and *LL* denote left arm, right arm, and left leg, respectively.

EEG

EEG is a recording of the electric potential of thousands or millions of neurons within the brain. The electrodes are placed on the scalp at certain anatomical locations. EEG was first systematically analyzed by Hans Berger in 1920, who introduced the term "electroencephalogram" to indicate fluctuations recorded from the brain. EEG waves are usually irregular and cannot be classified in the normal brain.

However, four characteristic frequencies have been identified: Alpha (8-13Hz), Beta (14-30Hz), Theta (4-7Hz), and Delta (below 3.5Hz) waves. Under pathological conditions, like epilepsy, distinct patterns can be observed and used to help predict the onset of the condition.

Using a simplified model of the brain and surrounding tissues as a sphere with several shells, it is possible to compute the EEG based on the measured intracerebral currents at the scalp. The field potential can be represented as a function of intracerebral currents or of the membrane potential. In an infinite, isotropic, and homogeneous medium, because of injected current densities $\vec{j}_i$ at a point $\vec{r}$, the electrical potential at a point $\vec{r}_0$ lying at a distance, R, from $\vec{r}$ ($R=|\vec{r}-\vec{r}_0|$)is the following:

$$V(\vec{r}_0) = -\frac{1}{4\pi\sigma}\int_{vol}\frac{\mathrm{div}\,\vec{j}_i}{R}\,d^3r$$

where σ is the conductivity of the medium; the operator div indicates differentiation of a vector. When the injected current densities originate at the cell membrane, by assuming that the neuronal membrane is equivalent to a double layer with an intracellular membrane potential V_m, the potential at a point $\vec{r}_0$ is given approximately by

$$V(\vec{r}_0) \approx -\frac{\sigma_i}{4\pi\sigma_e}\int_{surf} V_m(\vec{r})\,d\Omega(\vec{r}-\vec{r}_0)$$

where σ_i is the intracellular conductivity, σ_e the extracellular conductivity, and

$$d\Omega(\vec{r}-\vec{r}_0)$$

is the solid angle subtended by an infinitesimal surface on the membrane surface and seen from the extracellular point $\vec{r}_0$.

Further Reading

Malmivuo, Jaako, and Robert Plonsey. *Bioelectromagnetism: Principles and Applications of Bioelectric and Biomagnetic Fields.* New York: Oxford University Press, 1995.

Niedermeyer, Ernst, and Fernado Lopes da Silva. *Electroencephalography: Basic Principles, Clinical Applications, and Related Fields.* 5th ed. Philadelphia, PA: Lippincott Williams & Wilkins, 2004.

Sanei, Saeid, and J. A. Chambers. *EEG Signal Processing.* Hoboken, NJ: Wiley Interscience, 2007.

Yih-Kuen Jan
Fuyuan Liao
Robert D. Foreman

See Also: Medical Imaging; Nervous System; Pacemakers.

Egyptian Mathematics

Category: Government, Politics, and History.
Fields of Study: Connections; Geometry; Measurement; Number and Operations; Representations.
Summary: Ancient Egyptians were adept at engineering and geometry and deeply dependent on accurate measurements of the annual Nile flood.

Our knowledge of Egyptian mathematics (3000–1000 B.C.E.) is based on hieroglyphic writings found on stone or as script (hieratic and demotic) in multiple papyri. Preserved in tombs and temples in the Nile valley, a papyrus is a narrow scroll of paper, about 15 feet in length, made by interweaving tiny strips of a water reed called *papu*. The key documents are the Moscow, Rhind, Rollin, and Harris papyri. These works are generally thought to be textbooks used by scribes to learn mathematics and solve problems.

In ancient Egypt, mathematics was used for many purposes necessary to everyday life: measuring time, drawing straight lines, measuring and recording the level of the Nile floodings, calculating land areas, and managing money and taxes. The Egyptians were also one ancient culture that came closest to determining the true length of Earth's year with mathematics. Perhaps most well known to the modern world are the fantastic tombs, pyramids, and other architectural marvels constructed using mathematics. Though their knowledge ranged from arithmetic calculations to algebraic rules to geometrical formulas to numerical ideas, historians consider the Egyptians' mathematical achievements to be somewhat less advanced compared to the Babylonians.

Egyptian Number System

Egyptian numbers are written using a simple grouping system whose symbols denote powers of 10. Their symbols included a vertical staff (10^0), heel bone (10^1), scroll (10^2), lotus flower (10^3), pointing finger (10^4), tadpole (10^5), and astonished man (10^6):

𓏺	𓎆	𓍢	𓆼	𓂭	𓆐	𓁨
1	10	100	1000	10,000	100,000	1,000,000

Using these symbols, a number was expressed additively. For example, the base-10 number 4501 was represented by a visual collection of 4 lotus flowers, 5 scrolls, and 1 vertical staff. As no place-value system is involved, these symbols can be written in any order or arrangement visually—they equal a numerical value as a group. Though able to represent large values of numbers with these symbols, the Egyptians' lack of place values deterred their ability to calculate proficiently using algorithms.

Again represented by hieroglyphic symbols, Egyptian fractions were restricted to unit fractions (numerator of 1) except for the special fraction 2/3. For example, the unit fraction 1/3 was represented by an ellipse (or dot) placed visually over 3 vertical staffs. The Egyptians had no symbol for zero as a place holder but such was not really needed because of their simple grouping system and use of distinct symbols for each power of 10.

Egyptian Arithmetic

Addition and subtraction are quite easy using the Egyptian numbers, involving only the union or removal of the grouped symbols. In addition, a symbol that appeared 10 times was replaced by the next higher level symbol; for example, 10 vertical staffs could be replaced by 1 heel bone. Similarly, in subtraction, a symbol could be traded in for 10 of the next lesser symbol if such was necessary. For example, to perform $23-8$, a heel bone could be traded for 10 vertical staffs so that 8 vertical staffs could be taken from the 13 vertical staffs.

Egyptian multiplication involved repeated addition, using a doubling process along with a counter. For example, to multiply 23×13, their process (in modern notation) would look like the following, with the counter on the right:

23	1*
46	2
92	4*
184	8*

Using the starred counters $(1+4+8=13)$, the product is obtained by adding the associated numbers $(23+92+184=299)$. The key to this multiplication is the distributive process, since

$$23\times 13 = 23\times(8+4+1) = (23\times 8)+(23\times 4)+(23\times 1)$$
$$= 184 + 92 + 23 = 299.$$

Thus, base two notation also is the underlying principle, since

$$13 = (1)(2^3) + (1)(2^2) + (0)(2^1) + (1)(2^0).$$

These processes of duplation and mediation (doubling and halving) remained as standard algorithms in Western mathematics until the 1500s.

Division required an inversion of the multiplication process. For example, to divide 299 by 23, the Egyptian scribe determined what number times 23 would produce 299, using a process like the following (in modern notation):

23*	1
46	2
92*	4
184*	8

Using the starred sums, $23 + 92 + 184 = 299$, the desired factor (or quotient) is obtained by adding the associated numbers, or $1 + 4 + 8 = 13$. The division process becomes complicated when no combination of the starred numbers equals the desired sum (for example, 300 divided by 23), requiring the use of unit fractions:

23*	1
46	2
92*	4
184*	8
1*	1/23

For more difficult divisions (for example, 301 divided by 23), considerable creativity was needed.

To aid in their computations, the Egyptians created tables for doubling and halving numbers, complemented by special $2/n$ tables that would help avoid odd-number situations. For example, the Rhind papyrus had a $2/n$ table for the odd numbers 5–101.

Egyptian Algebra

Though without an algebraic notation, the Egyptians solved numerous types of algebraic equations, known as "aha" calculations. The majority of their problems were linear equations with one unknown (called the "heap"). Their solution process involved the method of false position, where an initial guess is made, examined, and then adjusted to obtain the correct solution. This same process is now fundamental to the area of numerical analysis and is used extensively for scientific computing using computers.

Consider this Egyptian problem, "Heap and a seventh of the heap together give 19." In modern notation, the associated linear equation is $x + x/7 = 19$, while their step-by-step solution was the following:

Make a guess for heap, for example, 7

$$\text{Then } 7+\frac{7}{7}=8$$

$$\text{But } \left(2+\frac{1}{4}+\frac{1}{8}\right)(8)=19$$

$$\text{Thus, heap } =(7)\left(2+\frac{1}{4}+\frac{1}{8}\right)=16+\frac{1}{2}+\frac{1}{8}.$$

The processes of multiplication and division, as well as the law of associativity, play very important roles:

$$\begin{aligned}&\left(2+\frac{1}{4}+\frac{1}{8}\right)(8)\\ &=\left(2+\frac{1}{4}+\frac{1}{8}\right)\left(7+\frac{7}{7}\right)\\ &=\left(2+\frac{1}{4}+\frac{1}{8}\right)\left[(7)\left(1+\frac{1}{7}\right)\right]\\ &=\left[\left(2+\frac{1}{4}+\frac{1}{8}\right)(7)\right]\left(1+\frac{1}{7}\right)\\ &=\left(16+\frac{1}{2}+\frac{1}{8}\right)\left(1+\frac{1}{7}\right)\\ &=19.\end{aligned}$$

The majority of the Egyptians' "aha" problems created practical situations requiring the use of ratios and proportions, such as determining feed mixtures or combinations of grains to make bread. In some instances, the Egyptians did use special hieroglyphic symbols as part of their algebraic work, including "plus" (legs walking left to right), "minus" (legs walking right to left) and other ideograms for "equals" and the "heap."

Egyptian Geometry

The Egyptians' geometry was rooted in an algebraic perspective, devoid of any evidence of generalization or proof. Approximately one-fourth of the problems found in the papyri are geometrical—focusing on practical measurements, such as the calculation of land areas, or volumes of storage containers. Similar to the Babylonians, the Egyptians used prescriptive formulas. For example, they viewed a circle's area as equal to that of a square erected on 8/9 of the diameter. That is,

$$A=\left(\frac{8}{9}(2r)\right)^2=\frac{256}{81}r^2$$

implying their value of π approximated 3.160493827.

Historians agree that the Egyptians knew key formulas for computing the area of a triangle, the volume of a cylinder, some curvilinear areas, and even the volume of the frustum of a square-based pyramid. These formulas were apparently put to great use by the Egyptians in their accurate construction of the pyramids, feats that required a solid understanding of ratios, proportions, dihedral angles, and even astronomy. No evidence suggests the Egyptians knew of the relationships described by the Pythagorean theorem. Some of their geometrical prescriptions were also incorrect. For example, the area of a general quadrilateral (with ordered side lengths a, b, c, d) was calculated by the formula

$$A = \frac{1}{4}(a+c)(b+d)$$

which is correct only if the quadrilateral is a rectangle or square.

Signs of Advanced Mathematical Thinking

Egyptian mathematics was utilitarian in its direct ties to the solution of practical problems. Also, because their numeration system involved simple grouping with no place values, it is not reasonable to expect that the Egyptians had explored ideas such as factors, powers, and reciprocals. This limitation perhaps explains why no record has been found of tables involving Pythagorean triples. Nonetheless, they did apparently use some number tricks; when multiplying a number by 10, they merely replaced each hieroglyphic symbol by the symbol representing the next higher power of 10 (that is, replacing each vertical staff with a heel bone, each heel bone with a scroll, and so forth).

Problem 79 in the Rhind Papyrus suggests that the Egyptians did some recreational mathematics that had no real-world applications. The problem states, "7 houses, 49 cats, 343 mice, 2401 ears of spelt, 16,807 hekats." Historians assume that the scribe was creating a problem involving seven houses, each with seven cats, each of which eats seven mice, each of which had eaten seven ears of grain, each of which had sprouted seven grains of barley…wanting to know the total number of houses, cats, mice, ears of spelt, and grains. Mathematically, the solution of this problem would require some knowledge of powers of 7 and geometric progressions.

Further Reading

Aaboe, Asger. *Episodes From the Early History of Mathematics*. Washington, DC: Mathematical Association of America, 1975.

Friberg, Jöran. *Unexpected Links Between Egyptian and Babylonian Mathematics*. Singapore: World Scientific Publishing, 2005.

Katz, Victor J., ed. *The Mathematics of Egypt, Mesopotamia, China, India, and Islam: A Sourcebook*. Princeton, NJ: Princeton University Press, 2007.

Van der Waerden, B. L. *Science Awakening*. Oxford, England: Oxford University Press, 1985.

———. *Geometry and Algebra in Ancient Civilizations*. Berlin: Springer, 1983.

Jerry Johnson

See Also: Arabic/Islamic Mathematics; Babylonian Mathematics; Chinese Mathematics; Greek Mathematics.

Einstein, Albert

Category: Space, Time, and Distance.
Fields of Study: Algebra; Geometry; Measurement; Representations.
Summary: One of the most well-known physicists, Albert Einstein's work continues to influence many fields.

During the twentieth century, research in the fields of mathematics, physics, chemistry, information technologies, and engineering exploded. People's perception concerning the world and the universe around them changed dramatically within a faster and faster changing world. If one were to choose a single influential scientist to represent this era, some might choose Albert Einstein. During "the Age of Einstein," he introduced many original concept widely used in various fields, such as mathematics, science and technology, world politics, economics, and philosophy.

Early Life and Education

One common myth about Einstein was that he failed mathematics as a child. Albert Einstein was born at Ulm (Württemberg, Germany) on March 14, 1879. He studied in various places, including Munich, Italy, and Switzerland. His uncle, an engineer, presented him with questions about mathematics, such as a challenge to find a proof of the Pythagorean theorem. Einstein noted, "After much effort I succeeded." After viewing a *Ripley's Believe It or Not* headline about his proposed failure in mathematics, biographers note that Einstein replied, "I never failed in mathematics. Before I was 15 I had mastered differential and integral calculus." In 1896, he entered the Swiss Federal Polytechnic School in Zurich to study physics and mathematics. In 1901, he began working at the Swiss Patent Office. In 1905, he obtained his doctorate degree. He was a professor at various universities in Europe until 1933, when he immigrated to America because of anti-Jewish laws in Germany.

Accomplishments

One notable quotation attributed to Einstein is: "Do not worry about your difficulties in mathematics, I assure you that mine are greater." During his life, Einstein published a great amount of papers in several fields of the sciences. Many equations and laws are named for him, including: Einstein's absorption coefficient, Einstein photoelectric law, Einstein frequency condition, Einstein diffusion equation, Einstein–Bohr equation and Einstein coefficients, Einstein frequency and Einstein elevator, Einstein–Planck law, Einstein mass–energy relation, and so on. In chemistry, a synthetic radioactive chemical element having the symbol Es, the atomic number 99, and atomic weight 252.08 is called "einsteinium." Einstein's principle of relativity, the basic postulate of Einstein's special relativity theory, states that the laws of nature have the same form in all inertial frames of reference. Einstein based his general theory of relativity on mathematical ideas like mathematician Bernhard Riemann's geometric formulations. Gravity was now described to be curved spacetime, "Matter tells spacetime how to curve and curved spacetime tells matter how to move." He was also philosophical about the applicability of mathematics, saying, "How can it be that mathematics, being after all a product of human thought which is independent of experience, is so admirably appropriate to the objects of reality? Is human reason, then, without experience, merely by taking thought, able to fathom the properties of real things?" Einstein's field equations from general relativity and their solutions have been a fruitful research area in mathematics and physics, leading to concepts like metrics for black holes and the notion of Einstein manifolds, named for him. However, he complained that he had difficulty understanding the theory of relativity after mathematicians "invaded" it.

Albert Einstein once said "A man should look for what is, and not for what he thinks should be."

Conservation of Mass and Energy

While twenty-first-century researchers continue to investigate Einstein's field equations, Einstein's most famous equation is probably $E = mc^2$. In his paper on the equivalence of matter and energy, he deduced the equation. It meant that conservation laws can be unified into a single law of the conservation of mass-energy. This equation also predicted the development of nuclear power. However, Einstein was an opponent of nuclear weapons. In 1939, Einstein wrote and signed a letter to U.S. President Franklin D. Roosevelt to warn him about research on uranium and the possibility of the development of an atomic bomb. The president took his words seriously, which was the beginning of the Manhattan Project—the effort to construct a nuclear bomb. In a 1954 letter to his friend, Linus Pauling, Einstein confessed that his letter to Roosevelt was the one great mistake of his life. During his life, he made many contributions for peace. Einstein stated, "We have to divide up our time like that, between our politics and our equations. But to me our equations are far more important, for politics are only a matter of present concern. A mathematical equation stands forever." Just before his death, in a letter to Bertrand Russell, he still urged all nations to give up nuclear weapons.

A Public Person

Einstein was in his whole life a public person. Being a good-humored speaker, he took part in a large number of conferences and traveled in many countries. His name became the brand for "genius," and a large number of sayings and anecdotes are told on his account, such as "Pure mathematics is, in its way, the poetry of logical ideas." He always seemed to have a clear view of the problems to solve and the will to solve them. He remained a very curious person and taught his pupils not to be afraid of asking, trying, and failing. Failures, together with achievements, are merely stepping stones for the next adventure of discovery, he said.

In the 1949 publication of *Autobiographical Notes*, he stated, "In the beginning (if there was such a thing), God created Newton's laws of motion together with the necessary masses and forces. This is all; everything beyond this follows from the development of appropriate mathematics methods by means of deduction." After his retirement from Princeton, he continued to work on a theory of unification of the basic concepts of physics, natural sciences, mathematics, and religion. Albert Einstein received a Nobel Prize in physics along with honorary doctorate degrees in science, medicine, and philosophy from many universities. A crater on the moon is named after him.

Further Reading

Cox, Brian, and Jeff Forshaw. *Why Does $E = mc^2$? (And Why Should We Care?)*. Cambridge, MA: Da Capo Press, 2009.

Cropper, William H. *Great Physicists: The Life and Times of Leading Physicists From Galileo to Hawking*. New York: Oxford University Press, 2001.

Geroch, Robert. *General Relativity From A to B*. Chicago: University of Chicago Press, 1978.

Isaacson, Walter. *Einstein: His Life and Universe*. New York: Simon & Schuster, 2007.

Staley, Richard. *Einstein's Generation: The Origins of the Relativity Revolution*. Chicago, IL: University of Chicago Press, 2009.

Troemei-Ploetz, Senta. "Mileva Einstein-Mari: The Woman Who Did Einstein's Mathematics." *Women's Studies International Forum* 13, no. 5 (1990).

Simone Gyorfi

See Also: Clocks; Elementary Particles; Geometry of the Universe; Gravity; Hawking, Stephen; Measuring Time.

Elections

Category: Government, Politics, and History.
Fields of Study: Communication; Data Analysis and Probability; Number and Operations.
Summary: Mathematics can help explain and predict elections.

Long the domain of economists, political scientists, and philosophers, systems of government has emerged as a field ripe for the application and study of mathematics. Elections are typically classified under an emerging branch of mathematics called "social choice theory," though there are historical connections and applications in a number of areas, such as combinatorics and probability theory. Economist Duncan Black's 1958 book *The Theory of Committees and Elections* is credited with helping to revive modern interest in using mathematics to study election questions.

In a democratic society, such as the United States, elections are the primary vehicle for providing citizens a fair and equal voice in the machinations of federal, state, and local governments. As such, it is fundamentally important that elections be conducted in a manner that is perceived to be fair by the citizenry; that is, a governing body derives its legitimacy from the equitable interpretation and application of the voting power of the public.

Beyond the widely known popular elections (electing the candidate with the most first-place votes) there are a number of alternative voting methods; many of these allow voters to express more information about their preferences of various candidates. Since it is possible for different methods to produce different winners given the same voter preferences, a number of

Townspeople lined up to vote in rural areas of Guatemala for the 2007 national elections. In a democratic society, elections provide citizens with a voice in the workings of federal, state, and local governments.

voting properties have been postulated. Each property states a desired outcome or effect that a voting system should express. For example, a voting system should be "anonymous" in that individual voters should be able to exchange ballots without affecting the outcome; in other words, one person's ballot should not have special significance. A more challenging property is "independence from irrelevant alternatives," which requires the relative outcome of an election to remain unaffected if candidates are added or removed from consideration (provided this addition or removal does not change the relative way voters feel about the other candidates). Economist Kenneth Arrow demonstrated mathematically in his doctoral dissertation that no voting system can satisfy all the desired properties. Arrow's Impossibility Theorem was later published in his 1951 book *Social Choice and Individual Values.*

A particular type of voting system, weighted voting, arises when voters are assigned different numbers of votes. This system is usually employed to reflect a situation where some voters should have greater say or representation than others. The Banzhaf Power Index, named after John Banzhaf, is a tool that elucidates the voting power enjoyed by the voters in a weighted voting scheme and reveals that voting power is not always commensurate with a voter's number of votes. It is also sometimes called the Penrose–Banzhaf Power Index to include its original inventor, Lionel Penrose.

The U.S. Electoral College, an example of a weighted voting system, is used to elect a winner in U.S. presidential elections. The U.S. Electoral College illustrates a drawback of weighted voting in that a winning presidential candidate may not have received a majority of popular votes. This has sparked much interest in replacing the U.S. Electoral College in favor of the popular vote method but smaller states that enjoy more voting power with the U.S. Electoral College are likely to block attempts at Constitutional reform.

Exit polling, invented by statistician Warren Mitofsky, allows social demographers to understand the dynamics of an election and to predict the winner. Exit polling has become an increasingly important tool for media and news outlets as they scramble to retain and inform viewers on the eve of an important election. A number of studies have investigated the influence of exit polling while an election is taking place; for instance, polls broadcast in real time may influence voters who have yet to vote and hence possibly change the outcome of an election. Exit polling has also garnered interest in recent presidential elections when erroneous predictions caused media sources to prematurely, or incorrectly, identify a winning candidate.

The Ballot Box Problem is an interesting mathematical puzzle, proposed by Joseph Bertrand, which seeks answers about how an election may unfold as ballots are removed from the ballot box and counted. The solution to Bertrand's theorem is a Catalan number, named for Eugène Catalan. An elegant proof was derived by Désiré André.

Types of Elections

Though most people are familiar with the plurality election (also known as "popular vote") in which the candidate with the most votes (most first-place votes) wins, there are a number of alternative election methods. One of the most prominent is the Borda method, named for Jean-Charles de Borda, where voters are required to rank all candidates from their first choice to their last; points are then assigned to each candidate based on the candidate's rank on the each ballot. The sum of a candidate's total points is used to determine the winner. This method allows voters to specify more information about how they view the candidates, other than merely selecting their favorite.

In the Sequential Pairwise method, two of the candidates vie in a head-to-head competition (an imaginary election with only the two candidates) where the losing candidate is eliminated and the winner proceeds forward to battle another candidate. Again, voters rank candidates in preference listings, which are used to determine the winner between a particular pair of candidates. The winner can be inferred from the preference lists by assuming each voter would select the candidate that is higher on his or her list. A drawback of this method is that the order in which the candidates are selected for the individual competitions can change the ultimate outcome of the election.

A Condorcet Winner is a candidate who beats every other candidate in a head-to-head election. When one exists, a Condorcet Winner will obviously win the Sequential Pairwise election but not all sets of voter preference rankings produce a Condorcet Winner. The method is named for Marie Jean Antoine Nicolas de Caritat, Marquis de Condorcet.

In an Instant Run-off election, a plurality vote is taken and the candidate with the least number of first-place votes is eliminated. Then the election is repeated with the remaining candidates until only one winner remains. Again, voter preference rankings can be used to simulate the repeated elections in order to determine the winner without holding a series of actual elections.

Weighted Voting

Much of rationale behind the U.S. system of government is based on the principle of "one person, one vote" (each citizen should have equal say in the system of government). There are times, however, when it is appropriate to give certain individuals (or groups) more voting power than others. This type of voting system, often called "yes–no voting" or "weighted voting," occurs when voters are assigned a different number of votes or "weights" to their votes. Elections are between two alternatives; the winner is selected if the vote total exceeds a predetermined threshold. Each voter must use all available votes toward the same candidate or choice—votes cannot be split between the candidates or choices.

An example of a weighted voting system was the European Economic Community (EEC) established in 1958 as a precedent to the current European Union. The original six members were assigned votes in proportion to their population size:

Country	# Votes
France	4
Germany	4
Italy	4
Belgium	2
Netherlands	2
Luxembourg	1

A threshold is established to determine the number of total votes necessary to win an election. Though this threshold is often simple majority, in the EEC example, a threshold of 12 (of the total 17 votes) was established to pass certain types of legislation.

An interesting question arises as to the dynamics of weighted voting systems and, more specifically, an entity's ability to influence the outcome of an election. Several theorists have shown that voting power is not necessarily proportional to an entity's vote count. For example, it would be misleading to assume that France enjoys 23.5% (4/17) of the voting power in the EEC example.

Banzhaf applied a power index to argue a landmark case in Nassau County, New York, in 1965. His voting power calculations demonstrated the disenfranchisement of certain entities within weighted voting schemes and thus questioned the system's constitutionality.

Banzhaf's computation is based upon the notion of a winning coalition (a collection of voters whose vote total exceeds the threshold). Such a coalition (or "voting block") can win an election by all voting the same way. A voter is critical to a winning coalition if by removing that voter, the coalition no longer exceeds the threshold. A voter's Banzhaf Power Index (BPI) is related to the number of times that voter is a critical member of a winning coalition.

In the EEC example, France, Germany, Italy, and Belgium form a winning coalition since their vote total of 14 exceeds the threshold of 12. France, Germany, and Italy are all critical members because the coalition ceases to win without their votes. However, Belgium is not a critical member since France, Germany, and Italy together still form a winning coalition. The number of times each voter appears as a critical member of some winning coalition is computed as follows:

Country	# Critical	BPI
France	10	10/42 = 23.8%
Germany	10	110/42 = 23.8%
Italy	10	10/42 = 23.8%
Belgium	6	6/42 = 14.3%
Netherlands	6	6/42 = 14.3%
Luxembourg	0	0/42 = 0%

Each country's BPI is the number of times it is critical compared to the total number of critical instances. Here, there are 42 total instances where an entity is critical; Belgium has 6 of them and thus 14.3% (6/42) of the voting power. Thus, Belgium commands 14.3% of the voting power even though it has 11.8% of the votes. In this scheme, Luxembourg has no voting power—it is not able to influence the outcome of any possible election. It is common in weighted voting schemes of smaller size (20 or fewer members) for entities with a greater number of votes to possess greater voting power, while small entities (with a fewer number of votes) possess less voting power. As the number of voters increase, voting power tends to better approximate the proportion of votes. But such weighted voting systems are subject to arbitrary swings of voting power as new voters are added or removed, or as seemingly subtle changes to the weights are made.

An equally popular voting power computation was proposed by Lloyd Shapely and Martin Shubik in 1954. Instead of critical members in winning coalitions, their system identifies pivotal voters as the ones who enter a coalition and cast the deciding vote by doing so. A similar calculation ensues in which voting power is correlated with the percentage instances in which each entity plays the pivotal role.

U.S. Electoral College

The voting system responsible for electing the president of the United States, the U.S. Electoral College, is essentially a weighted voting scheme. A state's electors (or "votes") arise from the sum of their congressional representation: one vote for each of a state's two senators and one vote for each representative to the House of Representatives. The District of Columbia receives three electors to form a total of 538 (100 senators, 435 representatives, and three from Washington, D.C.). A presidential candidate needs a majority of the electoral votes—at least 270—to claim victory.

Under such a system, it is possible that the winning candidate need not garner a majority of first-place votes. In fact, U.S. presidential elections in 1824, 1876, 1888, and 2000 all produced a winner who lost the popular vote total.

Those elections and other issues have created an endless interest in reforming or removing the U.S. Electoral College and replacing it with a popular vote system. As recently as 2004, the Every Vote Counts Amendment proposed to replace the U.S. Electoral College with a popular vote initiative. Such a reform requires a Constitutional change and thus approval of 75% of the states.

It is unlikely such a measure would ever be adopted because small states enjoy significantly more voting power in the U.S. Electoral College than they would in a popular vote system. A state with few votes, such as South Dakota, would likely be ignored by campaigners since the voting population is too small to make a difference under a popular vote election.

The National Popular Vote Compact is an alternative attempt at election reform. In this compact, individual states would cast their electoral votes according to the national popular vote, not simply the tallies within the state. This has the effect of choosing a president elected by popular vote within the Electoral College system and thus bypassing the hurdle of constitutional reform. To date, this compact has been adopted by five states (61 electoral votes) with a number of others considering the compact in state legislature—enough states to compile 270 electoral votes would have to sign on to the compact in order to have the intended effect of electing a president by popular vote.

Exit Polling

An important factor associated with elections is the attempt to predict election outcomes through the surveying of voters as they leave the voting areas, a procedure known as "exit polling." This procedure contrasts with pre-election polls in that actual voters who have (presumably) just cast a vote are being sampled and thus results are typically more accurate than surveying people prior to an election who are "likely" to vote, or who may change their mind between being polled and actually casting a vote.

Although the science of predicting election outcomes has been around as long as elections themselves, it is at the beginning of the twenty-first century—with widespread electronic media coverage and more sophisticated polling techniques—that exit polling has garnered more national attention. A number of papers have been written about the effects of exit polling being broadcast in real time; the researchers hypothesize that exit polling influences voter behavior primarily by making an election seem closer or not closer than was previously perceived. This effect is especially true in the United States where, as a function of different time zones, voters in western states have access to more complete results of a national election unfolding across the country.

Exit polling has garnered an additional spotlight with the controversial presidential elections of 2000 and 2004. In both cases, especially the 2004 election, exit polling differed significantly from the actual vote tally, causing many media outlets to incorrectly, or prematurely, announce a victor.

Ballot Problem

There are several interesting mathematical puzzles based on elections and voting; perhaps the most well known of them is the Ballot Problem, originally presented by Joseph Bertrand in the late nineteenth century. Consider an election between two people, Alice and Bob, where Alice has received A votes and Bob B votes. Let $A > B$ so that Alice wins the election. The puzzle arises from the counting of the votes: what is the probability that as the votes are pulled randomly from the ballot box and tallied one by one, that Alice and Bob are tied in their vote total at some point after the first vote is read?

Survivor!

The popular television series *Survivor* nicely illustrates a ballot-box type of problem. In individual tribal councils, as well as the final vote for an overall winner, ballots are drawn from a ballot box and read aloud. It is easy to hypothesize that the ballots are not drawn in a random order but instead are selected so as to maximize the suspense of the election outcome. Another interesting question, related to information theory, is that ballots are read only until the election outcome is certain; unread ballots are not presented to the remaining tribe members, thereby depriving them of strategic information about the voting behavior of their fellow competitors.

The puzzle's solution is a creative argument based on combinatorics and probability. Sequences, a listing of votes as they are pulled from the ballot box, can be identified as those with ties and those without. The following is a sequence from an election with nine voters (A = 5, B = 4):

$$b\,b\,a\,b\,a\,a\,b\,a\,a.$$

In this sequence, the first tie occurs with the reading of the sixth vote, though there is also a subsequent tie. There is also a "matching" partial sequence in which the *a*'s and *b*'s exchange places up through the point of the first tie:

$$a\,a\,b\,a\,b\,b\,b\,a\,a.$$

Every such sequence of strings that produces a tie somewhere in the intermediate vote tally comes in matching pairs as shown. Out of each pairing, one sequence must start with an *a* while its match starts with a *b*. Since Alice wins the election, some of the sequences starting with an *a* will result in a tie but not all of them. However, every sequence that starts with a *b* must at some point achieve a tie since ultimately there will be more *a*s than *b*s. There are three categories of sequences:

- sequences that start with an *a* but never have a tie
- sequences that start with an *a* and achieve a tie at some point
- sequences that start with a *b* and achieve a tie at some point

The probability that any sequence is found starting with a *b* is

$$\frac{B}{A+B}$$

since there are B ballots out of $A + B$ total ballots where a *b* can be the first vote drawn. There are exactly as many sequences that start with an *a* and also achieve a tie because each one is matched with exactly one *b*-starting sequence. Therefore, the probability of reading the votes and achieving a tie along the way is exactly

$$\frac{2B}{A+B}.$$

This problem has spawned a number of related problems with interesting ties to Catalan numbers.

Further Reading

Freeman, Steven F. *The Unexplained Exit Poll Discrepancy.* Philadelphia: Center for Organizational Dynamics, University of Pennsylvania. 2004.

Hodge, Jonathan K., and Richard E. Kilma. *The Mathematics of Voting and Elections.* Providence, RI: The American Mathematical Society, 2005.

Sudman, Seymour. "Do Exit Polls Influence Voting Behavior?" *Public Opinion Quarterly* 50, no. 3 (1986).

Taylor, Alan D. *Mathematics and Politics: Strategy, Voting Power, and Proof.* New York: Springer-Verlag. 1995.

Matt Kretchmar

See Also: Census; Congressional Representation; Game Theory; Gerrymandering; Government and State Legislation; Mathematics, Elegant; Voting Methods.

Electricity

Category: Architecture and Engineering.
Fields of Study: Algebra; Representations.
Summary: Electricity, arising from the flow of electrons, can be described mathematically.

Daily operations of modern industrial societies, including transportation, communication, heating, cooling, lighting, computing, and medical technology, rely on the use of electrical power. Power from batteries and electrical outlets is derived from the flow of electrons, known as "electric current." The term "electricity" refers to a variety of physical effects, both static and dynamic, that arise from electric charge. The mathematical description of electric and magnetic phenomena developed in the eighteenth and nineteenth centuries contributed to a rapid expansion of electrical technology, which is powered today by a vast grid of electric power stations and distribution systems.

Electric Charge and Coulomb's Law

Electric charge is a property of matter that can be negative (as in electrons), positive (as in protons), or zero.

Most matter has a net charge of zero, containing essentially the same number of electrons as protons. Two objects whose charges are both positive or both negative repel each other, while objects with opposite charges attract each other. Static electricity is created when electrons build up on or are depleted from the surface of a material, often by rubbing materials together. Effects of static electricity are seen, for example, in a rubbed balloon clinging to a wall, or in hair standing on end. In metals, electrons are not strongly bound to individual atoms but move freely through the lattice of protons. Materials with freely moving charges are known as "conductors." The force between two charged particles at rest is described by Coulomb's Law, named after French engineer Charles-Augustin de Coulomb (1736–1806). Coulomb's Law states that the magnitude F of the force exerted by one charged particle on the other is

$$F = \frac{kqq'}{r^2}$$

where q and q' are the magnitudes of the charges of the particles, r is the distance between the two particles, and k is a constant. This equation shows, for example, that if one charge is tripled, then the force is tripled, and if both charges are tripled, then the force becomes nine times as large. On the other hand, tripling the distance r between the particles multiplies the right-hand side of the equation by $1/3^2$, or $1/9$, reducing the force to a ninth of its previous value.

Electric Field and Electric Current

The presence of charged particles creates an electric field that exerts a force on other charged particles in the region. An electric power generator, usually driven by a steam turbine fueled by coal or a nuclear reactor, creates an electric field between two terminals by building an over-supply of electrons (negative charge) in one terminal and a deficit of electrons (positive charge) in the other. The flow of electrons from a negative toward a positive terminal along a conducting path, such as a wire, is an electric current. In lightning, electrons from negatively charged clouds in the atmosphere are attracted to positively charged objects on the ground beneath the cloud. Here the electric field is so strong that electric current passes through air, which usually acts as an insulator that prevents the flow of electrons. Batteries operate by producing an electric current between oppositely charged terminals of chemical cells. A battery produces direct current (DC), where electrons flow in one direction, while a power generator creates alternating current (AC), where the direction of electron flow alternates rapidly, typically at a frequency of 60 hertz (cycles per second). The hertz is named for German physicist Heinrich Hertz (1857–1894), who made important advances in understanding the connection between electric and magnetic fields.

Ohm's Law

The energy that an electric field imparts to a unit charge moving from one terminal to another is the number of volts (V) between the terminals, named after Italian physicist Alessandro Volta (1745–1827). On electric bills, energy usage is typically given in kilowatt hours (kWh). The watt, named for British engineer James Watt (1736–1819), is a unit of power, or energy per time, and 1 kilowatt is 1000 watts. Multiplying power (in kilowatts) by time (in hours) yields energy, in kilowatt-hours. In an electric current, the current intensity (I) is abbreviated as "current" and is the quantity of charge that moves past a cross-section of the conducting path per unit time. As electric current flows through a material, the motion of the electrons is hindered by positive ions, creating electrical resistance (R). Resistance in the path of a current creates heat and light, as in appliances, such as stoves and light bulbs. Electrical energy can be transformed into mechanical energy to power motors as in cars, airplanes, power tools, kitchen blenders, and hair dryers when electric current passes through a coil of wire, inducing a magnetic field that sets the coil in motion.

Ohms's Law, formulated by German physicist Georg Ohm (1789–1854), states that for a metal conductor at constant temperature, the voltage (V) is $V = IR$, where I is the current, and R is the resistance. This equation shows, for example, that if the resistance is cut in half, then to maintain the same voltage, the current must be doubled. If too little resistance is present, the current may become so strong as to damage electrical equipment. Circuit breakers then sever the path of the current to avoid damage.

Electric Power from Generator to Consumer

High voltage generated at power stations is propagated along power lines almost instantaneously, over many miles, to substations near cities and towns. At the substations, the voltage is reduced and transmitted to electric distribution centers that channel the voltage to homes, offices, and other facilities. In standard electrical outlets in the United States, there are 120 volts between the wires leading to the two vertical slots. When an appliance is plugged into the outlet, the vertical prongs of the plug make contact with these wires, creating a pathway of current through the appliance. The third slot in the outlet carries a protective ground wire. In appliances with a three-pronged plug, the ground wiring is designed to provide a preferred pathway for escaped current so that it will not travel through the body of the person holding the appliance.

Large appliances, including most drying machines and ovens, operate at 240 volts, using a different type of outlet. Touching one or more openings in an electrical outlet or touching the prongs of a plug as it is inserted into the outlet may pass an electric current through the body that can be harmful or even deadly. At electrical facilities, "High Voltage" signs warn of the danger of electric shock because of the presence of high voltage.

Further Reading

California Energy Commission. "What Is Electricity?" http://energyquest.ca.gov/story/chapter02.html.

Herman, Stephen L., and Crawford G. Garrard. *Practical Problems in Mathematics for Electricians*. 6th ed. Albany, NY: Delmar, 2002.

U.S. Energy Information Administration: Independent Statistics and Analysis. "Electricity." http://www.eia.doe.gov/fuelelectric.html.

BARBARA A. SHIPMAN

See Also: Elementary Particles; Light; Light Bulbs; Lightning; Microwave Ovens; Nanotechnology; Radiation.

Elementary Particles

Category: Space, Time, and Distance.
Fields of Study: Data Analysis and Probability; Number and Operations; Representations.
Summary: Various branches of mathematics are employed to study elementary particles, the smallest particles in the universe.

Particle physics is a branch of physics that seeks to describe and explain the universe on the smallest scales. The particles thought to be the fundamental building blocks of matter and force are called "elementary particles." Like all branches of physics, the study of elementary particles relies heavily upon many branches of mathematics, including calculus, geometry, group theory, algebra, and statistics. Particle physics also contributes to mathematical research by posing questions that give rise to new mathematical theories.

History

For thousands of years, scientists and philosophers have been asking the questions, "What is the universe made of?" and "Are there fundamental units that make up space, matter, energy, and time, or are these infinitely divisible?" As early as the fifth century B.C.E., Greek philosopher Democritus (c. 460–370 B.C.E.) hypothesized that all matter is made of indivisible, fundamental units called "atoms." Despite these early hypotheses, there was very little progress in this field until the dawn of the twentieth century.

The twentieth century saw the emergence of several new branches of physics. Among these was particle physics, a field that seeks to explore the universe on the smallest scales. Particle physicists try to identify the particles that form matter and force, describe their properties, and understand how these particles relate to each other. Some of these particles are not composed of any other particles and are therefore called "elementary particles." These elementary particles form the basic building blocks of the universe.

The understanding of particle physics at the beginning of the twenty-first century is embodied in the Standard Model of Particle Physics, an elaborate yet still incomplete model that attempts to list and describe all existing particles. Jokingly referred to as "The Particle Zoo," the Standard Model lists dozens of particles and includes elementary particles with exotic names such as "gluon," "muon," and "quark." Many of the particles in the Standard Model have yet to be detected experimentally, and their existence is conjectured based on theoretical work.

Mathematics Used in the Study of Particle Physics

Like all physical theories, particle physics relies heavily upon mathematics, which provides the theoreti-

NASA scientists detected a ring of dark matter that formed during a collision between galaxy clusters. Astronomers don't know what dark matter is made of; however, they believe it is a type of elementary particle.

cal framework physicists use to explain and describe physical phenomena. Mathematics also enables physicists to make predictions that can later be tested using modern tools, such as particle accelerators.

One of the most useful branches of mathematics is calculus, a field that has applications in practically all branches of the natural sciences, as well as in engineering and even in the social sciences. It is therefore not surprising that calculus occupies a central role in the theory of elementary particles. Differential calculus may be used to describe properties of particles at an instant, while integral calculus is used to describe cumulative effects of a particle or a system of particles over time and space.

Calculus is but one branch of the mathematical field of analysis that is useful in particle physics. Other branches of analysis—partial differential equations, complex analysis, and functional analysis—play important roles as well.

Geometry has traditionally been used to describe the universe on the grandest scales, those of galaxies, galaxy clusters, and the universe as a whole. Recently, geometry has found a place in elementary particle research as well. French mathematician Alain Connes (1947–) has described a theoretical model for particle physics that is based on noncommutative geometry, which is a geometrical representation of noncommutative algebras—systems in which the order of factors in an operation determines the value of the operation. For example, if a and b are real numbers, then it is always true that $a \times b = b \times a$, as multiplication is commutative for real numbers. However, if A and B are matrices, then generally $A \times B \neq B \times A$. Matrix multiplication is therefore noncommutative.

Symmetry, Group Theory, and Quantum Mechanics

One of the most fundamental mathematical concepts in elementary particles is symmetry. In mathematics, symmetry is defined as an operation on an object that leaves some of the object's properties unchanged. As an example, consider a square drawn in the plane and an axis of rotation that passes through the square's center, perpendicular to the plane. If the square is rotated by 90 degrees around that axis, the square will appear unchanged. Rotation by 90 degrees is thus called a "symmetry" of the square. The set of all symmetries of an object forms a mathematical construct called a group (a set with an operation that obeys several axioms). Group theory, a branch of algebra, plays an important role in particle physics, as properties of many elementary particles can be explained and described by the use of symmetry.

The chief group-theoretic structure in particle physics is the Lie (pronounced "Lee") group, named after Norwegian mathematician Sophus Lie (1842–1899). Lie groups are groups that posses the properties of geometric constructs known as "differentiable manifolds." Lie groups thus provide yet another connection between geometry and elementary particles.

One of the most important physical theories of the twentieth century is quantum mechanics, a theory that holds that, at the atomic and subatomic levels, behavior of particles is a statistical rather than a deterministic phenomenon. Since elementary particles obey quantum-mechanical laws, statistics and probability are invariably major components of the mathematical framework of elementary particles.

While physicists use mathematics as a tool for exploring the universe, the relationship between particle physics and mathematics is not one-directional. Research in particle physics drives the emergence of new mathematical theories, just as mechanics drove the emergence of calculus in the seventeenth century. In 1990, American theoretical physicist Edward Witten (1951–) won the Fields Medal, the highest honor in mathematics, for his many contributions to mathematics. He is the only non-mathematician ever to win the prestigious award. As both mathematicians and physicists continue to explore new horizons, the cross-fertilization of ideas will benefit both fields in decades to come.

Further Reading

Griffiths, David. *Introduction to Elementary Particles.* Weinheim, Germany: Wiley-VCH, 2008.

Hellemans, Alexander. "The Geometer of Particle Physics." *Scientific American* 295, no. 2 (2006).

Mann, Robert. *An Introduction to Particle Physics and the Standard Model.* Boca Raton, FL: CRC Press, 2010.

Or Syd Amit

See Also: Gravity; Relativity; Symmetry.

Elevation

Category Space, Time, and Distance.
Fields of Study: Geometry; Number and Operations.
Summary: Various aspects of elevation can be calculated using mathematical techniques.

Trigonometry has long been used to measure height. Elevation is often the height of a point relative to sea level, and its measurement is called "hypsometry." Elevation affects air pressure, temperature, and gravity, all of which have noteworthy effects on people. Astronomers and mathematicians such as Blaise Pascal and Edmund Halley investigated relationships between barometric pressure and elevation.

Historical surveys of elevation include those who used barometers, like John Charles Frémont, who was at one time professor of mathematics of the Navy, and physician Christopher Packe. However, this method is sensitive to a number of variables. In the twenty-first century, detailed elevation data are available. Mount Everest is known as Earth's highest elevation. Topographical maps represent elevation by using contour lines, each line following a path of constant elevation. Transits were developed in the nineteenth century, and they can be used to calculate changes in elevation. Contour integrals and generalized contours for functions of two variables are investigated in multivariable calculus classrooms. Mathematicians and computer scientists have helped create realistic computer models of land elevation, called "digital elevation models." They have explored ideas like irregular-mesh grids or shifting nested grids in surface reconstruction. Other types of elevation studies also benefit from mathematical techniques, like using the ocean wave spectrum to investigate sea surface elevation peaks, or statistical techniques to investigate the impacts of elevation changes. Mathematician and astronomer Nilakantha Somayagi investigated the elevation of lunar cusps in the sixteenth century. The term "angle of elevation" in high school classrooms represents the angle between where an observer is standing and the line of sight to an object. The angle of elevation is found in many contexts, including in the Pyramids of Egypt, in the astrolabe, and in global positioning systems.

Highest Elevations on Earth

Elevations are nearly always computed relative to sea level, the average height of the ocean's surface. Sea level is an inexact measure since tides, temperature, wind, salinity, and air pressure affect the oceans. Mount Everest (above) in the Himalaya Mountains near the border of Nepal and Tibet is the highest mountain on Earth at an elevation of 29,035 feet as of 2010. Everest gains more than two inches of elevation per year because of the collision of tectonic plates and there are discrepancies in its listed height.

Earth is not spherically symmetric; its radius near the equator is more than 13 miles greater than its radius near the poles. Consequently, Mount Chimborazo in Ecuador holds the distinction of having the summit farthest from Earth's center. Lying about one degree south of the equator, where Earth is widest, Mount Chimborazo is approximately 20,561 feet above sea level, enough to make its summit more than a mile farther from Earth's center than Mount Everest's summit.

Topographic Maps

A topographic map is a two-dimensional map that conveys elevation information as well as other features of an area. Contour lines are the key to capturing elevation changes from a three-dimensional world on a two-dimensional map. A contour line is a path that follows a constant elevation. Early uses of contours date to the eighteenth and nineteenth centuries and include the work of engineer Jean-Louis Dupain-Triel and astronomer and mathematician John Couch Adams.

A contour line is drawn each time a predetermined elevation change is achieved. For example, a map may use 100-foot elevation increments, with one contour line following points having an elevation of 100 feet and the next marking an elevation of 200 feet. Consecutive contour lines always differ by 100 feet in elevation. As the mapped terrain climbs more steeply, the contour lines on the map will be closer together. The lines can mark elevations that increase and decrease, representing terrain that rises and falls intermittently. Contour lines can represent elevations that are zero, or negative numbers as when mapping an ocean floor.

A topographic map of an area with constant elevation at its boundary, such as an island bounded by the sea, will not have contour lines extending off the map's edge. In such cases, all contour lines will appear as closed curves. A curve is closed if it loops back to where it started. Typically, contour lines appear as simple closed curves that do not cross themselves. The pattern of contour lines as nonintersecting rings lying one within another is common on topographic maps.

Also common is to have two separate sets of nonintersecting rings contained within a single contour line, as when two hills are surrounded by a larger path of constant elevation.

The U.S. Geological Survey (USGS) has created a complete large-scale topographic map of the United States in more than 56,000 pieces. The National Elevation Dataset is noted as the "the primary elevation product of the USGS." The data set is updated regularly, and historic data sets are also available for investigations.

There is an ever-growing growing need for digitized maps, which allow a computer user to read elevation at any spot on the map. Some digitized maps enable the user to view a landscape from different perspectives, creating a three-dimensional view of the area's elevation changes, similar to what would be seen at the actual location. Data from existing topographic maps and aerial photography are used to create digitized maps. Improvements in technology will continue to affect the science of map making.

Effects of High Elevation

As elevation increases, air temperature drops because of a decrease in air pressure. At about 18,000 feet above sea level, for example, the air pressure is half that at sea level. In the troposphere, the lowest layer of Earth's atmosphere, a general rule of thumb is that air temperature drops 6.5 degrees Celsius for every 1000 meters of elevation gain, or roughly one degree Fahrenheit for every 280 feet of elevation gain in standard conditions. This phenomenon, which can be modeled with an equation, can be seen directly when an observer standing at a low elevation on a warm day views a tall mountain covered with snow.

Another consequence of this cooling is that water vapor in the air condenses, sometimes causing increased rainfall on the windward side of a mountain range and a "rain shadow" downwind from the mountains. Many deserts lie just downwind from a mountain range. For example, sand dunes in Death Valley, California, lie in the rain shadow of Mount Whitney, the highest peak in the continental United States.

Because of these differences in temperature and precipitation, tall mountains can have multiple climatic zones, with different plant species thriving near the summit than at lower elevations. Some animal species, such as Roosevelt elk, migrate seasonally to take advantage of elevation effects, climbing to cooler locations in the summer and descending to warmer valleys in winter.

The lower atmospheric pressure at high elevations makes breathing more difficult. Mountain climbers at high elevations use special apparatus to breathe. Some competitive distance runners train at high elevations in order to challenge their cardiovascular systems. When they race at a lower elevation, the air feels relatively dense and oxygen-rich, giving them a competitive advantage.

With the less-dense atmosphere at high elevations, the sun's rays can penetrate more easily, making sunburn possible even on cold days. Engines of naturally aspirated cars get less horsepower at higher elevations. Projectiles travel farther, a phenomenon known to golfers and baseball players. Standard equations for projectile motion sometimes assume a sea-level location; adjustments must be made to account for elevation.

The effect of gravity is reduced with travel to high elevations; mass remains the same but weight decreases slightly, primarily because of the increase in distance from Earth's center of mass. A person's weight would be less atop Mount Chimborazo than anywhere else on Earth.

Further Reading

Smith, Arthur. "Angles of Elevation of the Pyramids of Egypt." *Mathematics Teacher* 75, no. 2 (1982) .

Thrower, Norman. *Maps & Civilization: Cartography in Culture and Society*. 3rd ed. Chicago: University of Chicago Press, 2008.

U.S. Geological Survey. "National Elevation Dataset." http://ned.usgs.gov.

David I. Kennedy

See Also: Curves; Gravity; Maps; Plate Tectonics; Temperature; Trigonometry; Weather Scales.

Elevators

Category: Architecture and Engineering.
Fields of Study: Algebra; Number and Operations.
Summary: Mathematics is used to quantify aspects such as the maximum speed and distance range of

elevators as well as model vibration and optimize traffic flow.

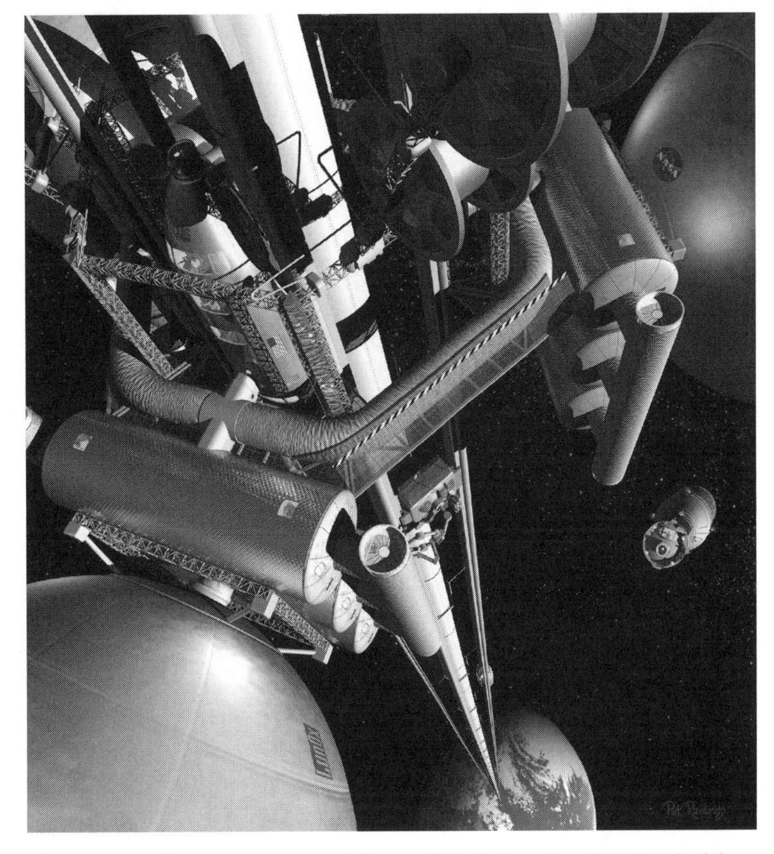

The National Aeronautics and Space Administration (NASA) holds an annual engineering competition to design a space elevator.

An elevator is a mechanism for vertical transport of persons or cargo. Mathematics is used to quantify aspects such as the maximum speed and distance range of elevators, determined by their purpose, such as lifting passengers, cars, or aircraft. Applied mathematical models focus on the dynamics and vibrations within different types of elevator mechanisms, such as hydraulic or rope systems. Mathematicians also investigate questions related to aspects such as waiting time, using probability models. Systems of multiple elevators are modeled as high-dimensional spaces using dynamical systems. The number of passengers in an elevator system constantly changes, making an optimal policy for what is referred to as an "elevator group control" mathematically interesting. At the end of the nineteenth century, scientist Konstantin Tsiolkovsky conceived of a space elevator. He was self-taught and worked as a mathematics teacher.

Hydraulic Elevators

The main concept related to why hydraulic elevators work is Pascal's Law, stating that when the pressure increases anywhere in a confined fluid, it equally increases everywhere. This, together with the fact that pressure (P) is equal to force (F) per unit area (A), can be exploited for an advantage of force. The elevator car stands on top of a piston ending in a wide shaft filled with oil, connected to a narrow shaft with oil. When a pump increases pressure in the narrow shaft, by applying a relatively small force, the equal pressure applies to the floor of the cabin, producing higher force because of the larger area: $P_1 = P_2$, and

$$\frac{F_1}{A_1} = \frac{F_2}{A_2}.$$

Hydraulic elevators are only used in relatively low buildings since the piston has to be as tall as the building to extend to the top floor but fully fit under the building when the elevator is on the ground floor. Digging as deep as a skyscraper is high to install an elevator is impractical. These elevators are mostly used for heavy loads in places such as car mechanic shops.

Roped Elevators

A mathematically interesting concept related to roped elevators is the conservation of energy. A roped elevator consists of two ends of a steel cable going around a pulley attached at the top, called a "sheave." The elevator car is attached to one end of the cable, and the counterweight, which weighs about the same, is attached to the other end.

When the elevator car is at the bottom of the shaft, the counterweight is at the top, and its potential energy converts to force, helping move the elevator car up. When the elevator car is higher than the counterweight, their roles are reversed. This way, it takes very little additional force to make the sheave rotate and the elevator car move up and down.

Logistics

In modern buildings with multiple elevators, computer programs determine how to dispatch elevators to minimize wait time and to save energy. For example, a sensor may detect that an elevator is near capacity and will not stop it for any additional passengers. An elevator going down may not open its doors for people who want to go up, avoiding carrying them back and forth. More sophisticated elevator software can take into account typical traffic patterns, directing elevators to the busiest floors.

Space Elevator

A space elevator is a structure for escaping the gravity well of a planet, transporting objects between the surface and a geostationary orbit. This proposed structure would consist of a large satellite counterweight in orbit and a cable connecting it to the ground. The inertia of the counterweight rotating around the planet will balance the gravitational pull on the cable, keeping the cable taut. The National Aeronautics and Space Administration (NASA) is working on several efforts related to construction of a space elevator, including an annual engineering competition. The technological problems include avoiding meteorites and dangerous atmospheric weather systems, developing materials strong enough for the cable, designing the counterweight, protecting passengers from radiation, and powering the elevator cars. In 2008, Japan announced plans to build a space elevator in the immediate future. Space elevators have frequently appeared in science fiction since the early twentieth century.

Further Reading

Bangash, M. Y. H., and T. Bangash. *Lifts, Elevators, Escalators and Moving Walkways/Travelators*. Leiden, The Netherlands: Taylor and Francis, 2007.

Van Pelt, Michel. *Space Tethers and Space Elevators*. New York: Copernicus Books, 2009.

Wuffle, A. "The Pure Theory of Elevators." *Mathematics Magazine* 55, no. 1 (January 1982).

Maria Droujkova

See Also: Engineering Design; Interplanetary Travel; Pulleys; Spaceships.

Encryption

See *Coding and Encryption*

Energy

Category: Space, Time, and Distance.
Fields of Study: Algebra; Measurement.
Summary: Mathematics is used to study energy and energy conservation as well as to develop new sources of energy.

The concept of energy and transportation of energy are central to the survival of any civilization. As mathematical physicist Ludwig Boltzmann noted, "Available energy is the main object at stake in the struggle for existence and the evolution of the world." At the start of the twenty-first century, human beings have accessed or created many forms of energy and power production, including coal-fired and oil-fired power plants, solar heating plants, wind farms, nuclear power plants, geothermal sources of heat, hydroelectric power produced by dams, biofuels that store solar energy, and tidal energy produced by gravitational interactions between Earth and the moon.

There are also potentially disruptive energy sources, including natural events, such as lightning, volcanoes, and earthquakes. Some global sources of energy and power that remain to be tapped by humans include the atmosphere's expansion and contraction, ocean currents, and sea level differences. Various calculations of energy, including chemical reactions and nuclear reactions, invoke the principle of conservation of energy. In relativistic or quantum terms, the conservation of mass-energy is also important. Energy, work, and quantity of heat are all expressed in "joules," a measure of work named for physicist James Joule. There is a vast array of energy problems that mathematicians research, and mathematics makes many contributions to energy issues.

Energy, Defined

Energy is found in nearly every system or process in the universe: mechanics, chemicals, heat, electricity, nuclear processes, and quantum effects. Mathematician and scientist René Descartes studied mechanics; centuries later, mathematician and philosopher Gottfried Leibniz criticized his ideas and developed what are referred to today as "kinetic energy," "potential energy," and "momentum." In mechanics, the kinetic energy (E) of an object is expressed as

$$E = \frac{1}{2}mv^2$$

where m is the object's mass and v is its velocity. Another form of energy found in mechanics is the energy of position called "potential energy." It has the units of joules. An example is the potential energy defined as work done in the compression of a coiled spring. The sum of all the kinetic and potential energies within a system comprises the mechanical energy of the system. Energy may be a conserved quantity within a closed system, or it may change forms, such as mechanical energy being converted to heat by friction. How energy in a system is measured is important. As noted, mechanical energy is measured as the sum of kinetic energy and potential energy, or energies of motion and position. Chemical energy is measured by the heat energy released in chemical reactions. Electrical energy is measured by work done in a system.

Energy Conservation

In general, the amount of energy of various types can be equated to an equivalent amount of heat energy. On an experimental scale, heat energy is the ability of work done to raise the temperature of water. The joule is a measure of thermodynamic energy and is the common unit of energy. James Joule is credited with experiments in the mid-1800s that demonstrated that work done on a system can be converted into heat. His experiments and those of others eventually led to the realization and statement of the "principle of conservation of energy" as a hypothesis, which was proved in certain restricted settings and generalized by induction. In 1865, mathematical physicist Rudolf Clausius worked on thermodynamics and stated his first law as, "The energy of the universe is constant." The principle of conservation of energy applies not only to certain mechanical systems but is also seen widely in systems where other forms of energy are considered. Thus, heat energy is produced by combustion and friction, radiant energy is from light and other forms of radiation, and chemical energy is stored in fuels and electrical energy. The principle is continually tested in new situations. This testing led to discoveries in the twentieth century in atomic physics. In the International System of Units, *Le Système International d'Unités* (SI), a joule is defined as a newton-meter, named for Isaac Newton. The systematic study of the relation of various physical quantities through an analysis of their dimensions is the subject of dimensional analysis. Richard Feynman noted, "For those who want some proof that physicists are human, the proof is in the idiocy of all the different units which they use for measuring energy."

One energy issue that has been important to mathematicians, philosophers, and physicists is the relationship between matter and energy. Some physicists wanted to assign matter-like properties to energy, such as Wilhelm Wien, who considered that energy might have a traceable motion. Mathematician William Clifford thought of matter and energy as types of curvatures. In the theory of special relativity of 1905, Albert Einstein proved an equivalence of mass and energy as expressed in his famous equation $E = mc^2$, where E is the energy equivalent of mass m, and c denotes the speed of light, 299,792,458 meters per second. There is no process available to human beings at the start of the twenty-first century in which matter can be converted completely into radiant energy.

For example, in a nuclear explosion, only a tiny fraction of nuclear material is converted into energy. The only known process of annihilating matter is to pair a particle of matter with a particle of anti-matter, with the result that two photons are formed with energies that are equivalent to the energies of the particles. This process is on a quantum scale. Fusion is one process for partially converting mass into energy and occurs naturally in stars. Many controlled fusion experiments have been performed but in the process of producing fusion, a greater amount of input energy is needed for the reaction than is ultimately released by the reaction. Only in uncontrolled thermonuclear explosions are large amounts of energy released by fusion.

Fusion

Scientists continue to explore novel sources of energy and power from sources that entail motion, heat, quantum uncertainty and other natural physical phenomena. One possible source of power is controlled fusion reactions, hot or cold. Controlled hot fusion reactions have not yet reached a break-even point where the energy of the reaction exceeds the energy input needed to trigger the reaction.

There are ongoing fusion experiments that use various solids and liquids with energy pumped into them by lasers in which fusion occurs but the fusion is not self-sustaining. The main problem is the energy input and inherent danger in heating suitable substances to temperatures at which fusion between atoms of hydrogen isotopes can occur. The hydrogen is in the form of deuterium or tritium, and the temperatures reached through compression must be on the order of millions of degrees, and there are often energetic byproducts that are dangerous to objects and people. In contrast to hot fusion, cold fusion (also known as "low-energy nuclear reactions" among the twenty-first-century research community) is the fusion of atoms at close to room temperature, generally through the use of supersaturated metal hydrides. These reactions produce heat, helium, and a very low level of neutrons. The energy output is greater than the input, leading many scientists and others to investigate this process as a viable solution to the energy needs of the future. Chemists Martin Fleischmann and Stanley Pons were the first, in 1989, to publicly announce that they had achieved cold fusion. Many competing scientific and mathematical models have been developed to explain how cold fusion works but many researchers and others remain skeptical regarding its existence or viability.

Other Mathematical Applications

Mathematicians and other scientists have long studied the various aspects of energy. The concept of energy is fundamental to many scientific and business theories, applications, and disciplines. For instance, mathematicians have modeled energy trading in financial markets, which is quantitatively interesting because, in such applications, energy possesses unique attributes as a non-storable and non-fungible commodity. They have also worked to design efficient shutdown schedules for electronic systems to address concerns related to energy conservation. Mathematics is important for explaining the cosmic phenomenon of dark energy. This type of energy, often modeled as a scalar field and inferred in large part from observation and mathematical analysis of gravitational fields, has implications for theories and measurement of universe expansion and dark matter. On the other hand, mathematicians such as Blake Temple have used mathematics to attempt to disprove the existence of dark energy and posit alternative explanations. Others have investigated the geometry of symplectic energy. Mathematicians are also influential in energy research and policy making via work at federal agencies like the U.S. Department of Energy. Mathematician J. Ernest Wilkins was a fellow at the Department of Energy's Argonne National Laboratory and physicist and mathematician Hermann Bondi was the chief scientific adviser to the Department of Energy. Mathematical analysis and computational methods have also been used to study energy problems related to equilibrium, stability, and energy transport.

Further Reading

Coopersmith, Jennifer. *Energy, the Subtle Concept: The Discovery of Feynman's Blocks from Leibniz to Einstein*. New York: Oxford University Press, 2010.

Gerritsen, Margot. "Mathematics Awareness Month—April 2009—Theme Essay: Mathematics in Energy Production." http://www.mathaware.org/mam/09/essays/Margot_EnergyMaths.pdf.

Greengard, Claude, and Andrzej Ruszczynski. *Decision Making Under Uncertainty: Energy and Power*. New York: Springer, 2010.

Society for Industrial and Applied Mathematics. "Fuel Cells, Energy Conversion, and Mathematics." http://www.siam.org/about/news-siam.php?id=1605.

Veigele, William. *How to Save Energy and Money at Home and on the Highway: The Mathematics and Physics of Energy Conservation and Reduction of Consumer Energy Costs*. Boca Raton, FL: Universal Publishers, 2009.

Julian Palmore

See Also: Einstein, Albert; Electricity; Geothermal Energy; Green Mathematics; Light Bulbs; Radiation; Solar Panels; Tides And Waves; Universal Constants; Wind and Wind Power.

Energy, Geothermal

See *Geothermal Energy*

Engineering Design

Category: Architecture and Engineering.
Fields of Study: Algebra; Geometry; Measurement.
Summary: Engineering design is a carefully regulated process to create optimal solutions for given problems.

Engineers design everything from automobiles and bridges to prosthetic limbs and sporting equipment. Designing is different than simply building in that it requires the adherence to a very systematic, yet iterative, process known as the "engineering design process." This process is to engineers what the scientific method is to scientists—guiding steps that help ensure that the end result is the best it can be. When a new product is created without following the steps of the engineering design process, there is a higher likelihood that the product designed will lack some important aspect: the end product may not appropriately account for the needs of its users, it may cost too much to manufacture, or it may not have been tested to ensure safety. Accordingly, the term "designing" refers to the entire process, such that an engineer "does design." The use of the term "design" as a noun may be used at different points in the process but may have very different meanings depending on what phase of the process the engineer is in. Design may really mean "design idea" during the brainstorming phase of the process or "model or prototype of the design" during the building phase of the process.

The engineering design process requires the application of mathematics in many of the steps. Throughout the process, engineers use basic mathematics concepts, including addition and multiplication to calculate costs; geometry to calculate surface areas for material needs; and measurements to ensure appropriate dimensioning. However, more sophisticated projects may require the application of higher-level mathematics, such as calculus and differential equations, to solve the technical engineering problems certain designs pose.

The Engineering Design Process

The engineering design process refers to the steps that are required to create the best possible solution to a problem. It is a process often undertaken by a team of engineers who work together, though it can be performed by an individual—trained or untrained as an engineer. Though there is no consensus as the exact breakdown and name of each step, the general design process is universally accepted.

In the first step of the engineering design process, the engineering team is presented with some type of problem or unmet societal need to be solved. Often, this problem is presented to the engineering team by a company that is trying to offer a product that better meets its customers' needs. The engineer must ask many questions to both the client and the user, as well as conduct background research, in an effort to establish the objectives and constraints of the design. The objectives are what the solution to the problem (the final designed product) should aim to accomplish. The constraints are the factors that limit the possible designs, such as time, money, or material restrictions. Time and money constraints are particularly important as they often drive the project and must be monitored throughout to ensure that the project is completed on time and within budget. At the end of this step of the design process, the engineering team fully understands the problem and has developed objectives and constraints to guide their possible solutions.

In the next step of the engineering design process, the engineers generate design ideas to solve the newly refined problem. Idea generation normally occurs through group brainstorming methods, with the goal of producing as many ideas as possible. There are a number of methods used to enhance the innovation and creativity of the ideas that come from the brainstorming session, including ensuring group diversity, drawing from existing stimulus and building off of each other's ideas. In this step of the process, some of the generated ideas will evolve into rough hand-drawn sketches. These sketches need to show perspective and relative size clearly.

The next step of the engineering design process is design selection. A method known as "decision analysis" is most commonly used for design selection. Decision analysis is a systematic process to objectively and logically choose the best idea to move forward with from the many generated through brainstorming. It is important because it reduces the likelihood of a designer's bias in selecting a design. As a first step, the brainstormed ideas must initially be narrowed down through discussion or other means to only the handful of ideas that appear to be most promising. These ideas are then compared through decision analysis. For the decision analysis, it is first necessary to create a list of design

Once the engineering team is satisfied with the final product, the design is executed through computer-aided design (CAD) drawings.

criteria and weight them based on their relative importance. As an example, as safety is paramount in design, the criteria of "safety" would be the most important criteria and would be weighted as 1.0 on a scale of 0–1. The criteria of "portability," on the other hand, might be desirable but not necessary, so it would be weighted as 0.5. There is no standard as to what weighting scale should be used but it is important to be consistent in its application. For each criterion, in addition to the determined weighted importance, a numerical range must also be established for rating each design with respect to the criterion. When possible, this range should be as objective and quantifiable as possible.

Each design being considered is then "scored" using the range for each criterion. The score is then multiplied by the relative criteria weight for a total score for each criterion and for each design. The total scores for each criterion are then summed for each design. The summed scores can be used to compare multiple designs, with the one scoring the highest being the one most likely to be successful.

After identifying a design to move forward with, refinement of the design is necessary. This step includes determining dimensions and materials that will be used to construct the chosen design. Detailed sketches, often drawn from multiple perspectives, are created and include the dimensions of each part to be made. Determining these dimensions often requires in-depth estimation and calculation. At the most simplistic level, dimensioning requires taking into account any necessary clearances or gaps in the design, especially when multiple parts need to be fitted together. It may also be necessary to determine the combinations of dimensions that ensure a specified surface area requirement is met, in which case algebra can be helpful. More in-depth designs may require that dimensions come from established tables of normative dimensions, such as anthropometric tables, providing typical measurements of different-sized people, or from engineering analysis, such as stress or buckling calculations. Deriving dimensions from engineering analysis methods often requires high-level mathematics and a technical background in engineering but ensures a stronger, safer product.

Once the design has been refined and the dimensions are known, building begins. For most designs, a scale model or a simplified prototype is created first to test for feasibility of the design before further time and money is invested. To create a scale model, all dimensions of the detailed sketches must be reduced by multiplying by some chosen scaling factor, often 1:2. Regardless of whether a full-size design or scale model is used, it is necessary to calculate the amount of each material that needs to be purchased to build the design. This requires thought and calculation, in particular when multiple parts could be cut from one piece of wood, metal, or fabric. Often, surface area is calculated according to the part's geometry to determine the total amount of material needed. Once material has been secured, building of the design can occur. Throughout building, it is essential to make careful measurements for all parts because almost all designs are made from multiple components that must fit together to function as one product. For example, if a piece of wood to be used for one leg of a chair is measured even ¼ inch shorter than the other legs, it will likely mean the finished chair will rock and wobble, and the design will be undesirable.

As a next step in the engineering design process, the constructed design is experimentally tested to determine its performance. This step helps to identify design strengths and weaknesses, which can be used to make recommendations for future refinement of the product. The specific experimental test performed is determined by the type of product designed and the design objectives. Regardless of the type of test conducted, measurements are taken throughout the experiment to record some aspect of the design's performance. Often, multiple trials will be taken, generating many data points. The data obtained from these measurements are then used to draw conclusions about the success of the design. Statistical analysis may also be employed to further assist in the interpretation of the data.

Almost always, the data collected during testing will suggest that the design could perform better if refined in some way. As such, it is common for the engineering team to return to the building stage and then iteratively cycle between it and testing steps until satisfied. At times, it may also be necessary to return to earlier steps in the engineering design process. Once the team is satisfied with the final product, final documentation is prepared to explain the design and share it with others. This is often done through computer-aided design (CAD) drawings and written technical reports.

Further Reading

Dym, Clive L. and Patrick Little. *Engineering Design: A Project Based Introduction.* Hoboken, NJ: John Wiley and Sons, 2009.

Eide, Arvide, et al. *Introduction to Engineering Design and Problem Solving.* New York: McGraw-Hill, 2001.

Pahl, Gerhard, et al. *Engineering Design: A Systematic Approach.* 3rd ed. New York: Springer, 2007.

Pilloton, Emily. *Design Revolution: 100 Products That Empower People.* Los Angeles, CA: Metropolis Books, 2009.

Ulrich, Karl, and Steven Eppinger. *Product Design and Development.* New York: McGraw-Hill/Irwin, 2007.

Kimberly Edginton Bigelow

See Also: Bridges; Green Design; Problem Solving in Society; Robots.

Equations, Polar

Category: History and Development of Curricular Concepts.

Fields of Study: Algebra; Communication; Connections; Geometry; Representations.

Summary: Polar coordinate systems were developed in the seventeenth century and have numerous modern applications.

The polar coordinate system is a coordinate system for the plane in which each point is determined by a distance from a fixed point, called the "pole," and an angle from a fixed direction, called the "polar axis." In normal usage, the pole is analogous to the origin in the Cartesian coordinate system, named for René Descartes. Both polar and rectangular (Cartesian) coordinates require two bits of data to place a point in the plane. While the Cartesian coordinate system requires knowing and placing two chosen lines to serve as axes, polar coordinates requires knowing one fixed point and one fixed ray. This characteristic makes polar coordinates useful in navigation. Students in twenty-first-century high schools are introduced to polar coordinate systems and the topic is further developed in college mathematics and physics classrooms.

History

The concept of using an angle and a radius may be dated to the first millennium B.C.E. There are references to Hipparchus of Rhodes (c. second century B.C.E.) using a type of polar coordinates to establish the positions of the stars that he studied. Archimedes of Syracuse describes his namesake spiral in the book *On Spirals*, as where the distance from a given point depends on the angle from a given radius.

In a number of articles about the development of polar coordinates, most notably the 1952 article "Origin of Polar Coordinates" by Julian Lowell Coolidge, further development of polar coordinates was generated by studying the Archimedean spiral. According to Coolidge's history, the first mention should go to Bonaventura Cavalieri in his 1635 treatise *Geometria indivisilibus continuorum* in which he studies the spiral of Archimedes. Cavalieri studies the area inside the spiral and relates it to other known areas.

Like all good stories in the history of mathematics, this assertion is not without disagreement. In 1647,

Grégoire de Saint-Vincent in his work *Opus Geometricum* claimed that he was familiar with the method and had sent his work to Christopher Grienberger in 1625. Grienberger had died in 1636, and the priority of the work was the subject of an article by Moritz Cantor in 1900.

Spiral curves were of interest to many mathematicians, including Gilles Personne de Roberval, James Gregory, Descartes, and Pierre Varignon. Gregory, Descartes, and Varignon all used a type of transformation of coordinates that heralded the complete development of polar coordinates. It appears to be Jacob Bernoulli and Isaac Newton who most completely developed these transformations. Bernoulli worked on the lemniscate and introduced the terms "pole" and "polar axis." Newton investigated transformations between coordinate systems, including polar coordinates, in his work *Method of Fluxions*, which was written in 1671 but not published until 1736.

Applications

Polar coordinates are the basis for navigation and radar, since the direction of travel can be given as an angle and distance from the origin. The radar screen that is used in air traffic control uses the location of the radar transmitter/receiver as the pole and magnetic north as the polar ray, zero degrees. This aspect and the fact that the angles continue in a clockwise direction instead of a counterclockwise direction are the major differences between a navigational use and the mathematical system. This same radar is the basis for all weather radar that is available for viewing either on television or from the Internet. Each radar location (there are 178 National Weather Service Doppler weather radar locations that cover the United States) sets a pole and covers a specific area. Storms are located and their paths are computed using the overlaps. This information must be transformed from the polar system (how far from the radar site and at what angle) into GIS coordinate system and then placed on a map to go to television or to the Internet. One well-known measuring device is the polar planimeter, created by mathematician and physicist Jacob Amsler in the nineteenth century. It measured the area enclosed by a curve. Amsler switched careers to focus on mathematical instruments, and he produced thousands of Amsler planimeters.

Other examples of the use of polar coordinates are very simplified uses in planning sprinkler systems in a building, as well as in irrigation systems in landscape and farming. Each of the sprinkler heads serves as a pole, and different walls, boundary lines and such serve as polar axes.

Different microphones have different recording patterns depending on the specific purpose. The omnidirectional microphone is used when sound from all directions is to be recorded, such as a choir or a large group. A cardioid microphone is a unidirectional microphone, which would be used to record a performer but not the crowd. Bidirectional microphones are used in an interview situation where the voices of both the interviewer and interviewee need to be recorded. The pattern of sounds that are picked up by the microphone are a lemniscate—the figure studied by Bernoulli.

Further Reading

Boyer, C. B. "Newton as an Originator of Polar Coordinates." *The American Mathematical Monthly* 56, no. 2 (1949).

Coolidge, J. L. "The Origin of Polar Coordinates." *The American Mathematical Monthly* 59, no. 2 (1952).

"A Periodic Shift in Polar Roses for Valentines Day." http://www.nikolasschiller.com/blog/index.php/archives/category/renderings/quilt/polar-coordinates.

David C. Royster

See Also: Climbing; Coordinate Geometry; Graphs; Maps; Transformations.

Escher, M.C.

Category: Arts, Music, and Entertainment.
Fields of Study: Geometry; Measurement; Representations.
Summary: The works of M.C. Escher are frequently used by mathematicians and mathematics educators to explore mathematical concepts.

Maurits Cornelis Escher (1898–1972) was a Dutch graphic artist perhaps best known for creating artwork with illusional and conceptual effects including woodcuts, lithographs, and mezzotints with meticulous detail. Despite the fact that he did poorly in mathematics in school, he accurately illustrated mathematical concepts in many of his works, which are frequently used by mathematicians and mathematics educators to illuminate and explore those concepts. He also wrote a paper called "Regular Division of the Plane with Asymmetric Congruent Polygons." About his own work and processes, Escher said:

> By keenly confronting the enigmas that surround us and by considering and analyzing the observations that I had made, I ended up in the domain of mathematics. Although I am absolutely without training or knowledge in the exact sciences, I often seem to have more in common with mathematicians than with my fellow-artists.

Early Work

As a student, Escher did not excel at any subject except drawing. After failing the final secondary school examinations, he enrolled in the Haarlem School of Architecture and Decorative Arts in 1919. Encouraged by one of his teachers, he shifted his interest from architecture to graphic arts. His first trip to the Mediterranean in 1922 made a strong impression upon him. He decided to leave the school and settle in Rome, where he married Jetta Umiker in 1924. In the following years, his fascination with Italian landscapes, combined with his passion for printmaking, resulted in a series of realistic woodcuts, *Castrovalva* being one of the most notable.

Hand with a Reflecting Globe (1935) marks the beginning of a period when the exploration of what he called an "inner vision" replaced his interest in the outward appearance of things. After fleeing from political turmoil to Switzerland, he testified that he was no longer inspired by his surroundings. During a visit to Spain in 1936, he worked extensively on copying the motifs from the Moorish mosaics in the Alhambra castle in Granada. The idea of creating patterns that would not only involve abstract shapes but also animal and human figures strongly obsessed him.

Art and Mathematics

In 1937 Escher created *Still Life and Street*, his first "impossible reality" image. In the same year, he moved with his family to Belgium, where he began to consider divisions of the plane using the work of mathematicians such as George Pólya regarding the 17 distinct plane symmetry groups. Using his own techniques, Escher explored questions such as the possible shapes for tiles that can produce a regular division of the plane, along with the various isometries that relate the edges of such tiles. Escher mapped adjacent tiles using translations, rotations and glide-reflections, all of which require the tiles' edges to be straight segments. This aspect became one of the central ideas of his art.

German occupation forced him to flee to Baarn, The Netherlands, in 1941, where he settled permanently. Two articles in *Life* and *Time* magazines in 1951 brought the world's attention to his work. Besides increasing demand for prints and numerous commissions, this recognition enabled him to start exchanges with many world-renowned scientists.

Escher's first ideas about infinity revolved around depicting decreasing figures as one moves toward the center of an image, as seen in his woodcut *Development II.* An article by a Canadian geometer Harold Scott MacDonald Coxeter made him reverse his point of view by creating a tessellation of a disc with tiles decreasing while moving toward the boundary of a disc. This approach produced some of his later prints, including *Circle Limit III*, and his last work, *The Snakes* (1969).

British mathematical physicist Roger Penrose, fascinated by Escher's lithograph *Relativity*, developed impossible objects known as the "Penrose tribar" and, together with his father, Lionel Penrose, the "Penrose staircase." After exchanging his ideas with the artist, these objects inspired lithographs *Waterfall* and *Ascending and Descending*. Mathematicians continue to investigate the mathematical details of Escher's work. Number theorist Hendrik Lenstra used the theory of elliptic curves and complex exponential functions to analyze aspects of Escher's *Print Gallery*.

Escher's most ambitious work, a 22-foot-wide woodcut, Metamorphosis III, was based on tessellations. Many other mathematical topics were also implemented in Escher's work: topology in depictions of the Möbius strip, the principle of self-reference in *Drawing Hands*, numerous polyhedra, concave and convex objects, irregular perspective, spherical geometry, optical illusions, and visual paradoxes, among others. Escher's creative interpretation of these subjects erased the boundaries between mathematics and art. He said,

> At first I had no idea at all of the possibility of building up my figures. I did not know any "ground

rules" and tried, almost without knowing what I was doing, to fit together congruent shapes that I attempted to give the form of animals. Gradually, designing new motifs became easier as a result of my study of the literature on the subject . . .especially as a result of my putting forward my own layman's theory, which forced me to think through the possibilities.

Legacy

M.C. Escher felt closer to mathematicians than to his peers. Although he frequently stated that he was a mere craftsman, not an artist, some of the images he created found their place in popular culture and mathematics, becoming icons of the twentieth century. Escher's son George noted that his father often did not seem to comprehend that his process of creation and exploration of the mathematical concepts he used in his work was in fact very much like a mathematician. His work tackles human understanding of the order of the universe and unveils it with unexpected beauty and refinement.

Further Reading

Ernst, Bruno. *The Magic Mirror of M.C. Escher.* New York: Random House, 1976.

Escher, M.C., *Escher on Escher: Exploring the Infinite.* New York: Harry N. Abrams, 1982.

Robinson, Sara. "M.C. Escher: More Mathematics Than Meets the Eye." *SIAM News* 35, no. 8 (2002).

Schattschneider, Doris. "The Mathematical Side of M.C. Escher." *Notices of the AMS* 57, no. 6 (2010).

Zoran Petrovic
Karim Salim

See Also: Crystallography; Geometry in Society; Optical Illusions; Symmetry; Transformations.

Ethics

Category: Mathematics Culture and Identity.
Fields of Study: Communication; Connections; Problem Solving.
Summary: Since the time of Plato, mathematicians have been analyzing and confronting ethical problems.

Mathematics and ethics have a long and tangled history. Philosophy has nurtured mathematical forms of thought that, in turn, have had a profound influence on ethical theorizing. For example, mathematics served as a model for Jeremy Bentham (1748–1832) whose goal in utilitarianism was to develop a calculus of pleasure and pain.

Several contemporary ethical theories are tied to the mathematics of game theory, especially the work of John Rawls (1921–2002). Ethical issues arise in mathematics teaching, research, industry, and government work. Mathematicians such as Lee Lorch challenge discriminatory practices and fight for human rights, justice, and equality. Other mathematicians have refused to work on projects they find ethically problematic. Ethical norms often change over time and for various contexts, leading to controversial applications of mathematics research, like the atomic bomb. In the face of increasing marketability of mathematical results, some have questioned the disparity between the academic tradition of making knowledge freely available and personal ownership of intellectual property. Many professional associations have developed, maintained, and revised ethical guidelines for their members, and mathematicians who wish to perform experiments must submit a proposal to an institutional review board for ethical review. In 2010, the National Science Foundation issued a program solicitation for an Ethics in Science, Mathematics, and Engineering Online Resource Center.

Mathematics and Ethics in Plato (429–347 B.C.E.)

Plato's *Republic* is the first systematic treatment of ethics. The best preparation for acquiring ethical knowledge is a firm foundation in mathematics. However, the connection between mathematics and ethics is much deeper. Methodologically, Plato develops his argument by building a simplified model of the state in the same manner in which a study of any geometrical figure is done in mathematics. Justice in the state is merely justice in the individual writ large. Thus, Plato appeals to similarity transformations. The argument is that, as a result of a uniform scaling operation, justice in the individual is similar to justice in the state. Further, within the Platonic tradition, mathematical and ethical knowledge have the same formal characteristics. They are both examples of purely intelligible objects grasped entirely by reason in an intellectual intuition and known as a result of a process of recollection. Thus, they are examples of immutable and unchangeable truths, which could not be other than what they are. Plato's very definition of justice contains a mathematical element, because justice is a type of equality. Justice is a matter of treating equal individuals equally and unequal individuals unequally. According to Plato, different political orders arise from the different conceptions of equality.

Mathematics and Ethics in Aristotle (384–322 B.C.E.)

For Aristotle, mathematics does not provide a model for ethics. However, mathematical concepts function in an analogical sense. Aristotle used a distinction between arithmetic and geometric proportion in his discussion of justice. Distributive justice is based on geometrical proportion, while rectificatory justice is based on arithmetical proportion. Issues of rectificatory justice arise when a judge must rectify a situation by attempting to restore equality to someone who has been injured. Issues of distributive justice arise when something has to be divided among two individuals.

A detail of Plato (left) and Aristotle from the painting The School of Athens by Italian artist Raphael.

Modern Moral Euclidian Philosophers

Both Thomas Hobbes (1588–1679) and Baruch Spinoza (1632–1677) incorporated the mathematical method of Euclid of Alexandria into their treatment of ethics. Hobbes thought that mathematical modes of thought could produce clarity in ethics and politics. However, it was Spinoza who most rigorously and consistently imitated Euclid's method. He begins each section of his *Ethics* with a set of definitions and axioms, which he then uses to demonstrate a series of propositions about the universe, human nature, and basic ethical precepts.

Mathematical Ethics

The guidelines of professional mathematical associations cover a wide range of topics. Creation, attribution, publication, and presentation of research, especially with regard to falsification and plagiarism, as well as skewed interpretations and one-sided "advertising" style arguments, are commonly addressed. These guidelines extend into the classroom, along with data sharing or loaning and responsible group work. Attention is also given to the nature of teacher-student and colleague relationships in which one individual has some level of authority over the other, especially when they involve professional decisions like hiring, granting tenure, issuing promotions, and conferring degrees.

Mathematician Philip Davis noted that ethics are typically derived from past experiences and so may do little good in addressing many future or even current dilemmas. Further, judging the past based on current criteria leads to additional difficulties. Arguments abound, for example, about whether statistical data gathered from Nazi medical experiments should be used or destroyed, or whether mathematicians can be held responsible for any future unanticipated uses of their work, such as computer viruses or code-breaking algorithms usurped by data thieves. The Manhattan Project exemplifies many of the moral dilemmas faced by mathematical scientists. Many participants have expressed profound regrets; others have not, citing the undeniable advances made in numerous fields and the need at the time to bring an end to the greater destruction of World War II. For example, the cyclotron was invented by Ernest O. Lawrence in 1931, who received the Nobel Prize in 1939 for this invention.

Further Reading

Ernest, Paul. "Values and the Social Responsibility of Mathematics." *Philosophy of Mathematics Education Journal* 22 (2007).

Hersh, Reuben. "Mathematics and Ethics." *The Mathematical Intelligencer* 12, no. 3 (1990).

Huff, Darrell. *How to Lie With Statistics*. New York: W. W. Norton, 1993.

Michael K. Green

See Also: Atomic Bomb (Manhattan Project); Game Theory; Genetics; Nanotechnology; Vietnam War.

Europe, Eastern

Category: Mathematics Around the World.
Fields of Study: All.
Summary: Eastern Europe has a long tradition of both mathematics research and education.

Throughout history, the countries of Europe have had shifting political and social boundaries. Eastern European mathematics evolved within the context of many mathematics traditions, including Soviet Union mathematics, over the past centuries. Historically, gifted young scholars from regions around the world completed their mathematical studies at Europe's well-known and respected universities. Studies of mathematicians' letters and scientific papers show that they often maintained connections with people in other countries who shared their fields of interest. The Soviet Union exercised broad social and political influence over most of eastern Europe and also impacted U.S. mathematics in the twentieth century. Within the Soviet Union, students from the far reaches of the nations within its boundaries were often brought to Russia for work or education, as well as sent to other parts of the Soviet Union to teach or to establish research centers. In the twenty-first century, students in the United States and around the worked attend study abroad programs, such as the Budapest Semesters in Mathematics. In the twenty-first century, the United Nations Statistics Division classified the following countries belonging to eastern Europe: Belarus, Bulgaria, Czech Republic, Hungary, Moldova, Poland, Romania, Russia, Slovakia, and Ukraine. The *CIA World Factbook* adds Estonia, Latvia, and Lithuania, which were among the member nations of the Soviet Union, though the United Nations classifies them as belonging to northern Europe. Geographical boundaries continued to change in the twentieth century because of post–World War II structures and, later, the breakup of the "Eastern Bloc" nations, which were once under the Soviet Union's political influence. Therefore, mathematics contributions of some people from eastern Europe may be included within the histories of other regions or countries.

History of Russian and Soviet Mathematics Education

When examining past and present states of mathematics in Belarus, Moldova, Russia, Ukraine, Estonia, Latvia, and Lithuania, it is pertinent to acknowledge that they share a common sociopolitical root: they are all former member states of the Soviet Union. Further, the broader Eastern Bloc of Soviet Union allies included Bulgaria, Romania, Hungary, East Germany, Poland, Albania (until the early 1960s), and Czechoslovakia (which later split into the Czech Republic and Slovakia). The Eastern Bloc is sometimes known historically as "eastern Europe," versus the "western Europe" countries allied with the United States, a rival of the Soviet Union. During its several decades of existence in the twentieth century, the Soviet Union included many mathematicians who made significant contributions to the body of modern mathematical knowledge. Further, Russian and Soviet mathematicians were influential on many other countries.

One important landmark in mathematics education in Russia is the creation in 1701 of the School of Mathematical and Navigational Sciences in Moscow. Peter the Great, who had traveled widely in other parts of Europe to study the state of mathematics and science as part of his effort to modernize Russia and expand the empire, founded this school. It educated students in basic mathematics as well as more specialized subjects, such as astronomy and navigation. Notably, students from all social classes except serfs were admitted, and financial assistance was available. Graduates worked in the navy, as engineers, and as teachers in a variety of settings, so the school had a multiplier effect in terms of spreading mathematics education throughout Russia. Peter the Great also founded the Saint Petersburg Academy of Sciences in 1724, influenced in part by correspondence with mathematician Gottfried Leibniz, who also purportedly recommended a three-tiered educational system of schools, universities, and academies. Many eminent foreign mathematicians, such as Leonard Euler, Christian Goldbach, and Daniel Bernoulli, worked at the Saint Petersburg Academy.

As part of her goal of modernizing Russia in the European style, Empress Catherine the Great, who was born in Germany, established the first gymnasiums in Russia. These gymnasiums were schools meant to prepare students for higher education and were created in most major Russian cities in the nineteenth century. Nicolai Ivanovich Lobachevsky, one of the first Russian mathematicians to achieve international recognition, was a beneficiary of this expanded educational opportunity. He graduated from Kazan Gymnasium and Kazan University (in Tatarstan) and is most noted for his work in hyperbolic geometry, a form of non-Euclidean geometry. However, despite this considerable expansion, access to education was far from universal until the Soviet era. The Soviet Union was founded by revolution in 1917, when the monarchy of the Russian Empire was overthrown, but was not made official until 1922. The Saint Petersburg Academy of the Sciences evolved into the Russian and then Union of Soviet Socialist Republics (USSR) Academy of the Sciences. It reverted to the Russian Academy of Sciences following the dissolution of the Soviet Union, and remains an influential organization in the twenty-first century. Academies of sciences were also founded in most of the states of the Soviet Union. Universal compulsory education was established in 1919. Soviet schools had both political and educational goals but the expectation that all children would attend school rapidly increased literacy and played a key role in modernizing and industrializing the country.

In the Soviet Union, the study of mathematics and the sciences was emphasized, a choice that not only fostered rapid economic growth but also became a point of national pride, as by mid-century the Soviet Union was frequently seen to rival or even surpass the United States in scientific and applied research. When the Soviet Union successfully launched the satellite Sputnik in 1957, it raised concern in the United States not only because of the possibility that the Soviet Union was developing weapons for which the United States had no counter but also because it put into question the common assumption that the United States was the world leader in mathematics and science. One result of Sputnik in the United States was a substantial increase in federal funding for scientific education and research in the hope of catching up and surpassing the Soviet Union in the "space race."

As part of this concern that the Soviet Union was surpassing the United States, many studies were commissioned of the Soviet educational system and how it differed from the American system. Among the differences noted by researchers were the facts that in Soviet schools, specialists taught mathematics from the fourth grade onward, a uniform curriculum was used across the entire country, and much greater emphasis was

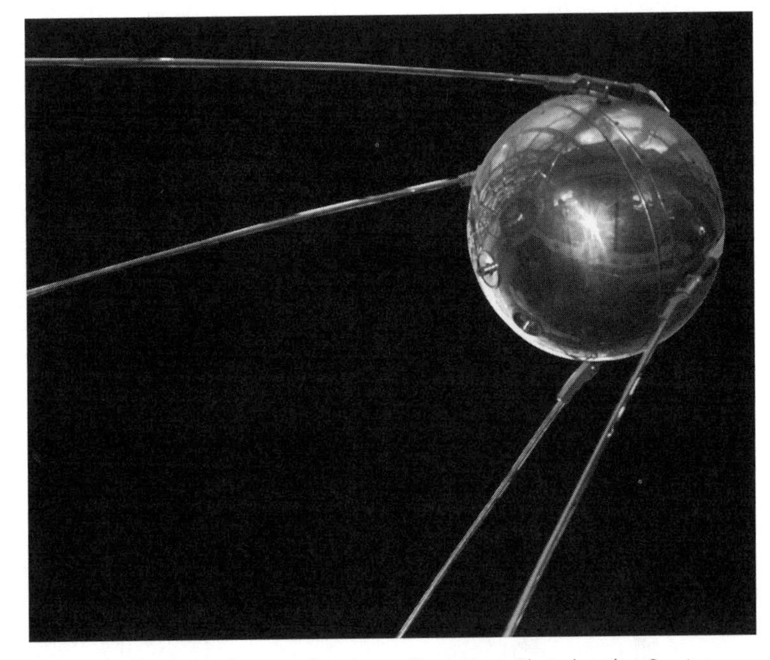

The launch of the first artificial satellite, Sputnik 1, by the Soviet Union on October 4, 1957, started the race to the moon.

placed on developing the talents of students who were identified as gifted in mathematics. The Soviet Union had "special schools," which were free boarding schools at high school level for gifted students and specialized in particular subjects. Four such schools were devoted to mathematics. Correspondence courses in advanced mathematics were also available to increase the number of students studying those subjects. American observers noted that the level of mathematics required for university admittance during the Soviet period was much higher than what would be expected for entering freshmen in the United States. At the same time, other authors have noted that English-language sources often do not reflect the full scope and influence of Russian and Soviet mathematics. These omissions may be because of Cold War influences and a period of Soviet isolationism from the United States and much of Europe, a policy that contrasts strongly with earlier Russian connections and the growing collaborations following the Soviet era.

Notable Soviet and Russian Mathematicians

Andrey Kolmogorov (1903–1987) is known for his work in the fields of probability theory and topology, including the Kolmogorov axioms, Kolmogorov's zero-one law, and Kolmogorov space.

Stefan E. Warschawski (1904–1989) studied at the University of Königsberg and Göttingen. His doctoral thesis was on the boundary behavior of conformal mappings.

Sergei Lvovich Sobolev (1908–1989) worked in mathematical analysis and partial differential equations. Sobolev spaces (named after him) can be defined by growth conditions on Fourier transforms.

Israel Moiseevich Gelfand (1913–2009) worked in the field of functional analysis. He is known for the Gelfand representation in Banach algebra theory; the representation theory of the complex classical Lie groups; contributions to distribution theory and measures on infinite-dimensional spaces; integral geometry; and generalized hypergeometric series. His name is linked to the development of mathematical education.

Igor Shafarevich (1923–) is the founder of the major school of algebraic number theory and algebraic geometry in the Soviet Union. He has also written well-known textbooks.

Grigori Perelman (1966–) declined the Fields medal, a prestigious award in mathematics often equated to the Nobel Prize, for his work on the Poincaré conjecture, named for Henri Poincaré. He cited inequities and reportedly noted, "If the proof is correct then no other recognition is needed."

Other well-known Soviet or Russian twentieth-century mathematicians include Boris Pavlovich Demidovich, who worked on problems in mathematical analysis, and Yakov Isidorovich Perelman, who was a science writer and author of many popular science books.

Czech Republic and Slovakian Mathematicians

Kurt Gödel (1906–1978) proved fundamental results about axiomatic systems. Gödel's Incompleteness Theorems are named for him.

Stefan Schwarz (1914–1996) studied semigroups, number theory, and finite fields and founded the Mathematico-Physical Journal of the Slovak Academy of Sciences in 1950.

Hungarian Mathematicians

Hungarian mathematicians of the twentieth century are well known in the mathematical world. Many of them immigrated to the United States after World War II.

Frigyes Riesz (1880–1956) was a founder of functional analysis. He produced representation theorems for functional on quadratic Lebesgue integrable functions, named for Henri Lebesgue, then introduced the space of q-fold Lebesgue integrable functions. He also studied orthonormal series and topology.

George Pólya (1887–1985) worked in probability, analysis, number theory, geometry, combinatorics, and mathematical physics. He wrote books about problem-solving methods, complex analysis, mathematical physics, probability theory, geometry, and combinatorics. He was regarded by many as a great teacher and influenced many mathematicians.

Cornelius Lanczos (1893–1974) worked on relativity and mathematical physics. He invented what is now called the Fast Fourier Transform, named for Joseph Fourier. He published more than 120 papers and books.

John von Neumann (1903–1957) worked in quantum mechanics, game theory, and applied mathematics, as well as helping pioneer computer science. His doctoral thesis was on set theory. His definition of ordinal numbers is the one commonly used in the early twenty-first century.

Rózsa Péter (1905–1977) is known for teaching, for her books on the history of mathematics, and for her series of theorems about primitive recursive functions.

Paul Erdos (1913–1996) is well known among mathematicians for his insatiable ability to pose and solve problems. It is often said that he lived on mathematics and coffee, touring the circle of his friends and pupils and giving lectures on combinatorics, graph theory, and number theory. He advocated for elegant and elementary proof. One of the most prolific mathematicians in history, he wrote more than 1500 papers.

Paul Richard Halmos (1916–2006) is known for his contributions to operator theory, ergodic theory, functional analysis (in particular Hilbert spaces, named for David Hillbert), and for his textbooks.

Alfréd Rényi (1921–1970) worked on probability theory, statistics, information theory, combinatorics, graph theory, number theory, and analysis.

László Lovász (1948–) published his first paper called *On graphs not containing independent circuits* when he was only 17 years old. He is a prominent figure of post–World War II mathematicians.

Notable Polish Mathematicians

Stefan Banach (1892–1945) worked on the theory of topological vector spaces, measure theory, integration, and orthogonal series. His doctoral thesis "On Operations on Abstract Sets and their Application to Integral Equations" (1920) marks the birth of modern functional analysis. He defined the "Banach space."

Benoit Mandelbrot (1924–2010) is known as the father of fractal geometry. The Mandelbrot set, a connected set of points in the complex plane, is named after him.

Mathematicians From Romania

János Bolyai (1802–1860) is perhaps the most famous Romanian mathematician because of his treatise on a complete system of non-Euclidean geometry in his book *Appendix*. In his own words, he created a new world out of nothing.

Caius Iacob (1912–1992) worked in the fields of analytic geometry, descriptive geometry, analysis, and complex functions.

Grigore C. Moisil (1906–1973) worked on differential equations, the theory of functions, and mechanics. He set up the first Romanian computer science course. Moisil was appreciated for his philosophy and humor.

Other important Romanian mathematicians include Dimitrie Pompeiu, Ferenc Radó, Isaac Jacob Schoenberg, Simion Stoilow, Gheorghe Titeica, Gheorghe Vranceanu, Octav Onicescu, Ion Colojoara, and Dan Barbilian.

Competitions and Contests

Building on eastern Europe's strong mathematics traditions, many mathematical contests are hosted frequently or entirely within the region, such as International Mathematical Olympiad, Romanian Master of Sciences (formerly called the Romanian Masters in Mathematics—it was expanded to include physics), Czech-Polish-Slovak Match, Bulgarian Competition in Mathematics and Informatics, Romanian National Olympiad, and the International Kangaroo Mathematics Contest (often called "Math Kangaroo") among others. Individuals from all over the world participate regularly in these competitions. There are also several winners of the Fields Medal who were born or worked in eastern Europe.

Further Reading

Davis, Robert B. "An Analysis of Mathematics Education in the Union of Soviet Socialist Republics." Report for the National Institute of Education. December 1979. http://www.eric.ed.gov/PDFS/ED182141.pdf.

Dickson, Paul. *Sputnik: The Shock of the Century*. New York: Walker Publishing, 2001.

Sinai, Iakov. *Russian Mathematicians in the 20th Century*. Singapore: World Scientific Publishing, 2003.

Vogeli, Bruce R. *Soviet Secondary Schools for the Mathematically Inclined*. Washington, DC: National Council of Teachers of Mathematics, 1968.

Simone Gyorfi
Sarah Boslaugh

See Also: Europe, Northern; Europe, Southern; Europe, Western.

Europe, Northern

Category: Mathematics Around the World.
Fields of Study: All.
Summary: Since the Enlightenment, Northern Europe has made considerable contributions to mathematics research and continues to do so.

Northern Europe has produced many outstanding mathematicians and scholars in related fields, from the development of calculus by Isaac Newton in the seventeenth century to the cosmological models developed by Stephen William Hawking in the twentieth and twenty-first centuries.

Northern Europe also led the way in developing many practical applications of mathematics and later statistics, including taking a national census like the Domesday Book undertaken in England in 1183 and developing mathematical ways to measure the influence of personal habits on health as in the studies of Richard Doll and Austin Bradford Hill on the relationship between smoking and disease. In the twenty-first century, the United Nations category of northern Europe includes the Åland Islands, the Channel Islands, Denmark, Estonia, Faeroe Islands, Finland, Guernsey, Iceland, Ireland, the Isle of Man, Jersey, Latvia, Lithuania, Norway, Svalbard and Jan Mayen Islands, Sweden, and the United Kingdom of Great Britain and Northern Ireland. However, the changing political boundaries in many of these countries throughout history, as well as the rise and fall of the Soviet Union, which included countries like Estonia, Latvia, and Lithuania, mean that mathematical contributions of some individuals may be included within the histories of other regions.

Sir Isaac Newton was one of the most influential mathematicians of the modern era. He shares credit with Gottfried Leibniz for developing integral and differential calculus, and he also made major contributions in the fields of physics and astronomy. Newton's 1687 book *Philosophiae Naturalis Principia Mathematica* laid the groundwork for classical mechanics including a description of the three laws of motion and remains one of the most influential books in the history of science. He also built the first reflecting telescope and developed a theory of color based on the visible spectrum displayed when visible light is refracted through a prism. Through his work with the laws of gravity and Kepler's laws of planetary motion, named for Johannes Kepler, Newton was able to demonstrate mathematically the validity of heliocentrism, which is the scientific principle that Earth and other planets revolve around the sun.

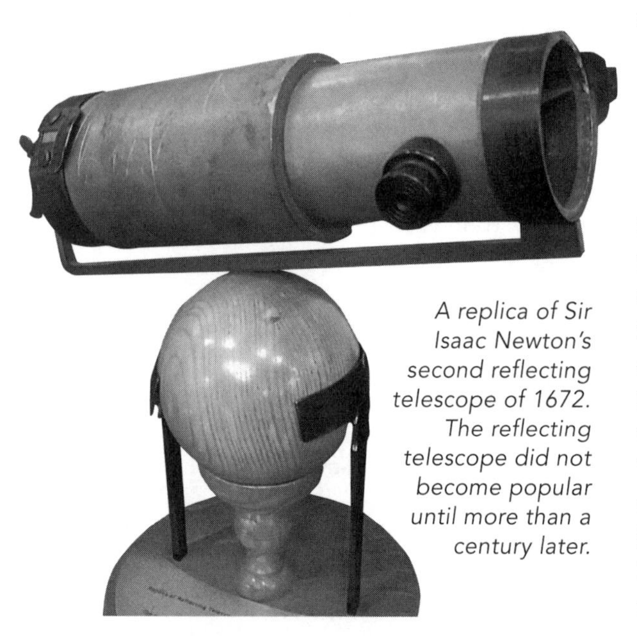

A replica of Sir Isaac Newton's second reflecting telescope of 1672. The reflecting telescope did not become popular until more than a century later.

The nineteenth century saw several major breakthroughs in mathematics by scholars from northern Europe. In England, philosopher and mathematician George Boole developed the system now known as "Boolean logic," which has many practical applications and was instrumental in the development of modern digital computers. His most famous works are *The Mathematical Analysis of Logic* (1847) and *The Laws of Thought* (1854). His slightly younger contemporary, Norwegian Niels Henrik Abel, invented the field of group theory (contemporaneously with Frenchman Evariste Galois), which has many applications in mathematics and physics. Abel is well known for a proof he wrote at age 19 that there can be no general algebraic solution of an equation greater than degree four. In Ireland, Sir William Rowan Hamilton provided an important reformulation of Newtonian mechanics and invented an extension of the number system called "quaternions."

In the period 1910–1913, the British scholars Bertrand Russell and Alfred North Whitehead wrote the influential *Principia Mathematica* in which they attempted to derive the foundations of mathematics from a set of axioms and inference rules. Russell was also a prominent writer and political activist who won the Nobel Prize for Literature in 1950, while Whitehead was also noted as a philosopher. More recently, Andrew Wiles, who was born and educated in the United Kingdom but immigrated to the United States, achieved fame for proving Fermat's Last Theorem (named for Pierre de Fermat), one of the most famous previously unsolved problems in mathematics.

Honors

There is no Nobel Prize for mathematics but several different international awards are offered that have been termed the "Mathematics Nobel Prize" because of their prestige. The Fields Medal is awarded every four years to one or more mathematicians of age 40 or younger by the International Mathematical Union. Winners from the United Kingdom have included Klaus Roth (1958), Michael Atiya (1966), Alan Baker (1970), Simon Donaldson (1986), Richard Borcherds (1988), and Timothy Gowers (1998). Lars Ahlfors of Norway won in 1936, the first year the medal was given; Atle Selberg of Norway won in 1950; and Lars Hormander of Sweden won in 1962. Another major mathematical prize, the Abel Prize, is named after Norwegian mathematician Niels Henrik Abel and is awarded annually by the Norwegian Academy of Science and Letters. The Abel Prize has been awarded since 2003. Northern European winners include Michael F. Atiyah of the United Kingdom and Lebanon in 2004 and Lennart Carleson of Sweden in 2006.

The Wolf Prize in Mathematics has been awarded almost annually by the Wolf Foundation since 1978 and more than one prize may be given per year. Northern European winners include Lars Ahlfors of Finland (1981), Atle Selberg of Norway (1986), Lars Hormander of Sweden (1988), Lennart Carleson of Sweden (1992), Andrew Wiles of the United Kingdom (1995/1996), and David B. Mumford of the United Kingdom (2008).

Northern European countries have been regular competitors in the International Mathematical Olympiad, an annual competition held since 1959 for high school students. Each competing country sends a team of six students who are assigned six questions to solve. Individual students are awarded medals based on their scores, and countries are also compared based on the total score for their team.

There have been many medal winners from northern European countries. The United Kingdom began participating in 1967 and even hosted the 1976 and 2002 competitions. Ireland first participated in 1988. The northern Europe countries from the former Soviet Union—Estonia, Latvia, and Lithuania—first participated in 1993, which coincided with the removal of Russian troops from the area and other political reorganization throughout the former Soviet Union. Among the Scandinavian countries, Sweden first participated in 1967, Norway in 1984, Finland in 1965, Denmark in 1991, and Iceland in 1985. Sweden hosted the 1991 competition, and Finland hosted it in 1985.

Further Reading

Knox, Kevin C., and Richard Noakes. *From Newton to Hawking: A History of Cambridge University's Lucasian Professors of Mathematics*. Cambridge, England: Cambridge University Press, 2003.

Krantz, Steven G. *An Episodic History of Mathematics: Mathematical Culture Through Problem Solving*. Washington, DC: Mathematical Association of America, 2010.

School of Mathematics and Statistics, St. Andrews University. "The MacTutor History of Mathematics." http://www-groups.dcs.st-andrews.ac.uk/~history/.

Westfall, Richard S. *Isaac Newton*. Cambridge, England: Cambridge University Press, 2007.

Sarah Boslaugh

See Also: Calculus and Calculus Education; Europe, Eastern; Hawking, Stephen; Lovelace, Ada; Wiles, Andrew.

Europe, Southern

Category: Mathematics Around the World.
Fields of Study: All.
Summary: Modern Western mathematics was developed in southern Europe and continues to thrive there.

The system of modern mathematics originated in southern Europe, with the ancient Greeks undoubtedly building on traditions already used in Egypt and by the Phoenicians. Like many areas of the world, the nations of southern Europe have had many different boundaries, names, and political alliances throughout history, and so the mathematical contributions of some individuals may be included within the histories of other regions. For example, many nations were member states of the former Soviet Union. The United Nations now includes Albania, Andorra, Bosnia and Herzegovina, Croatia, Gibraltar, Greece, Holy See, Italy, Malta, Montenegro, Portugal, San Marino, Serbia, Slovenia,

Spain, and the former Yugoslav Republic of Macedonia in Southern Europe.

Ancient Greeks and Romans

The earliest Greek school of mathematics is ascribed to Thales (c. 640–550 B.C.E.), who came from Miletus, in present-day Turkey, and Pythagoras (c. 569–500 B.C.E.) who hailed from the Mediterranean island of Samos and later moved to Sicily. Archytas, who subscribed to the Pythagorean philosophy and worked on the harmonic mean, was from Tarentum in modern-day Italy. One of the most well-known Greek mathematicians of the ancient world, Euclid of Alexandria (c. 330–260 B.C.E.), was also not from the Greek mainland. He lived in Alexandria, in modern-day Egypt, and his work proved hugely influential to subsequent mathematicians with his detailed hypotheses and proofs. The great mathematician Archimedes of Syracuse (c. 285–212 B.C.E.) also studied in Alexandria but was from Sicily, where he spent most of his life.

These early Greek mathematicians were undoubtedly an influence on the Romans but the Romans themselves were seemingly more interested in applied mathematics—especially how it related to engineering and building—than in the pure mathematics that was favored by the Greeks. Mathematics was certainly taught in Roman schools and historians have long pondered why Roman mathematicians did not have more influence. This dearth of mathematical advancement has generally been ascribed to the Romans' lack of a designation for "zero" and their awkward system of numbers, which may have prevented any great advances in theory. The Roman Empire did, however, see a continual flourishing of mathematics in Greece and the Greek diaspora, in particular the city of Alexandria. Anicius Manlius Severinus Boethius (c. 475–525) was a well-known Roman mathematician who worked during the declining years of the Roman Empire.

The Renaissance

The Bishop of Seville, Isidorus Hispalensis (570–636), helped develop mathematics in Spain and there were great advances made in arithmetic with the Moorish invasions of Spain and the incorporation of many of the advances made in the Muslim world. The great trading cities of Genoa and Venice soon established themselves as important centers of finance, as did Florence during the Renaissance. Venice, in particular, because of its geographical position and its connections with the Arab world, saw the importation of many books and manuscripts on Arab mathematics—at that stage well advanced in pure mathematics theories compared to Europe. This Arab influence saw Leonardo Pisano Bigollo (c. 1170–1250), the son of an Italian merchant in North Africa, develop theories—the most well-known being the Fibonacci numbers, which were termed after his assumed name.

Several centuries later, the advent of the printing press also led to a republication of the works of Greek mathematicians such as Euclid, albeit in Latin translation. Cardinal Bessarion, the former Archbishop of Nicaea, helped bridge the link between Byzantium and Rome, helping to preserve some of the Greek learning that was lost when the city of Constantinople was captured and sacked in 1453. Leonardo da Vinci (1452–1519) developed mathematics theories, testing out some of them in siege machines designed for Cesare Borgia and others. Girolamo Maggi (c. 1523–1572), another Italian mathematician, was involved in designing military defenses in Cyprus. He was captured by the Ottoman Turks and executed in Constantinople but not before writing two major treatises from memory while in prison there.

The Renaissance saw a new interest in mathematics in Italy, with Galileo Galilei (1564–1642) being a well-known mathematician and scientist. He was a great influence on many subsequent mathematicians, including Alessandro Marchetti (1633–1714). Evangelista Torricelli (1608–1647) invented a barometer; Giovanni Ceva (1647–1734) proved Ceva's theorem in elementary geometry; and the Jesuit Franceso Cetti (1726–1778) helped connect mathematics to other scientific discoveries. Later Italian mathematicians include Giulio Ascoli (1843–1896) who taught in Milan, and Carlo Emilio Bonferroni (1892–1960) who developed the theory of Bonferroni inequalities. The Italian Mathematical Union was established in 1922 by Salvatore Pincherle and others, and its journal, the *Bollettino dell'Unione Matematica Italiana,* is widely respected around the world.

Professional Associations

Professional associations in the region other than the Italian Mathematical Union include the Bosnian Mathematical Society; the Croatian Mathematical Society; the Cyprus Mathematical Society; the Montenegro Mathematical Society; the Portuguese Society of Mathematics; the Mathematical Society of Serbia; the Mathematics, Physics, and Astronomy Society of Slovenia; and the Royal Spanish Mathematical Society. Mathematicians also gather from all over Europe in the European Mathematical Society. The International Mathematical Olympiad is a competition for high school students that originated in 1959. Albania first participated in 1993, Bosnia and Herzegovina in 1993, Croatia in 1993, Greece in 1975, Italy in 1967, Montenegro in 2007, Portugal in 1989, Serbia in 2006, Slovenia in 1993, Spain in 1983, Yugoslavia in 1963, and the former Yugoslav Republic of Macedonia in 1993. Greece was a host of the competition in 2004, Slovenia in 2006, Spain in 2008, and Yugoslavia in 1967 and 1977.

Further Reading

Field, Judith Veronica. *The Invention of Infinity: Mathematics and Art in the Renaissance.* New York: Oxford University Press, 1997.

Hodgkin, Luke. *A History of Mathematics: From Mesopotamia to Modernity.* New York: Oxford University Press, 2005.

Manaresi, Mirella. *Mathematics and Culture in Europe: Mathematics in Art, Technology, Cinema, and Theatre.* New York: Springer, 2007.

Justin Corfield

See Also: Archimedes; Europe, Eastern; Greek Mathematics; Roman Mathematics.

Europe, Western

Category: Mathematics Around the World.
Fields of Study: All.
Summary: Western Europe has been home to many of the important astronomical and mathematical discoveries of the early modern age.

Historically, the term "western Europe" has had cultural and political definitions. For example, during the Cold War it was often used to designate a collection of noncommunist countries allied in some way with the United States. In the early twenty-first century, the United Nations Statistics Division for western Europe contains Austria, Belgium, France, Germany, Liechtenstein, Luxembourg, Monaco, the Netherlands, and Switzerland. There is a rich history of mathematics scholarship, education, and achievement in western Europe. Important work in a diverse array of mathematical areas like calculus, number theory, analytical geometry, probability, statistics, functional analysis, graph theory, logic, and number theory was produced by people from this geographic region, as well as many mathematical contributions to related disciplines like physics, astronomy, optics, engineering, and surveying.

Historical Contributions

Western European mathematicians have made major contributions to the development of mathematics and the application of mathematical theory to practical problems, from German mathematician and astronomer Johannes Kepler, who worked with Danish astronomer Tycho Brahe and helped established the laws of planetary motion, to French mathematician René Thom, who founded the study of catastrophe theory.

Much of modern science and mathematics has its roots in work done in Europe in the seventeenth century. Johannes Kepler studied at the University of Tubingen, where he learned both the geocentric model of astronomy (the view that Earth is the center of the universe, with the other planets revolving around it) and the heliocentric model of German astronomer Nicolaus Copernicus (the view that the sun is the center of the universe and the planets, including Earth, revolve around it). He later worked with Brahe and established the laws of planetary motion in several influential publications: *Astronomia Nova*, *Harmonices Mundi*, and *The Epitome of Copernican Astronomy*. Also in Germany, mathematician Gottfried Leibniz developed the field of calculus independent of Sir Isaac Newton in England.

In France, mathematician and philosopher René Descartes developed analytical geometry, including the development of Cartesian coordinates, did important work in optics, and was also one of the fathers of modern Western philosophy with influential books such as *Meditations on First Philosophy*, *Discourse on the Method* (which contains the oft-quoted statement

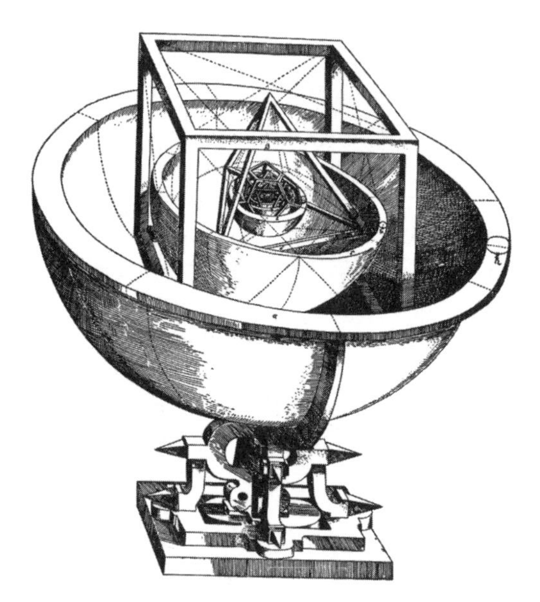

Johannes Kepler's platonic solid model of the solar system was published in 1596.

cogito ergo sum, or "I think, therefore I am"), and *Principles of Philosophy*. Also in France, the basics of probability theory were developed by mathematicians Pierre de Fermat and Blaise Pascal, while Fermat also did important work in number theory, analytic geometry, and optics. Fermat's Last Theorem, mentioned but not proved by Fermat in 1637 in the margin of a book, was among the unsolved problems in mathematics until British mathematician Andrew Wiles proved it in 1994. Pascal invented the mechanical calculator and the hydraulic press and is well known among middle school students for Pascal's Triangle, a presentation of binomial coefficients.

In the eighteenth century, Swiss mathematician and physicist Leonhard Euler spent much of his adult life working at the Russian Academy of the Sciences in St. Petersburg. He developed the concept of the function and the notation $f(x)$, one of several notation conventions he developed that are still used in the early twenty-first century (others include using the letter *e* for the natural logarithm, *i* for an imaginary unit, and the Greek letter *sigma* (Σ) for summation). He also made important contributions to calculus, number theory, graph theory (he solved the famous Seven Bridges of Konigsberg problem), and applied mathematics. French and Italian astronomer and mathematician Joseph-Louis Lagrange, who was born in Italy but worked primarily in France and Prussia, created the calculus of variations, developed a method of solving differential equations and transformed Newtonian mechanics into a branch of analysis, which facilitated the development of mathematical physics. He was also the first professor of analysis at the École Polytechnique, an elite engineering school founded in France in 1794. Also in France, mathematician and astronomer Pierre-Simon LaPlace played a key role in the development of Bayesian statistics, named for English minister and mathematician Thomas Bayes, and mathematical astronomy. He also posited the existence of black holes and gravitational collapse in the solar system.

In the nineteenth century, mathematician German Carl Friedrich Gauss made important contributions to several mathematical and physics fields including statistics, number theory, astronomy, surveying (he invented the heliotrope), and optics. The well-known normal distribution is sometimes referred to as the "Gaussian distribution" because he is often credited with discovering it. In France, Augustin-Louis Cauchy not only worked as an engineer but also pursued mathematical studies in his spare time and was appointed to the Académie des Sciences in 1816. He made numerous contributions to mathematics and physics, including his development of complex function theory, clarification of the principle of calculus, and development of the argument principle. In France, mathematician Evariste Galois proved, in parallel with the work of Norwegian mathematician Niels Henrik Abel, that there was no general method for solving polynomial equations of degree of greater than degree four.

In 1900, German mathematician David Hilbert gave an influential talk at the International Congress of Mathematicians in which he identified 23 unsolved problems in mathematics, which served as a spur for other mathematicians to focus on those problems (10 have been solved as of 2010). Hilbert is also well known for formulating the theory of Hilbert spaces, which are key to functional analysis, and did important work in mathematical logic and proof theory. Austrian mathematician Kurt Gödel, best known for his two incompleteness theorems, immigrated to the United States to escape World War II and spent his later years at Princeton University. A group of primarily French mathematicians, including Jean Dieudonne and André Weil, began publishing anonymously under the pseudonym "Nicolas Bourbaki." They are now known as the "Bourbaki Group" or "Association des collaborateurs de Nicolas Bourbaki" and have published several books in which they attempt to ground different areas of mathematics in set theory.

Awards and Honors

There is no Nobel Prize for mathematics but several different international awards are offered that have been termed the "Mathematics Nobel Prize" because of their prestige. The Fields Medal is awarded every four years to one or more mathematicians of age 40 or younger by the International Mathematical Union. Winners of the Fields Medal from western Europe include Laurent Schwartz of France (1950), Jean-Pierre Serre of France (1954), Rene Thom of France (1958), Pierre Deligne of Belgium (1978), Alain Connes of France (1982), Gerd Faltings of Germany (1986), Jean Bourgainof Belgium (1994), Pierre-Louis Lions of France (1994), Jean-Christophe Yoccoz of France (1994), Laurent Lafforgue of France (2002), Wendelin Werner of France (2006), Ngo Bao Chau of Vietnam and France (2010), and Cedric Villani of France (2010).

The Abel Prize, named after Norwegian mathematician Niels Henrik Abel, is awarded annually by the Norwegian Academy of Science and Letters. Western European winners include Jean-Pierre Serre of France (2003), Jacques Tits of Belgium and France (2008), and Mikhail Gromov of Russia and France (2009).

The Wolf Prize is awarded in several fields, including mathematics, by the Wolf Foundation. The first prizes were given in 1978 and it is awarded almost annually, with the possibility of more than one winner in a field in a given year. Western European winners include Carl L. Siegel of Germany (1978), Jean Leray of France (197), André Weil of France and the United States (1979), Henri Cartan of France (1980), Friedrich Hirzebruch of Germany (1988), Mikhail Gromov of Russia and France (1993), Jacques Tits of Belgium and France (1993), Jurgen Moser of Germany and the United States (1994/1995), Jean-Pierre Serre of France (2000), and Pierre Deligne of Belgium (2008).

Western European countries have been regular competitors in the International Mathematical Olympiad, held annually for students younger than 20 who have not yet begun tertiary education. There is both an individual and a team competition. Each country sends six students who are assigned six questions to solve. Countries are compared based on the total score for their team, while individual students may be awarded gold, silver, and bronze medals depending on how many problems they solve correctly. Germany has twice hosted the International Mathematical Olympiad and has participated since 1977.

East Germany also twice hosted the Olympiad and first participated in 1959, the year the Olympiad began. France began competing in 1967 and hosted the competition once. Belgium began participating in 1969. Austria began competing in 1970 and has served once as host. The Netherlands hosted the Olympiad in 2011 and has been competing since 1969. Luxembourg began competing in 1970, Switzerland began competing in 1991, and Liechtenstein began competing in 2005.

Further Reading

Bradley, Robert E., Lawrence A. D'Antonio, and C. Edward Sandifer, eds. *Euler at 300: An Appreciation.* Washington, DC: Mathematical Association of America, 2007.

Hahn, Robert. *Pierre Simon Laplace, 1749–1827: A Determined Scientist.* Cambridge, MA: Harvard University Press, 2005.

Joyce, David E. "The Mathematical Problems of David Hilbert." http://aleph0.clarku.edu/~djoyce/hilbert.

Krantz, Steven G. *An Episodic History of Mathematics: Mathematical Culture Through Problem Solving.* Washington, DC: Mathematical Association of America, 2010.

Mashaal, Maurice. *Bourbaki: A Secret Society of Mathematicians.* Translated by Anna Pierrehumbert. Providence, RI: American Mathematical Society, 2006.

Repcheck, Jack. *Copernicus' Secret: How the Scientific Revolution Began.* New York: Simon & Schuster, 2008.

School of Mathematics and Statistics, St. Andrews University. "The MacTutor History of Mathematics." http://www-groups.dcs.st-andrews.ac.uk/~history.

Segal, Sanford L. *Mathematicians Under the Nazis.* Princeton, NJ: Princeton University Press, 2003.

Sarah Boslaugh

See Also: Astronomy; Daubechies, Ingrid; Europe, Eastern; Europe, Northern; Europe, Southern; Mathematicians, Religious.

Expected Values

Category: History and Development of Curricular Concepts.

Fields of Study: Communication; Connections; Data Analysis and Probability.

Summary: The mathematical concept of "expected value" arose in the study of fairness in gambling but it has many scientific applications.

When people play lotteries or purchase insurance, they are investing money for a chance of some future financial return that may or may not occur. From the lottery or insurance company's perspective, money comes in from multiple purchasers and is paid out to the winners or claimants. Both sides may have questions regarding whether the investments are worthwhile or the payments are fair. These questions appear to date back to antiquity. Evidence of gambling games has been found in archaeological excavations of caves and in many ancient civilizations, including Egypt, Greece, and the Roman Empire. Babylonians used a form of maritime insurance and the Romans paid some investments in annuities.

A question concerning the fairness of certain gambling games spurred the development of probability theory in the seventeenth century. Mathematicians Blaise Pascal and Pierre de Fermat addressed fairness and related concepts while corresponding about a scenario in which two people wanted to quit playing a game and divide the winnings fairly, given that one player had a better chance of winning the game than the other. Mathematician Pierre-Simon Laplace seems to have first defined expected value in his 1814 work *Essai Philosophique sur les Probabilitiés*, writing, "This advantage in the theory of chance is the product of the sum hoped for by the probability of obtaining it.…We call this advantage mathematical hope." Expected value is the long-term average of the possible outcomes of a random variable or process, like tossing a six-sided

Expected Value

Consider a game in which a player rolls a standard six-sided die one time. If the result is a six, the player wins $4. If the result is any number from one to five, the player loses $1. If the player continues to play the game many, many times, will the overall outcome be a profit, a loss, or will the player break even? Mathematical calculations of expected value can be used to find an answer to this question and determine whether the game is fair to both sides.

Let x = the outcome of a single roll of a six-sided die, expressed as financial gain or loss

Die Roll	Outcome x	Probability $p(x)$
1	Lose $1	1/6
2	Lose $1	1/6
3	Lose $1	1/6
4	Lose $1	1/6
5	Lose $1	1/6
6	Win $4	1/6

The expected value would be

$$\text{Expected Value} = E(x) = \sum xp(x)$$

$$= -1\$\frac{1}{6} - 1\$\frac{1}{6} - \$1\frac{1}{6} - \$1\frac{1}{6} - \$1\frac{1}{6} + \$4\frac{1}{6} \approx -\$0.17.$$

This value means that over a large number of times playing the game, the player should expect to lose 17 cents per play, on average. However, in the short run, a player might win or lose more, since a winning a single roll could yield $4 or a series of losses could cost several dollars. If the amount received for winning were $5 instead of $4, the expected value would be $0 and the game would be "fair" in the sense that neither side would have a monetary advantage. This notion of fairness is different than the fairness or equal chances of the die rolls, which determine the probabilities and could also affect the expected value.

die. Mathematically, expected value is computed as the weighted sum of the outcomes, where the weights are the corresponding probabilities. For discrete random variables, expected value is a summation; for continuous variables, it is an integration. While computing means for data is very common beginning in middle school classrooms in the twenty-first century, finding expected values for random variables is more commonly part of high school and college curricula. Though initially motivated by notions of fairness, expected values have many important applications in probability and statistical theory and practice.

Applications

Scientific problems involving measurement were an inspiration for many mathematical advances in probability and applied data analysis. Astronomers in the eighteenth century often computed arithmetic means (or averages) for data to estimate parameters and describe distributions of "errors," like those they found when taking multiple measurements of the same astronomical distance. These averages were likely to be close to the true distance or value, or so they generally believed. This technique was used without proof for a long time, though mathematician Thomas Simpson had shown that an average was a better measure than a single observation in a very limited set of cases. Some issues in finding a suitable proof stemmed from the fact that probability distributions commonly used for describing errors at that time presented mathematical difficulties when trying to find expected values for averages versus expected values for individual observations. Work by mathematicians Abraham de Moivre and Laplace led to the Central Limit Theorem, derived by Laplace in the nineteenth century and later extended by other mathematicians such as Francis Edgeworth. This result is sometimes called the "DeMoivre–Laplace theorem" and was given its more common name in work by George Pólya in the early twentieth century. The primary impact of the Central Limit Theorem with regard to expected values is that it defined the expected value for the sampling distribution of the mean, given sufficiently large sample sizes. It established a theoretical basis for estimation and a later hypothesis testing for various parameters.

There are many different probability distributions that mathematicians, statisticians, and others have found, derived, named, and studied. For many years the normal distribution, credited to mathematician Carl Friedrich Gauss, played a central role in error modeling and other applications. However, approaching the twentieth century, increasing application of probability and statistics in a wide variety of fields, including biology, business, genetics, and psychophysics, led investigators like statistician Karl Pearson to research non-normal or skewed distributions to better represent phenomena they encountered. The problem then became to estimate parameters for these distributions and discover their mathematical properties. The method of moments estimates parameters like variance and skew using expected values. It primarily considers deviations of points from the distribution mean, called "central moments," which are conceptually related to the idea of moment or torque about a point in physics. Deviations are raised to various powers so that the k-th moment corresponds to the k-th power. The first central moment is zero, since it essentially sums all deviations from the mean or expected value. Variance is the second central moment, which is the expected value (the weighted sum) of all squared deviations from the mean. The third moment quantifies skew or asymmetry and is the expected value of all cubed deviations from the mean. A symmetric distribution has skew of zero. The fourth moment is called "kurtosis" and measures whether the distribution is taller or shorter and has thicker or thinner tails than a normal distribution with the same variance. Mixed moments can be found for two variables together to quantify the covariance and, by extension, correlation. Measures of skewness and kurtosis based on moments are credited to Pearson.

Further Reading

Aven, T. *Risk Analysis: Assessing Uncertainties Beyond Expected Values and Probabilities.* Hoboken, NJ: Wiley, 2008

Fey, James, Elizabeth Phillips, and Catherine Anderson. *What Do You Expect: Probability and Expected Value.* Palo Alto, CA: Dale Seymour Publications, 1997.

Hald, Anders. *A History of Parametric Statistical Inference from Bernoulli to Fisher, 1713–1935.* New York: Springer, 2007.

Carmen M. Latterell

See Also: Data Analysis and Probability in Society; Dice Games; Game Theory; Lotteries; Measures of Center; Probability.

Exponentials and Logarithms

Category: History and Development of Curricular Concepts.
Fields of Study: Algebra; Communication; Connections.
Summary: Exponential and logarithmic functions are used to study and analyze a variety of mathematical relationships.

Much of the language and notation of mathematics involves a very advanced shorthand. As ideas grow and become more complex, mathematicians seek ways to express highly condensed thought in relatively simple terms. Exponents are an elementary example: if one wants to multiply the number 2 times itself 10 times, rather than write "$2 \cdot 2 \cdot 2 \cdot 2 \cdot 2 \cdot 2 \cdot 2 \cdot 2 \cdot 2 \cdot 2$" one can write "$2^{10}$" instead. From these beginnings, which date to ancient Egypt and Babylon, the remarkable worlds of exponential and logarithmic functions emerge. When one develops the understanding of what it means to take 2 to any real number power, one naturally considers the function $f(x) = 2^x$, an example of what is called an "exponential function." For larger and larger positive x, the function grows amazingly fast: $2^{10} = 1024$, $2^{20} = 1{,}048{,}576$, and $2^{30} = 1{,}073{,}741{,}824$.

The exponential function $f(x) = e^x$, where e is the so-called "natural base," an irrational number whose decimal approximation is $e \approx 2.71828$, is an important exponential function. With e in homage to the great Swiss mathematician Leonhard Euler (1707–1783), this special exponential function $f(x) = e^x$ might rightly lay claim to the title of "the most important function in all of mathematics." Exponential growth and decay functions, along with the number e itself, have a wide variety of uses and applications.

In classrooms in the twenty-first century, the logarithm of a number is defined as the exponent or power to which a stated number, called the "base," is raised to obtain the given number. The development of logarithms in the seventeenth century led to a revolution in scientific calculation, especially when the slide rule replaced tables of logarithms. While the advent of calculators and computers eliminated the need for calculation by logarithms in the latter part of the twentieth century, logarithms remain important in order to understand financial and natural processes. For instance, the Richter scale to measure earthquakes, named for Charles Richter, is a logarithmic scale. In chemistry, the pH scale is based on the negative logarithm of the concentration of free hydrogen ions. Students in the middle grades investigate exponential notation while high school students explore exponential and logarithmic functions.

Archimedes of Syracuse investigated that the addition of what he called "orders" corresponded with their product, known today as the "first law of exponents." The number e may have first appeared in the early seventeenth century in an appendix to John Napier's work on logarithms. This number also arose in the work of Christiaan Huygens in the mid-seventeenth century when he was exploring the area under the hyperbola $xy = 1$. Finally, in the late seventeenth century through work involving continuous compound interest, Jacob Bernoulli was led to consider the expression

$$\left(1 + \frac{1}{n}\right)^n$$

for large values of n, and this expression approaches e as n grows without bound. Mathematicians explored many issues related to e and exponentials, including such people as Euler, Gotthold Eisenstein, and others, who investigated the convergence of sequences of iterated exponentials. Bernoulli may also have been the first mathematician to realize that the number e was intricately linked to emerging ideas with logarithms.

The Natural Exponential Function

Because any exponential function can be written in terms of e, one finds that functions of the form $P(x) = Me^{kx}$, where M and k are constants that depend on the context, arise in many natural settings. Exponential cell and population growth, as well as exponential decay in radioactive materials, are modeled by functions of this form. Once the values of M and k are identified, the function easily indicates the corresponding output for any input value x. For example, if a car is initially valued (at time $t = 100$) at \$10,000 that depreciates at a certain continuous rate, one might use the function $P(t) = 10000e^{-0.2t}$ to model the worth of the car in year t.

Functions like this generate very natural questions, including ones like "At what time t will the car's value be \$3,000?" Before trying to answer this more complicated question, consider some simpler ones. For instance, what value of t makes $10^t = 17$? Since $10^1 = 10$, while $10^2 = 100$, it seems like there ought to be a number between 1 and 2 such that 10 raised to that power is 17. But what is the number? Here, some very considerable mathematical ideas are involved: the function $y(t) = 10^t$ is continuous; the range of y is all positive real numbers; and $-y$ is always increasing, making it a one-to-one function. All these facts together combine to indicate that one can pick any positive real number y and know that there must be one and only one real number t that satisfies the equation $10^t = y$. In other words, there is a function h that takes any positive real number y, and to this value y associates the real number t so that 10 raised to the power t is y. This explanation is how teachers usually describe to students where logarithms come from—the logarithm is the very function that accomplishes this association. It is all a matter of perspective; if t is known and y is sought, the exponential function is used, while if y is known and t is sought, the logarithm function is used. Expressed in words, it is "y equals 10 to the power t" and "t is the power to which we raise 10 to get y." Babylonian clay tablets presented similar questions.

Historical Development

Historically, the further development of logarithms arose very differently. In the late fifteenth century and early sixteenth century, both John Napier and Jost Burgi, who were each interested in key problems in astronomy, developed logarithms for a much different use: as a new tool to help do arithmetic with large numbers. Their approach to logarithms was fundamentally geometric, as algebra was not yet sufficiently well developed to aid their work, although Napier's approach was more algebraic than Burgi's methods. Napier noted, "Seeing there is nothing that is so troublesome to mathematical practice, nor doth more molest and hinder calculators, than the multiplications, divisions, square and cubical extractions of great numbers, which besides the tedious expense of time are for the most part subject to many slippery errors, I began therefore to consider in my mind by what certain and ready art I might remove those hindrances." In 1624, Henry Briggs published logarithm tables in *Arithmetica Logarithmica* and he is noted by some as perhaps the man most responsible for popularizing logarithms among scientists. The development of the slide rule made logarithms easy to use, since they reduced the reliance on tables. In 1620, Edmund Gunter noted logarithms on a ruler by marking the position of numbers relative to their logarithms. William Oughtred placed two sliding logarithmic rulers next to each other and by 1630, the portable circular slide rule reduced multiplication computations to the act of lining up two numbers and reading a scale. Logarithms remain a useful way to deal with large numbers in the early twenty-first century, because the logarithm of a large number is a much, much smaller one. R. C. Pierce Jr. noted, "It has been postulated that logarithms literally lengthened the life spans of astronomers who had formerly been sorely bent and often broken early by the masses of calculations their art required." Modern mathematicians have also come to fully understand the connection between logarithms and the area under the curve $xy = 1$, which was explored by Huygens in the 1600s.

Using Logarithms to Solve Exponential Functions

Perhaps the most powerful property of logarithms is that they "undo" exponential functions. For example, for the natural logarithm of base e, denoted "ln," one obtains $\ln(e^5) = 5$. Remember, $\ln(e^5)$ means "the power to which one raises e to get e^5." This power, of course, is 5. The general property that holds here is that for any real number t, $\ln(e^t) = t$. This rule proves to be immensely useful in solving exponential equations. To see how, consider an earlier example: the function $P(t) = 10000e^{-0.2t}$ (the value of a car in year t). At what time t will the car's value be \$3,000? This question is equivalent to solving the equation:

$$0.3 = e^{-0.2t}.$$

Taking the natural logarithm of both sides of the equation "undoes" the effects of the exponential function and hence gains more direct access to the variable t: $\ln(0.3) = \ln(e^{-0.2t}) = -0.2t$.

Dividing both sides of the last equation above by -0.2, one finds that

$$t = \frac{\ln(0.3)}{-0.2} \approx 6.0199$$

so that the car's value will be $3,000 in just over six years. The natural logarithm of 0.3 is central to answering the question.

While the motivation for the need for logarithms can be seen in relatively elementary terms—solving exponential equations—the actual mathematics that explains what logarithms really are and how they work is deep and is best supported using some sophisticated ideas from calculus. Even with exponential functions, there are some big questions without answers: how is *e* to the 5th power calculated? How is the natural logarithm of 0.3 computed? Until the invention of personal computers in the 1970s, such computations were all done by hand, usually with the assistance of elaborate tables, or with slide rules. At one point in history, entire books were written that held nothing but tables of values for logarithms. People now use inexpensive handheld calculators, computer algebra systems like *Maple* or *Mathematica*, or even Google, and each returns a value almost immediately. These modern technological tools rely on a rich and beautiful mathematical theory of exponential and logarithmic functions. Beyond their interesting mathematical properties, exponential and logarithmic functions remain important for their many applications, such as the key role that exponential functions play in the study of differential equations, including those that model vibrations in bridges and buildings, thus forming a central component of modern civil engineering.

Further Reading

Maor, Eli. *e: The Story of a Number*. Reprint. Princeton, NJ: Princeton University Press, 1994.

Nahin, Paul J. *An Imaginary Tale: The Story of i (The Square Root of Minus One)*. Princeton, NJ: Princeton University Press, 2010

Pierce Jr., R. C. "A Brief History of Logarithms." *The Two-Year College Mathematics Journal* 8, no. 1 (1977).

Stoll, Cliff. "When Slide Rules Ruled." *Scientific American* 294, no. 5 (2006).

Strogatz Steven. "Power Tools—NYTimes.com" http://opinionator.blogs.nytimes.com/2010/03/28/power-tools.

Matt Boelkins

See Also: Calculators in Society; Carbon Dating; Earthquakes; Functions; Mathematics, Elegant.

Extinction

Category: Weather, Nature, and Environment.
Fields of Study: Algebra; Problem Solving.
Summary: Causes and factors of extinction can be quantified and modeled using mathematical and statistical techniques.

Extinction occurs when the last member of a species dies. A species survives for much longer than any of its members. For example, a human can live up to about 120 years, whereas the human species (*Homo sapiens*) is thought to have existed for hundreds of thousands of years. It is not known how long our species will endure and indeed most species on Earth have already become extinct. There are many causes of extinction, some natural and others as a result of human activities. Many factors influence whether an endangered species can avoid extinction. These factors can be quantified and modeled using mathemati-

Causes of Extinction

A species can become extinct for various reasons, including intense competition with other species, disease, or failure to adapt to changing climatic conditions, as well as the disappearance of a species' prey. Anthropomorphic reasons for extinction include over-hunting by humans, habitat loss from human activities such as deforestation, and social planning (the intentional eradication of smallpox).

cal and statistical techniques. A species can disappear in some parts of its habitat but not in others. Not all species have existed on Earth for the same length of time—some appear only briefly while others manage to persist for incredibly long periods of time. Human activities may be increasing the rate at which other species become extinct.

Rise of Extinction

A species is endangered when it consists of a small number of members. In such cases, individuals may have trouble finding each other because of geographical separation. For a species that is endangered, it is of interest to know whether the species is likely to become extinct. It is customary to let $N(t)$ represent the size of a population at time t. The fact that the species is endangered implies that $N(t)$ takes positive values close to zero. If $N(t)$ is eventually measured to be zero, then the species has become extinct. However, if $N(t)$ rebounds to larger positive values, then the species persists. In general, stochastic effects largely determine whether an endangered species will become extinct. Given population data $N(t)$ at different times t, one may compute the mean (μ) of the population growth rate.

$$R(t) = \ln\left(\frac{N(t)}{N(t-1)}\right).$$

For example, if $t = 10$ then

$$\mu = \frac{(R(1) + R(2) + \cdots + R(10))}{10}.$$

A positive (or negative) value of μ indicates that the population is growing (or declining) on average. Combining this information with the standard deviation (σ) of $R(t)$ allows one to assess the risk for extinction, which is typically highest when μ is negative and σ is small. Complex models of population dynamics exist to predict whether a species will persist or become extinct. These include geometric growth models in which a population multiplies at a fixed rate, logistic growth models in which populations slowly attain steady-state sizes, and Lotka–Volterra predator-prey models for interactions between multiple species, named for Alfred Lotka and Vito Volterra.

Local Extinction

A species can become extinct in one area (such as an island) and still persist elsewhere (such as a continent). If the species is able to recolonize the former area, then this is known as a "rescue effect." If local extinction events become synchronized—as a result of global climate change, for example—then the risk of a species becoming globally extinct is much higher.

Rate of Extinction

Scientists estimate that there may be 10 million species alive today and yet they account for fewer than 1 in 1000 species that have ever lived. The average time to extinction for a species, as measured from the time of its first appearance, is close to 10 million years. When the time to extinction for a species is much longer, such as more than 100 million years, then later members are said to be living fossils.

Mass Extinction

A mass extinction occurs when a large number of species become extinct in a short period of time. Although rare, the fossil record indicates that these events have occurred at least five times, the most famous being the mass extinction of non-flying dinosaurs 65 million years ago in what was probably a meteor impact. Many scientists believe that we are currently in the midst of a sixth mass extinction, with up to 40,000 species becoming extinct each year—a rate that is roughly 100–1000 times higher than in prehistoric times.

Further Reading

Allen, Linda J. S. *An Introduction to Mathematical Biology*. Upper Saddle River, NJ: Prentice Hall, 2007.

Bright, Michael. *Extinctions of Living Things* (*Timeline: Life on Earth*). Portsmouth, NH: Heinemann, 2008.

Erickson, J., and A. E. Gates. *Lost Creatures of the Earth.* New York: Facts on File, 2001.

Hallam, T. *Catastrophes and Lesser Calamities: The Causes of Mass Extinctions.* New York: Oxford University Press, 2005.

Hecht, J. *Vanishing Life: The Mystery of Mass Extinctions.* New York: Atheneum, 2009.

Thieme, Horst R. *Mathematics in Population Biology.* Princeton, NJ: Princeton University Press, 2003.

Andrew Nevai

See Also: Animals; Climate Change; Deforestation; Forest Fires; Mathematical Modeling; Predator–Prey Models.

Extreme Sports

Category: Games, Sport, and Recreation.
Fields of Study: Algebra; Geometry.
Summary: The emphasis on fast motion, tricks, and personal expression in extreme sports makes geometry especially relevant to athletes.

There is no single definition of extreme sports, though they generally include dangerous sporting activities that involve a substantial risk of injury, like Buildings, Antennae, Spans, and Earth (BASE) jumping, cliff diving, street luge, or even the traditional running of the bulls in Pamplona, Spain. Extreme sports are believed to be attractive to participants because of the challenge and adrenaline rush and to spectators because the results are typically unpredictable.

The popularity of extreme sports grew rapidly in the latter part of the twentieth century. The television network ESPN created the Extreme Games, now called the "X Games," in 1995, making extreme sports more visible to the general public. Other networks have also begun to televise these types of competitions and some extreme sports events have been included in the Olympic Games. Mathematics is important in extreme sports. Knowing and applying concepts from geometry and probability helps participants be safe and successful. Innovative equipment manufacturers use concepts and techniques from many areas, including geometry, statistics, modeling, and simulation, to prototype and refine their designs, resulting in greater safety and effectiveness.

Skateboarding

Skateboarders perform tricks using a wheeled board, either on a flat surface or using equipment like ramps or rails. Many stunts rely on differential pressure applied by the rider's feet to various parts of the skateboard to tilt or flip it, often rotating both board and rider in one or more axes. Lip tricks require a vertical orientation and transitional edge like the lip of a swimming pool or ramp. In aerial tricks, the rider leaves the ground completely, using counterpressure of hands and feet to maintain control of the board while spinning or flipping.

Tony Hawk is one of the most well-known extreme athletes and a vertical skateboarding pioneer. He was the first person to competitively perform an aerial turn of two and a half rotations, or 900 degrees, at the 1999 X Games. In the past, he has done 720 degree turns. For the 900, he exerted greater takeoff force in the direction of the turn, producing more rotational velocity. Tony Hawk's Project 8 video game used motion capture technology to smoothly animate professional skaters, while Tony Hawk Ride allowed players to simulate the sport using a skateboard-like controller.

Snowboarding

Snowboarding is similar to skateboarding and involves standing on a board and sliding down a snow-covered hill. Snowboarding became an Olympic sport in 1998, with giant slalom and half pipe competitions taking place. The giant slalom is a speed race in which athletes speed down a steep hill with gates that require them to zigzag between. Determining an optimal path from one gate to another without crashing or wasting time requires mathematics, especially geometry. A half-pipe consists of two quarter-cylinders connected by a flat space and topped by a small lip. The competition is a more artistic event, with athletes generating enough speed using the curves of the pipe to become airborne and do tricks. These may include multiple rotations, both twisting and somersaulting. At the 2010 Olympics, Shaun White executed a record-setting 1260-degree trick consisting of two flips and three and a half spins.

BMX Biking

In bicycle motocross (BMX), athletes ride specially designed smaller bicycles that enable them to shift their center of mass to make precision movements. BMX courses often use steep hills to launch the rider into the air to perform tricks. Other tricks and spins may be done on flat ground. The sport was added to the list of events for the 2012 Summer Olympic Games. Billy Gawrych is a professional BMX competitor who performs intricate routines, often set to music, with tricks linked together in a series of connected, flowing patterns.

When a skateboarder performs an ollie, the forces acting on the board are the weight of the rider, the force of gravity on the board, and the force of the ground pushing up on the board, which balance out to zero net force.

Sports Engineering and Equipment

Sports engineering is a growing interdisciplinary field that draws from mathematics, engineering, biology, physics, materials science, and many other disciplines to study characteristics of athletes and equipment, as well as their interaction. The focus is on performance and safety. For example, engineer Mont Hubbard described the motion of skateboards with riders using two mathematical models, and mathematicians develop new models using techniques and theories from areas like trigonometry, physics, differential equations, and probability. Quality function deployment is a method of quality control that attempts to translate often subjective customer requirements into mathematical engineering specifications. One research group studied the subjective perception of the "feel" of snowboards. They used field evaluations and laboratory data to create matrices of parameters. Snowboards for freeride and freestyle, the two primary types of snowboarding, have somewhat different designs; however, issues of flexibility, torsional stiffness, and curvature were the important factors affecting feel and performance for both styles. Equipment for sports of all kinds is subjected to statistically designed tests to evaluate safety, and data from accidents and failures helps fuel further research.

Further Reading

Clemson, Wendy, David Clemson, Oli Cundale, Laura Berry, and Matt King. *Using Math to Conquer Extreme Sports*. New York: Gareth Stevens Publishing, 2004.

Estivalet, Margaret, and Pierre Brisson. *The Engineering of Sport 7*. Vol. 1 New York: Springer, 2008.

Gutman, Bill. *Being Extreme: Thrills and Dangers in the World of High-Risk Sports*. New York: Citadel Press, 2003.

Sagert, Kelly Boyer. *Encyclopedia of Extreme Sports*. Westport, CT: Greenwood Press, 2008.

Thorpe, Holly. *Snowboarding Bodies in Theory and Practice (Global Culture and Sport)*. New York: Palgrave Macmillan, 2011.

Tyler, M., and K. Tyler. *Extreme Math: Real Math, Real People, Real Sports*. Waco, TX: Prufrock Press, 2003.

Michele LeBlanc
Nena Amundson

See Also: Mathematical Modeling; Probability; Trigonometry.